Lecture Notes in Artificial Intelligence 9956

Subseries of Lecture Notes in Computer Science

LNAI Series Editors

Randy Goebel
University of Alberta, Edmonton, Canada
Yuzuru Tanaka
Hokkaido University, Sapporo, Japan
Wolfgang Wahlster
DFKI and Saarland University, Saarbrücken, Germany

LNAI Founding Series Editor

Joerg Siekmann
DFKI and Saarland University, Saarbrücken, Germany

More information about this series at http://www.springer.com/series/1244

Toon Calders · Michelangelo Ceci
Donato Malerba (Eds.)

Discovery Science

19th International Conference, DS 2016
Bari, Italy, October 19–21, 2016
Proceedings

 Springer

Editors
Toon Calders
Universiteit Antwerpen
Antwerp
Belgium

Michelangelo Ceci
Università degli Studi di Bari Aldo Moro
Bari
Italy

Donato Malerba
Dipartimento di Informatica
University of Bari "Aldo Moro"
Bari
Italy

ISSN 0302-9743 ISSN 1611-3349 (electronic)
Lecture Notes in Artificial Intelligence
ISBN 978-3-319-46306-3 ISBN 978-3-319-46307-0 (eBook)
DOI 10.1007/978-3-319-46307-0

Library of Congress Control Number: 2016950906

LNCS Sublibrary: SL7 – Artificial Intelligence

Printed on acid-free paper

This Springer imprint is published by Springer Nature
The registered company is Springer International Publishing AG Switzerland

Preface

The 19th International Conference on Discovery Science (DS 2016) was held in Bari, Italy, during October 19–21, 2016. As in previous years, the conference was co-located with the International Conference on Algorithmic Learning Theory (ALT 2016), which was already in its 27th year. First held in 2001, ALT/DS has been one of the longest-running series of co-located events in computer science. The unique combination of latest advances in the development and analysis of methods for discovering scientific knowledge, coming from machine learning, data mining, and intelligent data analysis, as well as their application in various scientific domains, on the one hand, with the algorithmic advances in machine learning theory, on the other hand, made every instance of this joint event unique and attractive. This volume contains the papers presented at the 19th International Conference on Discovery Science, while the papers of the 27th International Conference on Algorithmic Learning Theory were published by Springer in a companion volume (LNCS Vol. 9925).

The 19th Discovery Science conference received 60 international submissions. Each submission was reviewed by at least three committee members. The committee decided to accept 30 papers. This resulted in an acceptance rate of 50 %. Following the tradition of the joint Discovery Science and the Algorithmic Learning Theory conferences, invited talks were shared between the two meetings. There was one shared invited speaker, John Shawe-Taylor from University College London, who contributed his talk "Margin Based Structured Output Learning." The invited talk for DS 2016 was "Collective Attention on the Web" by Kristian Kersting from the Technical University of Dortmund, while the ALT invited talk was "Learning About Agents and Mechanisms from Opaque Transactions" by Avrim Blum from Carnegie Mellon University. Furthermore, two tutorials, one from ALT and one from DS, were shared. For DS, the tutorial "Perspectives of Feature Selection in Bioinformatics: From Relevance to Causal Inference" was presented by Gianluca Bontempi from the Université Libre de Bruxelles, while for ALT the tutorial "How to Estimate the Mean of a Random Variable?" was given by Gábor Lugosi from Pompeu Fabra University. Abstracts of all five invited talks and tutorials are included in these proceedings.

We would like to thank all the authors of submitted papers, the Program Committee members, and the additional reviewers for their efforts in evaluating the submitted papers, as well as the invited speakers and tutorial presenters. We are grateful to Ronald Ortner and Hans Ulrich Simon for ensuring the smooth coordination with ALT. We would also like to thank all the members of the Organizing Committee: Annalisa Appice, Roberto Corizzo, Claudia d'Amato, Nicola Di Mauro, Stefano Ferilli, Corrado Loglisci, Gianvito Pio, Francesco Serafino. We are grateful to the people behind EasyChair for making the system available free of charge. It was an essential tool in the paper submission and evaluation process, as well as in the preparation of the Springer proceedings. We are also grateful to Springer for their continuing support of Discovery Science and for publishing the conference proceedings. A special issue on the topics of

Discovery Science has also been scheduled for the Springer journal *Machine Learning*, thus offering the possibility of publishing in this prestigious journal an extended and reworked version of papers presented at Discovery Science 2016.

The joint event ALT/DS 2016 was organized under the auspices of the University of Bari "Aldo Moro". Financial support was generously provided by the Fondazione Puglia and the Consorzio Interuniversitario Nazionale per l'Informatica (National Interuniversity Consortium for Informatics, CINI). The event was also supported by the FP7 project MAESTRA - Learning from Massive, Incompletely annotated, and Structured Data (grant number ICT-2013-612944) from the European Commission.

Finally, we are indebted to all conference participants, who contributed to make this momentous event a worthwhile endeavor for all involved.

October 2016   Toon Calders
                                                          Michelangelo Ceci
                                                            Donato Malerba

Organization

General Chair

Donato Malerba   University of Bari "Aldo Moro", Italy

Program Chairs

Michelangelo Ceci   University of Bari "Aldo Moro", Italy
Toon Calders   Universiteit Antwerpen, Belgium

Social Media Chair

Claudia d'Amato   University of Bari "Aldo Moro", Italy

Proceedings Chair

Stefano Ferilli   University of Bari "Aldo Moro", Italy

Web Chair

Francesco Serafino   University of Bari "Aldo Moro", Italy

Publicity Chair

Nicola Di Mauro   University of Bari "Aldo Moro", Italy

Local Arrangements Chairs

Annalisa Appice   University of Bari "Aldo Moro", Italy
Corrado Loglisci   University of Bari "Aldo Moro", Italy
Gianvito Pio   University of Bari "Aldo Moro", Italy
Roberto Corizzo   University of Bari "Aldo Moro", Italy

Program Committee

Annalisa Appice   University of Bari "Aldo Moro", Italy
Albert Bifet   Telecom ParisTech, France
Hendrik Blockeel   K.U. Leuven, Belgium
Bruno Cremilleux   Université de Caen, France
Ivica Dimitrovski   Ss. Cyril and Methodius University in Skopje, Macedonia
Saso Dzeroski   Jozef Stefan Institute, Slovenia

Additional Reviewers

Massimo Bilancia
Martin Breskvar
Sophie Burkhardt
Juan Colonna
Nicola Di Mauro
Nicola Fanizzi
Khaled Fawagreh
Esther Galbrun
Kambiz Ghazinour
Massimo, Guarascio

Andreas Karwath
Mehdi Kaytoue
Gaël Lejeune
Jurica Levatic
Regis Pires Magalhães
Yoshiaki Okubo
Panagiotis Papapetrou
Ettore Ritacco
Livia Ruback
Hiroto Saigo

Francesco Serafino
Safwan Shatnawi
Nikola Simidjievski
Ricardo Sousa
Martin Strazar
Antonio Vergari
Douglas Woodford
Bernard Zenko

Invited Talks

Margin Based Structured Output Learning

John Shawe-Taylor

Department of Computer Science, University College London Gower Street,
London WC1E6BT, UK
J.Shawe-Taylor@cs.ucl.ac.uk

Abstract. Structured output learning has been developed to borrow strength across multidimensional classifications. There have been approaches to bounding the performance of these classifiers based on different measures such as microlabel errors with a fixed simple output structure. We present a different approach and analysis starting from the assumption that there is a margin attainable in some unknown or fully connected output structure. The analysis and algorithms flow from this assumption but in a way that the associated inference becomes tractable while the bounds match those attained were we to use the full structure. There are two variants depending on how the margin is estimated. Experimental results show the relative strengths of these variants, both algorithmically and statistically.

Bio. John Shawe-Taylor is a professor at the University College London where he directs the Centre for Computational Statistics and Machine Learning and heads the Department of Computer Science. His research has contributed to a number of fields ranging from graph theory through cryptography to statistical learning theory and its applications. However, his main contributions have been in the development of the analysis and subsequent algorithmic definition of principled machine learning algorithms founded in statistical learning theory. He has co-authored two influential text books on kernel methods and support vector machines. He has also been instrumental in coordinating a series of influential European Networks of Excellence culminating in the PASCAL networks.

Collective Attention on the Web

Kristian Kersting

Computer Science Department, TU Dortmund University, 44221 Dortmund,
Germany
kristian.kersting@cs.tu-dortmund.de

Abstract. It's one of the most popular YouTube videos ever produced, having been viewed more than 840 million times. It's hard to understand why this clip is so famous and actually went viral, since nothing much happens. Two little boys, Charlie and Harry, are sitting in a chair when Charlie, the younger brother, mischievously bites Harry's finger. There's a shriek and then a laugh. The clip is called "Charlie Bit My Finger–Again!"

Generally, understanding the dynamics of collective attention is central to an information age where millions of people leave digital footprints everyday. So, can we capture the dynamics of collective attention mathematically? Can we even gain insights into the underlying physical resp. social processes? Is it for instance fair to call the video "viral" in an epidemiological sense?

In this talk I shall argue that computational methods of collective attention are not insurmountable. I shall review the methods we have developed to characterize, analyze, and even predict the dynamics of collective attention among millions of users to and within social media services. For instance, we found that collective attention to memes and social media grows and subsides in a highly regular manner, well explained by economic diffusion models. Using mathematical epidemiology, we find that so-called "viral" videos show very high infection rates and, hence, should indeed be called viral. Moreover, the spreading processes may also be related to the underlying network structures, suggesting for instance a physically plausible model of the distance distributions of undirected networks. All this favors machine learning and discovery science approaches that produce physically plausible models.

This work was partly supported by the Fraunhofer ICON project SoF-WIReD and by the DFG Collaborative Research Center SFB 876 project A6.

Bio. Kristian Kersting is an Associate Professor in the Computer Science Department at the Technical University of Dortmund. His main research interests are data mining, machine learning, and statistical relational artificial intelligence, with applications to medicine, plant phenotyping, traffic, and collective attention. He gave several tutorials at top conferences and co-chaired BUDA, CMPL, CoLISD, MLG, and SRL as well as the AAAI Student Abstract track and the Starting AI Research Symposium (STAIRS). Together with Stuart Russell (Berkeley), Leslie Kaelbling (MIT), Alon Halevy (Goolge), Sriraam Natarajan (Indiana) and Lilyana Mihalkova (Google) he cofounded the international workshop series on Statistical Relational AI. He served as area chair/senior PC for several top conference and co-chaired ECML PKDD 2013, the premier European venue for Machine Learning and Data Mining. Currently, he is an action editor of JAIR, AIJ, DAMI, and MLJ as well as on the editorial board of NGC.

Perspectives of Feature Selection in Bioinformatics: From Relevance to Causal Inference

Gianluca Bontempi

Machine Learning Group, Interuniversity Institute of Bioinformatics
in Brussels (IB)², Université libre de Bruxelles, Bld de Triomphe,
1050 Brussels, Belgium
mlg.ulb.ac.be

Abstract. A major goal of the scientific activity is to model real phenomena by studying the dependency between entities, objects or more in general variables. Sometimes the goal of the modeling activity is simply predicting future behaviors. Sometimes the goal is to understand the causes of a phenomenon (e.g. a disease). Finding causes from data is particular challenging in bioinformatics where often the number of features (e.g. number of probes) is huge with respect to the number of samples. In this context, even when experimental interventions are possible, performing thousands of experiments to discover causal relationships between thousands of variables is not practical. Dimensionality reduction techniques have been largely discussed and used in bioinformatics to deal with the curse of dimensionality. However, most of the time these techniques focus on improving prediction accuracy, neglecting causal aspects. This tutorial will introduction some basics of causal inference and will discuss some open issues: may feature selection techniques be useful also for causal feature selection? Is prediction accuracy compatible with causal discovery? How to deal with Markov indistinguishable settings? Recent results based on information theory, and some learned lessons from a recent Kaggle competition will be used to illustrate the issue.

Bio. Gianluca Bontempi is Full Professor in the Computer Science Department at the Université Libre de Bruxelles (ULB), Brussels, Belgium, co-head of the ULB Machine Learning Group and Director of (IB)², the ULB/VUB Interuniversity Institute of Bioinformatics in Brussels. His main research interests are big data mining, machine learning, bioinformatics, causal inference, predictive modelling and their application to complex tasks in engineering (forecasting, fraud detection) and life science. He was Marie Curie fellow researcher, he was awarded in two international data analysis competitions and he took part to many research projects in collaboration with universities and private companies all over Europe. He is author of more than 200 scientific publications, associate editor of PLOS One, member of the scientific advisory board of Chist-ERA and IEEE Senior Member. He is also co-author of several open-source software packages for bioinformatics, data mining and prediction.

Learning About Agents and Mechanisms from Opaque Transactions

Avrim Blum

School of Computer Science, Carnegie Mellon University, Pittsburgh,
PA 15213–3891, USA
avrim@cs.cmu.edu

Abstract. In this talk I will discuss some learning problems coming from the area of algorithmic economics. I will focus in particular on settings known as combinatorial auctions in which agents have preferences over items or sets of items, and interact with an auction or allocation mechanism that determines what items are given to which agents. We consider the perspective of an outside observer who each day can only see which agents show up and what they get, or perhaps just which agents' needs are satisfied and which are not. Our goal will be from observing a series of such interactions to try to learn the agent preferences and perhaps also the rules of the allocation mechanism.

As an example, consider observing web pages where the agents are advertisers and the winners are those whose ads show up on the given page. Or consider observing the input-output behavior of a cloud computing service, where the input consists of a set of agents requesting service, and the output is a partition of them into some whose requests are actually fulfilled and the rest that are not—due to overlap of their resource needs with higher-priority requests. From such input-output behavior, we would like to learn the underlying structure. We also consider a classic Myerson single-item auction setting, where from observing who wins and also being able to participate ourselves we would like to learn the agents' valuation distributions.

In examining these problems we will see connections to decision-list learning and to Kaplan-Meier estimators from medical statistics.

This talk is based on work joint with Yishay Mansour and Jamie Morgenstern.

Bio. Avrim Blum is Professor of Computer Science at Carnegie Mellon University. His main research interests are in Foundations of Machine Learning and Data Mining, Algorithmic Game Theory (including auctions, pricing, dynamics, and connections to machine learning), the analysis of heuristics for computationally hard problems, and Database Privacy. He has served as Program Chair for the IEEE Symposium on Foundations of Computer Science (FOCS) and the Conference on Learning Theory (COLT). He was recipient of the Sloan Fellowship, the NSF National Young Investigator Award, the ICML/COLT 10-year best paper award, and the Herbert Simon Teaching Award, and is a Fellow of the ACM.

How to Estimate the Mean of a Random Variable?

Gabor Lugosi

Department of Economics, Pompeu Fabra University,
Ramon Trias Fargas 25–27, 08005, Barcelona, Spain
gabor.lugosi@gmail.com

Abstract. Given n independent, identically distributed copies of a random variable, one is interested in estimating the expected value. Perhaps surprisingly, there are still open questions concerning this very basic problem in statistics. In this talk we are primarily interested in non-asymptotic sub-Gaussian estimates for potentially heavy-tailed random variables. We discuss various estimates and extensions to high dimensions. We apply the estimates for statistical learning and regression function estimation problems. The methods improve on classical empirical minimization techniques.

This talk is based on joint work with Emilien Joly, Luc Devroye, Matthieu Lerasle, Roberto Imbuzeiro Oliveira, and Shahar Mendelson.

Bio. Gábor Lugosi is an ICREA research professor at the Department of Economics, Pompeu Fabra University, Barcelona. His research main interests include the theory of machine learning, combinatorial statistics, inequalities in probability, random graphs and random structures, and information theory. He has co-authored monographs on pattern classification, on online learning, and on concentration inequalities.

Contents

Applications

Ensemble Learning

Classification

Pattern Mining and Rules

Exceptional Preferences Mining

Cláudio Rebelo de Sá[1,2(✉)], Wouter Duivesteijn[3], Carlos Soares[1],
and Arno Knobbe[2]

[1] INESCTEC, Porto, Portugal
claudio.r.sa@inesctec.pt, csoares@fe.up.pt
[2] LIACS, Universiteit Leiden, Leiden, The Netherlands
a.j.knobbe@liacs.leidenuniv.nl
[3] Data Science Lab and iMinds, Universiteit Gent, Ghent, Belgium
wouter.duivesteijn@ugent.be

Abstract. Exceptional Preferences Mining (EPM) is a crossover
between two subfields of datamining: local pattern mining and preference learning. EPM can be seen as a local pattern mining task that finds
subsets of observations where the preference relations between subsets of
the labels significantly deviate from the norm; a variant of Subgroup Discovery, with rankings as the (complex) target concept. We employ three
quality measures that highlight subgroups featuring exceptional preferences, where the focus of what constitutes 'exceptional' varies with the
quality measure: the first gauges exceptional overall ranking behavior,
the second indicates whether a particular label stands out from the rest,
and the third highlights subgroups featuring unusual pairwise label ranking behavior. As proof of concept, we explore five datasets. The results
confirm that the new task EPM can deliver interesting knowledge. The
results also illustrate how the visualization of the preferences in a Preference Matrix can aid in interpreting exceptional preference subgroups.

1 Introduction

Consider a survey where detailed preferences of sushi types have been collected,
along with demographic details of the respondents. For each example in the
dataset, we have personal details (age, gender, income, etc.) as well as a set of
sushi types, ordered by preference [16]. By mapping the demographic attributes
and unusual preferences, marketeers would be able to target key demographics
where specific sushi types have greater potential.

The study of preference data has been approached from a number of perspectives, grouped under the name *Preference Learning* (PL) (e.g., as Label Ranking
[3,5,24]). Typically, the aim is to build a global predictive model, such that the
preferences can be predicted for new cases. However, in several areas, such as
marketing, there is also great value in identifying subpopulations whose preferences deviate from the norm. If some sushi type is markedly underpreferred by
a certain age group or in a certain region, then the vendor can develop specific

© Springer International Publishing Switzerland 2016
T. Calders et al. (Eds.): DS 2016, LNAI 9956, pp. 3–18, 2016.
DOI: 10.1007/978-3-319-46307-0_1

strategies for those groups. Finding coherent groups of customers to focus on is an invaluable part of promotion strategies.

Arguably the most generic setting for discovering local, supervised deviations is that of Subgroup Discovery (SD) [18]. The aim of SD is to discover subgroups in the data for which the target shows an unusual distribution, as compared to the overall population [18]. SD is generic in the sense that the actual nature of the target variable can be quite diverse [1,15,23]. In this paper, we develop a Subgroup Discovery approach that focuses on a deviation target concept representing preferences over a fixed set of labels.

1.1   Main Contributions

This work provides focus specifically on the discovery of meaningful subgroups with exceptional preference patterns (see Sect. 4). We propose three quality measures for this purpose, reflecting different facets of interestingness one might have about the unusual preferences. All quality measures contrast the ranking of the labels in the subgroup with the ranking of the labels in the entire dataset; they differ in the granularity of the measured deviation. A subgroup is deemed interesting by the first quality measure if the overall ranking is exceptional, by the second quality measure if one particular label behaves exceptionally, and by the third quality measure if a single pair of labels displays exceptional behavior. Hence, Exceptional Preferences Mining provides subgroups displaying exceptional ranking behavior; different quality measures allow for this exceptional behavior to either encompass the entire label space, or focus on more local peculiarities.

2   Label Ranking

Label Ranking (LR) studies the problem of learning a mapping from instances to rankings over a finite number of predefined labels [13]. It can be considered a variant of the conventional classification problem [4]. However, in contrast to a classification setting, where the objective is to assign examples to a specific class, in LR we are interested in assigning a complete preference order of the labels to every example.

More formally, in classification, given an instance x from the instance space $\mathbb{X}$, the goal is to predict the label (or class) λ to which x belongs, from a predefined set $\mathcal{L} = \{\lambda_1, \ldots, \lambda_k\}$. In Label Ranking, the goal is to order the labels in $\mathcal{L}$ by their association with x. A ranking is a total order over $\mathcal{L}$ defined on the permutation space Ω. A total order can be represented as a permutation π of the set $\{1, \ldots, k\}$, such that $\pi(a)$ is the position of λ_a in π.

A total order

$$\lambda_{\pi(1)} \underset{x}{\succ} \lambda_{\pi(2)} \underset{x}{\succ} \ldots \underset{x}{\succ} \lambda_{\pi(k)}$$

is associated with every instance $x \in \mathbb{X}$, representing a ranking $\pi \in \Omega$. In cases where the orders are partial, they are represented as rankings with ties [11].

The goal in label ranking is to learn the mapping $\mathbb{X} \to \Omega$. The training data is defined as D, which is a bag of n records of the form $x = (a_1, \ldots, a_m, \pi)$, where $\{a_1, \ldots, a_m\}$ is set of values from m independent variables $\mathcal{A}_1, \ldots, \mathcal{A}_m$ describing instance x and π is the corresponding target ranking.

Pairwise comparisons have been used to decompose LR or Multi-Label problems into binary problems [13]. In LR, the most relevant approach is Ranking by Pairwise Comparisons (RPC) [10], which decomposes the LR problem into a set of binary classification problems. Then, a learning method is trained with all examples for which either $\lambda_i \succ \lambda_j$ or $\lambda_j \succ \lambda_i$ is known [10]. The resulting predictions are then combined to predict a total or partial ranking [3].

Recently, some approaches have been suggested for mining preferences and ranks [12, 19]. These approaches tackle different problems from the one we propose in this paper. In [12], the authors suggest an approach to mine the rankings with association rules that search for *subranking* patterns, while our approach relates the ranking patterns with descriptors (otherwise referred to as independent variables). From a different perspective, [19] suggests a *ranked tiling* approach to search for rank patterns, whereas we are interested in the preference relations derived from the ranks.

3 Subgroup Discovery and Exceptional Model Mining

Subgroup Discovery (SD) [18] is a data mining framework that seeks subsets (satisfying certain user-specified constraints) of the dataset where something exceptional is going on. In SD, we assume a flat-table dataset D, which is a bag of n records of the form $x = (a_1, \ldots, a_m, t_1, \ldots, t_\ell)$. We call $\{a_1, \ldots, a_m\}$ the *descriptors* and $\{t_1, \ldots, t_\ell\}$ the *targets*, and we denote the collective domain of the descriptors by $\mathcal{A}$. We are interested in finding interesting subsets, called *subgroups*, that can be formulated in a *description language* $\mathcal{D}$. In order to formally define subgroups, we first need to define the following auxiliary concepts.

Definition 1 (Pattern and coverage). *Given a description language $\mathcal{D}$, a pattern $p \in \mathcal{D}$ is a function $p : \mathcal{A} \to \{0, 1\}$. A pattern p covers a record x iff $p(a_1, \ldots, a_m) = 1$.*

Patterns induce subgroups, and subgroups are associated with patterns, in the following manner.

Definition 2 (Subgroup). *A subgroup corresponding to a pattern p is the bag of records $S_p \in D$ that p covers:*

$$S_p = \{x \in D \mid p(a_1, \ldots, a_m) = 1\}$$

For simplicity, we will loosely identify pattern and subgroup with each other.

The exact choice of the description language is left to the domain expert or analyst. A typical choice is the use of conjunctions of conditions on attributes.

Restricting the findings of SD from all subsets to only subgroups that can be defined in such a way, ensures results of the following form:

$$\text{Age} \geq 30 \ \wedge \ \text{Likes} = \text{Salmon Roe} \quad \text{is unusual}$$

Restricting the search from subsets to subgroups, combined with a sensible choice of description language, ensures that SD delivers subgroups that are defined in terms of attributes of the dataset. This means that the results are delivered in a form with which dataset domain experts are familiar. In other words, the focus of SD lies on delivering *interpretable* results.

Formally, the interestingness of a subgroup can be measured using all information available in its associated pattern. In practice, it depends on the task we are trying to solve. Therefore, we should define one or more *quality measures* to assess the interestingness we want to explore.

Definition 3 (Quality Measure). *A* quality measure *is a function* $\varphi : \mathcal{D} \to \mathbb{R}$.

In the most common form of pattern mining, *frequent itemset mining* [2], interestingness is measured by the frequency of the pattern. *Subgroup Discovery* [18], on the other hand, measures interestingness in a *supervised* form. One designated target t_1 is identified in the dataset, and subgroup interestingness is gauged by an unusual distribution of that target. Hence, considering that a poll revealed that the majority of Japanese people like *Fatty tuna* sushi, an interesting subgroup could refer to a group of people for which the majority prefers *Tuna roll*:

$$\text{Age} \geq 30 \ \wedge \ \text{Lives in region} = \text{Hokkaido} \quad \Rightarrow \quad \text{Likes} = \text{Tuna roll}$$

If instead of a single target, multiple targets $t_1, \ldots, t_\ell$ are available, and if we are not interested in finding unusual target distribution, but unusual target interaction, we can employ *Exceptional Model Mining* (EMM) [6,7] instead of SD. EMM is instantiated by selecting two things: a model class and a quality measure. Typically, a model class is defined to represent the unusual interaction between multiple targets we are interested in. A specific quality measure that employs concepts from that model class must be defined to express exactly when an interaction is unusual and, therefore, interesting.

The target concept at hand in this paper has only one target object t, which resembles SD. However, that target object is a label ranking $\pi \in \Omega$, as defined in Sect. 2. Hence it represents unusual interactions between multiple individual labels, which is more consistent with EMM.

3.1   Traversing the Search Space

Typically, *subgroups* are found by a level-wise search through attribute space [21]. We define constraints on single attributes and define the corresponding subgroups as those records satisfying each one of those constraints.

The actual phenomenon of the data that a given quality measure favors, depends on the target concept (binary, numeric, preferences, ...). For very small subgroups, one easily finds an unusual distribution of the target. To favor larger subgroups, one defines the quality measure such that it balances the exceptionality of the target distribution with the size of the subgroup.

SD approaches have been developed for binary, nominal [1] and numeric target variables [14,15], as well as for targets encompassing multiple attributes [23]. However, none of the previous approaches is able to capture all the sets of preferences that can be derived from rankings within an SD framework.

4 Exceptional Preferences Mining

Exceptional Preferences Mining (EPM) is the search for subgroups with deviating preferences. Exactly what constitutes an interesting deviation in preferences is governed by the employed quality measure, and can be inspired by the application at hand.

When the number of labels is large, the search for preference patterns can be hard to analyze and visualize. A real world example is the Sushi dataset [16], which represents the preferences of 5 000 persons over 10 types of sushi. Even this relatively modest number of sushi types can be ranked in a large number of combinations: more than 98 % of the 5 000 rankings present in this dataset are unique. This illustrates why it can be more difficult to directly learn a ranker that associates a reliable complete ranking for any subset in $\mathbb{X}$ when the number of labels is large.

In EPM, we want to search for strong preference behavior. However, in cases like the Sushi dataset, it is difficult to get strong total orders, due to the low number of ranking repetitions. In other words, searching for subgroups where all types of sushi are consistently ranked in this exact same order can be unfruitful. For this reason, we also propose lower-granularity measures that focus on one label versus the others (Labelwise). That is, we look for subgroups where at least one type of sushi is often preferred to all the others. As an example, if a subgroups ranks *tekka-maki* consistently in the top 3 while the majority in the dataset ranks it in the last 3, this measure will find it to be very interesting. We also propose a measure of even lower granularity, focusing on label versus label (Pairwise) preferences. This means that, if most people display a preference *tamago* $\succ$ *kappa-maki*, a subgroup where most people prefer *kappa-maki* $\succ$ *tamago* will be deemed interesting by this measure.

Our assumption is that, even though over 98 % of the total rankings in the Sushi dataset are unique, there is plenty of information present in these rankings: the partial orders and pairwise comparisons can reveal interesting subgroups.

4.1   Preference Matrix

Let us define a function, ω, assigning a numeric value to the pairwise comparison of the labels λ and $\hat{\lambda}$:

$$\omega\left(\lambda,\hat{\lambda}\right) = \begin{cases} 1 & \text{if } \lambda \succ \hat{\lambda} \; (\lambda \text{ preferred to } \hat{\lambda}) \\ -1 & \text{if } \lambda \prec \hat{\lambda} \; (\hat{\lambda} \text{ preferred to } \lambda) \\ 0 & \text{if } \lambda \sim \hat{\lambda} \; (\lambda \text{ indifferent to } \hat{\lambda}) \\ n/a & \text{if } \lambda \perp \hat{\lambda} \; (\lambda \text{ incomparable to } \hat{\lambda}) \end{cases}$$

Note that, by definition, $\omega\left(\lambda,\hat{\lambda}\right) = -\omega\left(\hat{\lambda},\lambda\right)$. We can use ω to represent a ranking π as a *Preference Matrix* (PM), M_π:

$$M_\pi\left(i,j\right) = \omega_\pi\left(\lambda_i,\lambda_j\right)$$

M_π is, by definition, an antisymmetric matrix with $tr\left(M_\pi\right) = 0$. PMs can natively represent partial or incomplete orders but can also be aggregated to represent sets of rankings from an entire dataset D or subgroup S. To aggregate the entries, the *mean* or the *mode* can be used.

Table 1. Example dataset $\hat{D}$. The first column is the only descriptor. The subsequent four columns represent the preferences among four labels, by providing their ranks. An alternative representation is presented in the rightmost section of the table.

$\mathcal{A}_1$	π				Alternative π
	λ_1	λ_2	λ_3	λ_4	
0.1	4	3	1	2	$\lambda_3 \succ \lambda_4 \succ \lambda_2 \succ \lambda_1$
0.2	3	2	1	4	$\lambda_3 \succ \lambda_2 \succ \lambda_1 \succ \lambda_4$
0.3	1	4	2	3	$\lambda_1 \succ \lambda_3 \succ \lambda_4 \succ \lambda_2$
0.4	1	3	2	4	$\lambda_1 \succ \lambda_3 \succ \lambda_2 \succ \lambda_4$

Aggregation of a PM for Sets of Rankings. The PM of a set of rankings from a dataset D with n rankings, M_D, aggregated with the mean is:

$$M_D\left(i,j\right) = \frac{1}{n}\sum_{\pi \in D} M_\pi\left(i,j\right)$$

The PM of the example dataset $\hat{D}$ (cf. Table 1) is the following:

$$M_{\hat{D}} = \begin{bmatrix} 0 & 0 & 0 & 0.5 \\ 0 & 0 & -1 & 0 \\ 0 & 1 & 0 & 1 \\ -0.5 & 0 & -1 & 0 \end{bmatrix}$$

This representation enables easy detection of strong partial order relations in a set. If row i has all the values very close to 1, then λ_i is highly preferred in this group. If entry $M_{\hat{D}}(i,j) = 1$ or $M_{\hat{D}}(i,j) = -1$, then all rankings in $\hat{D}$ agree that $\lambda_i \succ \lambda_j$ or $\lambda_i \prec \lambda_j$, respectively.

All the elements of $\hat{D}$ reveal distinct total preferences, but λ_3 is always preferred to λ_2, which is easily verified by checking that $M_{\hat{D}}(3,2) = 1$. In the ranking representation of $\hat{D}$, this fact follows from four distinct combinations of ranks: rank $3 > 1$, rank $2 > 1$, rank $4 > 2$ and rank $3 > 2$ (this information is found in the two columns below λ_2 and λ_3). Conversely, λ_4 is never preferred to λ_3, which is represented by $M_{\hat{D}}(4,3) = -1$. In some cases, the overall trend is not as clear (e.g. λ_1 is preferred to λ_4 but not always) and in other cases, there is no trend at all (e.g. λ_1 and λ_2).

Representing a set of rankings as a PM has another advantage over the traditional permutation representation: it enables simple measurement of *labelwise* (by rows/columns of the PM) and *pairwise* (by single entries of the PM) distances (see Sect. 4.2).

From the PM of a subgroup S, one can derive a new ranking π_S. How to do so is a non-trivial question, which has received a lot of attention in several research fields with similar types of matrix [13]. The straightforward way is to sum the rows of the PM and then assign a score to each corresponding label. Higher values correspond to a relatively more preferred label.

The generation of a PM is basically a pairwise decomposition problem. The complexity is $\mathcal{O}\left(sk^2\right)$ per subgroup, where s is the size of the subgroup and k the number of labels in the ranking. Even though any number of labels is theoretically permitted in label ranking, in practice the number of labels is usually smaller than 20. Hence, the generation of PMs should not be an issue in terms of computational time.

We use a visual representation of PM that is a set of colored tiles (cf. Fig. 1). Each tile represents an entry of the PM. The entries of a PM can vary from -1 to 1. The negative entries of the matrix are represented with red tiles, the positive with green tiles, and 0 is represented in white. The colored tiles fade out as they get closer to 0.

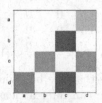

Fig. 1. PM representation of the set of rankings in $\hat{D}$ (cf. Table 1). Dark green tiles represent 1 and dark red tiles respresent -1. (Color figure online)

4.2   Characterizing Exceptional Subgroups

The table has now been set to formally define the quality measures for EPM, which will evaluate how exceptional the preferences are in the subgroups. A subgroup can be considered interesting both by the amount of deviation (distance) and by its size (number of records *covered* by the subgroup, cf. Sect. 3) [9]. Since, reasonable quality measures should take both these factors into account, we divide the quality measures into two parts: the *distance* component and the *size* component.

$$QM_S = size_S \cdot distance_S$$

In order to allow direct comparisons between different quality measures, both components are normalized to the interval $[0, 1]$. A common measure for the size in Subgroup Discovery is $\sqrt{s}$ [17]. To normalize, we use the square root of the fraction of the dataset covered by S: $size_S = \sqrt{s/n}$.

Before introducing the distance components, let us first define a distance matrix L_S, as the distance matrix between the PMs M_S and M_D:

$$L_S = \frac{1}{2}(M_D - M_S)$$

where $S \subseteq D$ (division by 2 limits the distance to the interval $[-1, 1]$). We can measure different properties of L_S and represent them with a numeric value. This way we get an indicator of the quality of the distance of preferences for a subgroup. Consider the subgroup $\hat{S}_1 : A_1 \geq 0.3$, which covers the last two cases from our example dataset $\hat{D}$. Its PM is:

$$M_{\hat{S}_1} = \begin{bmatrix} 0 & 1 & 1 & 1 \\ -1 & 0 & -1 & 0 \\ -1 & 1 & 0 & 1 \\ -1 & 0 & -1 & 0 \end{bmatrix}$$

The first row clearly reveals that λ_1 is always preferred to all other labels in this subgroup. If we compute the difference matrix $L_{\hat{S}_1}$ we get:

$$L_{\hat{S}_1} = \begin{bmatrix} 0 & -0.5 & -0.5 & -0.25 \\ 0.5 & 0 & 0 & 0 \\ 0.5 & 0 & 0 & 0 \\ 0.25 & 0 & 0 & 0 \end{bmatrix}$$

The difference matrix $L_{\hat{S}_1}$ shows that the behavior of λ_1 is exceptional in $\hat{S}_1$.

Only subgroups for which we can infer at least one pairwise preference are considered interesting in Exceptional Preferences Mining. That is, subgroups with a PM containing only zeros are not considered interesting.

As we are interested in subgroups with exceptional preferences, we use the distance matrix L_S to measure exceptionality. The distance measures we employ here typically consider a particular subset of the cells of the distance matrix L_S.

Norm. Maximizing the distance of preferences is also maximizing the magnitude of L_S. The most fundamental mathematical way to measure the magnitude of a vector or matrix is the *norm*. Hence we can use the *Frobenius norm* of L_S as a distance measure.

$$\text{Norm}(S) = \sqrt{s/n} \cdot \|L_S\|_F = \sqrt{s/n} \cdot \sqrt{\sum_{i=1}^{k}\sum_{j=1}^{k} L(i,j)^2}$$

If one is searching for preference deviations in general, one should use the *Norm* quality measure, as it considers all the PM entries at the same time. After the subgroups are found, ideally, we can derive a complete ranking from their PMs. The overall deviation can be due to one label deviating strongly or from multiple labels deviating less strongly.

Labelwise. An interesting task in the PL field is the *labelwise* analysis [3]. Instead of focusing on a whole ranking, it focuses on the preference behavior from the perspective of individual labels. A data analyst might be interested in finding if a particular label λ behaves substantially different according to most members in a subgroup S, compared to its behavior on the overall dataset. Hence, the fact that only one label behaves differently, disregarding the interaction between the other labels, can also be interesting. We can measure the distance of each label, in subgroup S, by computing the *norm* of the rows from L_S. Since in this case we are interested in exceptionality of only one label, we consider the maximum value found:

$$\text{Labelwise}(S) = \sqrt{s/n} \cdot \max_{i=1,\ldots,k} \frac{1}{(k-1)} \sum_{j=1}^{k} L(i,j)$$

Pairwise. Another well-studied task in PL is Pairwise Preferences [13] which decomposes the preferences into pairs label-vs-label. In situations where there are not even exceptional labelwise preferences, one can still search for localized preference strongholds. If we are interested in subgroups with, at least one pair with distinctive preference behavior, we can employ the following quality measure:

$$\text{Pairwise}(S) = \sqrt{s/n} \cdot \max_{i,j=1,\ldots,k} L(i,j)$$

This quality measure is the least restrictive of this set: a subgroup is interesting if one pair of labels interacts unusually, disregarding all other label interactions.

5   Experiments

We incorporate Exceptional Preferences Mining in the Cortana[1] software package [22]. This package delivers a generic framework for SD, implements several SD instances, and offers many generic features allowing for different SD

---

[1] http://datamining.liacs.nl/cortana.html.

approaches. The description language consists of logical conjunctions of conditions on single attributes.

Our experiments use a standard *beam search* approach. Since the Subgroup Discovery algorithm itself is not the topic of this paper, we will skip over the algorithmic details, but they can be found elsewhere: the relevant pseudo-code is given in [6, Algorithm 1]. The most influential parameters are set as follows: we use a relatively generous search width w (also known as beam width or beam size) of 100, allowing for a relatively broad (albeit heuristic) search, and a maximum search depth d of 2, which keeps the resulting subgroups interpretable. We explore some striking subgroups found with the quality measures on a variety of datasets, providing evidence of the versatility of our work.

All the findings we present in this paper have gone through the DFD validation procedure [8] with 100 copies, and all have been found significant at a significance level of $\alpha = 1\%$.

5.1   Datasets

Statistics regarding the datasets used in this work are shown in Table 2. The majority are Label Ranking datasets from the KEBI Data Repository at Philipps University of Marburg [4]. These datasets were adapted from multi-class and regression problems both from the UCI repository [20] and the Statlog collection [4]. In the process, the features were normalized, and their names were replaced by $A1, A2, \ldots, Am$. Therefore, on these datasets, the reported subgroups cannot be interpreted on the original dataset domain, whereas for general datasets, this interpretability is a key feature of Exceptional Preference Mining. We choose to experiment with these datasets anyway, since they are well-known in the preference learning community.

For illustrating domain-specific interpretation of the results, we experiment with two further datasets. We adapt the COIL 1999 Competition Data from UCI [20]. This dataset concerns the frequencies of algae populations in different environments. We refer to this dataset as *Algae*. The original COIL dataset consists of 340 examples, each representing measurements of a sample of water from different European rivers in different periods. The measurements include concentrations of chemical substances such as nitrogen (in the form of nitrates, nitrites and ammonia), oxygen and chlorine. Also the pH, season, river size and flow velocity are registered. For each sample, the frequencies of 7 types of algae are also measured. In this work, we consider the algae concentrations as preference relations by ordering them from larger to smaller concentrations. Those with 0 frequency are placed in last position and equal frequencies are represented with ties. Missing values are set to 0.

Our final dataset is the Sushi preference dataset [16], which is composed of demographic data about 5 000 people and sushi preferences. Each person sorts a set of 10 different sushi types by preference. The 10 types of sushi, are (a) shrimp, (b) sea eel, (c) tuna, (d) squid, (e) sea urchin, (f) salmon roe, (g) egg (h) fatty tuna, (i) tuna roll and (j) cucumber roll. Since the attribute names were not transformed in this dataset, we can make a richer analysis of it.

Table 2. Dataset details. The column U_π represents the percentage of unique rankings.

Datasets	#Examples	#Labels	#Attributes	U_π
Cpu-small	8 192	5	6	1 %
Elevators	16 599	9	9	1 %
Wisconsin	194	16	16	100 %
Algae (COIL)	316	7	10	72 %
Sushi	5 000	10	10	98 %

For all the experiments, all results and statistical tests are completed in less than 5 min on an Intel Core 2 Duo CPU @ 2.93 GHz with 4 GB RAM.

5.2   Results

We start this section by presenting a discovery which provides an exemplary demonstration of one advantage of the PM representation.

Elevators Dataset. Figure 2 shows the subgroup with highest score found with the *Norm* quality measure in the Elevators dataset. Considering the base matrix, which has information from all the rankings in the dataset, we conclude that e, f, g, h have fixed relative positions: $e \succ g \succ f \succ h$. This information is not easy to obtain with the usual representations of rankings, but is clearly revealed in the PM representation. In fact, 13 403 from a total of 16 599 rankings have $e \succ g \succ f \succ h$. This illustrates how the visual ranking representation in a PM can be very useful for supporting predictive methods and for data exploration. The subgroup, $A6 \geq 0.436$, covering 7 048 instances, had a norm of 0.0028. It shows a distinct behavior between the sets of labels a, b, c, d and the set e, f, g, h. In the whole data, labels a, b, c, d are a bit more desirable than e, f, g, h. However, in the subgroup, the latter are clearly preferred to a, b, c, d.

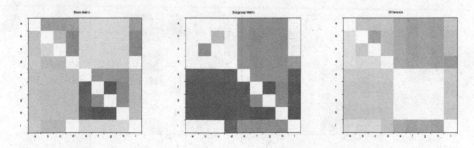

Fig. 2. PM representation of the dataset Elevators (base matrix), the subgroup $A6 \geq$ 0.436 (subgroup matrix) and the difference (difference matrix).

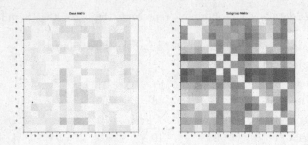

Fig. 3. PM representation of the dataset Wisconsin (Base Matrix) and the subgroup $A_5 \leq -0.527$ (Subgroup Matrix). (Color figure online)

Wisconsin. Using the *Norm* quality measure on the Wisconsin dataset, we obtain 30 subgroups, the 1st-ranked of which (it happens to occur at depth 1 in the search) is represented in Fig. 3. The base matrix reveals that the dataset has balanced preferences, by the low intensity of the colored tiles. The red rows of the PM of subgroup $A_5 \leq -0.527$ (Subgroup Matrix in Fig. 3) indicate a strong behavior of the labels f, h and i. The PM reveals that labels f, h, i are consistently ranked lower than the other labels in this specific subgroup. Since PMs are antisymmetric, the 3 green columns represent the same phenomena but from the perspective of the other labels. If we focus on these 3 labels, we can see that tile (f, h) is white, which means f and h are equivalent. On the other hand, tiles (i, f) and (i, h) are green, which means that $i \succ f$ and $i \succ h$. If one had to guess a reliable partial order from this subgroup using only the PM, a logical choice would be to say that $a, b, c, d, e, g, j, k, l, m, n, o, p \succ i \succ f, h$.

Algae. With the Algae dataset, we obtain results about the concentrations of algae with the *Norm* measure. One such example is that during *Spring*, the types of algae a, b and c are much more common in rivers than the others. This can be easily concluded by studying the PM representation of the subgroup

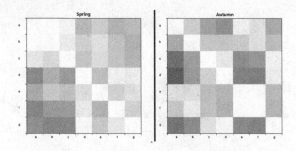

Fig. 4. PM representation of the subgroups *Season = Spring* (left subgroup matrix) and *Season = Autumn* (right subgroup matrix) from the Algae dataset.

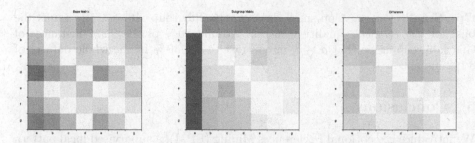

Fig. 5. PM representation of the dataset Algae (base matrix) and the subgroup $V10 \leq 59 \land V6 \leq 11.867$ (subgroup matrix), with difference matrix on the right.

(Fig. 4). This subgroup has a norm of 0.010647. On the other hand, we also see an interesting behavior during the *Autumn* season, with a norm of 0.01058.

With the *Labelwise* measure, we find more than 400 subgroups, the best of which is presented in Fig. 5. The PM clearly reveals the effect of the Labelwise quality measure: in the subgroup, the label a is strongly preferred over all others, while the image is much more nuanced over the whole dataset. If we ignore the label a, the PMs for both the overall dataset and the subgroup are rather bland, and their difference is not very pronounced. But for this one particular label a, the behavior on the subgroup is extremely clear-cut, and the Labelwise quality measure picks up on that effect.

Sushi. With the *Labelwise* measure, we find 149 subgroups on the Sushi dataset. We present the best subgroup using this measure in Fig. 6. The subgroup (Males over 30 years) shows a preference for Sea Urchin, since the majority of men rank this sushi type in the top 4. By contrast, in the whole population, more than half rate it between 5th to 10th, and every fifth person rate it in last place.

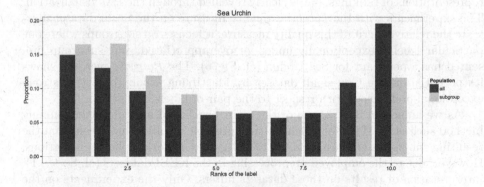

Fig. 6. Percentage of ranks for Sea Urchin (Sushi dataset) for all individuals in comparison to the subgroup (males older than 30 years).

Cpu-Small. On the Cpu-small dataset, the subgroup $A6 \geq 0.127$ ranks the best for the *Pairwise* quality measure. Around 80 % of the 2 221 instances of this subgroup agree that $a \succ d$, in contrast to the 30 % in the whole dataset of 8 192 instances.

6   Conclusions

We introduce Exceptional Preferences Mining (EPM), a supervised local pattern mining task where the target concept is a ranking of a fixed set of labels. The result of this task is a set of subgroups, which are coherent subsets of the dataset that can be described in terms of a conjunction of few conditions on an attribute, where the label preferences are exceptional in some sense.

The relevant statistics on a set of preference relations is collected in the cells of a Preference Matrix (PM). A PM is compiled for the entire dataset, and for each subgroup under consideration. A subgroup whose PM deviates significantly from the PM for the whole dataset is then considered to be interesting. We define three quality measures for EPM that instantiate this concept of 'interesting' to different levels of granularity. The *Norm* quality measure deems a subgroup interesting if the full set of preference relations is substantially displaced. The *Labelwise* quality measure highlights subgroups where any one label interacts exceptionally with the other labels, agnostic of how those other labels interact with each other. The *Pairwise* quality measure finds a subgroup interesting if any one pair of labels display exceptional preference relations. Hence, by choosing the appropriate quality measure, EPM delivers subgroups featuring preference relations that are exceptional at your preferred scope.

The experiments with the *Norm* quality measure on the Elevators dataset illustrate the value of the PM visualization. The PM, as displayed in Fig. 2, clearly indicates that there are strong relations between a subset of the available labels. We learn that quite frequently, labels e, f, g, h have fixed relative positions: $e \succ g \succ f \succ h$. This information is not easy to obtain with the usual representations of rankings, but is clearly revealed through the PM visualization. The experiments with the *Labelwise* quality measure on the Sushi dataset illustrate the relative merit of this quality measure: it focuses on subgroups where one particular label is exceptionally under- or overappreciated. The subgroup presented has a penchant for Sea Urchin (cf. Fig. 6). The *Pairwise* measure shows its potential on the Cpu-small dataset by identifying a subgroup with strong exceptional preferences with respect to the pair of labels a and d.

As we argued in Sect. 3, one of the main benefits of a local pattern mining method such as EPM is that it delivers interpretable results. That means that the resulting subgroups are ideally suited to instigate real-world policies and actions. However, due to the employed preprocessing in the KEBI datasets (cf. Sect. 5.1), interpretation of results on those datasets falters. Only the experiments on the Algae and Sushi datasets allow a more extensive exploration of interpretable results. In future work, we would be interested in evaluating EPM on more label ranking datasets that come with interpretable attributes.

Acknowledgments. This work was supported by the European Union through the ERC Consolidator Grant FORSIED (project reference 615517). This work was also financed by the ERDF - European Regional Development Fund through the Operational Programme for Competitiveness and Internationalisation - COMPETE 2020 Programme within project POCI-01-0145-FEDER-006961 and by National Funds through the FCT - Fundação para a Ciência e a Tecnologia (Portuguese Foundation for Science and Technology) as part of project UID/EEA/50014/2013.

References

1. Abudawood, T., Flach, P.: Evaluation measures for multi-class subgroup discovery. In: Buntine, W., Grobelnik, M., Mladenić, D., Shawe-Taylor, J. (eds.) ECML PKDD 2009, Part I. LNCS (LNAI), vol. 5781, pp. 35–50. Springer, Heidelberg (2009). doi:10.1007/978-3-642-04180-8_20

2. Agrawal, R., Mannila, H., Srikant, R., Toivonen, H., Verkamo, A.: Fast discovery of association rules. Adv. Knowl. Disc. Data Min. **12**, 307–328 (1996)

3. Cheng, W., Henzgen, S., Hüllermeier, E.: Labelwise versus pairwise decomposition in label ranking. In: LWA 2013, Lernen, Wissen & Adaptivität, Workshop Proceedings Bamberg, 7–9 October 2013, pp. 129–136 (2013)

4. Cheng, W., Huhn, J.C., Hüllermeier, E.: Decision tree and instance-based learning for label ranking. In: Proceedings of the 26th Annual International Conference on Machine Learning, ICML 2009, Montreal, Quebec, Canada, 14–18 June 2009, pp. 161–168 (2009)

5. de Sá, C.R., Soares, C., Knobbe, A.J.: Entropy-based discretization methods for ranking data. Inf. Sci. **329**, 921–936 (2016)

6. Duivesteijn, W.: Exceptional model mining. Ph.D. Thesis, Leiden University (2013)

7. Duivesteijn, W., Feelders, A., Knobbe, A.: Exceptional model mining. Data Min. Knowl. Disc. **30**, 47–98 (2016)

8. Duivesteijn, W., Knobbe, A.J.: Exploiting false discoveries - statistical validation of patterns and quality measures in subgroup discovery. In: Proceedings of ICDM, pp. 151–160 (2011)

9. Dzyuba, V., Leeuwen, M.: Interactive discovery of interesting subgroup sets. In: Tucker, A., Höppner, F., Siebes, A., Swift, S. (eds.) IDA 2013. LNCS, vol. 8207, pp. 150–161. Springer, Heidelberg (2013). doi:10.1007/978-3-642-41398-8_14

10. Fürnkranz, J., Hüllermeier, E.: Pairwise preference learning and ranking. In: Lavrač, N., Gamberger, D., Blockeel, H., Todorovski, L. (eds.) ECML 2003. Lecture Notes in Artificial Intelligence (LNAI), vol. 2837, pp. 145–156. Springer, Heidelberg (2003). doi:10.1007/978-3-540-39857-8_15

11. Fürnkranz, J., Hüllermeier, E.: Preference Learning, 1st edn. Springer, New York (2010)

12. Henzgen, S., Hüllermeier, E.: Mining rank data. In: Japkowicz, N., Matwin, S. (eds.) DS 2014. LNCS (LNAI), vol. 8777, pp. 123–134. Springer, Heidelberg (2014). doi:10.1007/978-3-319-11812-3_11

13. Hüllermeier, E., Fürnkranz, J., Cheng, W., Brinker, K.: Label ranking by learning pairwise preferences. Artif. Intell. **172**(16–17), 1897–1916 (2008)

14. Jin, N., Flach, P., Wilcox, T., Sellman, R., Thumim, J., Knobbe, A.J.: Subgroup discovery in smart electricity meter data. IEEE Trans. Industr. Inf. **10**(2), 1327–1336 (2014)

15. Jorge, A.M., Pereira, F., Azevedo, P.J.: Visual interactive subgroup discovery with numerical properties of interest. In: Japkowicz, N., Matwin, S. (eds.) DS 2006. LNCS (LNAI), vol. 4265, pp. 301–305. Springer, Heidelberg (2006). doi:10.1007/11893318_31

16. Kamishima, T.: Nantonac collaborative filtering: recommendation based on order responses. In: Proceedings of KDD, pp. 583–588 (2003)

17. Klösgen, W.: Explora: a multipattern and multistrategy discovery assistant. In: Advances in Knowledge Discovery and Data Mining, pp. 249–271 (1996)

18. Klösgen, W., Zytkow, J.M. (eds.): Handbook of Data Mining and Knowledge Discovery. Oxford University Press, New York (2002)

19. Van, T., Leeuwen, M., Nijssen, S., Fierro, A.C., Marchal, K., Raedt, L.: Ranked tiling. In: Calders, T., Esposito, F., Hüllermeier, E., Meo, R. (eds.) ECML PKDD 2014. LNCS (LNAI), vol. 8725, pp. 98–113. Springer, Heidelberg (2014). doi:10.1007/978-3-662-44851-9_7

20. Lichman, M.: UCI Machine Learning Repository (2013)

21. Mannila, H., Toivonen, H.: Levelwise search and borders of theories in knowledge discovery. Data Min. Knowl. Discov. 1(3), 241–258 (1997)

22. Meeng, M., Knobbe, A.: Flexible enrichment with cortana-software demo. In: Proceedings of BeneLearn, pp. 117–119 (2011)

23. Umek, L., Zupan, B.: Subgroup discovery in data sets with multi-dimensional responses. Intell. Data Anal. 15(4), 533–549 (2011)

24. Vembu, S., Gärtner, T.: Label ranking algorithms: a survey. In: Fürnkranz, J., Hüllermeier, E. (eds.) Preference Learning, pp. 45–64. Springer, Heidelberg (2010)

Local Subgroup Discovery for Eliciting and Understanding New Structure-Odor Relationships

Guillaume Bosc[1]([✉]), Jérôme Golebiowski[3], Moustafa Bensafi[4],
Céline Robardet[1], Marc Plantevit[2], Jean-François Boulicaut[1],
and Mehdi Kaytoue[1]

[1] Université de Lyon, CNRS, INSA-Lyon, LIRIS, UMR5205, 69621 Lyon, France
guillaume.bosc@insa-lyon.fr
[2] Université de Lyon, CNRS, Université Lyon 1, LIRIS, UMR5205,
69622 Lyon, France
[3] Université de Nice, CNRS, Institute of Chemistry, Nice, France
[4] Université de Lyon, CNRS, CRNL, UMR5292, INSERM U1028, Lyon, France

Abstract. From a molecule to the brain perception, olfaction is a complex phenomenon that remains to be fully understood in neuroscience. A challenge is to establish comprehensive rules between the physico-chemical properties of the molecules (e.g., weight, atom counts) and specific and small subsets of olfactory qualities (e.g., fruity, woody). This problem is particularly difficult as the current knowledge states that molecular properties only account for 30 % of the identity of an odor: predictive models are found lacking in providing universal rules. However, descriptive approaches enable to elicit local hypotheses, validated by domain experts, to understand the olfactory percept. Based on a new quality measure tailored for multi-labeled data with skewed distributions, our approach extracts the top-k unredundant subgroups interpreted as descriptive rules $description \rightarrow \{subset\ of\ labels\}$. Our experiments on benchmark and olfaction datasets demonstrate the capabilities of our approach with direct applications for the perfume and flavor industries.

1 Introduction

Around the turn of the century, the idea that modern, civilized human beings might do without being affected by odorant chemicals became outdated: the hidden, inarticulate sense associated with their perception, hitherto considered superfluous to cognition, became a focus of study in its own right and thus the subject of new knowledge. It was acknowledged as an object of science by Nobel prizes (e.g., [2] awarded 2004 Nobel prize in Physiology or Medicine); but also society as a whole was becoming more hedonistic, and hence more attentive to the emotional effects of odors. Odors are present in our food, which is a source of both pleasure and social bonding; they also influence our relations with others in general and with our children in particular. The olfactory percept encoded in odorant chemicals contribute to our emotional balance and wellbeing.

© Springer International Publishing Switzerland 2016
T. Calders et al. (Eds.): DS 2016, LNAI 9956, pp. 19–34, 2016.
DOI: 10.1007/978-3-319-46307-0_2

While it is generally agreed that the physicochemical characteristics of odorants affect the olfactory percept, no simple and/or universal rule governing this Structure Odor Relationship (SOR) has yet been identified. Why does this odorant smell of roses and that one of lemon? Considering that the totality of the odorant message was encoded within the chemical structure, chemists have tried to identify relationships between chemical properties and odors. However, it is now quite well acknowledged that structure-odor relationships are not bijective. Very different chemicals trigger a typical "camphor" smell, while a single molecule, the so-called "cat-ketone" odorant, elicit two totally different smells as a function of its concentration [4]. At best, such SOR rules are obtained for a very tiny fraction of the chemical space, emphasizing that they must be decomposed into sub-rules associated with given molecular topologies [5]. A simple, universal and perfect rule does probably not exist, but instead, a combination of several sub-rules should be put forward to encompass the complexity of SOR.

In this paper, we propose a data science approach with a view to advance the state of the art in understanding the mechanisms of olfaction. We create an interdisciplinary synergy between neuroscientists, chemists and data miners to the emergence of new hypotheses. Indeed, data-mining methods can be used to answer the SOR discovery problem, either through the building of predictive models or through rules discovery in pattern mining. One obstacle to this is that olfactory datasets are very complex (i.e., several thousand of dimensions, heterogeneous descriptors, multi-label, unbalanced classes, and non robust labelling) and, above all a lack of data-centric methods in neuroscience suitable for this level of complexity. The main aim of our study is to examine this issue by linking the multiple molecular characteristics of odorant molecule to olfactory qualities (fruity, floral, woody, etc.) using a descriptive approach (pattern mining). Indeed, a data science challenge was recently proposed by *IBM Research* and *Sage* [12]. Results suggest difficulties in the prediction of the data for *olfactory datasets* in general. The reason is that there is a strong inter- and intra-individual variability when individuals are asked about the quality of an odor. There are several explanations: geographical and cultural origins, each individual repertory of qualities (linguistic), genetic differences (determining olfactory receptors), troubles such as *anosmia* (see [3,10]). It appears that designing pure predictive models remains today a challenge, because it depends on the individual's genome, culture, etc. Most importantly, the most accurate methods generally never suggest a descriptive understanding of the classes, while fundamental neurosciences need descriptive hypotheses through exploratory data analysis, i.e., descriptions that partially explain SOR. For that, we develop a descriptive approach to make the results intelligible and actionable for the experts.

The discovery of (molecular) descriptions which distinguish a group of objects given a target (class label, i.e. odor quality(ies)) has been widely studied in AI, data mining, machine learning, etc. Particularly, supervised descriptive rules were formalized through subgroup discovery, emerging-pattern/contrast-sets mining, etc. [14]. In all cases, we face a set of objects associated to descriptions (which forms a partially ordered set), and these objects are related to one or several class labels. The strength of the rule (SOR in our application) is evaluated through a quality measure (F1-measure, accuracy, etc.). The issues of multi-labeled datasets have been deeply studied in the state of the art [15]. However, to the best of our knowledge, most of existing methods to explore multi-label data are learning tasks. The existing descriptive approach known as Exceptional Model Mining (EMM) deals with multi-label data but it only considers them together, and not separately. Indeed, this method extracts subsets of objects (e.g., odorants) which distribution on all labels (e.g., odors) is statistically different (i.e., exceptional) w.r.t. the distribution of the entire set of objects. However, we aim to focus on subsets of few labels at a time. Moreover, the experts expect rules highlighting which values of features result in a subset of labels, and not only to extract relevant features for some labels as feature selection does. Our contributions are as follows:

- We explain the main problems of existing descriptive rule discovery approaches for dataset such as olfactory datasets, that are (i) multi-labeled with (ii) an unbalanced label distribution (i.e., a high variance in the labels occurrences).
- For (i), we consider the enumeration of pairs consisting of a description and a subset of labels (a variant of redescription mining [9]).
- For (ii), we propose a new measure derived from the F-score but less skewed by imbalance distribution of labels and that can dynamically consider the label distributions. We show this fact both theoretically and experimentally.
- We devise an algorithm which explores the search space with a beam-search strategy. It comes with two major issues that we jointly tackle: Finding the best cut points of numerical attributes during the exploration, also overcoming a redundancy among the extracted patterns.
- We thoroughly demonstrate the actionability of the discovered subgroups for neuroscientists and chemists.

The rest of the paper is organized as follows. We formally define the SOR discovery problem in Sect. 2 and we show why state-of-the-art methods are not adapted for this problem. We present our novel approach in Sect. 3 while the algorithmic details are given in Sect. 4. We report an extensive empirical study and demonstrate the actionability of the discovered rules in Sect. 5.

2   Problem Formulation

In this section, we formally define our data model as well as its main characteristics before introducing the problem of mining discriminant descriptive rules in these new settings. Indeed, we recall after the two most general approaches that can deal with our problem although only partially, namely *subgroup discovery* [14] and *redescription mining* [9]. We demonstrate this fact and highlight their weaknesses through an application example.

Definition 1 (Dataset $\mathcal{D}(\mathcal{O}, \mathcal{A}, C, class)$). *Let $\mathcal{O}$ and $\mathcal{A}$ be respectively a set of objects (molecules) and a set of attributes (physicochemical properties). The value*

Table 1. Toy olfactory dataset.

ID	MW	nAT	nC	Quality	ID	MW	nAT	nC	Quality
1	150.19	21	11	{Fruity}	4	152.16	23	11	{Fruity}
2	128.24	29	9	{Honey, Vanillin}	5	151.28	27	12	{Honey, Fruity}
3	136.16	24	10	{Honey, Fruity}	6	142.22	27	10	{Fruity}

domain of an attribute $a \in \mathcal{A}$ is denoted by $Dom(a)$ where a is said numerical if $Dom(a)$ is embedded with an order relation, or nominal otherwise. Each object is described by a set of labels from the nominal set $Dom(C)$ by the function class : $\mathcal{O} \mapsto 2^{Dom(C)}$ that maps the olfactory qualities to each object.

Running example. Let us consider the toy olfactory dataset of Table 1 made of $\mathcal{O} = \{1, 2, 3, 4, 5, 6\}$ the set of molecules IDS and $\mathcal{A}$ the set of 3 of physicochemical attributes giving the molecular weight (MW), the number of atoms (nAT), and the number of carbon atoms (nC). Each molecule is associated to one or several olfactory qualities from $Dom(C) = \{Fruity, Honey, Vanillin\}$. The assignments of an odor to a molecule is made by domain experts.

Real-life olfactory datasets, instances of this model, show specific characteristics: (i) **high dimensions**, (ii) **multi-label**, (iii) **unbalanced classes**, and (iv) **non-robust labeling**. Indeed, (i) the number of attributes is large, up to several thousands of physicochemical attributes and a hundred of labels ; (ii) a molecule takes several labels and (iii) the label distribution is highly unbalanced, i.e., with a high variance in the frequency with which the labels occur in the dataset. Odors like *fruity* (resp. *powdery*) are strongly over-represented (resp. under-represented) (see Fig. 1). Then, (iv) labels (odors) attached to each molecule are given by experts based on their own vocabulary. However there is both a high inter- and intra-individual variability concerning the perception of odors [12], the latter involving more than 400 genes encoding molecular receptors (whose expressions differ between people). Perception is subject to the context of the data acquisition phases (questionnaires), cultural elements, etc.

Building an original dataset. One prominent methodological lock in the field of neuroscience concerns the absence of any large available database (>1000 molecules) combining odorant molecules described by two types of descriptors: perceptual ones such as olfactory qualities (scent experts defining a perceptual space of odors), and chemical attributes (chemical space). The dataset provided by the IBM challenge [12] is a clinical one: i.e., odorant molecules were not labeled by scent experts. To tackle this issue, the neuroscientists selected a list of 1,689 odorants molecules described by 74 olfactory qualities in a standardized atlas [1]. They then described using *Dragon 6 software* (available on talete.mi.it) all of these molecules at the physicochemical levels (each odorant molecule was described by more than 4,000 physicochemical descriptors). As such, and to the best of our knowledge, the present database, created by neuroscientists, is one of the very few in the field that enable quantification and qualification of more than 1,500 molecules at both, perceptual (neurosciences) and physicochemical (chemistry) levels. The distribution of the 74 olfactory qualities is illustrated in Fig. 1 (filled bars).

Problem 1 (SOR Problem). Given an olfactory dataset, the aim is to *characterize* and *describe* the relationships between the *physicochemical properties* of odorant molecules and their *olfactory qualities.*

A data-science approach. Answering this problem requires experts of different domains. The odor space is related to the study of olfaction in neuroscience, the understanding of the physicochemical space requires chemical skills, and finally, exploring jointly these two spaces requires data analysis techniques from computer science. In the latter, we cannot afford to use black box predictive models as we need intelligible patterns. Second, as olfactory datasets suffer from a poor label predictability, we cannot use global models to model the dataset but local models, i.e. subsets of data that are specific to some labels. These approaches are known as descriptive rule discovery methods [14], divided into two main trends: *subgroup discovery* and *redescription mining.* We introduce these methods and show their strengths and weaknesses to deal with our problem.

2.1   Subgroup Discovery

Subgroup Discovery (SD) has attracted a lot of attention for two decades under several vocables and research communities (subgroups, contrast sets, emerging patterns, etc.) [14,16]. The aim is to find groups of objects, called subgroups, for which the distribution over the labels is statistically different from that of the entire set of objects. A subgroup is defined (i) by its extent, i.e. the subset of objects it covers and (ii) by its intent, a description connecting restrictions on the attribute domains, such that the intent covers the extent. The intent can be defined on several languages, e.g. conjunctions of attribute domain restrictions.

Definition 2 (Subgroup). *The description of a subgroup is given by $d = \langle f_1, \ldots, f_{|\mathcal{A}|} \rangle$ where each f_i is a restriction on the value domain of the attribute $a_i \in \mathcal{A}$. A restriction is either a subset of a nominal attribute domain, or an interval contained in the domain of a numerical attribute. The set of objects covered by the description d is called the support of the subgroup $supp(d) \subseteq \mathcal{O}$. The set of all subgroups forms a lattice with a specialization/generalization ordering.*

Definition 3 (Quality measure). *The SD approach hence relies on a quality measure which evaluates the singularity of the subgroup within the population regarding a target class function: the class attribute. The choice of the measure depends on the dataset but also on the purpose of the application [8]. There are two main kind of quality measures: the first one is used with monolabeled dataset, e.g., the F-1 measure, the WRAcc measure, the Giny index or the entropy (the original SD [17]); and the second one is used with multilabeled dataset, e.g., the Weighted Kullback-Leibler divergence(WKL) as used in EMM [6]).*

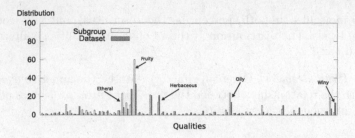

Fig. 1. Subgroup label distributions with WKL.

Running example. The support of the description $d_1 = \langle MW \leq 151.28, 23 \leq nAT \rangle$ is $\{2, 3, 5, 6\}$. For readability, we omit a restriction f_i in a description if there is no effective restriction on the attribute a_i. The description $d_2 = \langle MW \leq 151.28, 23 \leq nAT, 10 \leq nC \rangle$ is a specialization of d_1 (d_1 is a generalization of d_2). Moreover, considering Table 1, $WKL(d_1) = 4/6 \times ((3/4 \log_2 9/10) + (1/4 \log_2 3/2) + (3/4 \log_2 3/2)) = 0.31$. The WRAcc measure of the descriptive rule $d_1 \rightarrow Honey$ is $WRAcc(d_1, Honey) = 4/6 \times (3/4 - 1/2) = 0.25$.

The SD problem. Given a dataset $\mathcal{D}(\mathcal{O}, \mathcal{A}, C, class), minSupp, \varphi$ and k, the objective is to extract the k best subgroups w.r.t. the measure φ, with a support cardinality higher than a given $minSupp$.

Limits of SD addressing Problem 1. The existing methods of SD either target only a single label at a time, or all labels together depending on the choice of the quality measure. In our application, it is required that a subgroup could characterize several labels at the same time. Only the WKL measure can achieve this goal [6]. However, it suffers of the curse of dimensionality: in presence of a large number of labels (74 odors in our experiments), the subgroups cannot characterize a small set of odors. This is shown with real data on Fig. 1: the distribution of labels for the full dataset and for the best subgroup are displayed: clearly, the subgroup is not characteristic of a few odors. We need thus to consider not only all the possible subgroups, but all label subsets for each subgroup. In other settings, this search space is actually considered by a method called *Redescription Mining* [9].

2.2   Redescription Mining

Redescription mining (RM) [9] aims at finding two ways of describing a same set of objects. For that, two datasets are given with different attributes but the same set of object ID. The goal is to find pairs of descriptions (c, d), one in each dataset, where $supp(c)$ and $supp(d)$ are similar. The similarity is given by a Jaccard index between $supp(c)$ and $supp(d)$. The closer to 1 the better the redescription. If we consider the first dataset as the chemicophysical attributes and the second as the labels, we can apply RM to find molecular descriptions and label sets that cover almost the same set of objects.

Running example. Let us consider the dataset of the Table 1 and the redescription $r = (d_P, d_Q)$ with $d_P = \langle MW \geq 150.19 \vee nAT = 27 \rangle$, $d_Q = \langle Fruity \wedge (\neg Honey) \rangle$. Thus, $supp(d_P) = \{1, 4, 5, 6\}$ and $supp(d_Q) = \{1, 4, 6\}$ and $J(r) = \frac{3}{4} = 0.75$. Note that RM allows an expressive language with negations and disjunctions for defining a description.

The RM problem. Given a dataset $\mathcal{D}(\mathcal{O}, \mathcal{A}, C, class)$, $minSupp$ and k, the objective is to extract the k best redescriptions w.r.t. the Jaccard Index, with a support cardinality higher than $minSupp$.

Limits of RM addressing Problem 1. RM gives us an algorithmic basis to explore the search space composed of all pairs of subsets of objects and subsets of labels. However, the quality measure used in RM, the Jaccard index, does not fit exactly what Problem 1 expects. The Jaccard index is symmetric implying the discovery of almost bijective relationships. Yet, it is widely acknowledged that the structure-odor relationships are not bijective. Therefore, this measure is not relevant for unbalanced datasets, which is the case of olfactory datasets. Thus it is difficult to find descriptions related to an over-represented odor.

As a conclusion, answering Problem 1 can be achieved by exploring the search space of redescriptions of the form (d, L), with d a description and L a subset of labels, using any quality measure from subgroup discovery (F1-measure, WRAcc, KWL, etc.). This however, does not take into account the *unbalanced classes* problem. We make this point explicit in the next section and propose a solution.

3   An Adaptive Quality Measure

Existing discriminant descriptive rule methods cannot address Problem 1: the SD generic framework does not explore the correct search space whereas in RM the quality measure is not adapted for this problem. Problem 1 requires a data mining method that simultaneously explores both the description space and the search space of the odor labels. For that, we define *local subgroups*.

Definition 4 (Local subgroup). *Given a dataset $\mathcal{D}(\mathcal{O}, \mathcal{A}, C, class)$, a local subgroup (d, L) takes a description d characterizing a subset of few labels $L \subseteq Dom(C)$ of the class attribute C with $1 \leq |L| << |Dom(C)|$. The support of a local subgroup is the support of its description: $supp(d, L) = supp(d)$. Note that $supp(L) = \{o \in \mathcal{O} \mid L \subseteq class(o)\}$.*

The aim is to find out *local subgroups* (d, L) where the description d is characteristic of the subset of few olfactory qualities $L \subseteq Dom(C)$. For that, we develop a SD method, that simultaneously explores this double search space. This method relies on a adaptive quality measure that enables to evaluate the singularity of the local subgroup (d, L) only for the subset of labels L it targets. This measure is adaptive for each local subgroup, i.e., it is automatically adjusted according to the balance of the subset of labels in the dataset.

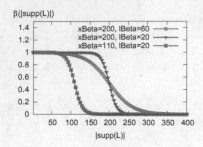

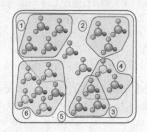

Fig. 2. The curves of $\beta(|supp(L)|)$.

Fig. 3. Necessity of an adaptive measure. (Color figure online)

The original F-Score. Complete surveys help understanding how to choose the right measure [8]. The generalized version of the WKL[1] considers the labels in the subset $L \subseteq Dom(C)$ as independent and does not look for their co-occurrences. The WRAcc measure is a gain measure on the precision of the subgroup and totally ignores the recall. However, we are interested in a measure that considers both precision ($P(d, L) = \frac{|supp(d) \cap supp(L)|}{|supp(d)|}$) and recall ($R(d, L) = \frac{|supp(d) \cap supp(L)|}{|supp(L)|}$) of a local subgroup. The F-Score does it:

$$F(d, L) = (1 + \beta^2) \times \frac{P(d, L) \times R(d, L)}{(\beta^2 \times P(d, L)) + R(d, L)} \tag{1}$$

Indeed, objects are described by both attributes and class labels, so F-score quantifies both the precision and the recall of the support of the description w.r.t. the support of the class labels.

The adaptive F_β. However, olfactory datasets involve unbalanced labels, i.e. the distribution of the labels is quite different from each other: some are over-represented, and other are under-represented. Thus, we decided to adapt the F-Score to unbalanced datasets considering the original constant β as a variable of $|supp(L)|$: the higher $|supp(L)|$, the closer to zero β is (the precision in the F-Score is fostered), and the lower $|supp(L)|$, the closer to one β is (the F-Score becomes the harmonic mean of precision and recall). Formally, given two positive real numbers x_β and l_β, we define the F_β measure derived from Eq. 1 with β a variable of $|supp(L)|$ as follows (see also Fig. 2)

$$\beta(|supp(L)|) = 0.5 \times \left(1 + \tanh\left(\frac{x_\beta - |supp(L)|}{l_\beta} \right) \right) \tag{2}$$

Intuitively, for over-represented labels, since it is difficult to find rules with high recall and precision, the experts prefer to foster the precision instead of the recall: they prefer extracting several small subgroups with a high precision than a huge local subgroup (d, L) with plenty of non-L odorants. In Fig. 3 the red odorants are over-represented in the dataset, but it is more interesting having the different local subgroups 1, 2, 3 and 4 with high precision, rather than a

---

[1] The generalized version of the WKL corresponds to the WKL measure restricted to the subset of labels $L \subseteq Dom(C)$ of the local subgroup (d, L).

single huge local subgroup 5 which precision is much lower. For odorants that are not over-represented, the measure considers both precision and recall: e.g., the local subgroup 5 is possible for the green molecules. The two real numbers x_β and l_β are set thanks to the characteristics of the dataset. In fact, due to the distribution δ_C of the classes in the dataset, fixing $x_\beta = E(\delta_L)$ and $l_\beta = \sqrt{\sigma(\delta_L)}$, where $E(X)$ and $\sigma(X)$ are respectively the average and the standard deviation of a random variable X, is sensible considering Problem 1.

The local subgroup discovery problem. Given a dataset $\mathcal{D}(\mathcal{O}, \mathcal{A}, C, class)$, the adaptive quality measure F_β, a minimum support threshold $minSupp$ and an integer $k \in \mathbb{N}^+$, the aim is to extract the k best local subgroups (d, L) w.r.t. the quality measure F_β with $1 \leq |L| \ll |Dom(C)|$, such that $supp(d, L) \geq minSupp$.

Running example. Considering the dataset of Table 1, with $x_\beta = 3$ and $l_\beta = 1.4$, let us discuss the local subgroup $(d_1, \{Fruity\})$ with $d_1 = \langle MW \leq 151.28, 23 \leq nAT \rangle$. First, $\beta(|supp(\{Fruity\}|) = 0.05$ since the $Fruity$ odor is over-represented ($|supp(\{Fruity\})| = 5$). We have that $F_\beta(d_1, \{Fruity\}) = 0.75$ and it fosters the precision rather than the recall. Now if we consider the local subgroup $(d_1, \{Honey, Fruity\})$: since $|supp(\{Honey, Fruity\})| = 2$, $\beta(|supp(\{Honey, Fruity\}) = 0.81$ from which it follows that $F_\beta(d_1, \{Honey, Fruity\}) = 0.41$ because it considers both recall and precision.

4   Mining Local Subgroups

Exploring the search space. The search space of local subgroups is structured as the product of the lattice of subgroups and the lattice of label subsets, hence a lattice. Let X be the set of all possible subgroups, and C the set of labels, the search space is given by $X \times 2^C$. Each element of this lattice, called *node* or local subgroup hereafter, corresponds to a local subgroup (d, L). Nodes are ordered with a specialization/generalization relation: the most general local subgroup corresponds to the top of the lattice and covers all objects, its set of labels is empty (or composed with labels that occur for –all– the objects). Each description d can be represented as a set of attribute restrictions: specializing a description is equivalent to add attribute domain restrictions (i.e. adding an element for nominal attributes, shrinking an interval to its nearest left or right value for a numerical attribute, see e.g. [11]).

Due to the exceptional size of the search space, we opt for a beam-search, starting from the most general local subgroup to the more specialized ones. This heuristic approach is also used in EMM and RM. It tries to specialize each local subgroup either by restricting an attribute or by extending subset of class labels with a new label it has also to characterize as long as the F_β score is improved. There are at most $|Dom(C)| + \sum_{a_i \in \mathcal{A}} |a_i|(|a_i| + 1)/2$ possibilities to specialize each local subgroup: we can proceed up to $|Dom(C)|$ extensions of the subset of labels to characterize L and $|\mathcal{A}|$ extensions of the description for which we can build up $|a_i|(|a_i| + 1)/2$ possible intervals for numeric attributes. We choose among those only a constant number of candidates to continue the exploration (the width of the beam: the $beamWidth$ best subgroups w.r.t. the quality measure). The search space is also pruned thanks to the anti-monotonic constraint on support.

Finding the attribute split points. When extending the description of a subgroup $s = (d, L)$ for a numerical attribute, the beam search exploration looks for the best cut points that optimize the F_β score of the resulting subgroup. Since the value domain of a numerical attribute a is finite (at most $|\mathcal{O}|$ different values), a naive approach would test all the possibilities to find the lower and the upper bounds for the interval that optimizes F_β ($O(|\mathcal{O}|^2)$ complexity). Our approach, inspired by a state-of-the-art approach [7], only searches for promising cut points. We define $r_i = \frac{|\{o \in supp(s)|a(o)=v_i, o \in supp(L)\}|}{|\{o \in supp(s)|a(o)=v_i, o \notin supp(L)\}|}$ for $v_i \in Dom(a)$. We say that a value v_i is a strict lower bound if $r_i > 1$ and $r_{i-1} \le 1$, and a value v_i is a strict upper bound if $r_i > 1$ and $r_{i+1} \le 1$. The algorithm searches for the best cut points among the strict lower and upper bounds.

Table 2. Characteristics of the datasets where $|\mathcal{O}|$ is the number of objects, $|\mathcal{A}|$ the number of attributes, $|\mathcal{C}|$ the number of labels, M_1 the average number of labels associated to an object, and M_2, min, max respectively the average, minimum and maximum number of objects associated to a label.

Dataset	$|\mathcal{O}|$	$|\mathcal{A}|$	$|\mathcal{C}|$	M_1	M_2	min	max
$\mathcal{B}_1$	7395	243	159	2.4	111.7	51	1042
$\mathcal{D}_1$	1689	43	74	2.88	67.26	2	570
$\mathcal{D}_2$	1689	243	74	2.88	67.26	2	570

However this approach can return an empty set of cut points, especially for under-represented subsets of class labels. Experimentally, the beam search exploration stops very quickly and only over-represented label sets can be output. For that, we consider the $maxBeginningPoints$ best lower bounds, i.e. value v_i such that $0 < r_i \le 1$, as possible cut points when the original method of Fayyad et al. [7] does not return any result. By default we set $maxBeginningPoints = 5$.

Mining diverse significant subgroups. Generally, when a method mixes a beam search and a top-k approach, the issue of redundancy is clearly an important thing to deal with. The risk is to extract a set of top-k local subgroups where several subgroups are redundant, i.e. that share same restrictions or support. For that, we implement a process to avoid redundancy during the exploration. Before adding a local subgroup s in the top-k resulting set, we quantify the redundant aspect of s w.r.t. each current top-k local subgroup by comparing the restrictions involved in its description but also the support of these restrictions. Formally, we compute a penalty score $pen(s_1, s_2) \in [0; 3]$ between two subgroups s_1 and s_2 by adding (i) the proportion of common attributes a_i involved in effective restrictions in both descriptions, and (ii) the values of the Jaccard index between the intervals $[l_1, u_1]$ and $[l_2, u_2]$ for each common attribute in the description, and (iii) the values of the Jaccard index between $supp(s_1)$ and $supp(s_2)$. The algorithm only adds a new local subgroup if the penalty score with all other subgroups is less than the threshold $maxRedundancy$, and if the penalty score is greater than $maxRedundancy$ the algorithm keeps the subgroup with the higher quality measure. By default, we fix $maxRedundancy$ to 2.2.

Finally, extracted subgroups have to be statistically significant: considering a local subgroup (d, L), the support of d and the support of L in the entire dataset have to be statistically meaningful. If we consider these distributions as independent, the probability that objects are included in both supports has to be low. To measure this, we compute the p-value: we test the distribution we face in the dataset against the null-model hypotheses.

5 Experiments

We experiment with the Bibtex dataset from the well-known MULAN[2] library for learning from multi-label datasets. The characteristics of this dataset are displayed in Table 2. The labels correspond to keywords the authors had chosen to their Bibtex entry. This Bibtex dataset is used to validate the method on both quantitative and qualitative sides because it does not require a deep expertise to interpret the results. We also used two real-world olfaction datasets. These datasets $\mathcal{D}_1$ and $\mathcal{D}_2$ have been derived from the dataset described in Sect. 2. Table 2 presents the characteristics of these datasets.

Performance study. To evaluate the efficiency of our algorithm, we consider the Bibtex dataset $\mathcal{B}_1$ and the two olfaction datasets $\mathcal{D}_1$ and $\mathcal{D}_2$. Experiments were performed on a 3.10 GHz processor with 8 GB main memory running Ubuntu 14.04.1 LTS. We vary $minSupp$, $beamWidth$ and $maxOutput$ separately and the non-varying parameters are fixed to $maxOutput = 100$, $beamWidth = 15$ and $minSupp = 30$. Surprisingly, in Fig. 4 (left), the runtime seems not to vary a lot when increasing the beam width. This is the same result when decreasing the minimum support threshold in Fig. 4 (right). This is due to the on-the-fly discretization method that is time-consuming. In

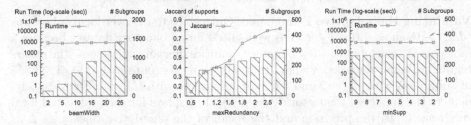

Fig. 4. The runtime and the number of output subgroups varying (left) the beam width, (middle) maxRedundancy and (right) the minimal support on $\mathcal{D}_1$.

---

[2] http://mulan.sourceforge.net.

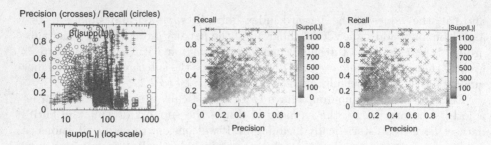

Fig. 5. (left) The precision and the recall of subgroups (d, L) as a function of $|supp(L)|$ on $\mathcal{B}_1$ with the value of $\beta(|supp(L)|)$. The precision and the recall of the output subgroups (d, L) on $\mathcal{B}_1$ according to $|supp(L)|$ (color scale) using $F1$ (middle), F_β (right).

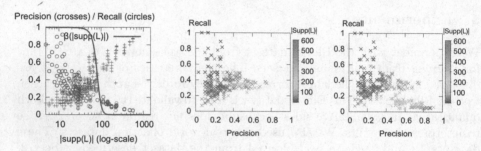

Fig. 6. The same experiments than those of Figs. 5 but on the dataseet $\mathcal{D}_1$.

Fig. 4 (left and right), we observe that even if $minSupp$ increases, the number of outputted subgroups is constant whereas when $beamWidth$ increases, the number of extracted subgroups is higher. This is due to the avoiding redundancy task: when $minSupp$ increases, the quality measure of the new generated local subgroups is improved, however, they may be redundant compared to other subgroups that are therefore removed. When $beamWidth$ increases, the diversity is increased so the subgroups are less redundant. Figure 4 (middle) depicts the impact of our avoiding redundancy step. The lower $maxRedundancy$, the less similar the support of subgroups, the fewer extracted subgroups.

Validating the adaptive F-measure. Our choice to discover local subgroups (d, L) with an adaptive F_β score is well-suited for an olfactory dataset because a molecule is associated to a few olfactory qualities. For an experiment (the others highlight similar remarks), we have that 60.6 % of subgroups with $|L| = 1$, 33.8 % of subgroups with $|L| = 2$ and 5.6 % of subgroups with $|L| = 3$. Figure 6 (left) depicts the impact of the factor $\beta(|supp(L)|)$. It displays for each extracted local subgroup (d, L) the precision and the recall of the descriptive rules $d \to L$ as a function of $|supp(L)|$, with the curve of the factor $\beta(|supp(L)|)$. Clearly, it works as expected: the subgroups for which $\beta(|supp(L)|)$ is close to 0 foster the precision rather than the recall, and the subgroup for which $\beta(|supp(L)|)$ is close to 1 foster both recall and precision. Figure 6 (right) shows this point in a different way: it displays the precision and the recall of each output subgroup (d, L). A color code highlights the size of $supp(L)$: for over-represented labels, the precision is fostered at the expense of the recall whereas in other cases both precision and recall are fostered. Comparing to Fig. 6 (middle) which displays this result with the F_1 score, we see that few output subgroups are relative to over-represented labels (the same applies for the Bibtex dataset $\mathcal{B}_1$, see Fig. 5).

Building a dataset for analyzing the olfactory percept. We worked on our original dataset presented in Sect. 2. For this subsection, we derived 3 datasets by changing the following conditions. As our approach cannot handle 4,000 molecular descriptors: we filter out correlated attributes with the Pearson product-moment correlation coefficient. As a result, attributes with a correlation higher than 90 % (resp 60 % and 30 %) were removed leaving only 615 (resp. 197 and 79) attributes. We ran our algorithm on these three datasets with the combinations of different parameters: standard F_1 score versus our adaptive measure F_β; $minSupp = 15$ (1 %) versus $minSupp = 30$ (2 %): and finally, we experiment with three different thresholds for the $maxRedundancy$ parameter (0.5, 1.5 and 2.5). All results are available at http://liris.cnrs.fr/olfamining/.

Identification of relevant physicochemical attributes. We consider the experiment on the dataset with 79 physicochemical properties, when we use the F_β score, $minSupp = 30$, and $maxRedundancy = 2.5$. A relevant information for neuroscientists and chemists concerns the physicochemical attributes that were identified in the descriptive rules. As showed in [13], the sum of atomic van der Waals volumes, denoted as Sv, is discriminant with regard to the hedonism of an odor, and especially the higher Sv, the more pleasant an odor. Moreover, the higher the rate of nitrogen atoms ($N\%$), the less pleasant an odor, consistent with the idea that amine groups ($-NH_2$) are associated with bad odors (such as cadaverine or putrescine). Based on this observation, we find subgroups related to either the *Floral* or *Fruity* quality that are characterized by a special range of values with regard to Sv and $N\%$. For example, $s_5 = \langle [27 \leq nOHs \leq 37]\ [6.095 \leq Sv \leq 7.871]\ [4 \leq N\% \leq 8]\ [25 \leq H\% \leq 28], \{Floral\}\rangle$ and $s_6 = \langle [1 \leq nCsp2 \leq 1]\ [2.382 \leq TPSA(Tot) \leq 2.483]\ [4 \leq N\% \leq 10], \{Fruity\}\rangle$ are output subgroups. The quality measure of s_5 is 0.91 with a precision of 0.91 and a low recall of 0.06. For s_6, its quality measure is up to 0.87, the same as its precision and its recall is 0.05. Each of these subgroups contains in its description the $N\%$ attribute associated to a very low percentage, and s_5 also includes the Sv attributes with a range of values that corresponds to its higher values. Note that, due to the F_β score, the recall of these subgroups is low because the odors *Fruity* and *Floral* are over-represented in the dataset. In general, the quality *Musk* is associated with large and heavy molecules: the molecular weight (MW) of these molecules is thus high. In the output subgroups, most of those associated to the musk quality include in their description the MW attribute with high values. For example, $s_7 = \langle [5 \leq nCar \leq 6]\ [3.531 \leq Ui \leq 3.737]\ [224.43 \leq MW \leq 297.3], \{Musk\}\rangle$ with a quality measure of 0.46 (precision: 0.48, recall: 0.37) is about molecules with a molecular weigh between 224.43 and 297.3. Moreover, when the quality *Musk* is combined with the quality *Animal*, we still have a high molecular weight but there are other attributes with specific range of values: $s_8 = \langle[3.453 \leq Ui \leq 3.691]$ $[238 \leq MW \leq 297.3]$ $[32 \leq nR = Cp \leq 87]$ $[1 \leq nCsp2 \leq 6]$, $\{Musk, Animal\}\rangle$. This latter topological attribute is consistent with the presence of double bonds (or so-called $sp2$ carbon atoms) within most musky chemical structure, that provides them with a certain hydrophilicity.

Providing relevant knowledge to solve a theoretical issue in the neuroscience of chemo-sensation. We consider the experiment on the dataset with 615 physicochemical properties, when we use the F_β score, $minSupp = 15$, and $maxRedundancy = 0.5$. Another important information brought by these findings to experts lies in the fact the SOR issue should be viewed and explored through a "multiple description" approach rather than "one rule for one quality" approach (i.e., bijection). Indeed, a number of odor qualities were described by very specific rules. For example, 44 % of the molecules described as *camphor* can be described by 3 rules physicochemical rules, with a very low rate of false positives (0.06 %; molecules being described by the physicochemical rule, but not described perceptively as *camphor*). Similar patterns were observed for other qualities: e.g., *mint* (3 descriptive rules; 32 % of the molecules described as *mint*; 0.06 % of false positives), *ethereal* (3; 35 %; 0 %), *gassy* (3; 36 %; 0.36 %), *citrus* (3; 42 %; 0.24 %), *waxy* (3; 43 %; 0 %), *pineapple* (3; 48 %; 0 %), *medicinal* (3; 49 %; 0.30 %), *honey* (4; 54 %; 0.06 %), *sour* (3; 56 %; 0.36 %). Focusing on these qualities, this confirms, as stated above, that a universal rule cannot be defined for a given odorant property, in line with the extreme subtlety of our perception of smells. For example, looking in more details on the produced rules for Camphor (see Fig. 7), it appears that one rule is mostly using topological descriptors, while the second rather uses chemical descriptors. The third rule has a combination of these two to fulfill the model.

Perspectives in neurosciences and chemistry. The present findings provide two important contributions to the field of neurosciences and chemo-sensation. First, although the SOR issue seems to be illusory for some odor qualities, our approach suggests that there exist descriptive rules for some qualities, and they also highlight the relevance of some physicochemical descriptors (Sv, MW, etc.). Second, the present model confirms the lack of bijective (one-to-one) relationship between the odorant and the odor spaces and emphasizes that several sub-rules should be taken into account when producing structure-odor relationships. From these findings, experts in neurosciences and chemistry may generate the

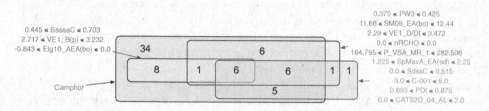

Fig. 7. Size of the support of three groups involving the *camphor* odor.

following new and innovative hypotheses in the field: (i) explaining inter-individual variability in terms of both behavioral and cognitive aspects of odor perception, (ii) explaining stability in odor-evoked neural responses and (iii) correlating the multiple molecular properties of odors to their perceptual qualities.

6 Conclusion

Motivated by a problem in neuroscience and olfaction, we proposed an original subgroup discovery approach to mine descriptive rules characterizing specifically subsets of class labels, as well as an adaptive quality measure to be able to characterize both under- and over- represented label subsets. We implemented its algorithmic counterpart and experimented it with real olfactory datasets. The powerful interpretability of the results and the information they bring, can improve the knowledge about the complex phenomenon of olfaction. Applying such structure/odor model in a dedicated olfactory data-analytics platform will improve understanding of the effects of molecular structure on the perception of odorant objects (foods, desserts, perfumes, flavors), enabling product formulation to be optimized with respect to consumers' needs and expectations.

Acknowledgments. This research is partially supported by the *CNRS* (Préfute PEPS FASCIDO) and the *Institut rhônalpin des systémes complexes* (IXXI).

References

1. Arctander, S.: Perfume and Flavor Materials of Natural Origin, vol. 2. Allured Publishing Corp., Carol Stream (1994)
2. Buck, L., Axel, R.: A novel multigene family may encode odorant receptors: a molecular basis for odor recognition. Cell **65**(1), 175–187 (1991)
3. Castro, J.B., Ramanathan, A., Chennubhotla, C.S.: Categorical dimensions of human odor descriptor space revealed by non-negative matrix factorization. PLoS ONE **8**(9), 09 (2013)
4. de March, C.A., Ryu, S., Sicard, G., Moon, C., Golebiowski, J.: Structure-odour relationships reviewed in the postgenomic era. Flavour Fragr. J. **30**(5), 342–361 (2015)
5. Delasalle, C., de March, C.A., Meierhenrich, U.J., Brevard, H., Golebiowski, J., Baldovini, N.: Structure-odor relationships of semisynthetic β-santalol analogs. Chem. Biodivers. **11**(11), 1843–1860 (2014)
6. Duivesteijn, W., Feelders, A., Knobbe, A.J.: Exceptional model mining - supervised descriptive local pattern mining with complex target concepts. Data Min. Knowl. Discov. **30**(1), 47–98 (2016)
7. Fayyad, U.M., Irani, K.B.: Multi-interval discretization of continuous-valued attributes for classification learning. In: IJCAI (1993)
8. Fürnkranz, J., Gamberger, D., Lavrač, N.: Foundations of Rule Learning. Springer, Heidelberg (2012)
9. Galbrun, E., Miettinen, P.: From black and white to full color: extending redescription mining outside the Boolean world. Stat. Anal. Data Min. **5**(4), 284–303 (2012)

10. Kaeppler, K., Mueller, F.: Odor classification: a review of factors influencing perception-based odor arrangements. Chem. Senses **38**(3), 189–209 (2013)
11. Kaytoue, M., Kuznetsov, S.O., Napoli, A.: Revisiting numerical pattern mining with formal concept analysis. In: IJCAI, pp. 1342–1347 (2011)
12. Keller, A., Vosshall, L., Meyer, P., Cecchi, G., Stolovitzky, G., Falcao, A., Norel, R., Norman, T., Hoff, B., Suver, C., Friend, S.: Dream olfaction prediction challenge (2015). www.synapse.org/#!Synapse:syn2811262. Sponsors: IFF, IBM Research, Sage Bionetworks and DREAM
13. Khan, R.M., Luk, C.-H., Flinker, A., Aggarwal, A., Lapid, H., Haddad, R., Sobel, N.: Predicting odor pleasantness from odorant structure: pleasantness as a reflection of the physical world. J. Neurosci. **27**(37), 10015–10023 (2007)
14. Novak, P.K., Lavrač, N., Webb, G.I.: Supervised descriptive rule discovery: a unifying survey of contrast set, emerging pattern and subgroup mining. J. Mach. Learn. Res. **10**, 377–403 (2009)
15. Tsoumakas, G., Katakis, I., Vlahavas, I.P.: Mining multi-label data. In: Data Mining and Knowledge Discovery Handbook, 2nd edn., pp. 667–685 (2010)
16. van Leeuwen, M., Knobbe, A.J.: Diverse subgroup set discovery. Data Min. Knowl. Discov. **25**(2), 208–242 (2012)
17. Wrobel, S.: An algorithm for multi-relational discovery of subgroups. In: Komorowski, J., Zytkow, J. (eds.) PKDD 1997. LNCS, vol. 1263, pp. 78–87. Springer, Heidelberg (1997). doi:10.1007/3-540-63223-9_108

InterSet: Interactive Redescription Set Exploration

Matej Mihelčić[1,2(✉)] and Tomislav Šmuc[1]

[1] Ruđer Bošković Institute, Bijenička cesta 54, 10000 Zagreb, Croatia
{matej.mihelcic,tomislav.smuc}@irb.hr
[2] Jožef Stefan International Postgraduate School, Jamova cesta 39,
1000 Ljubljana, Slovenia

Abstract. We propose a novel approach for interactive redescription set exploration and redescription analysis realized through the tool Inter-Set. The tool is developed for interaction with possibly large redescription sets, produced on large datasets, and it enables better understanding of the underlying data and relations between attribute sets. New insights from redescription sets can be obtained through three different interaction modes based on: (i) similarity of entity occurrence in redescription support sets, (ii) attribute co-occurence in redescriptions and (iii) redescription quality measures. These modes provide additional contextualization, which is a major advantage compared to current state of the art approaches that allow interactive redescription set exploration, enabling users to obtain new knowledge in the form of interesting redescription subsets which can be analysed further on the level of individual redescriptions.

Keywords: Knowledge discovery · Redescription mining · Redescription set · Interactive exploration · Self organising map · Heatmap · Crossfilter

1 Introduction

We focus our research on redescription mining [16], a field of data mining with a specific goal of finding different descriptions (called redescriptions) of similar groups of entities. These entities are described by one or more sets of Boolean, categorical or numerical attributes called views which are usually disjoint if more than one view is used. The benefits of using redescription mining are twofold: it provides information about groups of entities and means of observing connections between attributes from one or more different attribute spaces.

1.1 Notation and Definition

Although redescription mining is not limited by the number of views, all current approaches (including InterSet) work with maximally two distinct views W_1 and

© Springer International Publishing Switzerland 2016
T. Calders et al. (Eds.): DS 2016, LNAI 9956, pp. 35–50, 2016.
DOI: 10.1007/978-3-319-46307-0_3

W_2 with the corresponding sets of variables V_1, V_2 and the set of entities E. In this setting, a redescription R is a pair of queries $R = (q_1, q_2)$ where each query describes a set of entities by using variables from the set of variables corresponding to one view. Variables in the queries are logically connected with conjunction, negation and disjunction operators.

We present one redescription obtained on our use case dataset describing world countries by using general country information and information about their trading patterns (fully described in Sect. 3). The redescription $R = (q_1, q_2)$ contains two queries q_1 and q_2 defined as:

$q_1 :\ 23.8 \leq$ UN_YOUTH_M $\leq 54.4 \land 66.4 \leq$ STOCKS ≤ 166.6

$q_2 :\ 5.0 \leq E_{66} \leq 6.0 \land 4.0 \leq E_{88} \leq 5.0$

Variables (UN_YOUTH_M - percentage of unemployed male youth, STOCKS - turnover ratio of traded stocks) in q_1 and (E_{66} - the percentage of total export obtained with medicinal and pharmaceutical products, E_{88}- the percentage of total export obtained with electrical machinery, apparatus and appliances) in q_2 are connected with the conjunction (AND) operator.

A single redescription is typically characterized by three quality measures: the support, the Jaccard index and the p-value.

The support of a query q_i ($supp(q_i)$) is a set of all entities satisfying its condition. The redescription $R = (q_1, q_2)$ describes the entity if this entity is in a support of all queries forming the redescription. All entities described by a redescription compose a redescription support set ($supp(R) = supp(q_1) \cap supp(q_2)$).

The intuition behind redescription mining is that queries describing similar sets of entities provide information about the shared properties of these entities. Higher similarity among sets of entities represents higher association between the queries. Thus, it is appropriate to use Jaccard index, defined as $J(R) = \frac{|supp(q_1) \cap supp(q_2)|}{|supp(q_1) \cup supp(q_2)|}$, as a measure of redescription accuracy.

The p-value (p_{val}), also used by Galbrun and Miettinen [6], reflects statistical significance of individual redescription and is computed from the binomial distribution: $p_{val}(R) = \sum_{n=|supp(R)|}^{|E|} \binom{|E|}{n} (p_1 \cdot p_2)^n \cdot (1 - p_1 \cdot p_2)^{|E|-n}$. $|E|$ equals the number of entities in the dataset and p_1, p_2 correspond to marginal probabilities of obtaining the query q_1 and q_2. For a given redescription $R = (q_1, q_2)$, $p_{val}(R)$ represents a probability of obtaining a set of a size equal to or larger than that of $supp(R)$, by combining two random queries with marginal probabilities corresponding to the marginal probabilities of queries q_1 and q_2.

We define $attr(R)$ as a set of attributes used in redescription queries and the attribute Jaccard index of two redescriptions as: $attJ(R_1, R_2) = \frac{|attr(R_1) \cap attr(R_2)|}{|attr(R_1) \cup attr(R_2)|}$. The average attribute Jaccard index of a redescription R_i is defined as: $AvgAJ(R_i) = \frac{2 \cdot \sum_{j \neq i} attJ(R_i, R_j)}{n \cdot (n-1)}$. By analogy, the entity Jaccard index of two redescriptions is defined as $elemJ(R_1, R_2) = \frac{|supp(R_1) \cap supp(R_2)|}{|supp(R_1) \cup supp(R_2)|}$ and the average entity Jaccard index as: $AvgEJ(R_i) = \frac{2 \cdot \sum_{j \neq i} elemJ(R_i, R_j)}{n \cdot (n-1)}$. These measures provide information about the redundancy of a redescription with respect to entities and attributes.

1.2   Related Work

Redescription mining is an unsupervised descriptive task, closely related to multi-view clustering [3]. Redescription mining in addition to finding interesting groups of entities also provides interpretable rules describing these groups. It is also related to association rule mining [1,10,22] because both approaches search for relations between attributes. The main difference is that redescription mining searches for equivalence relations whereas association rule mining finds implication relations.

The first approaches developed for redescription mining [15,16,21] used redescription support and Jaccard index as sole constraints to limit redescription creation. Statistical significance was later incorporated into redescription mining process by Gallo et al. [8] to further constrain redescription creation. They proposed to compute the p-value of redescriptions from the binomial distribution. Recent approaches [6,14,23] incorporate information about statistical significance and mostly return smaller sets of redescriptions to the final user, though the exact number of produced redescriptions varies depending on different algorithm parameters. The goal is to make redescription queries understandable and non-redundant. These approaches are able to mine redescriptions containing Boolean, categorical and numerical attributes which extends the capabilities of previous approaches that only worked with Boolean attributes. Despite the efforts to create smaller sets of accurate and understandable redescriptions, it is still hard and time consuming to analyse all produced redescriptions and their properties. It is even harder to notice potential connections between different redescriptions, their support sets and different attributes only by observing algorithm output files.

For this reason and to allow more customizable exploration process, it is necessary to develop interactive applications that respond to user inputs and provide the required information. Zaki and Ramakrishnan [21] developed a console based application that allows limited user interventions such as finding attributes describing a given set of entities, finding entities described by a given set of attributes. It also allows placing constraints on entities and attributes, Jaccard index and redescription support to allow interaction with exploration process. Siren [7] is fully interactive redescription mining environment. It allows mining redescriptions and contains several visualizations of individual redescriptions. The parallel coordinates plot, allows visualizing values of entities from redescription support, those not contained in the support but described by some redescription query and other entities not described by either redescription query. The visualization is useful to observe potential regularities of entity values for some attribute contained in redescription support. The decision tree visualization shows interactions between different queries contained in the redescription (especially useful to understand redescriptions created by some decision tree based approach). The entity scatter plot allows comparing values of described entities based on two different attributes which enables correlation analysis. If geographical locations are described by redescriptions, the tool is able to represent the locations described by redescription queries on a map. Each generated redescription can be expanded by allowing the tool to improve the accuracy by adding new attributes to the redescription queries. Redundant redescriptions can be filtered based on redescription support.

The Siren tool offers many visualizations aimed at analysis of individual redescriptions. Although very useful, the approach requires users to scroll through the list of redescriptions examining each individually to get some context about the described entities and used attributes. By performing such exploration it is hard to place each redescription in a bigger context (determine redescription relation with respect to described entities and attributes used in queries). Besides filtering, there are no mechanisms that allow grouping of different redescriptions based on their properties that allow exploring parts of redescription space that are of immediate interest to the user.

Several tools for visualizing and exploring association rules are related to our work. The Self Organizing Map (SOM) [11] is used by Rojas et al. [5] to display the spatial distribution of the entities associated with the association rule on a map. The association rule visualization system for exploratory data analysis [13] uses scatter plot to visualize rules from the association rule set. The rules can be explored on the individual level by modifying the rules and observing comparative bar charts prior and post attribute addition or deletion. The MIRAGE [20] is a framework for mining, exploring and visualising minimal association rules. It uses lattice-based visualisation and exploration of minimal association rules. A user driven and quality oriented visualization for mining association rules [4] embeds association rules to 3D landscape. It allows users to select rule subsets, navigate among different subsets and to filter them on several interestingness measures by using sliders. Multi level spatial association rules mined by the tool ARES [2] are visualized by using graphs.

1.3   Contributions

We describe the InterSet (Fig. 1), a tool aimed at interactive, comprehensive exploration and interpretation of redescription sets. The InterSet uses large diversity and potentially higher level of granularity in the redescription set to increase the usefulness of the exploration. The exploration can be done based on: (i) entities described in redescriptions from the redescription set through the SOM visualization (EC-View), (ii) attributes used in redescriptions to describe different entities through the heatmap visualization (AI-View) and (iii) quality measures assigned to individual redescriptions by using cross-filter on multiple redescription quality criteria (RQ-View). The proposed views allow contextualization, grouping and targeted exploration of different redescriptions.

The tool uses the intuition that the high overlap of entities described by redescription queries indicates existence of shared properties and possible associations between the used attributes. This property is used to build a SOM map that groups entities based on their membership in support sets of redescriptions contained in a redescription set. Resulting groups potentially share many common properties and are interesting for exploration. In addition, we obtain a spatial map of entities based on similarity of their shared properties across

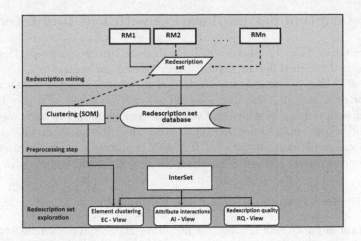

Fig. 1. Schematic description of the process that leads to interactive exploration of redescription sets: (i) Redescription set with one (mandatory) or more (optional) redescription mining algorithms must be generated, (ii) preprocessing step involves the use of Self organizing Maps clustering algorithm (optional) and database preparation (mandatory), necessary to perform (iii) interactive redescription set exploration with the InterSet tool.

both views. Attribute co-occurrence in redescription queries is used to create a heatmap of frequent cross-view pairwise attribute associations (as these are commonly observed when more than one view is used). The cross-filter visualization allows obtaining smaller set of redescriptions with some desired properties that can be explored in more detail. Redescription set obtained in each exploration view can contain a number of similar redescriptions which can be used to enhance the analysis. Comparing similar redescriptions allows understanding interactions between attributes and redescription support or detecting groups of entities with many common properties. If such level of granularity is not needed, the set can be reduced by eliminating redescriptions with large entity or attribute overlap, thus obtaining diverse and compact set with some desired properties. Except for the general insight into redescription set, the tool allows obtaining specific knowledge on the level of individual redescriptions. This includes value distribution analysis of entities contained in redescription support across all attributes contained in redescription queries and comparison with value distributions of all entities in the dataset or in some more specific groups, such as attribute numeric interval. Violin plot, used in the tool, allows visualization of irregular distribution shapes obtained when disjunction and negation operators occur in queries. The analysis allows understanding complex queries by transforming them to the Disjunctive normal form (DNF) which allows exploring parts of queries represented as clauses. These features complement and deepen the level of insight provided by the parallel coordinates plot.

2   Redescription Set Exploration with the Tool InterSet

The InterSet tool (Fig. 1) allows obtaining insight into some properties of the original data through different visualizations based on redescription set properties and selecting potentially interesting, non-redundant set of redescriptions suitable for detailed analysis. The selected set can be saved to a .csv file containing redescription queries and the values of corresponding quality measures.

The following subsections motivate and describe components used in redescription set exploration process for each exploration view. Tool capabilities are demonstrated on the use case data describing world countries in Sect. 3.

2.1   Entity Based Redescription Set Exploration

Redescription support sets usually do not have strictly hierarchical structure. Rather, they can be highly overlapping with the level of overlap depending on the underlying data, number of redescriptions in the redescription set and the algorithm used to create redescriptions. With such general structure of redescription supports, we decided to use the Self Organising Map [11] as it groups entities based on similarities and embeds similar groups closer together on a 2D visualization map. It allows representing entities from potentially large datasets in a compact form where each entity is member of only one SOM cluster.

Rather than exploiting entity similarity in the original dataset representation, as in work from Rojas et al. [5], we utilise the matrix of entity occurrence in a support set of individual redescriptions from the redescription set to obtain a map of entities sharing many cross-view properties. For a given redescription set $\mathcal{R}$ such that $|\mathcal{R}| = n$ and the original dataset containing m entities, we construct a $m \times n$ matrix A. The rows of A correspond to the entities from the original dataset, and columns correspond to the redescriptions from the redescription set. Thus, $A_{ij} = 1$ if and only if $R_j \in \mathcal{R}$ describes an entity $e_i \in E$.

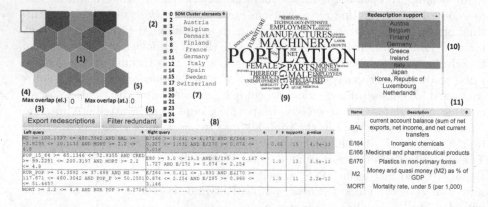

Fig. 2. The entity based interface of the InterSet tool.

The entity based redescription set exploration starts with the SOM map depicted in Fig. 2, Control (1). The layout of SOM is customizable and can be easily experimented with. Each hexagon contains a distinct group of entities and the color of each hexagon reflects the number of entities contained in each group, displayed on the legend (Control (2)). The average homogeneity of a cluster, defined as the average Jaccard index between redescription support and the entities contained in the cluster, can be used as an additional cluster selection criteria. Selecting a hexagon, red square in Fig. 2, provides more detailed information about it's content (Table (7)) and additional controls for in depth exploration (Table 8, Control 9). By observing information about all entities contained in the selected hexagon (Table (7)) and a Word net (9) displaying words most commonly used in attribute descriptors contained in redescription queries describing at least one entity from the SOM cluster, users can determine if it is of interest to explore all redescriptions describing at least one entity contained in the selected cluster (Table 8). Redescriptions can be analysed further on the query and the attribute level (described in Sect. 2.4) where Table (10) provides information about the entities described by the selected redescription (members of the SOM cluster being highlighted in green color) and Table (11) provides additional descriptions of compact attribute codes. It is possible to export (Control (3)) or filter (Controls (4), (5) and (6)) redescriptions contained in Table (8). Filtering process (described in Algorithm 1) allows obtaining a set of redescriptions with user defined maximal entity and attribute overlap.

2.2 Attribute Based Redescription Set Exploration

Research in many scientific fields such as biology, pharmacy and medicine requires discovering relevant associations between variables. Such associations can be explored with the InterSet by observing frequently co-occurring attributes in redescription queries (Fig. 3).

The SOM based representation can be applied to attributes used in redescription queries similarly as it was applied to entities. The main advantage of using the SOM is that it reveals interactions of more than two different variables. However, it is not possible to distinguish between views in such visualization and it is hard to explore associations between the neighbouring groups of attributes. The heatmap visualisation enables exploring interactions between all cross-view attribute pairs and arranging rows and columns based on different criteria which is the main reason for our choice. The focus is on cross-view relations since these are usually interesting when exploring similarities based on different contexts, though the tool can also be used to show all co-occurrence frequencies.

The heatmap (Control (1) in Fig. 3) is a starting point of the attribute based redescription set exploration. It is represented as a $k \times s$ matrix where rows represent k attributes from the first view and columns s attributes from the second view. Three initial row-column layouts can be chosen with Control (2): (1) Ordered by name, (2) Ordered by frequency and (3) Ordered by co-occurrence. When ordered by name (useful for domain experts), the rows and columns are sorted by the attribute code, Ordered by frequency layout arranges

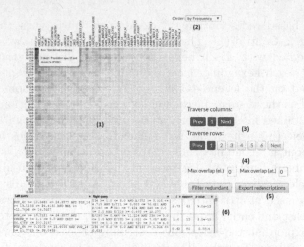

Fig. 3. The attribute based interface of the InterSet tool.

rows and columns of the heatmap to place frequently occurring attributes in redescription queries closer to the top left corner of the heatmap. Ordered by co-occurrence layout arranges rows and columns so that it sorts the heatmap diagonal in descending order by attribute co-occurrence frequency. This layout allows finding potentially larger groups of highly connected attributes if co-occurring in redescription queries. The heatmap is adopted to be used with large number of attributes by loading smaller submatrices of the potentially large cross-view attribute matrix, whose rows are all attributes from the first view and columns all attributes from the second view. The attributes are sorted in descending order by frequency when loaded into heatmap so that the (row,column) page combination $(1,1)$ (Control (3)) contains most frequently occurring attributes from both views. The user can scroll on two dimensions, visualizing parts of cross-view attribute space which can be explored further. The gray color denotes the co-occurrence level of the attribute pair and the table (Control (6)) equivalent to that from Fig. 2 lists all redescriptions from the redescription set containing the selected attribute pair in their queries. Analysis of selected redescription is described in Sect. 2.4 while redescription filtering (Control (4)), (Algorithm 1) and redescription export (Control (5)) work as described in Sects. 2.1 and 2.3. Combination of various attribute pair arrangements with redescription exploration and filtering allows better understanding of the attribute interactions.

2.3   Property Based Redescription Set Exploration

The last redescription set exploration view provided in the InterSet tool is based on redescription properties (quality measures). It enables users to filter the original redescription set by using one or more user-defined criteria which results in smaller, more interesting set, that is easier to explore. The exploration view uses sliders as in [4], with the important addition of a crossfilter (Control 1 in Fig. 4), which allows instantaneous display of distribution of the filtered set (Control 4) for all measures and corresponding redescriptions (Table 5). Sliding through the values of one or more different criteria (Control 4), allows observing changes in distribution of other criteria which provides information about the underlying data. The visualization can be efficiently used with redescription sets containing large number of redescriptions which are much harder to represent with some other visualization techniques such as parallel coordinate plots.

We use several redescription quality criteria currently used in the literature and allow adding new ones, such as different interestingness or unexpectedness criteria, to be used in the filtering process. The crossfilter (Fig. 4, Control 1), consists of histograms showing value distribution for each criterion used in the exploration process. Besides previously defined standard redescription quality measures, we have computed two additional measures: the average entity Jaccard index (Control (2)) and the average attribute Jaccard index (Control (3)) presented in Sect. 1.1. These criteria can be used to extract redescriptions describing (in)frequently described sets of entities or that containing (in)frequent combination of attributes depending on the crossfilter setting. The selection provides no guarantees on entity or attribute overlap between pairs of redescriptions in the newly constructed set. However, filtering (described in Algorithm 1) reduces this overlap to user defined level (Control 6).

Algorithm 1. The filtering algorithm

Input: Redescription set $\mathcal{R}$, max entity overlap ε_{el}, max attribute overlap ε_{at}
Output: Filtered redescription set $\mathcal{R}'$
1: **procedure** FILTER
2:     $criteria \leftarrow ((J, desc), (attJ, asc), (supp, desc), (elemJ, asc))$
3:     $\mathcal{R} \leftarrow \text{sort}(\mathcal{R}, criteria)$
4:     **for** $i = 0;\ i < |\mathcal{R}| - 1;\ i + + $ **do**
5:         **for** $j = i + 1;\ j < |\mathcal{R}|;\ j + + $ **do**
6:             **if** $elemJ(\mathcal{R}[i], \mathcal{R}[j]) \geq \varepsilon_{el}$ OR $attJ(\mathcal{R}[i], \mathcal{R}[j] \geq \varepsilon_{at})$ **then**
7:                 $\mathcal{R} \leftarrow \mathcal{R}.\text{delete}(\mathcal{R}[j])$
8:     **return** $\mathcal{R}$

Criteria array from line 2 in Algorithm 1 contains pairs of redescription quality criteria and sorting direction: desc - descending, asc - ascending. Redescription with preferred values of quality criteria is used to eliminate all redescriptions with unacceptably high entity or attribute Jaccard with the selected redescription.

2.4   Analysing Individual Redescriptions

For detailed redescription analysis, the tool requires redescription queries to be transformed in the Disjunctive normal form (DNF). This decomposed form allows analysing distribution of a subset of redescription support described by each clause in this representation. Since the general goal is to produce short,

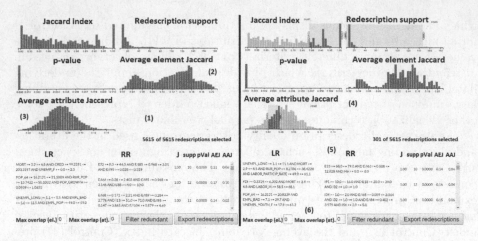

Fig. 4. The InterSet interface based on redescription properties. Initial configuration is shown in the left and the filtering step in the right part of the Figure.

understandable queries and every formula in Propositional logic can be transformed to an equivalent DNF [12], it is a reasonable and mostly feasible requirement aimed at increasing understandability. Depending on the query complexity, the analysis contain (i) three comparative violin plots if the DNF representation doesn't contain disjunction operators (explanation 1, 2, 3 in Fig. 5), (ii) four plots if the DNF representation contains disjunction operators (explanation 1, 2, 3, 5 in Fig. 5) and (iii) five plots if an attribute occurs in more than one clause in the DNF representation of a query (explanation 1, 2, 3, 4, 5 in Fig. 5).

Compared to parallel coordinates plot in Siren, our approach allows analysing value distributions and decomposed queries. Major benefit is it's invariance to dataset or redescription support size whereas parallel coordinates plot tend to have increasing number of possibly overlapping lines making analysis difficult. If Boolean or categorical values are used, violin plots are replaced with piecharts.

Redescription analysis process is demonstrated on redescription $R' = (q'_1, q'_2)$:
$q'_1 : \neg (44.1 \leq M_2 \leq 198.5 \wedge 15.5 \leq POP_{64} \leq 20.8 \wedge 2.0 \leq AGR\_EMP \leq 20.5)$
$q'_2 : (0.0 \leq E/I_{80} \leq 0.6 \wedge 0.0 \leq I_{34} \leq 5.0 \wedge 0.0 \leq E/I_{83} \leq 0.6) \vee (0.0 \leq E/I_{80} \leq 0.3 \wedge 6.0 \leq I_{34} \leq 24.0) \vee (0.0 \leq E_{47} \leq 1.0 \wedge 0.7 \leq E/I_{80} \leq 28.8 \wedge 99.0 \leq E_{97} \leq 100.0)$. It describes 155 countries with $J(R') = 0.88$ and $p_{val}(R') = 0.0023$. The analysis of its queries is demonstrated in Fig. 5.

Query q'_1 is transformed to the equivalent formula in DNF: $\neg (44.1 \leq M_2 \leq 198.5) \vee \neg (15.5 \leq POP_{64} \leq 20.8) \vee \neg (2.0 \leq AGR\_EMP \leq 20.5)$. Using negations splits the value distribution of entities in two parts. For example: $\neg (15.5 \leq POP_{64} \leq 20.8)$ is equivalent to $POP_{64} < 15.5 \vee POP_{64} > 20.8$. Query q'_2 contains three disjunctions, two of which contain attribute I_{34} (denoted as Cl1 and Cl2). Value distributions show that first two clauses describe orthogonal entities with respect to the values of this attribute. It also shows that clause 1 constitutes a backbone of the query describing the largest amount of entities.

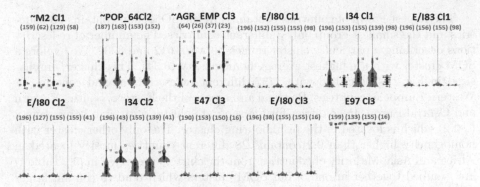

Fig. 5. Comparative violin plots showing entity value distribution for all attributes occurring in the selected redescription. The violin plots show entity distribution in order: (1) entity value distribution for the selected attribute for all entities containing non-missing values in the dataset, (2) entity distribution for all entities containing non-missing values in a numeric interval, defined in redescription query, for this attribute, (3) entity value distribution for all entities contained in redescription support set for a given attribute, (4) entity value distribution of all entities contained in redescription support set that are described by at least one clause containing the attribute under investigation, (5) entity value distribution of all entities contained in the redescription support set described by a particular clause which contains the analysed variable.

3 Exploring Redescriptions Obtained on the Country Data

The tool's capabilities are demonstrated on dataset describing 199 world countries [9,17,19] by using general country descriptors (demographic descriptors, unemployment, etc.) as one view and country trading patterns (with the values of percentage of export (E) and import (I) that a given commodity forms in total country export or import and the information on the ratio of these values (E/I)) as the other view. All attributes in the dataset contain numerical values.

We used the redescription mining algorithm, presented in [14], to create a set containing 5448 different redescriptions where some of them, by design, had high level of similarity. For all $R \in \mathcal{R}$, $J(R) \geq 0.5$, $p_{val}(R) \leq 0.01$ and $supp(R) \geq 10$.

The SOM needs to be precomputed and the obtained cluster membership represents the input to the tool[1]. In the experiments, we used the R package *kohonen* [18] to create the SOM with the 4×4 layout. To train it, we used 1000 iterations with the learning rate linearly declining from 0.05 to 0.01.

3.1 Redescription Set Analysis

The results and analysis presented in this Section are obtained with the tool InterSet, available at www.zel.irb.hr/interset.

---

[1] zel.irb.hr/interset.

The largest group of countries contained in the SOM map contains 28 countries and the smallest group only 3 different countries. The number of redescriptions describing a particular cluster ranges between 612 and 3002. We explore a SOM cluster with the highest average homogeneity (0.34). This cluster, emphasized in Fig. 2 is described with 2737 different redescriptions and contains 11 Western European countries. Portugal and a part of the Eastern, South - Eastern and Central European countries form a separate cluster with high homogeneity (>0.2) which is located in the neighbouring cluster. The only other cluster with homogeneity higher than 0.2 contains 28 different countries mostly located in Africa and Asia. Majority of countries from the clusters presented in [9] (Table 4) are grouped together in one of our SOM clusters which tend to contain larger groups of countries spatially ordered by similarity of their shared properties.

The table containing redescriptions from the selected SOM cluster was sorted in descending order by Jaccard index and in ascending order by redescription support to search for accurate redescriptions with possibly homogeneous support. We selected the third result ($R_{ex} = (q_1'', q_2'')$) for in depth analysis:

q_1'' : $-6.9 \le \text{MON\_GR} \le 6.6 \ \wedge \ 17.1 \le \text{POP}_{64} \le 21.1 \ \wedge \ 41.9 \le \text{STOC} \le 166.6$
q_2'' : $4.0 \le E_{24} \le 26.0 \ \wedge \ 3.0 \le I_{95} \le 5.0 \ \wedge \ 1.1 \le E/I_{85} \le 3.2$. This redescription describes 10 countries with $J(R_{ex}) = 1.0$ and $p_{val}(R_{ex}) = 6.3 \cdot 10^{-12}$. It describes Austria, Denmark, Finland, France, Germany, Italy, Portugal, Spain, Sweden and United Kingdom. Portugal, not a member of selected cluster, is contained in the neighbouring cluster (below). We compare value distribution of countries described by a redescription to a value distribution of all countries in the dataset with respect to attributes used in its queries. The described set of countries tend to have higher values in POP_{64} (percentage of population aged 65+). The difference is significant, as computed with the Mann-Whitney U test, with the p-value of $9.05 \cdot 10^{-7}$. These countries tend to have smaller values in MON_GR (money and quasi money growth - $p = 6.7 \cdot 10^{-5}$), higher values in STOC (stock trade - $p = 0.0002$), E_{24} (labour-intensive and resource-intensive manufactures - $p = 0.011$), I_{95} (articles of apparel and clothing accessories - $p = 1.54 \cdot 10^{-5}$) and E/I_{85} (industrial machinery and parts - $p = 2.3 \cdot 10^{-6}$). Countries from the selected SOM cluster, with the exception of Switzerland, tend to have higher values of E/I_{69} (plastics in primary form - $p = 0.0003$), E/I_{83} (specialised machinery - $p = 1.2 \cdot 10^{-6}$) and tend to have smaller values in AGR_M (percentage of mail employees working in agriculture - $p = 0.0024$). A part of discoveries related to export of technology, manufactures and population parameters match those reported in [9], here we present additional observations, which are a very small subset of information obtainable by the tool.

Attribute associations are, due to finding equivalence relations, an important and distinguishing feature of redescription mining. Attributes are ordered by co-occurrence which reveals high co-occurrence between attributes in the top left corner of the heatmap (Fig. 3). We explore correlations between all pairs of attributes contained in the top - left 5×5 submatrix. Since the Kolmogorov-Smirnov test shows that entity values for many attributes contained in this submatrix are not normally distributed, we compute correlation values (presented in Table 1) by using Spearman's rho and Kendall's tau correlation coefficients.

Table 1. Spearman's ρ and Kendall's τ correlation coefficient of a selected 5×5 cross-view attribute associations. For each attribute pair, a upper bound of p-values for both correlation coefficients is displayed below the correlation values.

Sp. ρ/Ken. τ	I_{34}	E/I_{66}	E/I_{83}	E/I_{85}	E/I_{93}
MORT	0.62/0.46 $< 10^{-15}$	$-0.66/-0.48$ $< 10^{-15}$	$-0.65/-0.47$ $< 10^{-15}$	$-0.63/-0.46$ $< 10^{-15}$	$-0.43/-0.29$ $< 10^{-8}$
CRED_COVER	$-0.52/-0.41$ $< 10^{-11}$	0.56/0.45 $< 10^{-15}$	0.43/0.32 $< 10^{-8}$	0.48/0.35 $< 10^{-10}$	0.48/0.36 $< 10^{-10}$
POP$_{14}$	0.69/0.52 $< 10^{-15}$	$-0.66/-0.46$ $< 10^{-15}$	$-0.61/-0.42$ $< 10^{-15}$	$-0.64/-0.45$ $< 10^{-15}$	$-0.48/-0.33$ $< 10^{-10}$
POP$_{64}$	$-0.64/-0.48$ $< 10^{-15}$	0.68/0.48 $< 10^{-15}$	0.62/0.44 $< 10^{-15}$	0.66/0.47 $< 10^{-15}$	0.49/0.33 $< 10^{-10}$
POP_F	$-0.36/-0.25$ $< 10^{-5}$	0.28/0.18 $< 10^{-3}$	0.35/0.23 $< 10^{-5}$	0.41/0.27 $< 10^{-7}$	0.37/0.26 $< 10^{-6}$

Results in Table 1 show that associations between view 1 attributes: mortality rate under 5 (MORT), private credit bureau coverage (CRED_COVER), percentage of population aged 0 to 14 (POP$_{14}$), percentage of population aged 65+ (POP$_{64}$) and percentage of female population (POP_F) contain statistically significant correlations with the view 2 attributes: I_{34} (cereals and cereal products), E/I_{66} (medicinal and pharmaceutical products), E/I_{83} (specialized machinery), E/I_{85} (other industrial machinery and part) and E/I_{93} (furniture and parts thereof). Many of this correlations might be caused by differences in country development. A number of developed countries have a high exports of all mentioned groups of commodities whereas developing countries either completely rely on the imports of this products or produce it for their own market. Since these two groups of countries differ in the population characteristics, it is possible for the correlation patterns as shown in Table 1 to emerge.

The third view is used to locate highly accurate redescriptions describing subsets of countries that are: (a) different than a majority of other redescriptions with respect to described entities and attributes used in redescription queries ($AvgAJ \leq 0.05$, $AvgEJ \leq 0.1$, $JS > 0.94$), (b) similar to larger number of other redescriptions with respect to described entities and attributes used in redescription queries ($AvgAJ \geq 0.1$, $AvgEJ \geq 0.15$, $JS \geq 0.94$). After obtaining each set with the cross-filter, we removed redundant redescriptions sharing more than 20 % entities and 20 % attributes. First experiment revealed 11 different redescriptions from which we present several interesting discoveries. The analysis revealed that 9 different African countries and a neighbouring Asian country Yemen have a smaller median of percentage of population aged 15 to 64 compared to the median of all countries, these countries also have high export contribution of textile fibres and their wastes in their total export. A subset of countries located in different SOM clusters show high export to import contribution ratio of iron, steel and chemical products. Very heterogeneous group of countries from various continents that share higher export contribution to import contribution ratio of prefabricated buildings, sanitary, heating and lighting fixtures was also discovered. Finally, a group of countries, largely comprised of eastern and south eastern European countries share a large import contribution of precious stones and non-monetary gold to their total import. Strict non-redundancy requirements left only one redescription in the second experiment describing 15 world countries that share many different socio-demographic and trading properties. The described group shows many characteristics of highly developed countries: large percentage of population older than 64, smaller percentage of rural population, higher money and quasi money (as percentage of GDP). Part of these countries have larger ratio of export to import contribution ratio of medicinal and pharmaceutical products, rubber manufactures, and a part of described countries has a large E/I ratio for other industrial machinery and parts.

Finding examples as presented in this section would be very time consuming with Siren because extensive redescription list exploration is required. Support for reasoning at the level of groups of entities, attribute associations or selecting groups of redescriptions with specific properties is not available in Siren.

4 Conclusions and Future Work

We have presented a tool that allows exploring potentially large redescription set obtained by one or more redescription mining approaches. It provides analytics mechanisms aimed at understanding individual redescriptions and uses the redescription set to obtain information about the underlying data - revealing connections and interactions between different entities and attributes. Potentially overlapping redescriptions are used as a tool to enhance the visualizations and allow high granularity exploration. Entity and property based exploration supports arbitrary number of data views while the attribute based view is easily extended by performing pairwise view exploration.

In future work we plan to increase exploration abilities of the tool by enabling different interactions between exploration views which is needed when faced with potentially large sets of redescriptions. On the technical side, we will aim to incorporate the SOM map in exploration process thus removing the need to train it separately and calling external scripts to feed the data into the database.

Acknowledgement. The authors acknowledge the European Commissions support through the MAESTRA project (Gr. no. 612944), the MULTIPLEX project (Gr.no. 317532) and support of the Croatian Science Foundation (Pr. no. 9623: Machine Learning Algorithms for Insightful Analysis of Complex Data Structures).

References

1. Agrawal, R., Mannila, H., Srikant, R., Toivonen, H., Verkamo, A.I.: Fast discovery of association rules. In: Advances in Knowledge Discovery and Data Mining, pp. 307–328. American Association for Artificial Intelligence (1996)
2. Appice, A., Buono, P.: Analyzing multi-level spatial association rules through a graph-based visualization. In: Ali, M., Esposito, F. (eds.) IEA/AIE 2005. LNCS (LNAI), vol. 3533, pp. 448–458. Springer, Heidelberg (2005). doi:10.1007/11504894_63

3. Bickel, S., Scheffer, T.: Multi-view clustering. In: Proceedings of the Fourth IEEE International Conference on Data Mining, ICDM 2004, pp. 19–26. IEEE Computer Society, Washington, DC (2004)
4. Blanchard, J., Guillet, F., Briand, H.: A user-driven and quality-oriented visualization for mining association rules. In: Proceedings of the 3rd IEEE International Conference on Data Mining (ICDM), Melbourne, Florida, USA, pp. 493–496 (2003)
5. Casstilo-Rojas, W., Peralta, A., Meneses, C.: Augmented visualization of association rules for data mining. In: Eight Alberto Mendelzon Workshop on Foundations of Data Management, AMW 2014, Cartagena de Indias, Colombia (2014)
6. Galbrun, E., Miettinen, P.: From black and white to full color: extending redescription mining outside the Boolean world. Stat. Anal. Data Min. **5**(4), 284–303 (2012)
7. Galbrun, E., Miettinen, P.: Siren: an interactive tool for mining and visualizing geospatial redescriptions. In: Proceedings of the 18th ACM SIGKDD International Conference on Knowledge Discovery and Data Mining, KDD 2012, pp. 1544–1547. ACM, New York (2012)
8. Gallo, A., Miettinen, P., Mannila, H.: Finding subgroups having several descriptions: algorithms for redescription mining. In: Proceedings of the SIAM International Conference on data mining (SDM), pp. 334–345. SIAM (2008)
9. Gamberger, D., Mihelčić, M., Lavrač, N.: Multilayer clustering: a discovery experiment on country level trading data. In: Japkowicz, N., Matwin, S. (eds.) DS 2015. Lecture Notes in Artificial Intelligence (LNAI), vol. 9356, pp. 87–98. Springer, Heidelberg (2014). doi:10.1007/978-3-319-11812-3_8
10. Hipp, J., Güntzer, U., Nakhaeizadeh, G.: Algorithms for association rule mining - a general survey and comparison. SIGKDD Explor. Newsl. **2**, 58–64 (2000)
11. Kohonen, T., Schroeder, R.M., Huang, T.S.T. (eds.): Self-Organizing Maps, 3rd edn. Springer-Verlag New York Inc., Secaucus (2001)
12. Kroening, D., Strichman, O.: Decision Procedures: An Algorithmic Point of View, 1st edn. Springer Publishing Company, Incorporated (2008)
13. Liu, G., Suchitra, A., Zhang, H., Feng, M., Ng, S.K., Wong, L.: Assocexplorer: an association rule visualization system for exploratory data analysis. In: KDD, pp. 1536–1539. ACM (2012)
14. Mihelčić, M., Džeroski, S., Lavrač, N., Šmuc, T.: Redescription mining with multi-target predictive clustering trees. In: Ceci, M., Loglisci, C., Manco, G., Masciari, E., Ras, Z.W. (eds.) NFMCP 2015. Lecture Notes in Artificial Intelligence (LNAI), vol. 9607, pp. 125–143. Springer, Heidelberg (2016). doi:10.1007/978-3-319-39315-5_9
15. Parida, L., Ramakrishnan, N.: Redescription mining: structure theory and algorithms. In: AAAI, pp. 837–844. AAAI Press/The MIT Press (2005)
16. Ramakrishnan, N., Kumar, D., Mishra, B., Potts, M., Helm, R.F.: Turning cartwheels: an alternating algorithm for mining redescriptions. In: Proceedings of the 10Th ACM SIGKDD International Conference on Knowledge Discovery and Data Mining, KDD 2004, pp. 266–275. ACM, New York (2004)
17. UNCTAD: Unctad database (2014). http://unctadstat.unctad.org/
18. Wehrens, R., Buydens, L.M.C.: Self and super-organising maps in R: the kohonen package. J. Stat. Softw. **21**(5) (2007). http://www.jstatsoft.org/v21/i05
19. WorldBank: World bank database (2014). http://data.worldbank.org/
20. Zaki, M.J., Phoophakdee, B.: MIRAGE: A framework for mining, exploring and visualizing minimal association rules. Technical Report 03-4, Computer Science Department, Rensselaer Polytechnic Institute (2003)
21. Zaki, M.J., Ramakrishnan, N.: Reasoning about sets using redescription mining. In: Proceedings of the 11th ACM SIGKDD International Conference on Knowledge Discovery in Data Mining, KDD 2005, pp. 364–373. ACM, New York (2005)

22. Zhang, M., He, C.: Survey on association rules mining algorithms. In: Luo, Q. (ed.) Advancing Computing, Communication, Control and Management. LNEE, vol. 56, pp. 111–118. Springer, Heidelberg (2010)
23. Zinchenko, T.: Redescription mining over non-binary data sets using decision trees. Master's thesis, Universität des Saarlandes Saarbrücken, Germany (2014)

Expect the Unexpected – On the Significance of Subgroups

Matthijs van Leeuwen[1](✉) and Antti Ukkonen[2](✉)

[1] Leiden Institute of Advanced Computer Science, Leiden, The Netherlands
m.van.leeuwen@liacs.leidenuniv.nl
[2] Finnish Institute of Occupational Health, Helsinki, Finland
antti.ukkonen@ttl.fi

Abstract. Within the field of exploratory data mining, subgroup discovery is concerned with finding regions in the data that stand out with respect to a particular target. An important question is how to validate the patterns found; how do we distinguish a true finding from a false discovery? A common solution is to apply a statistical significance test that states that a pattern is real iff it is different from a random subset.

In this paper we argue and empirically show that this assumption is often too weak, as almost any realistic pattern language specifies a set of subsets that strongly deviates from random subsets. In particular, our analysis shows that one should *expect the unexpected* in subgroup discovery: given a dataset and corresponding description language, it is very likely that high-quality subgroups can —and hence will— be found.

1 Introduction

Subgroup Discovery (SD) [7,19] is concerned with finding regions in the data that stand out with respect to a given target. It has many closely related cousins, such as Significant Pattern Mining [17] and Emerging Pattern Mining [1], which all concern the discovery of patterns correlated with a Boolean target concept. The Subgroup Discovery task is more generic, i.e., it is agnostic of data and pattern types. For example, the target can be either discrete or numeric [5].

A large number of SD algorithms and quality measures have been proposed and it is easy to mine thousands or millions of patterns from data. One question that naturally arises is how to validate these (potential) discoveries: how do we distinguish a 'true' pattern from a 'fluke' that is present in the data by chance and therefore does not represent any true correlation between description and target? In other words, how do we distinguish true from false discoveries?

Statistical testing for pattern mining has received quite some attention over the past decade. Webb was one of the pioneers of this topic and pointed out already in 2007 that the size of the entire pattern space under consideration should be used as Bonferroni correction factor for multiple hypothesis testing [18]. He also observed that when the null hypothesis supposes that the consequent of a rule/pattern is independent of its antedecent, Monte Carlo sampling of the target variable can be used to generate a null distribution. Almost simultaneously, Gionis et al. introduced randomisation testing on Boolean matrices [3].

© Springer International Publishing Switzerland 2016
T. Calders et al. (Eds.): DS 2016, LNAI 9956, pp. 51–66, 2016.
DOI: 10.1007/978-3-319-46307-0_4

Permutation tests [4] are a non-parametric approach to hypothesis testing, with the main advantage that p-values are computed from simulations instead of formulas of parametric distributions. This makes them especially suitable for situations where the null-hypothesis cannot be assumed to have a known parametric form. As a result, permutation tests have since become fairly popular in the data mining community and have been used for, e.g., studying classifier performance [15] and statistical testing for subgroup discovery [2].

More recently, several efficient statistical pattern mining methods have been proposed, mostly for rule discovery in Boolean matrices. Hämäläinen, for example, proposed Kingfisher [6], for efficiently searching for both positive and negative dependency rules based on Fisher's exact test. Terada et al. [17] introduced LAMP, which considers a similar setting but employs a smaller Bonferroni correction based on the notion of 'testable patterns'. Recently improvements have been introduced that make LAMP more efficient [13] and versatile [12].

Expect the Unexpected. The technical assumption underlying most statistical tests for pattern mining is that of *exchangeability* of the observations under the null hypothesis [4]. Simply put, *a priori* the selection of any subset of the data is deemed to be equally likely and the observed patterns are tested against this assumption. In this paper we argue and empirically show that this assumption is often too weak: *in practice almost any pattern language and dataset specify a set of data subsets that strongly deviates from this assumption.*

In particular, we will investigate how likely it is to observe a completely random subset having a large effect size, i.e., a large deviation from the global distribution. As we will see, this probability tends to be very small. Next, we will investigate how likely it is that a given description language, i.e., a set of possible patterns, contains descriptions having large effect size. As we will see, this probability tends to be substantially larger. As a result, we conclude that one should *expect the unexpected* in pattern mining in general and in subgroup discovery in particular: given a dataset and a description language, it is very likely that high-quality subgroups can —and hence will— be found.

One possible conclusion to draw from this is that null-hypothesis significance testing for individual patterns should be used with caution: although it helps to eradicate some very obvious false discoveries, the often-used null hypothesis based on exchangeability may be too weak for the p-values to be useful. Note that we are not the first to warn for the use of significance-based filtering in pattern mining [11], but we are the first to analyse and empirically investigate the effect of the description languages used in pattern mining and their relation to the null hypothesis that assumes random data subsets.

Section 2 introduces the basics of Subgroup Discovery, after which Sect. 3 presents our approach to quantifying the significance of description languages. More precisely, we will formalise the odds of observing a pattern having a large effect size versus observing a random subset having such a large effect size. As computing these odds exactly is clearly infeasible, we introduce the machinery required for estimating its two components in Sects. 4 and 5. We present the empirical analysis in Sect. 6 and conclude in Sect. 7.

2   Subgroup Discovery

A dataset $\mathcal{D}$ is a bag of tuples t over the attributes $\{A_1, \ldots, A_m, Y\}$, where the A_i are the *description attributes* A and Y is the target attribute. Each attribute has a Boolean, nominal or numeric domain.

A *subgroup* consists of a *description* and a corresponding *cover*. That is, a *subgroup description* is a pattern S, i.e., a conjunction of conditions on the description attributes. Its corresponding *subgroup cover* is the bag of tuples that satisfy the pattern defined by predicate S, i.e., $C(S) = \{t \in \mathcal{D} \mid t \vDash S\}$. We slightly abuse notation and refer to either the description or its cover using S (depending on context). We use $|S|$ to denote the size of the cover, also called *coverage*. Further, a *description language* L consists of all possible subgroup descriptions, parametrised by maximum depth *maxdepth*, which imposes a maximum on the number of conditions that are allowed in a description.

The Subgroup Discovery task is to find the top-k ranking subgroups according to some quality measure $\varphi : 2^Y \mapsto \mathbb{R}$, which assigns a score to any individual subgroup based on its target values. We consider the well-known Weighted Relative Accuracy (WRAcc), defined as $\varphi_{WRAcc}(S) = \sqrt{|S|}(\mu(S) - \mu)$, where μ is the mean of the target variable (restricted to the tuples in the subgroup cover in case of $\mu(S)$). WRAcc is well-defined for Boolean target attributes by interpreting the proportion of ones as the mean of Boolean values.

To discover high-quality subgroups, top-down search through the pattern space is commonly used. Several parameters influence the search, e.g., a minimum coverage threshold requires subgroup covers to consist of at least *mincov* tuples and the *maxdepth* parameter allows to restrict the size of the search space. During search, the overall top-k subgroups are usually kept as final result.

3   Estimating the Significance of a Description Language

In this section we motivate and formalise the approach that we will take to empirically investigate the significance of subgroups or, more accurately, their description languages. For this, we start by making three important observations.

The first observation, which we already mentioned in the Introduction, is that *the null hypotheses of most statistical tests for pattern mining are based on random subsets of the data*. This is also the case for, for example, the target permutation test proposed by Duivesteijn and Knobbe [2]. The rationale for permuting the target attribute is that this will result in datasets with no meaningful functional relationship between the description attributes and the target. Subgroups found in the actual data should have a higher quality than those found on the permuted data, or otherwise the subgroup is there "by chance".

We will empirically show, however, that given some 'non-random' dataset, almost any description language corresponds to a set of subsets that is very different from the set of random subsets, trivially resulting in very low p-values. Hence, this type of null hypothesis is often too weak and any approach based on this assumption will render many patterns significant; see, e.g., [2,18].

The second observation is that *pattern mining techniques search for the best patterns present in a given description language.* This observation has two immediate consequences. First, one cannot only consider the best found pattern and treat this as 'the result', i.e., as if this were the only observation. Instead, as Webb observed [18], one has to take into account that all patterns in the description language are (implicitly or explicitly) considered and hence apply multiple hypothesis correction, for example Bonferroni correction. Second, this also means that there is essentially no difference between testing (1) a description language in its entirety and (2) the best patterns that were found using search. We therefore deviate from the common approach and investigate the 'significance' of entire description languages rather than that of individual patterns.

Finally, the third observation is that *it is key to distinguish —using appropriate statistical terms— sample size on one hand and effect size on the other hand.* In case of Subgroup Discovery, this implies distinguishing subgroup coverage from the difference in target distribution between the subgroup and the dataset. These two quantities are usually combined by the quality measure for the purpose of ranking the patterns, but —unlike Duivesteijn and Knobbe [2]— we will treat these strictly separately for our analysis. There are two reasons for this: (1) quality measures such as Weighted Relative Accuracy are somewhat arbitrary combinations of coverage and relative accuracy, and using a global quality threshold may therefore result in somewhat arbitrary decisions, (2) larger effect sizes are more likely for smaller sample sizes.

Thus, we need to define sample size and effect size. Sample size is simply subgroup coverage as defined in the previous section. For the purpose of this paper we base the effect size on WRAcc and hence define it as $q(S) = (\mu(S) - \mu)/\sigma$, where μ and σ are the mean and standard deviation of the target variable.

3.1 Description Languages and Accessible Subsets

Although patterns are specified by the descriptions that make up a description language, what really matters for our analysis are the subsets of the data that these descriptions imply, i.e., their corresponding covers. We therefore introduce the notion of *accessible subsets.*

Given a description language L and data $\mathcal{D}$, denote by $X_{L,\mathcal{D}}$ the *set of all possible non-empty subgroup covers* in $\mathcal{D}$ that are *accessible* using L. More formally, we define the set of accessible subsets as

$$X_{L,\mathcal{D}} = \{E \subseteq \mathcal{D} \mid \exists S \in L \text{ s.t. } C(S) = E, |E| > 0\}. \tag{1}$$

While the precise size of $X_{L,\mathcal{D}}$, i.e., $|X_{L,\mathcal{D}}|$, is data dependent, in general it will be (much) smaller than 2^N, with $N = |\mathcal{D}|$. This leads to another observation: *in most practical cases only a fraction of subsets of $\mathcal{D}$ are accessible with L.* For instance, in this paper we define L as all conjunctions of conditions on the description attributes up to a given number of conditions (*maxdepth*). This language can only represent subgroups with a cover that is an intersection of half-spaces in the feature space of $\mathcal{D}$, the number of which is substantially lower than 2^N for most practical datasets.

In general, the size of $X_{L,\mathcal{D}}$ can be regarded as a measure of the *expressiveness* of the language L in the database $\mathcal{D}$; it is more informative than the number of descriptions, as multiple descriptions may imply the same cover. Even more important for our purposes, however, is that it fully specifies the set of subsets that one needs to consider in order to determine the significance of a given combination of description language and dataset.

3.2   On the Significance of Description Languages

Although the task of Subgroup Discovery is to find descriptions of 'large' covers having a 'large' effect size, in general all subsets of the data can be associated with an effect size. Thus, we can introduce notions that quantify how probable it is to observe a certain effect size in either (1) an accessible or (2) a random subset and compare these two probabilities. As noted before, this strongly depends on the sample size, i.e., the number of tuples in the subset. We therefore propose to *compare the collection of accessible subsets with all (random) subsets having the same size distribution to determine if the description language and data together specify subgroups having relatively high effect size.*

Let us first consider the accessible subsets. Let $X_{k,L,\mathcal{D}}$ denote the set of all accessible subsets with coverage at least k, i.e., $|S| \geq k$. Then, we are interested in the probability that one such accessible subset has an effect size larger than θ, i.e., how likely is that event? We will empirically estimate this probability as

$$\Pr(q(S) \geq \theta \mid X_{k,L,\mathcal{D}}) = \frac{|\{S \in X_{k,L,\mathcal{D}} \mid q(S) \geq \theta\}|}{|X_{k,L,\mathcal{D}}|}.$$

Next we are interested in the same probability, but computed for all possible subsets while taking the coverage distribution of the accessible subsets into account. We denote this probability by $\Pr(q(S) \geq \theta \mid \mathcal{D}_k)$. Based on our observations we expect the former probability to be much larger than the latter, as this would indicate that the description language is 'significant' with regard to the target attribute: it contains subgroups that have larger effect sizes than are likely to be observed in random subsets.

We formalise the ratio between these two probabilities as the *odds*, a function of coverage threshold k, minimum effect size θ, description language L and dataset $\mathcal{D}$:

$$\mathrm{odds}(k, \theta, L, \mathcal{D}) = \frac{\Pr(q(S) \geq \theta \mid X_{k,L,\mathcal{D}})}{\Pr(q(S) \geq \theta \mid \mathcal{D}_k)}. \tag{2}$$

Exactly computing the odds is clearly infeasible for any realistic dataset. In the following two sections we will therefore introduce techniques to estimate $\Pr(q(S) \geq \theta)$ for accessible subsets and for all subsets, respectively.

4   Estimating $\Pr(q(S) \geq \theta)$ for Accessible Subsets

We will now introduce a method for estimating the size of $X_{L,\mathcal{D}}$ for $|S| \geq k$, with or without constraint on the effect size θ, so that we can estimate $\Pr(q(S) \geq \theta)$

for accessible subsets. For this we build upon the SPECTRA algorithm [10]; we next provide a brief description but refer to [10] for the details, as we will focus on the modifications needed for our current purposes.

The SPECTRA Algorithm. SPECTRA is a modification of Knuth's seminal algorithm [8] for estimating the size of a combinatorial search tree, which is based on the idea of sampling random paths from the root of the tree to a leaf. Let $\mathbf{d} = (d_0, \ldots, d_{h-1})$ denote the sequence of branching factors observed along a path from the root (on level 0) to a leaf (on level h). The estimate produced by $\mathbf{d}$, which we call the *path estimate*, is defined as $\hat{e}(\mathbf{d}) = \sum_{i=1}^{h} \prod_{i=0}^{i-1} d_i$.

Since search trees are rarely regular in practice, Knuth's algorithm samples a number of different paths, and uses the average of the $\hat{e}(\mathbf{d})$ values as an estimate of the size of the tree. To use this method for estimating the number of patterns having coverage $\geq k$, we sample a number of paths from the root of the pattern lattice up to (and including) the frequent/infrequent border, compute the path estimates, and take the average of these. At every step the branching factor is given by the number of extensions to the current pattern that are still frequent. We need to make a small modification though, because the patterns do not form a tree but a *lattice*, i.e., each node can be reached via multiple paths. We therefore end up with the following estimator:

$$e(\mathbf{d}) = \sum_{i=1}^{h} \frac{1}{i!} \prod_{i=0}^{i-1} d_i, \tag{3}$$

which includes the normalisation term $1/i!$ to account for the $i!$ possible paths that can reach every node on the ith level.

The SPECTRA$^+$ Algorithm. We describe two extensions to SPECTRA that allow to estimate both $|X_{k,L,\mathcal{D}}|$ and $|\{S \in X_{k,L,\mathcal{D}} \mid q(S) \geq \theta\}|$.

First, observe that, in the context of frequent itemset mining, the number of accessible subsets is equivalent to the number of closed itemsets [16]. We exploit this observation to estimate $|X_{k,L,\mathcal{D}}|$: we estimate the number of closed subgroups, where a subgroup is closed iff there exists no extension that does not change its cover. Let c_i denote an indicator variable for the pattern at level i that is 1 iff it is closed and 0 otherwise. Then, a variant of Knuth's original method can be used to estimate the desired number[1]:

$$e(\mathbf{d}) = \sum_{i=1}^{h} c_{i+1} \frac{1}{i!} \prod_{i=0}^{i-1} d_i. \tag{4}$$

Second, observe that $q(S) \geq \theta$ can be considered as another constraint that can be evaluated for each individual subgroup. Hence, to estimate $|\{S \in X_{k,L,\mathcal{D}} \mid q(S) \geq \theta\}|$ we use $c_i = \mathbb{I}\{S_i \text{ is closed} \wedge q(S_i) \geq \theta\}$ in combination with Eq. 4. Note that this simple yet effective constraint-based estimator can be used for many other types of constraints as well.

---

[1] Note that we here use a simpler yet more versatile approach for estimating the number of closed patterns compared to the one proposed in [10].

Although these modifications maintain the desirable property that the expected estimate is correct, they do lead to an increase in the variance. This is due to the larger differences between the possible paths, which would require many more samples to mitigate. Another potential solution to this problem, one that is computationally less expensive, is to *bias* the path samples: instead of choosing an extension uniformly at random, one can design an alternative sampling distribution for each pattern extension. We consider two such biased samplers:

Frequency-Biased sampler. From all frequent extensions, one is randomly chosen proportional to coverage. To adjust the estimates accordingly, each branching factor is now computed as the inverse probability of choosing this extension, i.e., $d_i = \frac{\sum_{S \in freq} |C(S)|}{|C(S_{i+1})|}$, where *freq* is the set of all frequent extensions considered at level i and S_{i+1} is the chosen one.

Quality-Biased sampler. The previous sampler has a preference towards sampling larger covers, but this not necessarily imply large effect size. Therefore, in an effort to more accurately estimate $|\{S \in X_{k,L,\mathcal{D}} \mid q(S) \geq \theta\}|$ for large θ, we propose to randomly choose proportional to subgroup quality, i.e., WRAcc. The branching factors are modified accordingly: $d_i = \frac{\sum_{S \in freq} \varphi(S)}{\varphi(S_{i+1})}$.

5 Estimating $\Pr(q(S) \geq \theta)$ for All Subsets

Next we turn our attention to estimating $\Pr(q(S) \geq \theta)$ for arbitrary data subsets. To compute this estimate under the constraint where the random subsets follow the same size distribution as the accessible subsets, as required above, we first use SPECTRA$^+$ to estimate the coverage distribution of accessible subsets using a number of different values of k. Then, we use the estimators discussed in this section separately for every value of k, and re-weight the estimates by the probabilities of the respective k. The final estimate is simply the sum of these.

It is thus enough to define the estimators for a fixed coverage k. First, note that the probability we are interested in is simply the expected value of the indicator function $\mathbb{I}\{q(S) \geq \theta\}$:

$$\Pr(q(S) \geq \theta) = \mathrm{E}_{\mathrm{unif}}[\mathbb{I}\{q(S) \geq \theta\}], \qquad (5)$$

where $\mathrm{E}_{\mathrm{unif}}$ denotes that the expected value is computed over the uniform distribution of all subsets with coverage k. Next we discuss three different approaches resulting in four different estimators for $\Pr(q(S) \geq \theta)$.

Asymptotic – **Normal Approximation.** Since we have defined $q(S) = (\mu(S) - \mu)/\sigma$, where μ and σ are the mean and standard deviation of the target variable, respectively, a simple but efficient heuristic for $\Pr(q(S) \geq \theta)$ is given by the tail of the standard normal distribution. According to the central limit theorem, we have $\sqrt{k}q(S) \sim \mathcal{N}(0,1)$, when $|S| = k$. Let $\mathrm{cdf}(x)$ denote the cumulative density of the standard normal distribution at x. Now we can use $1 - \mathrm{cdf}(\sqrt{k}\theta)$ as an estimate of $\Pr(q(S) \geq \theta)$. In the experiments we will call this the *asymptotic* estimator.

Naive – **Sample Mean.** We can also use the sample mean to estimate the expected value in Eq. 5. Given $S_1, \ldots, S_R$, a uniformly drawn sample of R subsets of size k, we have

$$\Pr(q(S) \geq \theta) \approx R^{-1} \sum_{i=1}^{R} \mathbb{I}\{q(S_i) \geq \theta\}. \tag{6}$$

This is guaranteed to be an unbiased and consistent estimate of $\Pr(q(S) \geq \theta)$, but it may require a prohibitively large R when estimating very small probabilities, because in such cases we are unlikely to find random subsets S with effect size above θ in a sample of realistic size. In particular, this will happen for large values of θ. We call this estimator *naive* when reporting experimental results.

Weighted – **Importance Sampling.** To obtain good estimates also for large minimum effect size θ, we resort to *importance sampling*. The idea of importance sampling (see e.g. [14]) is to increase the probability of observing the event of interest, in our case $\mathbb{I}\{q(S) \geq \theta\}$, and re-weight these when computing the sample mean so that the resulting expected value is equal to the desired probability. It is easy to show (see Appendix A.1[2]) that the expected value of the indicator function under the uniform distribution (Eq. 5) is equal to the expected value of the indicator function weighted by the coefficient $W(S)$ under a biased sampling distribution, that is,

$$E_{\text{unif}}[\mathbb{I}\{q(S) \geq \theta\}] = E_{\text{biased}}[W(S)\mathbb{I}\{q(S) \geq \theta\}]. \tag{7}$$

Here $W(S) = \frac{p(S)}{p'(S)}$, where $p(S)$ and $p'(S)$ are the probabilities to draw S under the uniform and biased distributions, respectively. Now we can estimate $E_{\text{biased}}[W(S)\mathbb{I}\{q(S) \geq \theta\}]$ by drawing $S_1, \ldots, S_R$ according to p' and computing

$$R^{-1} \sum_{i=1}^{R} W(S_i)\mathbb{I}\{q(S_i) \geq \theta\}. \tag{8}$$

Unlike the basic sample mean estimator of Eq. 6, the weighted variant in Eq. 8 can give good estimates with a much smaller R, i.e., we need fewer samples, provided we have chosen the biased sampling distribution p' appropriately. We must thus define both p', as well as compute the weighting factor $W(S)$.

The Biased Sampling Distribution. We first discuss p'. Since we want to mainly draw random subsets that have a high $q(S)$, in an ideal situation $p'(S) \propto q(S)$. A very simple way to achieve an effect similar to this is to draw S using condition-specific probabilities that are proportional to the target values of the possible extensions, i.e., all conditions that are considered to be added to the current description. That is, we use a *weighted sampling without replacement* scheme, where the condition weights are proportional to their target values.

---

[2] Non-essential derivations can be found in Appendix at https://anon.to/NA9I9A.

More formally, let $p'(u)$ denote the probability to draw the condition u (before other conditions have been drawn), and define

$$p'(u) = \frac{t(u)}{\sum_v t(v)}, \tag{9}$$

where $t(\cdot)$ is the target value of an item. After drawing an item u we set the probability $p'(u)$ to zero, and renormalise the remaining probabilities to sum up to 1. This is repeated until we have drawn k items. This will have the effect that items having a high target value are more likely to be selected into S, and hence $q(S)$ will be biased towards higher values.

Next we discuss the weighting factor $W(S)$. The problem is that to compute $W(S)$ exactly we must compute $p'(S)$ (see also Appendix A.2), but no closed form expression for $p'(S)$ exists under weighted sampling without replacement, and computing $p'(S)$ exactly is infeasible. Hence we consider two heuristics.

Approximating $W(S)$ by Assuming a Sampling with Replacement Scheme. As a first approximation, we can consider a scheme that corresponds to sampling *with replacement*. In practice we still draw samples without replacement as usual, but compute $W(S) = p(S)/p'(S)$ as if we had used sampling with replacement. This has the upside that all required probabilities have simple and easy to compute closed form expressions. In particular, we have $\bar{p}(S) = \frac{1}{n^k}$ for uniform sampling with replacement, and $\bar{p}'(S) = \prod_{u \in S} p'(u)$ for weighted sampling with replacement. Given these we can approximate $W(S)$ as

$$\bar{W}_1(S) = \left(n^k \prod_{u \in S} p'(u) \right)^{-1} \tag{10}$$

where $p'(u)$ is defined as in Eq. 9. Our first importance sampling estimator, called W_1 below, is thus defined by replacing $W(S)$ in Eq. 8 with $\bar{W}_1(S)$ of Eq. 10 above. This estimator is not guaranteed to converge to the correct expected value, but the experiments show that it still produces reasonable results.

Approximating $W(S)$ by Sampling Permutations of S. Our second approach to approximate $W(S)$ relies on a different technique. As mentioned above, the basic problem of computing $W(S)$ exactly under weighted sampling without replacement is that it would require us to consider the probabilities of all possible permutations in which S could have been drawn. Doing this is clearly infeasible in practice, as there are $k!$ permutations for sets of size k. However, we can compute an approximation of $W(S)$ by using only a small sample of permutations (see also Appendix A.3). Concretely, given a sample of Q permutations of the set S, denoted $\pi^1, \ldots, \pi^Q$, we can estimate $W(S)$ as

$$\bar{W}_2(S) = \left(\frac{1}{Q} \sum_{i=1}^{Q} \Pr(\pi^i) \right)^{-1} \frac{(n-k)!}{n!}, \tag{11}$$

where $\Pr(\pi^i)$ is the probability to draw the permutation π^i under the weighted sampling without replacement scheme. Our second importance sampling estimator, called W_2, can now be defined by replacing $W(S)$ in Eq. 8 with $\bar{W}_2(S)$ of Eq. 11.

6 Experiments

We here present three experiments, of which the first two concern an assessment of the sampling accuracy of the estimators introduced in the previous two sections. After that, we present our empirical analysis of the odds of observing high effect size subgroups on a selection of datasets.

Table 1 presents the datasets that we use for Experiments 2 and 3, which all have either a Boolean or a numeric target. Except for Crime and Elections, which were described in [9], all were taken from the UCI Machine Learning repository[3]. On-the-fly discretisation of numeric description attributes was applied, meaning that 6 equal-size intervals were created upon pattern extension.

Table 1. Datasets according to target type. For each dataset the number of tuples and the number of discrete resp. numeric description attributes are given.

Dataset	Properties			Dataset	Properties		
	$\|\mathcal{D}\|$	|disc|	|num|		$\|\mathcal{D}\|$	|disc|	|num|
Boolean target				Numeric target			
Adult	48842	8	6	Abalone	4177	1	7
Breast cancer	699	0	9	Crime	1994	1	101
Mushroom	8124	22	0	Elections	1846	71	2
Pima	768	0	8	Helsinki	8337	1	22
Spambase	4601	57	0	Housing	506	1	12
Tic-tac-toe	958	9	0	RedWine	1599	0	11
				Wages	534	7	3

6.1 Experiment 1: Estimating $\Pr(q(S) \geq \theta)$ in All Subsets

We discuss an experiment to compare the four estimators of $\Pr(q(S) \geq \theta)$ in all subsets. One difficulty with this is to generate a target variable so that an exact ground truth probability can be calculated. It should also be possible to vary the skew of the resulting distribution, as the estimators may be affected by this.

In this experiment we generate the target variable as follows. Let $\mathbf{x}$ denote a vector of length n. We set the first m elements of $\mathbf{x}$ equal to $v \geq 1$, and let the remaining $n - m$ elements of $\mathbf{x}$ be equal to 1. That is, let $\mathbf{x}[1 : m] = v$, and $\mathbf{x}[(m + 1) : n] = 1$. By varying m and v we can adjust the skew, with

---
[3] http://archive.ics.uci.edu/ml/.

lower values of m and higher values of v implying a higher skew. Let μ and σ denote the mean and standard deviation of $\mathbf{x}$, respectively. The ground truth probability for a given θ is now given by

$$\Pr(q(S) \geq \theta_l) = \binom{n}{k}^{-1} \sum_{i=k-l}^{k} \binom{m}{i} \binom{n-m}{k-i}, \qquad (12)$$

where $l = \lfloor \frac{k(v-\theta\sigma-\mu)}{v-1} \rfloor$, and $\theta_l = (((k-l)v+l)/k - \mu)/\sigma$ (Appendix A.4). Notice that since we take the floor when computing l, the resulting probability does not correspond exactly to the given θ, but matches a value θ_l that is slightly larger than θ.

Table 2. Parameter values for synthetic target generation used in Experiment 1. Notice that values for skew parameter m and subgroup coverage k are given as fractions of n (the size of the generated target variable). For example, one combination that these give for n, m and k is $n = 2000$, $m = 0.3 \times 2000 = 600$, $k = 0.05 \times 2000 = 100$.

Parameter	Values	Parameter	Values
n	$1000, 2000, 3000, 4000$	k	$0.05 \times n, 0.10 \times n$
m	$0.1 \times n, 0.3 \times n, 0.6 \times n$	θ	$0.2, 0.4, 0.6, 0.8, 1.0$
v	$3, 6, 9$	R	500

We run all our estimators with all possible parameter combinations shown in Table 2. For the naive estimator we use $10 \times R$ samples, while W_1 and W_2 use both exactly R samples. When estimating with W_2 we set $Q = 200$. To get a fair comparison between the importance sampling estimators, we use the same set of R samples for both W_1 and W_2. Notice that computing the exact probability may fail due to problems with numerical computation. Such cases are omitted from consideration. This experiment was run in R 3.3^4 on Linux.

As estimating small probabilities accurately is difficult, we are mainly interested in calculating the correct order of magnitude, i.e., the estimates we report are always base-10 logarithms of $\Pr(q(S) \geq \theta)$. We also report the mean absolute error (MAE) of these over some subsets of the parameter combinations. Moreover, for some inputs one or more of the estimators fail to produce a positive, non-zero estimate. To get an overview of how prevalent this is, we compute the *recall* over all 360 parameter combinations (the two values of R are handled separately) by considering the fraction of cases where an estimate was obtained of those where it was possible to compute the exact probability.

An overview of the results is given in Table 3 for $R = 500$. On the left we show both MAE (as well as the minimum and maximum error) and recall when considering all parameter combinations, and on the right we report the same numbers for cases with $\theta \geq 0.6$. Larger values of θ are often more interesting in

---

[4] https://www.R-project.org.

Table 3. Mean absolute error (MAE), minimum and maximum error, and recall of the estimators for low θ (left side) as well as high values of θ (right side) when $R = 500$.

Estimator	$\theta \geq 0$			$\theta \geq 0.6$		
	MAE	(min, max) error	Recall	MAE	(min, max) error	Recall
Asymptotic	2.55	(0.004 19.2)	1.00	3.96	(0.035, 19.22)	1.000
Naive	0.18	(0.00, 1.5)	0.26	0.81	(0.677, 0.95)	0.014
W_1	2.76	(0.00, 25.8)	0.84	1.61	(0.00, 11.19)	0.743
W_2	2.67	(0.00, 30.3)	0.84	0.75	(0.00, 15.73)	0.743

practice, as we will show below when studying $\Pr(q(S) \geq \theta)$ in accessible subsets of real data. We observe that the naive estimator has the lowest MAE in every case. However, it also has a very low recall, meaning that it only rarely produces a useful estimate, but if it does, the estimate is rather accurate. (This is of course what one would expect.) But especially for $\theta \geq 0.6$ the naive estimator is in practice useless due to having almost zero recall. The asymptotic estimator, on the other hand, always has the highest possible recall (=1), while the importance sampling estimators W_1 and W_2 fail to calculate a positive non-zero estimate in about 15 to 25 % of the cases. When $\theta \geq 0$, the importance sampling estimators perform roughly at the same level as the asymptotic estimator in terms of MAE. However, when considering performance for $\theta \geq 0.6$, we find that the importance sampling estimators, especially W_2, have substantially lower MAE without losing that much in recall. We conclude that for high values of θ the W_2 estimator gives very good results.

6.2   Experiment 2: Estimating $\Pr(q(S) \geq \theta)$ in Accessible Subsets

As explained in Sect. 3, we estimate $\Pr(q(S) \geq \theta)$ from two components: (1) the number of accessible subsets having $(q(S) \geq \theta)$, and (2) the total number of accessible subsets. We estimate both components using SPECTRA$^+$. To evaluate the estimators we compare them against probabilities based on exhaustive counts. Although these can be obtained through simply enumerating all patterns and then counting the number of distinct accessible subsets, this is clearly only feasible for relatively small datasets and description languages. In the following we therefore limit our evaluation to datasets and maximum depth settings for which it was possible to enumerate and count all accessible subsets within a day.

We have introduced three variants of SPECTRA$^+$: the *uniform* variant that uses uniform path sampling, the *frequency* variant that uses the frequencies of the refinements to bias the sampling, and the *quality* variant that uses Weighted Relative Accuracy to bias the sampling. On the left side of Table 4 we present a comparison of these three variants when estimating $\Pr(q(S) \geq \theta)$ in the datasets. Top part of the table shows datasets for which the true probability can be computed, bottom part shows cases where only estimates are available. Note that values of θ vary across datasets, as for some datasets there are fewer high effect

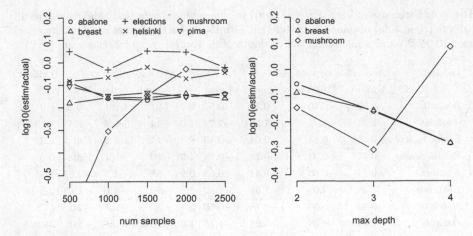

Fig. 1. Estimation accuracy of $\Pr(q(S) \geq 0)$ in accessible subsets using uniform sampling both as a function of the number of path samples (left) and search depth (right) with those datasets and parameter combinations that we were able to compute the exact counts by enumerating all accessible subsets.

size subsets than for others. Datasets having a numeric target attribute are shown in *italics*. The description language corresponding to a maximum depth of three is used, and the estimates are computed from 1000 path samples.

We observe that none of the estimators consistently has the highest accuracy. All tend to catch the correct order of magnitude, with the exception of the quality-biased sampler that is clearly the least accurate. Based on this we choose to use the uniform sampling variant for the remaining experiments, as it is the simplest to implement, and the other approaches do not seem to lead to substantial practical benefits[5] when estimating $\Pr(q(S) \geq \theta)$.

Next we investigate the effect of the number of path samples as well as the maximum search depth on the accuracy of the estimates. For this we compare estimates obtained using $\{500, 1000, 1500, 2000, 2500\}$ path samples, as well as maximum depths of 2, 3, and 4. The estimation accuracy is shown in Fig. 1 for both cases. As we can see from the results, using more path samples does not affect the quality of the estimates in most cases. For most data sets the estimator underestimates the true probability slightly, but the effects are negligible. Given that we are only interested in correctly estimating the order of magnitude, we use 1000 path samples for all remaining experiments. Similar results can be found for maximum depth. Increasing maximum depth may increase the factor by which the path sampling estimator underestimates, but the results are not consistent across all data sets.

---

[5] We note, however, that for the simpler problem of merely counting accessible subsets using the frequency-biased sampler may give more accurate results. These results are omitted due to length restrictions.

Table 4. Estimates of $\log_{10}(\Pr(q(S) \geq \theta))$ in accessible subsets (left side) as well as all subsets (right side) obtained with different estimators for the lowest support threshold used $(0.05|\mathcal{D}|)$ and a maximum search depth of 3. Target: *Numeric*/Boolean.

Dataset	Accessible subsets				θ	All subsets			Odds
	Uniform	Frequency	Quality	Exact		Asymp	Naive	W_2	
Abalone	−0.52	−0.50	−0.19	−0.52	0.4	−10	−Inf	**−9.5**	9.0
Abalone	−1.71	−1.71	−1.56	−1.74	1.0	**−44**	−Inf	−Inf	42.3
Breast cancer	−0.15	−0.13	−0.04	−0.12	0.3	**−1.3**	**−1.3**	Fail	1.2
Breast cancer	−0.37	−0.34	−0.16	−0.29	1.0	**−10**	−Inf	Fail	9.6
Elections	−0.44	−0.60	−0.44	−0.63	0.2	−1.9	**−1.7**	−22.2	1.3
Elections	−1.46	−2.09	−1.51	−1.82	0.5	−6.1	**−4.0**	−19.3	2.5
Helsinki	−0.49	−0.60	−0.16	−0.50	0.4	**−20**	−Inf	**−20**	19.5
Helsinki	−1.08	−1.08	−0.66	−1.16	1.0	**−100**	−Inf	−Inf	98.9
Housing	−0.35	−0.34	−0.15	−0.38	0.4	−1.5	**−1.5**	−1.6	1.2
Housing	−1.23	−1.30	−0.94	−1.24	1.0	−8.1	−Inf	**−6.9**	5.7
Mushroom	−0.74	−0.26	-	−0.27	0.1	**−0.9**	−0.9	−134	0.2
Mushroom	−0.53	−0.76	-	−0.55	0.2	**−5.1**	−Inf	−134	4.6
Pima	−0.44	−0.47	−0.15	−0.50	0.2	−1.4	**−1.6**	−Inf	1.2
Pima	−1.88	−1.51	−1.36	−1.82	0.8	**−7.5**	−Inf	−Inf	5.6
Adult	−0.33	−0.28	−0.19		0.1	**−1.0**	**−1.0**	NA	0.7
Adult	−1.89	-	−2.43		0.2	**−32**	−Inf	NA	30.1
Crime	−0.39	−0.35	-		0.1	−1.0	**−1.0**	−12.4	0.6
Crime	−0.76	−0.73	−0.38		0.4	−5.5	**−4.5**	−11.8	3.7
RedWine	−0.33	−0.34	-		0.1	−0.3	**−0.3**	−0.85	0.0
RedWine	−1.25	−1.22	−0.75		0.6	**−8.3**	−Inf	−Inf	7.1
Spambase	−0.22	−0.28	−0.02		0.2	−3.4	**−3.4**	−97	3.2
Spambase	−0.86	−1.06	−0.59		1.0	**−54**	−Inf	−97	53.1
Tic-tac-toe	−0.28	−0.27	−0.05		0.1	−0.3	**−0.3**	−9.2	0.0
Tic-tac-toe	−1.33	−1.43	−1.06		0.4	−2.8	**−3.1**	−9.2	1.8
Wages	−0.83	−0.78	−0.36		0.4	−2.0	**−1.9**	−1.9	1.1
Wages	−1.99	−2.27	−1.51		1.0	−7.1	−Inf	**−5.5**	3.5

6.3   Experiment 3: The Odds of Finding Patterns Having Large Effect Sizes

Lastly, we run the estimators for arbitrary subsets on the real data sets, and compute the odds as defined in Eq. 2. The probability estimates in arbitrary subsets are shown on the right side of Table 4. We have indicated the most reliable estimator in bold. Notice that the importance sampling estimator W_2 tends to produce bad estimates for small values of θ, as we also observed above. In general, if the naive estimator produces a non-zero positive estimate, we take this to be the most reliable value. For small probabilities, however, this will fail, and we must choose between the asymptotic estimator and W_2. We choose W_2 over the asymptotic in situations where the estimators do not deviate substantially.

Finally, we compute the base-10 log odds between (1) the uniform estimator in accessible subsets and (2) the estimator shown in bold for all subsets. We can observe that, for *all datasets*, the probability of observing a high effect size subgroup is *several orders of magnitude* higher in the accessible subsets than what it is in arbitrary subsets having the same size distribution.

7   Conclusions

We have shown that realistic pattern languages contain large numbers of high-quality subgroups; much more so than can be expected from randomly drawn subsets. Although maybe not a surprising result on itself, this does have implications for approaches that aim to eliminate false discoveries. Specifically, statistical tests often rely on the null hypothesis that any subset of the data is equally likely, but this assumption is clearly too weak. That is, one should *expect the unexpected* in pattern mining: given a dataset and a description language, it is very likely that high-quality subgroups can —and hence will— be found.

Note that we do not claim that significance testing is unusable in the context of pattern mining; existing approaches can certainly help to identify obvious false discoveries. Our analysis does demonstrate, however, that it is of interest to investigate significance tests for description languages (rather than individual patterns) that take the inherent structure of accessible subsets into account.

References

1. Dong, G., Zhang, X., Wong, L., Li, J.: CAEP: classification by aggregating emerging patterns. In: Japkowicz, N., Matwin, S. (eds.) DS 1999. LNCS (LNAI), vol. 1721, pp. 30–42. Springer, Heidelberg (1999). doi:10.1007/3-540-46846-3_4
2. Duivesteijn, W., Knobbe, A.: Exploiting false discoveries - statistical validation of patterns and quality measures in subgroup discovery. In: Proceedings of the ICDM 2011, pp. 151–160 (2011)
3. Gionis, A., Mannila, H., Mielikäinen, T., Tsaparas, P.: Assessing data mining results via swap randomization. ACM Trans. Knowl. Discov. Data 1(3), 14 (2007)
4. Good, P.I.: Permutation, Parametric and Bootstrap Tests of Hypotheses, 3rd edn. Springer, New York (2005)
5. Grosskreutz, H., Rüping, S.: On subgroup discovery in numerical domains. Data Min. Knowl. Disc. 19(2), 210–226 (2009)
6. Hämäläinen, W.: Kingfisher: an efficient algorithm for searching for both positive and negative dependency rules with statistical significance measures. Knowl. Inf. Syst. 32(2), 383–414 (2012)
7. Klösgen, W.: Explora: a multipattern and multistrategy discovery assistant. In: Advances in Knowledge Discovery and Data Mining, pp. 249–271 (1996)
8. Knuth, D.: Estimating the efficiency of backtrack programs. Math. Comput. 29(129), 122–136 (1975)

9. van Leeuwen, M., Knobbe, A.: Non-redundant subgroup discovery in large and complex data. In: Gunopulos, D., Hofmann, T., Malerba, D., Vazirgiannis, M. (eds.) ECML PKDD 2011. LNCS (LNAI), vol. 6913, pp. 459–474. Springer, Heidelberg (2011). doi:10.1007/978-3-642-23808-6_30

10. van Leeuwen, M., Ukkonen, A.: Fast estimation of the pattern frequency spectrum. In: Calders, T., Esposito, F., Hüllermeier, E., Meo, R. (eds.) ECML PKDD 2014. LNCS (LNAI), vol. 8725, pp. 114–129. Springer, Heidelberg (2014). doi:10.1007/978-3-662-44851-9_8

11. Lemmerich, F., Puppe, F.: A critical view on automatic significance-filtering in pattern mining. In: Proceedings of ECMLPKDD 2014 Workshop on Statistically Sound Data Mining (2014)

12. Llinares-López, F., Sugiyama, M., Papaxanthos, L., Borgwardt, K.M.: Fast and memory-efficient significant pattern mining via permutation testing. Proc. KDD **2015**, 725–734 (2015)

13. Minato, S., Uno, T., Tsuda, K., Terada, A., Sese, J.: A fast method of statistical assessment for combinatorial hypotheses based on frequent itemset enumeration. In: Calders, T., Esposito, F., Hüllermeier, E., Meo, R. (eds.) ECML PKDD 2014. LNCS (LNAI), vol. 8725, pp. 422–436. Springer, Heidelberg (2014). doi:10.1007/978-3-662-44851-9_27

14. Motwani, R., Raghavan, P.: Randomized Algorithms. Cambridge University Press, New York (1995)

15. Ojala, M., Garriga, G.C.: Permutation tests for studying classifier performance. J. Mach. Learn. Res. **11**, 1833–1863 (2010)

16. Pasquier, N., Bastide, Y., Taouil, R., Lakhal, L.: Discovering frequent closed itemsets for association rules. In: Beeri, C., Buneman, P. (eds.) ICDT 1999. LNCS, vol. 1540, pp. 398–416. Springer, Heidelberg (1999). doi:10.1007/3-540-49257-7_25

17. Terada, A., Okada-Hatakeyama, M., Tsuda, K., Sese, J.: Statistical significance of combinatorial regulations. Proc. Natl. Acad. Sci. **110**(32), 12996–13001 (2013)

18. Webb, G.I.: Discovering significant patterns. Mach. Learn. **68**(1), 1–33 (2007)

19. Wrobel, S.: An algorithm for multi-relational discovery of subgroups. In: Komorowski, J., Zytkow, J. (eds.) PKDD 1997. LNCS, vol. 1263, pp. 78–87. Springer, Heidelberg (1997). doi:10.1007/3-540-63223-9_108

Min-Hashing for Probabilistic Frequent Subtree Feature Spaces

Pascal Welke[1]([✉]), Tamás Horváth[1,2], and Stefan Wrobel[1,2]

[1] Department of Computer Science, University of Bonn, Bonn, Germany
welke@uni-bonn.de
[2] Fraunhofer IAIS, Schloss Birlinghoven, Sankt Augustin, Germany

Abstract. We propose a fast algorithm for approximating graph similarities. For its advantageous semantic and algorithmic properties, we
define the similarity between two graphs by the Jaccard-similarity of
their images in a binary feature space spanned by the set of frequent
subtrees generated for some training dataset. Since the feature space
embedding is computationally intractable, we use a probabilistic subtree
isomorphism operator based on a small sample of random spanning trees
and approximate the Jaccard-similarity by min-hash sketches. The partial order on the feature set defined by subgraph isomorphism allows for
a fast calculation of the min-hash sketch, without explicitly performing
the feature space embedding. Experimental results on real-world graph
datasets show that our technique results in a fast algorithm. Furthermore, the approximated similarities are well-suited for classification and
retrieval tasks in large graph datasets.

1 Introduction

A common paradigm in distance-based learning is to embed the instance space
into some appropriately chosen feature space equipped by a metric and to define
the dissimilarity between two instances by the distance of their images in the
feature space. In this work we deal with the special case of this paradigm that the
instances are vertex and edge labeled graphs (without any structural restriction),
the feature space is the set of vertices of the d-dimensional *Hamming-cube* (i.e.,
$\{0, 1\}^d$) spanned by the elements of a feature set of cardinality d for some $d > 0$,
and the metric is defined by the *Jaccard-distance*. This problem class is valid
because any vertex of the d-dimensional Hamming-cube can be regarded as a
set over a universe of cardinality d.

The crucial step of the generic approach above is the appropriate choice of the
feature space. Indeed, the quality (e.g., predictive performance) of this method
applied to some particular problem strongly depends on the semantic relevance of
the features selected, implying that one might be interested in feature languages
of large expressive power. If, however, the features are arbitrary graph patterns
and the (binary) value of a pattern P on a graph G is defined by 1 if and only
if P is a subgraph of G then this approach suffers from severe computational
limitations. Indeed, even in the simple case that P is a path, it is NP-complete

T. Calders et al. (Eds.): DS 2016, LNAI 9956, pp. 67–82, 2016.
DOI: 10.1007/978-3-319-46307-0_5

to decide whether P is a subgraph of G (i.e., whether P is subgraph isomorphic to G). This implies that the embedding of a graph into the feature space is computationally intractable if the pattern size is not bounded by some constant.

Our goal in this paper is to use *frequent* subgraphs as features, without any structural restriction on the graph class defining the instance space. This is motivated, among others, by the observation that frequent subgraph kernels [3] are of remarkable predictive performance e.g. on the ligand-based virtual screening problem [5]. The features, i.e., the set of frequent subgraphs are generated initially for a training set of graphs. To arrive at a practically fast algorithm, we restrict the pattern language to trees. This restriction alone is, however, not sufficient for a polynomial time algorithm for the following reasons: Mining frequent subtrees from arbitrary graphs is computationally intractable [6] and, as mentioned above, deciding subgraph isomorphism from a tree into a graph is NP-complete. To overcome these computational limitations, we give up the demand on the correctness of the pattern matching operator (i.e., subgraph isomorphism). More precisely, for each training graph we first take a set of k spanning trees generated uniformly at random, where k is some user specified parameter, replace each training graph with the random forest obtained by the vertex disjoint union of its k random spanning trees, and calculate finally the set of frequent subtrees for this forest database for some user specified frequency threshold. Clearly, the output of this probabilistic technique is always *sound* (any tree found to be frequent by this algorithm is a frequent subtree with respect to the original dataset), but *incomplete* (the algorithm may miss frequent subtrees). Since frequent subtrees in forests can be generated with polynomial delay [7], our frequent pattern generation algorithm runs in time polynomial in the combined size of the training dataset $\mathcal{D}$ and the set of frequent subtrees generated, as long as the number k of random spanning trees is bounded by a polynomial of the size of $\mathcal{D}$.

We follow a similar strategy for the embedding step: For an unseen graph G and a frequent tree pattern T, we generate a set F of k random spanning trees of G with the same method as for the frequent pattern mining algorithm and return 1 if T is subgraph isomorphic to F; 0 otherwise. On the one hand, in this way we decide subgraph isomorphism from a tree into a graph with one-sided error, as only a negative answer may be erroneous, i.e., when T is subgraph isomorphic to G but not to F. On the other hand, this subgraph isomorphism test with one-sided error can be performed in polynomial time. In a recent paper [12], we have empirically demonstrated that remarkable predictive performance can be obtained by the method sketched above. We show in this paper that our probabilistic algorithm decides subgraph isomorphism from T into G correctly with high probability if k is chosen appropriately.

Though the method sketched above runs in polynomial time, it is still impractical for large graphs and/or massive graph datasets for the following reasons: The fastest known algorithm for subtree isomorphism into forests runs in $O\left(n^{2.5}/\log n\right)$ time [9]. Although this is polynomial, it is prohibited even for datasets with a few hundred thousands of small graphs [12]. A second reason is the high dimensionality of the feature space, resulting in practically infeasible time and space complexity. Running time and memory can, however, be significantly reduced by using *min-hashing* [1], an elegant probabilistic technique for the approximation of the Jaccard-similarity. Given a binary feature vector f and a permutation π of f, the method is based on calculating the min-hash value $h_\pi(f)$, i.e., the position of the first occurrence of 1 in the permuted order of f.

For the feature set formed by the set of all paths up to a *constant* length, minhashing has already been applied for graph similarity estimation by performing the embedding *explicitly* [11]. We show for the more general case of tree patterns of *arbitrary* length that for a feature vector f and permutation π, $h_\pi(f)$ can be computed *without* calculating f. On the one hand, we can utilize the fact that we are interested in the first occurrence of a 1 in the order of π; once we have found it, we can stop the calculation, as all patterns after $h_\pi(f)$ are irrelevant for minhashing. Beside this straightforward speed-up of the algorithm, the computation of the min-hash can further be accelerated utilizing the facts that a tree pattern T need not be evaluated if T or a subtree of T has already been considered for this or another permutation. These facts allow us to define a linear order on the patterns to be evaluated and to avoid redundant subtree isomorphism tests.

Our experimental results clearly demonstrate that using our technique, the number of subtree isomorphism tests can dramatically be reduced with respect to the min-hash algorithm performing the embedding explicitly. It is natural to ask how the predictive performance of the approximate similarities compares to the exact ones. We show that even for a few random spanning trees per chemical compound, remarkable precisions of the active molecules can be obtained by taking the k nearest neighbors of an active compound for $k = 1, \ldots, 100$ and that these precision values are close to those obtained by the *full* set of frequent subtrees. In a second experimental setting, we analyze the predictive power of support vector machines using our approximate similarities and show that it compares to that of state-of-the-art related methods. The stability of our incomplete probabilistic technique is explained by the fact that a subtree generated by our method is frequent not only with respect to the training set, but, with high probability, also with respect to the set of spanning trees of a graph.

The rest of the paper is organized as follows. In Sect. 2 we collect the necessary notions and sketch the min-hashing technique. In Sect. 3 we present our algorithm for calculating min-hashing in probabilistic tree feature spaces. In Sect. 4 we report our empirical results and conclude finally in Sect. 5 along with mentioning some interesting problems for future work.

2 Notions

In this section we collect the necessary notions and fix the notation. The set $\{1, \ldots, n\}$ will be denoted by $[n]$ for all $n \in \mathbb{N}$. The following basic concepts from graph theory are standard (see, e.g., [4]). An *undirected* (resp. *directed*) *graph* G is a pair (V, E), where V (vertex set) is a finite set and E (edge set) is a subset of the family of 2-subsets of V (resp. $E \subseteq V \times V$). Unless otherwise stated, by graphs we mean undirected graphs. An *unrooted* (or *free*) tree is a connected graph that contains no cycle. For simplicity, we restrict the description of our method to unlabeled graphs, by noting that all concepts can naturally be generalized to labeled graphs.

Among the classical embedding (or pattern matching) operators, subgraph isomorphism is the most widely used one in pattern mining. For this reason, in the next section we will present our method for *subgraph isomorphism* and discuss potential generalizations to other embedding operators in Sect. 5. Let $G_1 = (V_1, E_1)$ and $G_2 = (V_2, E_2)$ be graphs. They are *isomorphic* if there exists a bijection $\varphi : V_1 \rightarrow V_2$ with $\{u, v\} \in V_1$ if and only if $\{\varphi(u), \varphi(v)\} \in V_2$ for all $u, v \in V_1$. G_1 is *subgraph isomorphic* to G_2, denoted $G_1 \preccurlyeq G_2$, if G_2 has a subgraph isomorphic to G_1; $G_1 \prec G_2$ denotes that $G_1 \preccurlyeq G_2$ and G_1 is not isomorphic to G_2. It is a well-known fact that subgraph isomorphism is NP-complete. This negative result holds even for the case that the patterns are restricted to trees.

For any graph class $\mathcal{H}$ containing no two isomorphic graphs, $(\mathcal{H}, \preccurlyeq)$ is a *poset*. Since $\mathcal{H}$ will be finite for our case, we represent $(\mathcal{H}, \preccurlyeq)$ by a directed graph $(\mathcal{H}, E)$ with $(H_1, H_2) \in E$ if and only if $H_1 \preccurlyeq H_2$ and there is no $H \in \mathcal{H}$ with $H_1 \prec H \prec H_2$ for all $H_1, H_2 \in \mathcal{H}$.

We will also use concepts and algorithms from frequent subgraph mining. For any graph class $\mathcal{H}$ (the *pattern* class), finite set $\mathcal{D}$ of graphs, and for any frequency threshold $\theta \in (0, 1]$, a pattern $H \in \mathcal{H}$ is *frequent* if $|\{G \in \mathcal{D} : H \preccurlyeq G\}| \geq \theta |\mathcal{D}|$. Given $\mathcal{H}$, $\mathcal{D}$, and θ, the problem of frequent subgraph mining is to *generate* all patterns from $\mathcal{H}$ that are frequent. This listing problem is computationally intractable [6]. It follows from the proof of this negative result that the problem remains intractable if $\mathcal{H}$ is restricted to trees.[1] If, however, $\mathcal{D}$ is restricted to forests then frequent subgraphs (i.e., subtrees) can be generated with polynomial delay [7].

We will measure the similarity between two graphs by the Jaccard-similarity of their images in the Hamming-cube $\{0, 1\}^{|\mathcal{F}|}$ spanned by the elements of some finite feature set $\mathcal{F}$. The binary feature vectors can then be regarded as the characteristic vectors of subsets of $\mathcal{F}$. Given two feature vectors $\boldsymbol{f_1}$ and $\boldsymbol{f_2}$ representing the sets S_1 and S_2, respectively, we define their similarity by the *Jaccard-similarity* of S_1 and S_2, i.e., by

$$\text{SIM}_{\text{Jaccard}}(\boldsymbol{f_1}, \boldsymbol{f_2}) := \text{SIM}_{\text{Jaccard}}(S_1, S_2) = \frac{|S_1 \cap S_2|}{|S_1 \cup S_2|}$$

with $\text{SIM}_{\text{Jaccard}}(\emptyset, \emptyset) := 0$ for the degenerate case. As long as the feature vectors are low dimensional (i.e., $|\mathcal{F}|$ is small), the Jaccard-similarity can quickly be calculated. If, however, they are high dimensional, it can be approximated with

---

[1] We note that the crucial property implying the negative complexity result in [6] is not necessarily the intractability of subgraph isomorphism; there are cases when efficient frequent subgraph mining is possible even for NP-hard pattern matching operators [7].

the following fast probabilistic technique based on *min-hashing* [1]: For a permutation π of $\mathcal{F}$ and feature vector $\boldsymbol{f}$, define $h_\pi(\boldsymbol{f})$ to be the index of the first entry with value 1 in the permuted order of $\boldsymbol{f}$. One can show that the following correspondence holds for the feature vectors $\boldsymbol{f_1}$ and $\boldsymbol{f_2}$ above (see [1] for the details):

$$\text{SIM}_{\text{Jaccard}}(S_1, S_2) = \mathbf{Pr}\left[h_\pi(\boldsymbol{f_1}) = h_\pi(\boldsymbol{f_2})\right],$$

where the probability is taken by selecting π uniformly at random from the set of permutations of $\mathcal{F}$. This allows for the following approximation of the Jaccard-similarity between $\boldsymbol{f_1}$ and $\boldsymbol{f_2}$: Generate a set $\pi_1, \ldots, \pi_K$ of permutations of the feature set uniformly at random and return K'/K, where K' is the number of permutations π_i with $h_{\pi_i}(\boldsymbol{f_1}) = h_{\pi_i}(\boldsymbol{f_2})$. The approximation of the Jaccard-distance with min-hashing results in a fast algorithm if the embedding into the feature space can be computed quickly.

3 Efficient Min-Hash Sketch Computation

In this section we present our method for *approximating* the similarity between two graphs. We define this similarity by the Jaccard-similarity of their binary feature vectors for its advantageous semantic and algorithmic properties. The underlying feature space is spanned by a certain set of frequent tree patterns fixed in advance. More precisely, given a finite training set $\mathcal{D}$ of graphs and a frequency threshold $\theta \in (0, 1]$, in a *preprocessing step* we first generate a subset $\mathcal{F}$ of the set of frequent subtrees of $\mathcal{D}$. Clearly, $\mathcal{F}$ is finite. It will span the Hamming feature space $\{0,1\}^{|\mathcal{F}|}$ and serve as the universe for the Jaccard-similarity. Given $\mathcal{F}$ and two graphs G_1, G_2, the similarity $\text{SIM}(G_1, G_2)$ between G_1 and G_2 is then defined by

$$\text{SIM}(G_1, G_2) := \text{SIM}_{\text{Jaccard}}(\boldsymbol{f_1}, \boldsymbol{f_2}),$$

where $\boldsymbol{f_i}$ is the characteristic vector of the set $\{T \in \mathcal{F} : T \preccurlyeq G_i\}$. The preprocessing step and the above definition of similarity raise the following three computational problems:

(P1) *The Frequent Subtree Mining Problem: Given* a finite set $\mathcal{D}$ of graphs and a frequency threshold $\theta \in (0, 1]$, *generate* the set of all *trees* that are frequent in $\mathcal{D}$.

(P2) *The Subtree Isomorphism Problem: Given* a tree T and a graph G, *decide* whether or not $T \preccurlyeq G$.

(P3) *Computing the Jaccard-similarity: Given* two binary feature vectors $\boldsymbol{f_1}$ and $\boldsymbol{f_2}$ as defined above, *compute* $\text{SIM}_{\text{Jaccard}}(\boldsymbol{f_1}, \boldsymbol{f_2})$.

Since we have no restrictions on $\mathcal{D}$ and G, problems (P1) and (P2) are computationally intractable. In particular, unless $P = NP$, the frequent subtree mining problem cannot be solved in output polynomial time [6] and deciding whether a tree is subgraph isomorphic to a graph is NP-complete. Regarding (P3), as $|\mathcal{F}|$ is typically some large set, computing the Jaccard-similarity makes the above algorithmic definition practically infeasible if a huge number of similarity queries must be evaluated.

3.1   Probabilistic Tree Patterns

We first focus on problems (P1) and (P2). To overcome the complexity limitations of these two problems, we follow our approach described in [12] and give up the demand on the completeness of (P1) and on the correctness of the subtree isomorphism test for (P2), and use the following *probabilistic* technique: We replace each graph in $\mathcal{D}$ for problem (P1) and G for problem (P2) by a set of its k spanning trees generated uniformly at random. For an empirical evaluation of this method, the reader is referred to [12]; a probabilistic justification together with an appropriate choice of k is given below. According to this probabilistic technique, a tree T will be found as frequent in $\mathcal{D}$ (P1) if it is a subtree of at least $\theta|\mathcal{D}|$ forests, each formed by the vertex disjoint union of the k random spanning trees. Similarly, T will be found to be subgraph isomorphic to G (P2) if T is subgraph isomorphic to the forest of G formed by the vertex disjoint union of the k random spanning trees generated for G.

Regarding (P1), the probabilistic technique above results in a *sound*, but *incomplete* algorithm. Indeed, any tree found to be frequent by the mining algorithm in the forest dataset obtained is a frequent tree in $\mathcal{D}$, but we have no guarantee that all frequent trees of $\mathcal{D}$ will be generated. Thus, our technique may ignore frequent tree patterns. Regarding (P2), we decide subtree isomorphism by one-sided error: If T is subgraph isomorphic to any of the k spanning trees of G then T is subgraph isomorphic to G; otherwise T may or may not be subgraph isomorphic to G.

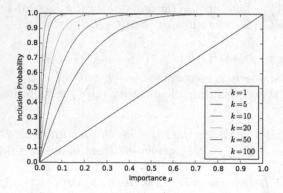

Fig. 1. The function $1 - (1 - \mu)^k$ for different values of k.

The rationale behind our probabilistic technique is as follows. For a graph G, let $\mathfrak{S}(G)$ be the set of spanning trees of G. Let $\mu \in (0, 1]$. A tree T is μ-*important* in G if

$$\frac{|\{S \in \mathfrak{S}(G) : T \preccurlyeq S\}|}{|\mathfrak{S}(G)|} \geq \mu.$$

Thus, the probability that a μ-important tree in G is subtree isomorphic to a spanning tree of G generated uniformly at random is at least μ. Notice that

$\mu = 1$ for any subtree of the forest formed by the set of bridges of G (i.e., by the edges that do not belong to any cycle in G). Let $\mathfrak{S}_k(G)$ denote a sample of k spanning trees of G generated independently and uniformly at random. Then

$$\mathbf{Pr}\left[\exists S \in \mathfrak{S}_k(G) \text{ such that } T \preccurlyeq S\right] \geq 1 - (1 - \mu)^k. \qquad (1)$$

The bound in (1) implies that for any graph G and μ-important tree pattern T in G for some $\mu \in (0, 1]$, and for any $\delta \in (0, 1)$,

$$\mathbf{Pr}\left[\exists S \in \mathfrak{S}_k(G) \text{ such that } T \preccurlyeq S\right] \geq 1 - \delta$$

whenever

$$k \geq \frac{\ln \delta}{\ln(1 - \mu)}$$

(see, also, Fig. 1 for the function $1 - (1 - \mu)^k$ for different values of k). Thus, if k is appropriately chosen, we have a probabilistic guarantee in terms of the confidence parameter δ that all μ-important tree patterns will be considered with high probability. As an example, we need less than 20 random spanning trees to correctly process a 0.15-important tree pattern with probability at least 0.95. Clearly, a smaller value of μ results in a larger feature set.

3.2   Fast Min-Hashing for Tree Patterns

We now turn to problem (P3) of computing the Jaccard-similarity between two binary feature vectors corresponding to the set $\mathcal{F}$ of tree patterns generated in the preprocessing step. More precisely, given a set $\mathcal{F}$ of trees that index our full feature space and two graphs G_1, G_2, we want to answer similarity queries of the form $\mathrm{SIM}(G_1, G_2)$ by calculating $\mathrm{SIM}_{\mathrm{Jaccard}}(\boldsymbol{f}_1, \boldsymbol{f}_2)$, where $\boldsymbol{f}_i$ is the binary feature vector representing the set of trees from $\mathcal{F}$ that are subgraph isomorphic to the forest formed by the vertex disjoint union of the k random spanning trees generated for G_i ($i = 1, 2$). Instead of using the naive brute-force algorithm, i.e., calculating first the explicit embedding of G_i into the feature space and computing then the exact value of $\mathrm{SIM}_{\mathrm{Jaccard}}(\boldsymbol{f}_1, \boldsymbol{f}_2)$, we follow Broder's probabilistic *min-hashing* technique [1] sketched in Sect. 2.

Min-hashing was originally applied to text documents using q-shingles as features (i.e., sequences of q contiguous tokens for some $q \in \mathbb{N}$), implying that one can calculate the explicit embedding in linear time by shifting a window of size q through the document to be embedded. In contrast, a naive algorithm embedding a graph with n vertices into the feature space corresponding to $\mathcal{F}$ would require $O\left(|\mathcal{F}|n^{2.5}/\log n\right)$ time by using the fastest subtree isomorphism algorithm [9], which is practically infeasible for large-sized feature sets. Another important difference between the two applications is that while the set of q-shingles for text documents forms an anti-chain (i.e., the q-shingles are pairwise incomparable), subgraph isomorphism induces a natural *partial order* on $\mathcal{F}$. The transitivity of subgraph isomorphism allows us to safely ignore features from $\mathcal{F}$ that do not influence the outcome of min-hashing, resulting in a much faster algorithm.

More precisely, in the preprocessing step, directly after the generation of $\mathcal{F}$, we generate K random permutations $\pi_1, \ldots, \pi_K : \mathcal{F} \to [|\mathcal{F}|]$ of $\mathcal{F}$ (see [2] for the details) and fix them for computing the min-hash values that will be used for similarity query evaluations (cf. Sect. 2). Since the computation of the embeddings is the most expensive operation, we can allow preprocessing time and space that is polynomial in the size of the pattern set $\mathcal{F}$. Therefore, we explicitly compute and store $\pi_1, \ldots, \pi_K$, and do not apply any implicit representations of them. This is particularly true, as we compute $\mathcal{F}$ explicitly in the preprocessing step and spend time that is polynomial in $\mathcal{F}$ anyway.

For a graph G and permutation π of $\mathcal{F}$, let

$$h_\pi(G) = \operatorname*{argmin}_{T \in \mathcal{F}} \{\pi(T) : T \preccurlyeq G\}.$$

The *sketch* of G with respect to $\pi_1, \ldots, \pi_K$ is then defined by[2]

$$\mathrm{SKETCH}_{\pi_1, \ldots, \pi_K}(G) = (h_{\pi_1}(G), \ldots, h_{\pi_K}(G)).$$

The rest of this section is devoted to the following problem: Given $\pi_1, \ldots, \pi_K$ and G as above, compute $\mathrm{SKETCH}_{\pi_1, \ldots, \pi_K}(G)$. The first observation that leads to an improved algorithm computing $\mathrm{SKETCH}_{\pi_1, \ldots, \pi_K}(G)$ is that for any $i \in [K]$, π_i may contain trees that can never be the first matching patterns according to π_i, for any query graph G. Indeed, suppose we have two patterns $T_1, T_2 \in \mathcal{F}$ with $T_1 \preccurlyeq T_2$ and $\pi_i(T_1) < \pi_i(T_2)$. Then, for any query graph G, either

1. $T_1 \preccurlyeq G$ and hence T_2 cannot be the first matching pattern in π_i or
2. $T_1 \npreccurlyeq G$ and hence, by the transitivity of subgraph isomorphism, we have $T_2 \npreccurlyeq G$ as well.

For both cases, T_2 will never be the first matching pattern according to π_i and can therefore be omitted from this permutation. Algorithm 1 implements this idea for a single permutation π of $\mathcal{F}$. The proof of the following result is immediate from the remarks above:

Lemma 1. *Let $\sigma = \langle T_1, \ldots, T_l \rangle$ be the output of Algorithm 1 for a permutation π of $\mathcal{F}$. Then, for any graph G,*

$$h_\pi(G) = \operatorname*{argmin}_{T_i \in \sigma}\{i : T_i \preccurlyeq G\}.$$

Algorithm 1 runs in time $O(|\mathcal{F}|)$. Loop 5 can be implemented by a DFS that does not recurse on the visited neighbors of a vertex. In this way, each edge of F is visited exactly once during the algorithm.

We now turn to the computation of $\mathrm{SKETCH}_{\pi_1, \ldots, \pi_K}(G)$. A straightforward implementation of calculating $\mathrm{SKETCH}_{\pi_1, \ldots, \pi_K}(G)$ for the evaluation sequences

---

[2] In practice, we do not store the patterns in $\mathrm{SKETCH}_{\pi_1, \ldots, \pi_K}(G)$ explicitly. Instead, we define some arbitrary total order on $\mathcal{F}$ and represent each pattern by its position according to this order.

Input: directed graph $F = (\mathcal{F}, E)$ representing a poset $(\mathcal{F}, \preccurlyeq)$ and permutation π of $\mathcal{F}$

Output: evaluation sequence $\sigma = \langle T_1, \ldots, T_l \rangle \in \mathcal{F}^l$ for some $0 < l \leq |\mathcal{F}|$ with $\pi(T_i) < \pi(T_j)$ for all $1 \leq i < j \leq l$

1: init $\sigma :=$ empty list
2: init $visited(T) := 0$ for all $T \in \mathcal{F}$
3: **for all** $T \in \mathcal{F}$ in the order of π **do**
4:     **if** $visited(T) = 0$ **then**
5:         **for all** $T' \in \mathcal{F}$ (including T) that are reachable from T in F **do**
6:             set $visited(T') := 1$
7:         append T to σ
8: **return** σ

Algorithm 1: Poset-Permutation-Shrink

$\sigma_1, \ldots, \sigma_K$ computed by Algorithm 1 for $\pi_1, \ldots, \pi_K$ just loops through each evaluation sequence, stopping each time the first match is encountered. This strategy can further be improved by utilizing the fact that a pattern T may be evaluated redundantly more than once for a graph G, if T occurs in more than one permutation before or as the first match. Lemma 2 below formulates necessary conditions for avoiding redundant subgraph isomorphism tests.

Lemma 2. *Let G be a graph, $F = (\mathcal{F}, E)$ be a directed graph representing a poset $(\mathcal{F}, \preccurlyeq)$, and let $\sigma_1, \ldots, \sigma_K$ be the evaluation sequences computed by Algorithm 1 for the permutations $\pi_1, \ldots, \pi_K$ of $\mathcal{F}$. Let $\mathfrak{A}$ be an algorithm that correctly computes $\text{SKETCH}_{\pi_1, \ldots, \pi_K}(G)$ by evaluating subgraph isomorphism in the pattern sequence $\Sigma = \langle \sigma_1[1], \ldots, \sigma_K[1], \sigma_1[2], \ldots, \sigma_K[2], \ldots \rangle$. Then $\mathfrak{A}$ remains correct if for all $i \in [K]$ and $j \in [|\sigma_i|]$, it skips the evaluation of $\sigma_i[j] \preccurlyeq G$ whenever one of the following conditions holds:*

1. *$\sigma_i[j'] \preccurlyeq G$ for some $j' \in [j-1]$,*
2. *there exists a pattern T before $\sigma_i[j]$ in Σ such that $\sigma_i[j] \preccurlyeq T$ and $T \preccurlyeq G$,*
3. *there exists a pattern T before $\sigma_i[j]$ in Σ such that $T \preccurlyeq \sigma_i[j]$ and $T \not\preccurlyeq G$.*

Proof. If Condition 1 holds then the min-hash value for permutation π_i has already been determined. If $\sigma_i[j] \preccurlyeq T$ and $T \preccurlyeq G$ then $\sigma_i[j] \preccurlyeq G$ by the transitivity of subgraph isomorphism. For the same reason, if $T \preccurlyeq \sigma_i[j]$ and $T \not\preccurlyeq G$ then $\sigma_i[j] \not\preccurlyeq G$. Hence, if Condition 2 or 3 holds then $\mathfrak{A}$ can infer the answer to $\sigma_i[j] \preccurlyeq G$ without explicitly performing the subgraph isomorphism test.

Algorithm 2 computes the sketch for a graph G along the conditions formulated in Lemma 2. It maintains a state for all $T \in \mathcal{F}$ defined as follows: 0 encodes that $T \preccurlyeq G$ is unknown, 1 that $T \preccurlyeq G$, and -1 that $T \not\preccurlyeq G$.

Theorem 1. *Algorithm 2 is correct, i.e., it returns $\text{SKETCH}_{\pi_1, \ldots, \pi_K}(G)$. Furthermore, it is non-redundant, i.e., for all patterns $T \in \mathcal{F}$, it evaluates at most once whether or not $T \preccurlyeq G$.*

Input: graph G, directed graph $F = (\mathcal{F}, E)$ representing a poset $(\mathcal{F}, \preccurlyeq)$ and K evaluation sequences $\sigma_1, \ldots, \sigma_K$ computed by Algorithm 1 for the permutations $\pi_1, \ldots, \pi_K$ of $\mathcal{F}$

Output: $\text{SKETCH}_{\pi_1, \ldots, \pi_K}(G)$

```
1: init sketch := [⊥, ..., ⊥]
2: init state(T) := 0 for all T ∈ F
3: for i = 1 to |F| do
4:     for j = 1 to K do
5:         if |σ_j| ≥ i ∧ sketch[j] = ⊥ then
6:             if state[σ_j[i]] ≠ 0 then
7:                 if state[σ_j[i]] = 1 then sketch[j] = σ_j[i]
8:             else if σ_j[i] ≼ G then
9:                 sketch[j] = σ_j[i]
10:                for all T' ∈ F (including T) that can reach T in F do
11:                    set state(T') := 1
12:            else
13:                for all T' ∈ F (including T) that are reachable from T in F do
14:                    set state(T') := -1
15: return sketch
```

Algorithm 2: Min-Hash Sketch

Proof. The correctness is immediate from Lemmas 1 and 2. Regarding non-redundancy, suppose $T \preccurlyeq G$ has already been evaluated for some pattern $T \in \mathcal{F}$ with $T = \sigma_i[j]$. Then, as $T \preccurlyeq T$, for any $\sigma_{i'}[j'] = T$ after $\sigma_i[j]$ in Σ either Condition 2 or 3 holds and hence $T \preccurlyeq G$ will never be evaluated again.

Once the sketches are computed for two graphs G_1, G_2, their Jaccard similarity with respect to $\mathcal{F}$ can be approximated by the fraction of identical positions in these sketches. (The similarity of G_1 and G_2 with $\text{SKETCH}_{\pi_1, \ldots, \pi_K}(G_1) = \text{SKETCH}_{\pi_1, \ldots, \pi_K}(G_2) = (\bot, \ldots, \bot)$ is defined by 0.)

4   Experimental Evaluation

We have conducted experiments on several real-world datasets. Since our method is restricted to connected graphs, disconnected graphs have been omitted. To obtain the feature sets of probabilistic tree patterns, we have applied the method in [12] to a randomly sampled subset of 10 % of the graphs in each dataset. We have restricted the maximum size of a tree pattern to 10 vertices, as this bound seemed optimal for the predictive/ranking performance for all chemical datasets used in our experiments by noting that our method is not restricted to constant-sized tree patterns. The same observation is reported in [11]. We have empirically investigated (i) the *speed* measured by the number of subtree isomorphism tests performed, (ii) the *quality* of our method for the retrieval of positive molecules in the highly skewed NCI-HIV dataset, and (iii) the *predictive performance* of support vector machines using a kernel function based on similarity measure.

Datasets: We have used the chemical graph datasets MUTAG, PTC, DD, NCI1, and NCI109 obtained from http://www.di.ens.fr/~shervashidze/, and NCI-HIV from https://cactus.nci.nih.gov/. MUTAG is a dataset of 188 connected compounds labeled according to their mutagenic effect on Salmonella typhimurium. PTC contains 344 connected molecular graphs, labeled according to the carcinogenicity in mice and rats. DD consists of $1,187$ protein structures, of which $1,157$ are connected. Labels differentiate between enzymes and non-enzymes. NCI1 and NCI109 contain $4,110$ resp. $4,127$ compounds of which $3,530$ resp. $3,519$ are connected. Both are balanced sets of chemical molecules labeled according to their activity against non-small cell lung cancer (resp. ovarian cancer) cell lines. NCI-HIV consists of $42,687$ compounds of which $39,337$ are connected. The molecules are annotated with their activity against the human immunodeficiency virus (HIV). In particular, they are labeled by "active" (A), "moderately active" (M), or "inactive" (I). While the first five datasets have a balanced class distribution, the class distribution of NCI-HIV is heavily skewed: Only 329 molecules (i.e., less than 1 %) are in class A, 906 in class M, and the remaining $38,102$ in class I.

Before going into the details, we first note that the similarities obtained by our method approximate the *exact* Jaccard-similarities quite closely on average. On a sample of roughly 1,500 graphs from the NCI-HIV dataset, the exact Jaccard similarities based on the full set of frequent trees and the similarities based on our min-hashing method showed a mean-squared-error of at most 0.005 for sketch size $K = 32$ and for an average Jaccard-similarity of 0.1396. For space limitations, we omit a detailed discussion of these results.

4.1 Speed-Up

The main goal of our method is to reduce the number of subgraph isomorphism tests during the computation of the min-hash sketch for a graph. We now show the effectiveness of our method from this aspect. To this end, we have compared our method given in Algorithm 2 not only with the *brute-force* explicit embedding, but also with the following *naive* embedding algorithm utilizing the partial order on the feature set $\mathcal{F}$: Given the poset $(\mathcal{F}, \preccurlyeq)$ as a directed graph F and a graph G, we traverse F starting at the vertices with in-degree 0. If a pattern T does not match G, we prune away all patterns reachable from T in F. In this way, we obtain the complete feature set of G with respect to $\mathcal{F}$ and compute the min-hash sketch accordingly. It is important to note that our algorithm may perform more subgraph isomorphism tests than the naive algorithm; this is due to the fact that, in contrast to the naive algorithm, we do not traverse F systematically. We leave a detailed discussion for the long version of this paper. We have compared our algorithm also with the naive method above in terms of the number of subgraph isomorphism tests performed. Table 1 shows the average number of subtree isomorphism tests per graph together with the pattern set size, for different datasets and parameters. The last four columns are the results of our algorithm for sketch size $K = 32, 64, 128, 256$, respectively. It can be seen that our algorithm (MH32–MH256) performs dramatically less subtree isomorphism tests than the brute-force one requiring $|\mathcal{F}|$ and outperforms also the naive algorithm in almost all cases. For example, on DD for $k = 10$ and $\theta = 10\,\%$, the naive algorithm evaluates subtree isomorphism for 11,006 patterns per graph on average, which is roughly one third of the total pattern set $(|\mathcal{F}|)$, while our method evaluates subtree isomorphism 345 times on average for sketch size 32, ranging up to 2190 times for sketch size 256. In general, the naive algorithm performs 4 (resp. 1.7) times as many subtree isomorphism tests as our method for $K = 32$ (resp. $K = 256$). This is a significant improvement in light of the speed $O\left(n^{2.5}/\log n\right)$ of the fastest subtree isomorphism algorithm [9].

Table 1. Average number of subtree isomorphism test per graph for several datasets with varying number k of sampled spanning trees and frequency thresholds θ. The table reports $|\mathcal{F}|$ and the average number of subtree isomorphism tests evaluated by the naive method and by Algorithm 2 for $K = 32, 64, 128, 256$ (last four columns).

Dataset	k	θ	$|\mathcal{F}|$	naive	MH32	MH64	MH128	MH256
MUTAG	5	10 %	452	206.38	49.93	68.24	96.12	127.42
MUTAG	10	10 %	543	244.11	42.77	63.77	90.57	125.39
MUTAG	15	10 %	562	254.86	45.39	65.96	94.87	133.91
MUTAG	20	10 %	573	260.18	55.34	76.32	105.15	135.11
PTC	5	10 %	1,430	321.04	70.07	102.62	121.12	156.12
PTC	5	1 %	9,619	734.79	236.31	327.27	475.35	611.92
PTC	10	10 %	1,566	354.20	79.63	108.59	109.44	147.91
PTC	20	10 %	1,712	376.65	17.60	25.81	31.49	39.62
DD	5	10 %	8,111	3,547.22	260.47	486.09	846.09	1,374.76
DD	10	10 %	18,137	6,670.93	317.82	568.23	1,072.58	1,936.42
DD	20	10 %	33,100	11,005.49	344.59	653.66	1,242.03	2,190.15
NCI1	5	10 %	1,819	431.19	89.12	137.75	185.22	221.21
NCI1	5	1 %	21,306	900.68	615.62	920.17	1,227.52	1,378.18
NCI1	20	10 %	2,441	557.70	115.07	183.54	220.14	255.58
NCI109	5	10 %	2,182	462.62	115.62	170.43	206.23	254.70
NCI109	5	1 %	19,099	886.06	532.38	727.15	1057.18	1,348.27
NCI109	20	10 %	2,907	598.36	110.42	175.76	226.07	284.92

4.2   Positive Instance Retrieval

In this section we evaluate the performance of our approach in terms of precision for retrieving similar molecules for a given active compound in the NCI-HIV dataset. We use a simple setup to evaluate the quality of the min-hash based similarity in comparison to the exact Jaccard similarity as well as to the similarities obtained by the path min-hash kernel [11].

For each molecule of class A (i.e., "active"), we retrieve its i nearest neighbors (excluding the molecule itself) from the dataset and take the fraction of the neighbors of class A. This measure is known in the Information Retrieval community as *precision at i*. As a baseline, a random subset of molecules from NCI-HIV is expected to contain less than 1 % of active molecules due to the highly skewed class distribution. In contrast, all methods show a drastically higher precision for the closest up to 100 neighbors on average.

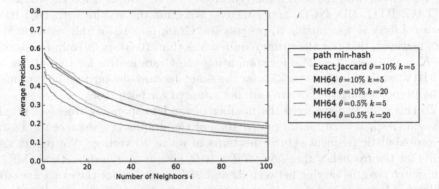

Fig. 2. Average fraction of "active" molecules among the i nearest neighbors of positive molecules in NCI-HIV dataset for path min-hash [11], exact Jaccard-similarity for frequent probabilistic tree patterns, and for our method with $K = 64$.

Figure 2 shows the average precision at i (taken over all 329 active molecules) for i ranging from 1 to 100. The number k of sampled spanning trees per graph, as well as the frequency threshold θ has a strong influence on the quality of our method. To obtain our results, we have sampled 5 (resp. 20) spanning trees for each graph and used a random sample of 4,000 graphs to obtain pattern sets for thresholds $\theta = 10\%$ and $\theta = 0.5\%$ respectively. We plot the min-hash-based precision for the four feature sets obtained in this way by our algorithm as a function of i for sketch size $K = 64$. We have compared this to the precision obtained by the exact Jaccard-similarity for $\theta = 10\%$ and $k = 5$ as well as to the precision obtained by path min-hash [11], both for the same sketch size $K = 64$.

The average precision obtained by using the exact Jaccard-similarities is slightly better than that of path min-hash. While our method performs comparably to path min-hash for $\theta = 0.5\%$ and $k = 5$, for $\theta = 0.5\%$ and $k = 20$ spanning trees it outperforms all other methods.

We were able to compute the precisions for the exact Jaccard-similarity neither for $\theta = 1\%$ nor for $k = 20$ sampled spanning trees. The Python implementation we used to calculate the similarity computations for exact Jaccard-similarity was not able to deal with the high dimensionality of the feature space, independently of the sparsity of the feature vectors. This indicates that the space required to compute the Jaccard-similarity is crucial for high-dimensional feature spaces.

4.3   Predictive Performance

In this section we empirically analyze the predictive performance of our method. The Jaccard-similarity induces a positive semi-definite kernel on sets, also known as a special case of the Tanimoto kernel (see, e.g., [8]). Interestingly, its approximation based on min-hashing is a kernel as well. Hence, we can use the tree pattern sets, resp. the min-hash sketches together with these two kernels in a support vector machine to learn a classifier. We have used 5-fold cross-validation and report the average area under the ROC curve obtained for the datasets MUTAG, PTC, DD, NCI1, and NCI109. We omit the results with NCI-HIV because LibSVM was unable to process the Gram matrix for this dataset. We note, however, that our algorithm required less than 10 (resp. 26) min for sketch size $K = 32$ (resp. $K = 256$) for computing the Gram matrix for the full set of NCI-HIV, while this time was 5.5 h for the exact Jaccard-similarity. The runtimes of the preprocessing step (3 min) are not counted for both cases.

To this end, we have fixed the number of random spanning trees per graph to $k = 5$ (resp. $k = 20$) and sampled 10 % of the graphs in a dataset to obtain the probabilistic frequent subtree patterns of up to 10 vertices. We report the results for the frequency threshold of $\theta = 10$ % for our min-hash method (MH*) with sketch sizes K varying between 32 and 256, as well as for the exact Jaccard similarity (Jaccard) based on the same feature set. A lower frequency threshold is practically unreasonable e.g. for MUTAG, as it contains only 188 compounds. We have compared the results obtained with our previous results in [12] that use a radial basis function kernel on the probabilistic subtree features (PSK) and with the frequent subgraph kernel (FSG) [3] using the full set of frequent connected subgraphs of up to 10 vertices with respect to the *full* datasets (i.e., not only for a sample of 10 %). We also report results of the Hash Kernel (HK) [10], which uses count-min sketching on sampled induced subgraphs up to size 9 to answer similarity queries.

Table 2. AUC values for our method (MH) for sketch sizes $K = 32, 64, 128, 256$, $k = 5$ spanning trees per graph, and frequency threshold $\theta = 10$ % to obtain the feature set. "n/a" indicates that the authors of [10] did not provide results for the respective datasets.

θ	Method	MUTAG	PTC	DD	NCI1	NCI109
10 %	$MH32$	87.84	58.97	77.58	77.36	77.48
10 %	$MH64$	87.73	58.68	79.91	78.04	79.54
10 %	$MH128$	87.59	56.97	82.07	79.94	79.94
10 %	$MH256$	87.78	57.18	83.58	80.76	81.72
10 %	Jaccard	89.04	57.72	85.38	82.28	82.41
10 %	PSK	84.22	54.17	84.67	79.09	78.05
10 %	FSG	87.34	56.76	82.20	81.66	81.55
	HK	93.00	62.70	81.00	n/a	n/a

Table 2 shows the results for frequency threshold $\theta = 10\%$ and $k = 5$ sampled spanning trees per graph. Jaccard and MH256 outperform PSK and even FSG on all datasets. On all datasets, except for PTC, the quality increases with the sketch size K. While HK is the best by a comfortable margin on MUTAG, the contrary is true for MH256, PSK, and Jaccard on DD. Most notably, MH usually outperforms PSK even for smaller sketch sizes and in many cases even the full frequent subgraph kernel FSG, while still being computable on DD, which has an enormous number of frequent patterns (compare Sect. 4.1). Note that the full datasets were used to generate the feature sets for PSK and FSG.

We also conducted experiments for $k = 20$ sampled spanning trees. For identical frequency threshold, the AUC improved by 3% on MUTAG, while only slightly changing for the other datasets.

5 Concluding Remarks

Algorithms 1 and 2 computing the min-hash sketches for a given tree pattern set and for subtree isomorphism can easily be adapted to any finite pattern set and pattern matching operator (e.g., homomorphism) if the pattern matching operator induces a pre-order on the pattern set. While the number of evaluations of the pattern matching operator can drastically be reduced in this way, the complexity of the algorithm depends on that of the pattern matching operator.

The one-sided error of our probabilistic subtree isomorphism test seems to have no significant effect on the experimental results. This raises the question whether we can further relax the correctness of subtree isomorphism obtaining an algorithm that runs in at most subquadratic time without any significant negative effect on the predictive/retrieval performance.

References

1. Broder, A.Z.: On the resemblance and containment of documents. In: Proceedings of the Compression and Complexity of Sequences, pp. 21–29. IEEE (1997)
2. Broder, A.Z., Charikar, M., Frieze, A.M., Mitzenmacher, M.: Min-wise independent permutations. J. Comput. Syst. Sci. **60**(3), 630–659 (2000)
3. Deshpande, M., Kuramochi, M., Wale, N., Karypis, G.: Frequent substructure-based approaches for classifying chemical compounds. Trans. Knowl. Data Eng. **17**(8), 1036–1050 (2005)
4. Diestel, R.: Graph Theory. Graduate Texts in Mathematics, vol. 173, 4th edn. Springer, Heidelberg (2012). http://dblp.dagstuhl.de/rec/bib/books/daglib/0030488
5. Geppert, H., Horváth, T., Gärtner, T., Wrobel, S., Bajorath, J.: Support-vector-machine-based ranking significantly improves the effectiveness of similarity searching using 2D fingerprints and multiple reference compounds. J. Chem. Inf. Model. **48**(4), 742–746 (2008)
6. Horváth, T., Bringmann, B., Raedt, L.: Frequent hypergraph mining. In: Inoue, K., Ohwada, H., Yamamoto, A. (eds.) ILP 2006. LNCS (LNAI), vol. 4455, pp. 244–259. Springer, Heidelberg (2007). doi:10.1007/978-3-540-73847-3_26

7. Horváth, T., Ramon, J.: Efficient frequent connected subgraph mining in graphs of bounded tree-width. Theor. Comput. Sci. **411**(31–33), 2784–2797 (2010)
8. Ralaivola, L., Swamidass, S.J., Saigo, H., Baldi, P.: Graph kernels for chemical informatics. Neural Netw. **18**(8), 1093–1110 (2005)
9. Shamir, R., Tsur, D.: Faster subtree isomorphism. J. Algorithms **33**(2), 267–280 (1999). doi:10.1006/jagm.1999.1044
10. Shi, Q., Petterson, J., Dror, G., Langford, J., Smola, A.J., Vishwanathan, S.V.N.: Hash kernels for structured data. J. Mach. Learn. Res. **10**, 2615–2637 (2009)
11. Teixeira, C.H.C., Silva, A., Meira Jr., W.: Min-hash fingerprints for graph kernels: a trade-off among accuracy, efficiency, and compression. J. Inf. Data Manag. **3**(3), 227–242 (2012)
12. Welke, P., Horváth, T., Wrobel, S.: Probabilistic frequent subtree kernels. In: Ceci, M., Loglisci, C., Manco, G., Masciari, E., Ras, Z.W. (eds.) NFMCP 2015. LNCS (LNAI), vol. 9607, pp. 179–193. Springer, Heidelberg (2016). doi:10.1007/978-3-319-39315-5_12

Structured Output Prediction

STIFE: A Framework for Feature-Based Classification of Sequences of Temporal Intervals

Leon Bornemann[1]([✉]), Jason Lecerf[2], and Panagiotis Papapetrou[3]

[1] Freie Universität Berlin, Berlin, Germany
leon.bornemann13@gmail.com
[2] Institut National des Sciences Appliquées Lyon, Lyon, France
[3] Department of Computer and Systems Sciences,
Stockholm University, Stockholm, Sweden

Abstract. In this paper, we study the problem of classification of sequences of temporal intervals. Our main contribution is the STIFE framework for extracting relevant features from interval sequences to build feature-based classifiers. STIFE uses a combination of basic static metrics, shapelet discovery and selection, as well as distance-based approaches. Additionally, we propose an improved way of computing the state of the art IBSM distance measure between two interval sequences, that reduces both runtime and memory needs from pseudo-polynomial to fully polynomial, which greatly reduces the runtime of distance based classification approaches. Our empirical evaluation not only shows that STIFE provides a very fast classification time in all evaluated scenarios but also reveals that a random forests using STIFE achieves similar or better accuracy than the state of the art k-NN classifier.

1 Introduction

Sequences of temporal intervals are ubiquitous in a wide range of application domains including sign language transcription [12], human activity monitoring, music informatics [10], and healthcare [3]. Their main advantage over traditional discrete event sequences is that they comprise events that are not necessarily instantaneous, but may have a time duration. Hence, sequences of temporal intervals can be encoded as a collection of labeled events accompanied by their start and end time values. It becomes apparent that in such sequences events may overlap with other events forming various types of temporal relations [12].

Examples. An example of such a sequence is depicted in Fig. 1, consisting of seven event intervals that have various time durations. Note that each event may occur several times in the sequence (e.g., events 0 and 1). Hence, a sequence of temporal intervals can be seen as a series of event labels (y axis) that can be active or inactive at a particular time point (x axis). Such sequences can appear in various application areas. One example is sign language [12]. A sentence expressed in signs consists of multiple, different gestures (e.g., head-shake, eyebrow-raise) or speech tags (e.g., noun, wh-word), which may have a time

© Springer International Publishing Switzerland 2016
T. Calders et al. (Eds.): DS 2016, LNAI 9956, pp. 85–100, 2016.
DOI: 10.1007/978-3-319-46307-0_6

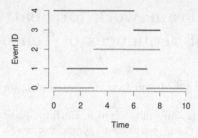

Fig. 1. Example of a sequence of temporal intervals: on the y axis we have five events, labeled as $\{0, 1, 2, 3, 4\}$, while on the x axis we can see the time points measured in seconds.

duration and can start at potentially different points in time. Another example is healthcare [3]. A sequence may correspond to a series of different types of treatments (events) for a particular patient. Treatments typically have a time duration and a patient could potentially be exposed to multiple treatments concurrently. An interesting and at the same challenging task involving sequences of temporal intervals is that of classification. For example, the correct classification of sign language videos can lead to the discovery of associations of certain types of expressions (labels) with various temporal combinations of gestural or grammatical events (features). Moreover, the extraction of certain combinations of treatments (features) may assist in the proper identification of adverse drug events (labels).

Previous research in the area of classification of sequences of temporal intervals has been limited to k-NN classifiers. Towards this direction, two state-of-the-art distance measures have been proposed, `Artemis` [5] and `IBSM` [6]. The first one quantifies the similarity between two sequences by measuring the fraction of temporal relations shared between them using a bipartite graph mapping, while ignoring their individual time duration. The second measure maps sequences of temporal intervals to vectors, where each time point is characterized by a binary vector indicating, which events are active at that particular time point. While the results obtained by both k-NN classifiers are promising, such classifiers still suffer from the fact that they consider only global trends or features in the data while ignoring local distinctive properties that may have a detrimental effect in predictive performance. Additionally the classification time of any k-NN classifier will always be at least linear to the size of the training set, while the computational cost of the chosen distance measure may severely impact the total runtime (e.g., `Artemis` is a distance measure with cubic computational complexity).

In this paper, we approach the problem of classification of sequences of temporal intervals by focusing on feature-based classifiers. Hence, the challenge is to identify and extract useful features from the sequences that could be then be used as input to traditional feature-based classifiers.

Contributions. Our main contributions can be summarized as follows: (1) we propose STIFE (Sequences of Temporal Intervals Feature Extraction

Framework), a novel framework for feature extraction from sequences of temporal intervals, and discuss its runtime complexity; (2) we present an improved method for calculating the IBSM distance, hence substantially reducing both runtime and memory requirements; (3) we provide an extensive empirical evaluation using eight real datasets as well as synthetic data, in which we compare our novel methods against the state-of-the-art.

2   Related Work

While arguably an understudied research area, sequences of temporal intervals have attracted some attention within the areas of data mining and databases. The first attempts at using sequences of temporal intervals mainly focused on simplifying the data without losing too much information. For example Lin et al. [8] show a way to mine maximal frequent intervals, but while doing so the different dimensions of the intervals were discarded. Another common form of simplification is to map sequences of temporal intervals to temporally ordered events without considering the actual duration of the intervals [1].

A large variety of Apriori-based techniques [2,7,9] for finding temporal patterns, episodes, and association rules on interval-based event sequences have been proposed. In addition, more advanced candidate generation techniques and tree-based structures have been employed by various methods [11–13], which apply efficient pruning techniques, thus reducing the inherent exponential complexity of the mining problem, while a non-ambiguous event-interval representation is defined by [14] that considers start and end points of event sequences and converts them to a sequential representation. The main weakness of performing such mapping is the fact that the candidate generation process becomes more cumbersome while introducing redundant patterns.

An approach for mining patterns of temporal intervals without performing any mapping to instantaneous events has been proposed by Papapetrou et al. [12]. The authors of this paper applied unsupervised learning methods to sequences of temporal intervals. In particular the Apriori algorithm for mining frequent item-sets had been adapted to fit sequences of temporal intervals. Subsequent to that, similarity of sequences of temporal intervals became more popular and has been looked at quite a bit. Robust similarity measures allow the use of sequences of temporal intervals as a data-basis for a lot of applications, some of them being similarity search, clustering and classification through k-NN classifier. It started with [5], where the authors propose two different similarity measures. The first one maps sequences of temporal intervals to time series data, the second one uses the temporal relations to construct and use a bipartite graph. This approach has been improved through different data representation and more robust similarity measures in [6].

With the transformation from sequences of temporal intervals to time series data being explored, it is unsurprising that also finding the longest common subpattern (LCSP) in sequences of temporal intervals has recently been considered. Finding the longest common subsequence (LCS) is a classic problem for time series data and finding the LCSP can be seen as its equivalent counterpart for sequences of temporal intervals. The problem of finding the LCSP was introduced in [4], where the authors prove its NP-hardness, introduce approximation algorithms as well as upper bounds.

3   Background

Let Σ define the alphabet of all possible events, i.e., the different types of intervals. A temporal interval is defined as: $I = (d, s, e)$ where $d \in \Sigma$ is an event label, and $s, e \in \mathbb{N}^+$ are the start and end times of the event interval, with $e \geq s$. Given an event interval $I = (d, s, e)$, we will sometimes denote d, s and e as $I.d, I.s$ and $I.e$, respectively.

A sequence of temporal intervals S is defined as an ordered multi-set of temporal intervals: $S = \{I_1, ..., I_m\}$. Note that it is allowed for multiple event intervals of the same label to overlap in a sequence. Further, a dataset of sequences of temporal intervals is denoted as $\mathcal{D}$.

The original IBSM method [6] represents a sequence S in a $|\Sigma| \times length(S)$ matrix called event tables, where $length(S)$ is the duration of the sequence (e.g. the time value at which the last interval stops). We briefly repeat the most important definitions next.

Definition 1. Active Interval. *Given a sequence S, an Interval $I \in S$ and a point of time t, I is called active at point of time t if $I.s \leq t \leq I.e$.*

Definition 2. Event Table. *Given a sequence S, its Event Table ET is defined as a $|\Sigma| \times length(S)$ matrix. The value of $ET(d, t)$ is the number of Intervals in S of dimension d that are active at point of time t. When we speak of the length of an event table we refer to the length of its corresponding sequence: $length(ET) = length(S)$.*

For the rest of this paper we assume that all sequences are of the same length, since this makes the definition of the distance measure easier. Note that this is not a big constraint, since sequences of smaller length can be interpolated to sequences of bigger length, by using linear interpolation. This was also suggested in the original definition of IBSM. The original distance between two event tables of the same length (number of columns) is called the IBSM-Distance.

Definition 3. IBSM-Distance. *Given two event tables A and B where $length(A) = length(B) = z$ the IBSM-Distance is defined as*

$$IBSM(A, B) = \sqrt{\sum_{d=1}^{|\Sigma|} \sum_{t=1}^{z} (A(d, t) - B(d, t))^2}$$

To counteract the large size of the event tables the authors suggest sampling methods which improve computation time but come at the cost of accuracy for the 1-NN classifier.

4   Compressed IBSM

The key idea of compressing IBSM without losing information is to reduce the size of the event table by only considering the points during which the value of a row can change, which are the start and end times of an event interval $I \in S$.

Definition 4. *Time Axis.* *Given a sequence S, let $T = \{t_1, ..., t_k\}$ be the sorted set of the start and end times of all intervals $I \in S$. We call T the time axis of S.*

Given a time axis $T = \{t_1, t_2, ..., t_k\}$ of sequence S we know according to the definition that $t_i < t_{i+1}$ for $i \in \{1, ..., k-1\}$. It is clear that for all $t \in \{1, .., length(S)\}$ where $t_i < t < t_{i+1}$: column t of the event table is equal to column t_i. Note that $k \leq 2m$ always holds, since k can at most be $2m$ but may be less since intervals in S can have the same start or end time. This allows us to just store the columns for all $t \in T$ and the other columns are implicitly given. We call the optimized form compressed event tables.

Definition 5. *Compressed Event Table (CET).* *Given a sequence S and its time axis $T = \{t_1, t_2, ..., t_k\}$ we define CET as the compressed event table of S as a $|\Sigma| \times |T|$ matrix where $CET(d, t_i)$ is the number of intervals in S of dimension d that are active at point of time t_i.*

Table 1. Uncompressed event table

	1	2	3	4	5	6	7	8	9
D_1	1	1	1	0	0	0	0	0	0
D_2	0	0	0	0	1	1	1	1	0

Table 2. Compressed event table

	1	4	5	9
D_1	1	0	0	0
D_2	0	0	1	0

Tables 1 and 2 present the different representations for a simple, small example. The distance between two compressed event tables can be calculated as follows:

Definition 6. *IBSM distance for Compressed Event Tables.* *Given two compressed event tables A and B with time axis $T_A = \{ta_1, ..., ta_k\}$ and $T_B = \{tb_1, ..., tb_p\}$, where $ta_k = tb_p$ (sequences have the same length) let $T = \{t_1, ..., t_r\} = T_A \cup T_B$ be the merged time axis (still ordered). Then we define the distance between the two compressed event tables as*

$$Dist(A, B) = \sqrt{\sum_{d=1}^{|\Sigma|} \sum_{j=1}^{|T|} E(A(d, I_A(t_j)), B(d, I_B(t_j))) \cdot \delta(j)} \qquad (1)$$

where

$$I_A(t) = max(\{i|t_i \in T_A \ t_i \le t\})$$
$$I_B(t) = max(\{i|t_i \in T_B \ t_i \le t\})$$
$$E(a,b) = (a-b)^2$$
$$\delta(j) = \begin{cases} t_{j+1} - t_j & if \ j < |T| \\ 1 & otherwise \end{cases}$$

The distance calculation now looks more complicated but the approach is straightforward. The squared error E is calculated for each cell of the table and is multiplied by the amount of time that the value would have been repeated in the old IBSM representation (δ). I_A and I_B are functions that map a point of time of the merged time axis T to the correct column index of their respective compressed event tables.

Given this definition, it is clear that given two sequences the event tables can be computed in $\Theta(m \cdot (log(m) + |\Sigma|))$. We need $\Theta(m \cdot log(m))$ to create the sorted time axis. Given two event tables the distance computation is linear to the number of cells in each table, which is $\Theta(|\Sigma| \cdot m)$. This is a clear improvement compared to the old $\Theta(|\Sigma| \cdot length(S))$, which is, as already mentioned, pseudo-polynomial.

5   Feature-Based Classification Through STIFE

While improving the performance of the distance measure is the key to improve the classification time of k-NN classifiers it can not address their overall disadvantage in classification time, which is that it will always be at least linear to the size of the training database. This can become problematic if the database is a huge size and classification of new instances is time critical. The rather broad application domain of real-time analysis of data-streams would be such an example.

Many feature based classifiers offer a classification time that is better than linear to the size of the database, such as decision trees or random forests. Thus if one is able to extract informative features from sequences of temporal intervals one could use these feature based classifiers to further improve classification time. An additional motivation besides time efficiency is that k-NN classifiers also have other drawbacks compared to feature based classifiers, such as sensitivity to outliers or units of measurements. Feature based classifiers might also yield better accuracy in some cases, depending of course on the usefulness of the extracted features.

In order to extract useful features we propose a novel method which we call the STIFE (**S**equences of **T**emporal **I**ntervals **F**eature **E**xtraction) framework. The rest of this section gives a detailed explanation of the framework.

5.1 STIFE Framework Components

Given a number of sequences as a training database, the main challenge of the framework is to explore and find features, which help to classify the training database. To do so we propose the STIFE Framework, which consists of three parts: (I) Static metrics, (II) Shapelet extraction and selection, and (III) Distance to class-cluster center.

Static metrics are simple, basic mappings that map one sequence to a set of features independent of the other sequences in the database S. The other two parts are dynamic, which means they consider the whole (training) database to extract the features that are particularly helpful to classify the sequences of that specific database. Subsects. 5.2, 5.3 and 5.4 describe the parts of the framework in detail. Afterwards Subsect. 5.5 summarizes the framework's time and memory complexity for training and classification.

5.2 I - Static Metrics

Let $S = \{I_1, ..., I_m\}$ be a sequence in which the intervals are sorted by start time, and in case of a tie by end time. We define the following basic metrics that will serve as static features:

- Duration: $I_m.e$
- Earliest start: $I_1.s$
- Majority dimension: The dimension d that occurs in most intervals $I \in S$.
- Interval count: $|S|$
- Dimension count: $|\{I.d | I \in s\}|$
- Density: $\sum_{I \in S} I.e - I.s$
- Normalized density: Density divided by the duration of the sequence.
- Max. overlapping intervals: Maximum number of overlapping intervals.
- Max. overlapping interval duration: The duration of the period with the highest number of overlapping intervals.
- Normalized max. overlapping interval duration: Max. overlapping interval duration divided by the duration of the sequence.
- Pause time: The total duration with no active dimension interval.
- Normalized pause time: Pause time divided by the duration of the sequence.
- Active time: The reverse of pause time.
- Normalized active time: Active time divided by the duration of the sequence.

These static metrics provide a very basic method to obtain some features. They are simple to understand, fast to compute and require little memory compared to the original training database. After sorting the intervals of the sequence all of these metrics can be calculated in either $\Theta(1)$ or $\Theta(m)$. Thus the overall runtime complexity of extracting static features from the database is $\Theta(n \cdot m \cdot log(m))$. Only $\Theta(n)$ additional memory is needed, since the number of static features is constant. The time to extract the features for an unseen sequence S_{new} is $\Theta(m \cdot log(m))$.

5.3   II - Shapelet Extraction and Selection

Shapelets are commonly defined as interesting or characteristic small subsequences of a larger sequence. The idea of shapelets has already been explored in the context of time-series data and has also been used as a tool for classification of time series data. Thus it is natural to also consider shapelets as candidates for features of sequences of temporal intervals. In this paper we will restrict ourselves to the shapelets of size 2 which are in the following referred to as 2-shapelets. To be able to define a 2-shapelet of a sequence of temporal intervals we must first define a few prerequisites such as temporal relationships between two intervals:

Definition 7. *Time Equality Tolerance. We define $\epsilon \in \mathbb{N}^+$ as the maximum tolerance which time values may differ from each other to still be considered as equal from a view point of temporal relationships. Since the value of ϵ can be quite domain specific we do not specify a fixed value here.*

Given the time equality tolerance we can define temporal relationships between temporal intervals:

Definition 8. *Temporal Relationship. Let A and B be two intervals with the following property: $A.s - \epsilon \leq B.s$ (B does not start before A). Then we define the set of possible temporal relationships as $R = \{meet, match, overlap, leftContains, contains, rightContains, followedBy\}$. Their individual definition is visualized in Fig. 2.*

These temporal relations for event intervals have already been used in the context of distance measures for sequences of temporal intervals on multiple occasions [5,6]. Note that for an ordered pair of event intervals exactly one of these relations applies, meaning the temporal relationship of two event intervals is unambiguous. Based on this, a 2-shapelet can be defined.

Definition 9. *2-shapelet. Given a sequence S and two temporal intervals $A, B \in S$ we define a 2-shapelet as $sh = (A.d, B.d, r)$ where $r \in R$ is the*

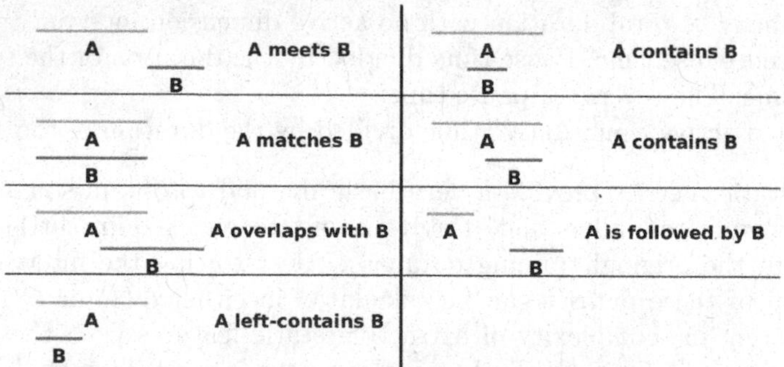

A	A meets B	A	A contains B
B		B	
A	A matches B	A	A contains B
B		B	
A	A overlaps with B	A	A is followed by B
B		B	
A	A left-contains B		
B			

Fig. 2. The 7 temporal relations between an ordered pair of event intervals (A,B)

Algorithm 1. Extract and select all shapelet features

Require: Let $S = \{S_1, ..., S_n\}$ be the sequences, Σ be the set of dimensions and k the number of features to keep.

$SM \leftarrow$ new $n \times (7 \cdot |\Sigma|^2)$ matrix

for $i = 1$ **to** n **do**

    for $(A, B) \in \{(A, B) \mid A, B \in S \land A.s - \epsilon \leq B.s\}$ **do**

        $r \leftarrow$ temporal relationship of (A, B)

        $j \leftarrow$ compute column index of shapelet $(A.d, B.d, r)$

        $SM[i, j] \leftarrow SM[i, j] + 1$

    end for

end for

$gains \leftarrow$ new $(7 \cdot |\Sigma|^2)$ array

for $j = 1$ **to** $7 \cdot |\Sigma|^2$ **do**

    $gains[col] \leftarrow$ information gain of column j of SM

end for

$featureIndices \leftarrow$ new List

for $j = 1$ **to** $7 \cdot |\Sigma|^2$ **do**

    if $gains[j]$ is in the top k of $gains$ **then**

        $featureIndices.add(j)$

    end if

end for

delete all columns of SM except for $featureIndices$

return SM

---

temporal relationship between the two intervals. In other words a two-shapelet (d_1, d_2, r) says, that there are two intervals in S of the respective dimensions d_1 and d_2 that have the temporal relationship r.

All 2-shapelets of a sequence S can be found by simply determining the relationships of all pairs of intervals (A, B), where $A, B \in S$ and B does not occur before A. The idea for the resulting features is simply to treat the number of occurences of each 2-shapelet as a feature of the sequence. This will result in exactly $|\Sigma| \cdot |\Sigma| \cdot |R| = 7 \cdot |\Sigma|^2$ possible features which is a swiftly increasing function of the number of dimensions. Thus it is necessary to perform feature selection afterwards which we do by information gain. Information gain is a measure of how much information is stored in an attribute with regard to the class label distribution and is commonly used when building decision trees. The formula is explained in detail in [15]. Since information gain is defined on categorical features and the number of shapelet occurrences in a sequence are numeric attributes, it is necessary to discretize them. We use the information gain of the best binary split (meaning the feature $\overrightarrow{a}$ is discretized to a vector of boolean values according to $\overrightarrow{a} \leq x$ for the $x \in \mathbb{N}$ that yields the highest information gain). The algorithm for the shapelet feature extraction is roughly summarized in Algorithm 1. To count all 2-shapelet occurences a $n \times 7 \cdot |\Sigma|^2$ matrix is used (one row per sequence). For each sequence all correctly ordered pairs need to be looked at, which amounts to the runtime $\Theta(m^2)$ per sequence, thus the runtime for the shapelet occurence counting is $\Theta(n \cdot m^2)$. Memory requirement is $\Theta(n \cdot |\Sigma|^2)$.

Calculating the information gain of a numeric attribute needs $\Theta(n \cdot log(n))$. This is done for each feature, which means the total runtime of feature selection via information gain is $\Theta(n \cdot log(n) \cdot |\Sigma|^2)$. Memory remains at $\Theta(n \cdot |\Sigma|^2)$. Thus, putting the two steps together we arrive at $\Theta(n \cdot (m^2 + log(n) \cdot |\Sigma|^2))$ for runtime and $\Theta(n \cdot |\Sigma|^2)$ memory to execute shapelet extraction and select the best shapelets as features. Calculating the occurences of the selected 2-shapelets for a new sequence takes $\Theta(m^2)$ time in the worst case since once again all its correctly ordered interval pairs need to be considered. Note that this is always independent of $|\Sigma|$, since a constant number of shapelets are selected in the feature selection step.

5.4   III - Distance to Class-Cluster Center

Our approach here is inspired by the k-medoids clustering algorithm. Since clustering is an approach that is used in unsupervised learning and we are in the supervised case (e.g. we have data with class labels) it is unnecessary to actually execute the clustering algorithms. Instead we can just assume that we have the clusters given by the class-labels of the training data and simply extract the medoids of each class-cluster.

Given the medoids of each class-cluster, these will then be used as reference points and the distance to them will result in features. As a distance measure we choose the IBSM distance over ARTEMIS, since the compressed way of calculating it as introduced by us has a better runtime than ARTEMIS and 1-NN classifiers using IBSM yield better accuracy which leads us to believe that it is the more suitable distance measure. Given the distance measure we formulate the algorithm for distance based feature extraction in Algorithm 2.

Since the class labels (and thus cluster labels) are given, the clustering takes $\Theta(n)$ time. Afterwards we need to calculate the medoid of each cluster and subsequently calculate the distance to those for all training sequences. Assuming the number of classes is constant we know that the size of each cluster can be $\Theta(n)$ but the number of clusters is constant. For each cluster all compressed event tables (see Sect. 3) and their pairwise distances ($\Theta(n^2)$) need to be computed and stored. Thus the runtime and memory complexity of finding the distances to all class-cluster medoids is $\Theta(n^2 \cdot m \cdot (|\Sigma| + log(m)))$ time and $\Theta(n^2 \cdot m \cdot |\Sigma|)$ memory. The online feature extraction requires $\Theta(m \cdot (|\Sigma| + log(m)))$ time and $\Theta(m \cdot |\Sigma|)$ memory.

5.5   Runtime and Memory Complexity Overview

When analyzing runtime and memory complexity of the STIFE framework, the two interesting measures are training time and classification time. Extracting and selecting the features based on the training data adds to the classifier's training time. Since the framework can be used with any feature based classifier

---

Algorithm 2. Calculate all medoids and extract the distance to those as features

---

Require: Let $S = \{S_1, ..., S_n\}$ be the sequences, k the number of classes and $D :$
$S \times S \mapsto \mathbb{R}^+$ a distance measure for sequences
  $FM \leftarrow$ new $n \times k$ matrix
  for $c = 1$ **to** k **do**
    $S^{(c)} \leftarrow \{S_j \mid S_j \in S \wedge class(S_j) = c\}$
    $DM \leftarrow$ new $|S^{(k)}| \times |S^{(k)}|$ matrix
    for $(S_i, S_j) \in \{(S_i, S_j) \mid S_i, S_j \in S^{(k)} \wedge i \leq j\}$ **do**
      $DM[i, j] \leftarrow D(S_i, S_j)$
      $DM[j, i] \leftarrow DM[i, j]$
    end for
    $min \leftarrow \infty$
    $minI \leftarrow -1$
    for $i = 1$ **to** $|S^{(c)}|$ **do**
      $dist \leftarrow$ row i of DM
      if $sum(dist) \leq min$ **then**
        $min \leftarrow sum(dist); minI \leftarrow i$
      end if
    end for
    for $i = 1$ **to** n **do**
      $d \leftarrow D(S[i], S^{(c)}[minI]); FM[i, c] \leftarrow d$
    end for
  end for
  return FM

---

we will use $CTT(n)$ to describe the classifier training time and $CTM(n)$ to describe the classifier memory need for training.

For an unseen sequence, feature extraction is performed before the classifier can be applied. We will use the term $CCT(n)$ to describe the classifier classification time and $CCM(n)$ for the classifier classification memory need. The exact training and classification runtime and memory complexities have already been mentioned in the respective subsections. Table 3 presents upper bounds for the whole framework.

Table 3. Upper bounds for memory and runtime complexity of STIFE.

Task	Upper bound for complexity
Training Time	$O(m^2 \cdot n^2 \cdot |\Sigma|^2 + CTT(n))$
Training Memory	$O(m \cdot n^2 \cdot |\Sigma|^2 + CTM(n))$
Classification Time	$O(m \cdot log(m) + m \cdot |\Sigma| + CCT(n))$
Classification Memory	$O(m \cdot |\Sigma| + CCM(n))$

It can be observed that the biggest influencing factor besides the size of the database is the number of dimensions $|\Sigma|$. How many dimensions actually exist in a data-set is once again dependent on the domain. If the number of dimensions is very high, the memory requirement of the shapelet extraction and selection step might not be practical (it is using an $n \times 7 \cdot |\Sigma|^2$ matrix). Since however the matrix is usually sparse, memory need could be reduced by using appropriate implementations.

6   Empirical Evaluation

Our evaluation consists of two parts. In Subsect. 6.1 we analyze classification time and accuracy for real-life data-sets and in Subsect. 6.2 we conduct experiments with synthetic data to analyze the individual performance of the proposed methods for specific parameter settings. The STIFE framework, classifiers and distance measures were implemented in java[1]. When evaluating STIFE, we used the random forest implementation of Weka.

6.1   Real Data-Sets

For our empirical evaluation we used eight publicly available data sets. Some basic information about each data set is given in Table 4. Note that many of these data-sets come from different domains, which is very relevant when judging the general applicability of classification algorithms based on the evaluation results.

Table 4. Basic properties of the data sets

Data-set	Size test & Training	# of classes	max # of intervals (m)	$|\Sigma|$	duration
ASL-BU	873	9	40	216	5901
ASL-BU-2	1839	7	93	254	14968
AUSLAN2	200	10	20	12	30
BLOCKS	210	8	12	8	123
CONTEXT	240	5	148	54	284
HEPATITIS	498	2	592	63	7555
PIONEER	160	3	89	93	80
SKATING	530	6	143	41	6829

The data-sets were evaluated for three classifiers using 10-fold cross validation. The three evaluated classifiers are 1-NN using the uncompressed (original) IBSM distance [6], 1-NN using our novel method of calculating the IBSM distance, in the following called compressed IBSM, and a random forest using the STIFE framework, in the following called STIFE-RF. For STIFE-RF the Time

---

[1] Implementation available at: https://github.com/leonbornemann/stife.

Table 5. Mean accuracy for 1-NN using the IBSM distance measure and a random forest using STIFE for feature extraction

Data-set	STIFE + Random forest accuracy [%]	IBSM accuracy [%]
ASL-BU	91.75	89.29
ASL-BU-2	**87.49**	76.92
AUSLAN2	**47.00**	37.50
BLOCKS	100	100
CONTEXT	99.58	96.25
HEPATITIS	**82.13**	77.52
PIONEER	98.12	95.00
SKATING	96.98	96.79

Table 6. Mean classification time for 1-NN using the IBSM and compressed IBSM distance as well as a random forest using STIFE for feature extraction

Data-set	STIFE + Random forest [ms]	Compressed IBSM [ms]	IBSM [ms]
ASL-BU	0.48	8.04	331.33
ASL-BU-2	0.47	22.46	1968.85
AUSLAN2	0.46	0.16	0.23
BLOCKS	0.15	0.07	0.15
CONTEXT	0.33	1.69	2.47
HEPATITIS	0.38	9.55	154.96
PIONEER	0.10	0.68	0.78
SKATING	0.18	4.36	97.31

Equality Tolerance (ϵ) as defined in Subsect. 5.3 was set to 5 and the amount of shapelet features to keep was set to 75. Furthermore the number of trees was set to 500 and the number of features per tree was set to $\sqrt{f}$, where f is the number of extracted features.

The results for the accuracy are presented in Table 5. Since both IBSM and compressed IBSM calculate the exact same distance value, both 1-NN classifiers also return the same accuracy which is why we only report one of them. The results for accuracy show that the random forest using STIFE is on par or better than the state of the art 1-NN classifier. Especially on data-sets that seem to be harder to classify (bold in the table) our novel method clearly beats the state of the art IBSM classifier.

When evaluating accuracy the ASL-BU and ASL-BU-2 were treated in a special manner, since they are multi-labeled data-sets, which means that each sequence can have multiple class-labels. This presents a difficulty when evaluating classifier accuracy. Since we introduce a novel method (Random forest + STIFE) we want to show that it is at least on par with the state of the art 1-NN classifiers. Thus we chose a method of evaluation that is more lenient towards the 1-NN classifiers. For both classifiers we eliminated all sequences from the training database that have no class label. Subsequently we modified the training database for the random forest: we copy each sequence once for each of its class-labels and assign each copy exactly one class label. Example: If the sequence S has class labels $\{1, 2, 3\}$, the training database for the random forest will contain three instances of S with different class labels: $\{(S, 1), (S, 2), (S, 3)\}$. The training database of the 1-NN classifier remains unaltered (except for the removal of unlabeled sequences). Subsequently we redefine accuracy in the following: If a test sequence S has class labels A and a classifier predicts a set of class labels P, we say that the sequence was correctly classified, if $A \cap P \neq \emptyset$. Note that this is a definition that favors the 1-NN classifiers, since they will output all class labels of the nearest neighbour, while the random forest can only output exactly one class label. The fact that the random forest using SITFE still achieves better accuracy for both data-sets, although being at a disadvantage gives strong evidence that it may be superior to the 1-NN classifiers.

Table 6 reports the classification time of each of the three classifiers. The results show that compressed IBSM is always faster than IBSM. As expected due to the nature of the algorithms, the speedup is most significant for data-sets that contain high-duration sequences, namely ASL-BU, ASL-BU-2, HEPATITIS and SKATING. The runtime of our second approach, the random forest, while not always being faster is a lot more stable. It never exceeded a classification time of 1 millisecond for all of the data-sets.

6.2   Synthetic Data

There are four different parameters that are relevant for the classification runtime of the three studied classifiers. These are the size of the training database (n), the number of intervals per sequence (m), the number of dimensions $(|\Sigma|)$ and the maximal duration of a sequence. In order to study their individual effects on the classification runtime we randomly generated sequences with fixed values for 3 of the four parameters while varying the fourth one. In order to study the impact of a parameter in a scenario close to reality we set each of the fixed parameters to the upper median of the eight data-sets described in 6.1. That way the fixed parameters that are kept constant reflect a "normal" task. The upper medians are: $n = 498$, $m = 93$, $|\Sigma| = 63$, $duration = 5901$. The results are depicted in Fig. 3. The plotted curves confirm that both compressed IBSM and STIFE-RF are independent of the sequence duration, as opposed to the original IBSM distance. Furthermore, compressed IBSM is faster than IBSM in all evaluated scenarios except for a very high number of intervals (given a fixed duration). On top of that STIFE-RF scales much better with the size of the training database (n) and the number of dimensions $(|\Sigma|)$ than both 1-NN classifier. Lastly the plots show clearly that STIFE-RF is extremely fast in all scenarios: It's classification time never exceeds 3 ms, which makes its plotted curves look constant.

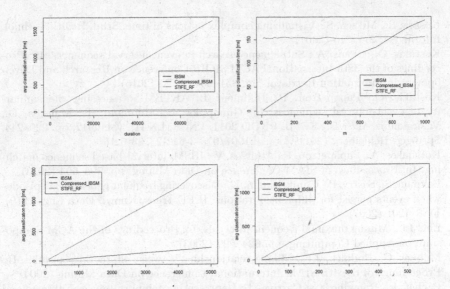

Fig. 3. Classifier performance for different parameters

7 Conclusions

Our main contribution in this paper is the formulation of the STIFE framework, a novel method that maps a sequence to a constant number of features, which can be used for classification. In addition, we presented an improved way of calculating the IBSM distance measure that reduces runtime and memory from the original pseudo-polynomial $\Theta(|\Sigma| \times length(S))$ to the fully polynomial $\Theta(m \cdot (log(m) + |\Sigma|))$. Our experimental evaluation on real and synthetic datasets showed that the STIFE framework using the random forest classifier outperforms the state-of-the-art 1-NN classifier using IBSM and compressed IBSM in terms of both classification accuracy and classification runtime. Directions for future work include the investigation of more elaborate feature selection techniques for selecting shapelets. Another direction is to compare the simple clustering by class of the distance based part of the framework to actual clustering methods and see if actually executing the k-medoids clustering algorithm results in medoids to which the distance is a more discriminative feature.

References

1. Giannotti, F., Nanni, M., Pedreschi, D.: Efficient mining of temporally annotated sequences. In: Proceedings of the 6th SIAM Data Mining Conference, vol. 124, pp. 348–359 (2006)
2. Höppner, F., Klawonn, F.: Finding informative rules in interval sequences. In: Hoffmann, F., Hand, D.J., Adams, N., Fisher, D., Guimaraes, G. (eds.) IDA 2001. LNCS, vol. 2189, pp. 125–134. Springer, Heidelberg (2001). doi:10.1007/3-540-44816-0_13

3. Kosara, R., Miksch, S.: Visualizing complex notions of time. Stud. Health Technol. Inform. **84**, 211–215 (2001)
4. Kostakis, O., Gionis, A.: Subsequence search in event-interval sequences. In: Proceedings of the 38th International ACM SIGIR Conference on Research and Development in Information Retrieval, pp. 851–854. ACM (2015)
5. Kostakis, O., Papapetrou, P., Hollmén, J.: ARTEMIS: assessing the similarity of event-interval sequences. In: Gunopulos, D., Hofmann, T., Malerba, D., Vazirgiannis, M. (eds.) ECML PKDD 2011. LNCS (LNAI), vol. 6912, pp. 229–244. Springer, Heidelberg (2011). doi:10.1007/978-3-642-23783-6_15
6. Kotsifakos, A., Papapetrou, P., Athitsos, V.: IBSM: interval-based sequence matching. In: Proceedings of SIAM Conference on Data Mining, pp. 596–604 (2013)
7. Laxman, S., Sastry, P., Unnikrishnan, K.: Discovering frequent generalized episodes when events persist for different durations. IEEE Trans. Knowl. Data Eng. **19**(9), 1188–1201 (2007)
8. Lin, J.L.: Mining maximal frequent intervals. In: Proceedings of the ACM Symposium on Applied Computing, pp. 624–629 (2003)
9. Mooney, C., Roddick, J.F.: Mining relationships between interacting episodes. In: Proceedings of the 4th SIAM International Conference on Data Mining (2004)
10. Pachet, F., Ramalho, G., Carrive, J.: Representing temporal musical objects and reasoning in the MusES system. J. New Music Res. **25**(3), 252–275 (1996)
11. Papapetrou, P., Kollios, G., Sclaroff, S., Gunopulos, D.: Discovering frequent arrangements of temporal intervals. In: Proceedings of IEEE International Conference on Data Mining, pp. 354–361 (2005)
12. Papapetrou, P., Kollios, G., Sclaroff, S., Gunopulos, D.: Mining frequent arrangements of temporal intervals. Knowl. Inf. Syst. **21**, 133–171 (2009)
13. Winarko, E., Roddick, J.F.: Armada - an algorithm for discovering richer relative temporal association rules from interval-based data. Data Know. Eng. **63**(1), 76–90 (2007)
14. Wu, S.Y., Chen, Y.L.: Mining nonambiguous temporal patterns for interval-based events. IEEE Trans. Knowl. Data Eng. **19**(6), 742–758 (2007)
15. Yang, Y., Pedersen, J.O.: A comparative study on feature selection in text categorization. ICML **97**, 412–420 (1997)

Approximating Numeric Role Fillers via Predictive Clustering Trees for Knowledge Base Enrichment in the Web of Data

Giuseppe Rizzo[✉], Claudia d'Amato, Nicola Fanizzi, and Floriana Esposito

LACAM – Universitá degli Studi di Bari Aldo Moro,
Via Orabona 4, 70125 Bari, Italy
{giuseppe.rizzo1,claudia.damato,nicola.fanizzi,
floriana.esposito}@uniba.it

Abstract. In the context of the Web of Data, plenty of properties may be used for linking resources to other resources but also to literals that specify their attributes. However the scale and inherent nature of the setting is also characterized by a large amount of missing and incorrect information. To tackle these problems, learning models and rules for predicting unknown values of numeric features can be used for approximating the values and enriching the schema of a knowledge base yielding an increase of the expressiveness, e.g. by eliciting SWRL rules. In this work, we tackle the problem of predicting unknown values and deriving rules concerning numeric features expressed as datatype properties. The task can be cast as a regression problem for which suitable solutions have been devised, for instance, in the related context of RDBs. To this purpose, we adapted learning predictive clustering trees for solving multi-target regression problems in the context of knowledge bases of the Web of Data. The approach has been experimentally evaluated showing interesting results.

1 Introduction

The Web of Data represents a novel vision of the Semantic Web [1], where a wealth of data and vocabularies offers properties used for linking resources to other resources but also to literals that specify their attributes. However, the Web of Data is also affected by a large amount of missing and the wrong information owing to its scale and its distributed nature. In particular, numeric features that are exploited to model various aspects of the real world (such as physical measurements, ratings and scores, etc.), usually represented in RDF/OWL as *datatype properties*. When their values for given resources are not explicitly provided in the datasets, they can hardly be derived even resorting to reasoning services. Moreover, such values may be merely incorrect.

Data mining approaches to the specific data representation and semantics (e.g. see [2]) can both fill these information gaps and detect the errors through the induction of predictive models to extend the factual knowledge and to provide so called non standard reasoning services [3]. In addition, when methods

T. Calders et al. (Eds.): DS 2016, LNAI 9956, pp. 101–117, 2016.
DOI: 10.1007/978-3-319-46307-0_7

derived from symbolic approaches are employed, it is possible to directly target the representation languages that are the theoretical formalism for the Web of Data. As a result, such methods allow to obtain new axioms/rules that can be potentially integrated in the knowledge bases increasing their expressiveness and by avoiding the complexity problems deriving from the employment of *Inductive Logic Programming* (ILP) methods [4].

In this context, it should be noted that an individual in a knowledge base like DBPEDIA is characterized by lots of properties whose values can be approximated by inducing models like *terminological regression trees* [5]. However, learning separate models may be a bottleneck in terms of execution times because it requires more training and prediction steps: the same set of instances has to be considered more times for solving each learning problem. To overcome these limitations, solutions based on a multi-target approach [6] to regression problems can be investigated. Specifically, in this paper, we propose a solution devised along the *predictive clustering* framework [7], which combines elements of the clustering and predictive models. Specifically, we are interested in the so-called predictive clustering trees [8], which generalizes the decision tree models to seek for homogeneous clusters of observations for which a predictive model can be associated. Such models estimate the values of the target properties at the same time but no one targets Description Logics representation languages.

Therefore, the main contribution of this paper is a framework for the induction of a new extension of tree models which combines predictive clustering trees and terminological regression trees that can be exploited to determine inductively the value of datatype properties with numeric ranges. This is a fundamental additional service which can complement other non-standard ones and can be the basis for solutions to more complex problems, e.g. ranking based on multiple criteria and detection of wrong numeric fillers. An empirical evaluation proves the effectiveness of such models.

The rest of the paper is organized as follows: the next section gives an overview about the literature concerning the datatype properties value prediction problem and the multi-target regression problem; Sect. 3 introduces the basics concerning Description Logics representation languages; Sect. 4 illustrates the algorithms for inducing predictive clustering trees for multi-target regression, estimating the target values and deriving rules; in Sect. 5 the empirical evaluation performed for assessing the feasibility of the approach is reported; finally, Sect. 6 concludes by proposing further extensions.

2   Related Works

The problem of numeric prediction in the Web of Data have been solved by inducing *terminological regression trees* [5], which derive from *terminological decision trees*, proposed instance-classification models [9] in *Description Logics* (DLs) [3], and also from *First Order Logic Regression Trees*, implemented in the TILDE system [10] (for solving regression problems with clausal knowledge bases). However, a terminological regression tree is a model adopted to solve a *single-target regression problem* in the context of the Web of Data: similarly to terminological decision trees, each inner node contains a concept description in Description Logics while the leaf nodes contain the approximated property values. The induction of such trees exploits a refinement operator to generate the concepts that will be installed as a node [5,9].

So far, to the best of our knowledge, terminological regression trees represented the only example of a model for solving the problem being compliant to the Web of Data representation languages while the literature regarding solutions for datasets in attribute-value or equivalent propositional representation languages is rich (see [11] for a survey). The proposed solutions can be roughly categorized as: (a) *problem transformation* methods (or *local* methods) that transform the multi-output problem into independent single-target problems to be solved using a simpler regression algorithm; (b) *algorithm adaptation* methods (or *big-bang* methods) that adapt a specific single-target method to directly handle multi-output data sets. These methods are more challenging since they usually aim at predicting the multiple targets and at modeling and interpreting the dependencies among these targets.

Other methods proposed in the literature which inspired the approach proposed in this paper are the *predictive clustering methods* [12], which have been employed successfully in the context of bioinformatics for solving *structured prediction* problems [13], *hierarchical classification* [8] and *image retrieval* [14].

3 Basics

The basic notions about Description Logics are shortly recalled to introduce the notation adopted in the illustration of the learning problem and of the methods (which are intended for the datasets and vocabularies in the *Web of Data*).

In Description Logics (DLs) [3], a domain is modeled through primitive *concepts* (classes) and *roles* (relations), which can be used to build complex descriptions regarding *individuals* (instances, objects), by using specific operators that depend on the adopted language. A *knowledge base* is a couple $\mathcal{K} = (\mathcal{T}, \mathcal{A})$ where the *TBox* $\mathcal{T}$ contains axioms concerning concepts and roles (typically subsumption axioms such as $C \sqsubseteq D$) and the *ABox* $\mathcal{A}$ contains assertions, i.e. axioms regarding the individuals ($C(a)$, resp. $R(a, b)$). The set of individuals occurring in $\mathcal{A}$ is denoted by $\mathsf{Ind}(\mathcal{A})$. The specific representation language can be extended in order to support *concrete domains* $\mathbf{D}$ such as boolean, string and, to our purpose, numerical domains. They can be used to define role assertions in the form $R(a, v)$ where $v \in \mathbf{D}$.

The semantics of concepts/roles/individuals is defined through interpretations. An *interpretation* is a couple $\mathcal{I} = (\Delta^{\mathcal{I}}, \cdot^{\mathcal{I}})$ where $\Delta^{\mathcal{I}}$ is the *domain* of the interpretation and $\cdot^{\mathcal{I}}$ is a *mapping* such that, for each individual a, $a^{\mathcal{I}} \in \Delta^{\mathcal{I}}$, for each concept C, $C^{\mathcal{I}} \subseteq \Delta^{\mathcal{I}}$ and for each role R, $R^{\mathcal{I}} \subseteq \Delta^{\mathcal{I}} \times \Delta^{\mathcal{I}}$. The semantics of complex descriptions descends from the interpretation of the primitive concepts/roles and of the operators employed, depending on the adopted language. $\mathcal{I}$ satisfies an axiom $C \sqsubseteq D$ (C is subsumed by D) when $C^{\mathcal{I}} \subseteq D^{\mathcal{I}}$ and an assertion $C(a)$ (resp. $R(a, b)$) when $a^{\mathcal{I}} \in C^{\mathcal{I}}$ (resp. $(a^{\mathcal{I}}, b^{\mathcal{I}}) \in R^{\mathcal{I}}$). $\mathcal{I}$ is a *model* for $\mathcal{K}$ iff it satisfies each axiom/assertion α in $\mathcal{K}$, denoted with $\mathcal{I} \models \alpha$. When α is satisfied w.r.t. these models, we write $\mathcal{K} \models \alpha$.

We will be interested in the *instance-checking* inference service: given an individual a and a concept description C determine if $\mathcal{K} \models C(a)$. Due to the *Open World Assumption* (OWA), answering to a class-membership query is more difficult w.r.t. *Inductive Logic Programming* (ILP) settings where the closed-world reasoning is the standard. Indeed, one may not be able to prove the truth of either $\mathcal{K} \models C(a)$ or $\mathcal{K} \models \neg C(a)$, as there may be possible to find different interpretations that satisfy either cases. For representing the rules, we adopt the Semantic Web Rule Language (SWRL) [15], extending the set of OWL axioms of a given ontology with Horn-like rule.

Definition 1 (Atoms and SWRL Rule). *Given a KB $\mathcal{K}$, An atom is a unary or binary predicate of the form $P_c(s)$, $P_r(s_1, s_2)$, $\mathsf{sameAs}(s_1, s_2)$ or $\mathsf{differentFrom}(s_1, s_2)$, where the predicate symbol P_c is a concept in $\mathcal{K}$, P_r is a role name in $\mathcal{K}$, s, s_1, s_2 are terms. A term is either a variable (denoted by x, y, z) or a constant (denoted by a, b, c) standing for an individual name or data value.*

A SWRL rule is an implication between an antecedent (body) and a consequent (head) of the form: $B_1 \wedge B_2 \wedge \dots B_n \to H_1 \wedge \dots \wedge H_m$, where $B_1 \wedge \dots \wedge B_n$ is the rule body and $H_1 \wedge \dots \wedge H_m$ is the rule head. Each $B_1, \dots, B_n, H_1, \dots H_m$ is an atom.

The rules can be generally called *multi-relational* rules since multiple binary predicates $P_r(s_1, s_2)$ with different role names of $\mathcal{K}$ could appear in a rule. The intended meaning of a rule is: whenever the conditions in the antecedent hold, the conditions in the consequent must also hold. A rule having more than one atom in the head can be equivalently transformed, due to the *safety condition* into multiple rules, each one having the same body and a single atom in the head. We will consider, w.l.o.g., only SWRL rules (hereafter just "rules") with one atom in the head. Differently from the Horne clauses adopted in logic programming, note that the OWA holds.

4   Predictive Clustering Trees for Multi-target Regression in DL Knowledge Bases

4.1   The Problem

In a regression problem, the objective is to approximate analytic function on the grounds of available training instances and predict approximately their correct values.

An even more general problem aims at predicting the correct values w.r.t. various properties: this is known as *multi-target regression*. Formally, in the context of DL knowledge bases the problem can be defined as follows:

Given:
- a knowledge base $\mathcal{K} = (\mathcal{T}, \mathcal{A})$;
- the target functional roles R_i, $1 \le i \le t$, ranging on the domains $\mathbf{D}_i$,
- a training set $\mathsf{Tr} \subseteq \mathsf{Ind}(\mathcal{A})$ for which the numeric fillers are known,
  $\mathsf{Tr} = \{a \in \mathsf{Ind}(\mathcal{A}) \mid R_i(a, v_i) \in \mathcal{A},\ v_i \in \mathbf{D}_i,\ 1 \le i \le t\}$

Build a regression model for $\{R_i\}_{i=1}^{t}$, i.e. a function $h : \mathsf{Ind}(\mathcal{A}) \to \mathbf{D}_1 \times \cdots \times \mathbf{D}_t$ such that it minimizes a loss function over Tr.

A possible loss function may be based on the *mean square error*: $L(h, \mathsf{Tr}) = \sum_{i=i}^{t} MSE[\hat{v}_i, v_i]$, with $\hat{\boldsymbol{v}} = h(a)$ and $\forall\, i \in \{1, \ldots, t\} : \mathcal{K} \models R_i(a, v_i)$.

4.2   The PCT Model

In this section, we aim at tackling the regression problem introduced in the Sect. 4.1 by solving a predictive clustering problem through a *predictive clustering tree* (PCT). A PCT can be derived from a hierarchical clustering model [16], where each cluster is associated to a (multi-target) local model that minimizes a loss function w.r.t. the training instances in the cluster. The nodes of the PCT can be regarded as quadruples with: (1) a (conjunctive) DL test concept; (2) a prototype vector $\hat{\boldsymbol{v}}$ where each $\hat{v}_i$ represents the value (computed via a local regression model) predicting the filler v_i for R_i; (3) the references to the two sub-trees $T.left$ and $T.right$ (possibly null) corresponding to the branches for the test outcomes. Figure 1 reports a simple PCT that might be induced from a knowledge base concerning movie domain. Each intermediate node contains a concept description that is used as a test in the prediction phase. Each leaf contains a prediction made using a local regression model, i.e. a prototype vector with the target values. In particular, the first element of the vector stands for the rating of a movie while the second value denotes the estimate of the taking of same movie. Each edge corresponds to the result of an instance check-test either on a concept or its complement.

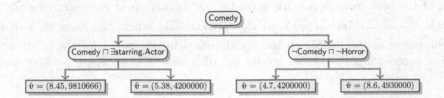

Fig. 1. A simple regression tree for multiple predictions in the movie domain

4.3   The Methods

The procedure for inducing PCTs and for estimating the values of some target functional roles ranging on concrete domains (datatype properties) are described in the following.

Training. Given the set of individuals Tr, a PCT is produced by means of a recursive strategy, which is implemented by the procedure INDUCEPCT shown in Algorithm 1 (to be started with INDUCEPCT(Tr; ⊤)). The algorithm resembles the procedure adopted for inducing similar tree-based models for classification [9] and regression [5] problems. The procedure for PCTs employs a generic stop condition (base case of the recursion) that may be customized according to the specific learning problem (function STOP). In the sequel, we will consider two stop conditions to our purpose: (1) *the number of individuals routed to the node is below some threshold*; (2) *the maximum number of levels was exceeded*, to avoid complex models. If either condition is satisfied, the node will be a leaf and a local regression model w.r.t. the prototype $\hat{v}$ is determined according to values of the instances sorted to that node (e.g. by averaging on each component or more complex ones). If no individuals are routed to the child node, the regression models are inherited from the parent node and used to compute the prototype.

The recursive case requires the generation of a new concept description that will be used for the tests at that node. As subsumption is a quasi-ordering over the space of possible concepts, a downward refinement operator ρ can be used to map the current concept C onto a set **S** of specializations C', $C' \sqsubseteq C$ [17]. In this case the refinement operator derives from the one adopted in [9] and it specializes a concept C in conjunctive normal form in various possible ways: (1) adding an atomic concept (or its complement) as a conjunct; (2) adding a general existential restriction (or its complement) as a conjunct; (3) adding a general universal restriction (or its complement) as a conjunct; (4) replacing a sub-description C_i in the scope of an existential restriction in C with one of its refinements; (5) replacing a sub-description C_i in the scope of a universal restriction with one of its refinements; Note that the cases of (4) and (5) are recursive. The proposed refinement operator performs a sort of random sampling from the (very large) space of specializations and it is biased towards small numbers of recursive calls. Then, the best test concept $E^* \in$ **S** is determined by SELECTBESTTESTCONCEPT according to a purity measure and finally it is installed in the current node. Stemming from the original TRT models, the selection of the best concept is made according to a criterion of *variance reduction*:
$$\text{VR}(E, \text{Tr}) = \text{Var}(\text{Tr}) - \left(\frac{|\text{Ps}|}{|\text{Tr}|} \text{Var}(\text{Ps}) + \frac{|\text{Ns}|}{|\text{Tr}|} \text{Var}(\text{Ns}) \right)$$ where the variance $\text{Var}(\cdot)$ is computed as the RMSE on the standardized target values for the input set of instances, while Ps and Ns are the set of instances with, resp., positive and

---

Algorithm 1. The routines for inducing a PCT

```
 1  INDUCEPCT(Tr : training set; C : concept): PCT
 2  begin
 3      T ← new PCT
 4      v̂ ← LOCALREGMODEL(D)  {local prototype}
 5
 6      if STOP(Tr) = true then
 7          {base case − leaf node}
 8          T ← ⟨null, v̂, null, null⟩;
 9      else
10          {recursive case}
11          S ← ρ(C)
12          E* ← SELECTBESTTESTCONCEPT(S, Tr)
13          ⟨Ps, Ns, Us⟩ ← PARTITION(Tr, E*)
14          T.left ← INDUCEPCT(Ps ∪ Us, E*)
15          T.right ← INDUCEPCT(Ns ∪ Us, ¬E*)
16          T ← ⟨E*, v̂, T.left, T.right⟩
17
18      return T
19  end
```

Algorithm 2. Prediction of target values

```
 1  PREDICT(a : individual; T : PCT) : D₁ × ··· × Dₜ      9    else if 𝒦 ⊨ ¬C(a) then
 2  begin                                                 10        N ← ROOT(T.right)
 3    N ← ROOT(T);                                        11    else
 4                                                        12        return v̂; {internal node prototype}
 5    while ¬LEAF(N, T) do                                13
 6      ⟨C, v̂, T.left, T.right⟩ ← INODE(N)                14    ⟨_, v̂, null, null⟩ ← INODE(N) {leaf case}
 7      if 𝒦 ⊨ C(a) then                                  15    return v̂;
 8        N ← ROOT(T.left)                                16  end
```

---

negative outcomes w.r.t. the candidate test concept E. After the assessment of the best concept E^*, the individuals are partitioned (by PARTITION) to follow the left or right branch according to the result of the instance-check w.r.t. E^*. Note that a training example a is replicated in both children in case both $E^*(a)$ and $\neg E^*(a)$ are satisfiable w.r.t. $\mathcal{K}$. The set of such individuals is denoted by Us. This strategy is applied recursively until the instances routed to a node satisfy the stopping condition.

Prediction. After a PCT is produced, predictions can be made relying on the resulting model. The related procedure, sketched in Algorithm 2, works as follows.

Given the individual for which the prediction is made and the tree T, the procedure PREDICT returns a target variable according to the leaf reached from the root in a path down the tree. Specifically, the algorithm traverses recursively the PCT performing an instance check w.r.t. the test concept contained in each visited node: let $a \in \mathsf{Ind}(\mathcal{A})$ the input individual and C the test concept installed in the current node, if $\mathcal{K} \models C(a)$ (resp. $\mathcal{K} \models \neg C(a)$) then the left (resp. right) branch is followed. If neither branch can be followed i.e. a definite classification w.r.t. C cannot determined, the algorithm will return the output of the regression model installed into the current node, the prototype $\hat{v}$. Alternatively, to simplify, this case can be assimilated with the non-membership case, that is, equivalent to considering $K \not\models C(a)$ for the right-branch test. The service routines ROOT, LEAF and INODE are employed, respectively to get the root node of a given PCT, to test whether a node is a leaf for a tree, and to select the node content.

Deriving Rules from PCTs. Similarly to decision trees, where each node in a path from the root to a leaf-node may be used to build a concept description, it is possible to derive a list of rules from a PCT. This trade-off between comprehensibility and effectiveness of the regression models may be regarded as an advantage over purely statistical models that are black boxes, not suitable for human inspection (or even modification). The procedure DERIVERULES (see Algorithm 3), explores all the paths leading to leaf-nodes which will constitute the consequents; for each path, an antecedent is formed by tracking back the paths from each leaf and collecting the intermediate test concepts. In this way, each path yields a different rule with a conjunctive antecedent that represents a local (partial) definition of the target functions.

5   Experiments

In this section, we illustrate the experiments aiming at assessing the effectiveness
of the proposed approach both on some small publicly available ontologies and
knowledge bases extracted from the Linked Data Cloud.

5.1   Experimental Settings

Algorithms Setup. We compared the predictive models induced according to
the approach described in this paper to some other approaches: *terminological
regression trees* (TRTs), induced for solving the target regression problems separately and two further methods, the *k-nearest neighbor regressor* and the *multi-
target linear regression model* [11]. The algorithm for growing PCTs requires the
value of the depth of the trees as a parameter. In our experiments this value was
set to 10, 15, 20, and 30. In the case of the TRTs two parameters control the
growth of the trees are required, namely θ and m [5], resp. a threshold for the
purity criterion and the number of instances that are routed to a leaf node. We
used the default settings for them, i.e. $\theta = 0.5$ and $m = 5$. For the propositional
models, the choice of a feature set characterizing the individuals of the knowledge
base is required to determine a similarity measure. In this case, the vectors were
built by using the membership w.r.t. atomic concepts in the knowledge base are
employed as feature set. Binary features were considered: the i-th value denotes
if for an individual $a \in \mathsf{Ind}(\mathcal{A})$, $C(a)$ is satisfiable. For the k-NN regression, the
Euclidean distance is employed and the neighborhood size k was set to $\log |\mathsf{Tr}|$.

Preparing Learning Problems on Small Ontologies. The effectiveness of
PCTs was evaluated on the datatype properties prediction tasks[1]. We considered various datasets extracted from publicly available Web ontologies[2]: BCO,
a medical ontology; GEOPOLITICAL that models information concerning countries in terms of latitude, longitude, population etc.; MONETARY, an ontology for
modeling domain related currencies and MUTAGENESIS, an OWL porting of the

---

Algorithm 3. Deriving rules from PCTs

1 DERIVERULES $(T : \text{PCT}, C : \text{parent concept})$: rule	8	$L \leftarrow [r]$
set	9	**else**
2 Let x be a variable name	10	$\langle D, M, T.\text{left}, T.\text{right} \rangle \leftarrow \text{INODE}(N)$
3 $N \leftarrow \text{ROOT}(T)$	11	$L.\text{left} \leftarrow \text{DERIVERULES}(T.\text{left}; C(x) \wedge D(x))$
4 **if** LEAF(N) **then**	12	$L.\text{right} \leftarrow \text{DERIVERULES}(T.\text{right}; C(x) \wedge \neg D(x))$
5 $\langle \text{null}, \hat{v}, \text{null}, \text{null} \rangle \leftarrow \text{INODE}(N)$	13	$L \leftarrow L.\text{left} + L.\text{right}$
6 **for** $i \leftarrow 1$ **to** t **do**	14	**return** L
7 $r \leftarrow (C(x) \rightarrow R_i(x, \hat{v}_i(x)))$		

---

[1] The source code and the benchmarks are available at: http://github.com/
Giuseppe-Rizzo/DLPredictiveClustering.

[2] The ontologies are available also at: http://www.inf.unibz.it/tones/index.php
(BCO, MONETARY), https://datahub.io/dataset (GEOPOLITICAL), https://github.
com/AKSW/DL-Learner/tree/develop/examples/mutagenesis (MUTAGENESIS).

Table 1. Small ontologies employed in the experiments

Ontology	DL language	Concepts	Roles	Data properties	Individuals
BCO	$\mathcal{ALCIF}(D)$	196	22	3	112
GEOPOLITICAL	$\mathcal{SHIN}(D)$	12	6	89	312
MONETARY	$\mathcal{ALCIF}(D)$	323	247	74	2466
MUTAGENESIS	$\mathcal{AL}(D)$	86	5	6	14145

well-known dataset employed for testing ILP algorithms. The details concerning the datasets, in terms of the numbers of classes, object and datatype properties, and individuals are reported in Table 1. The learning task for each dataset was the prediction of the (approximate) fillers for numeric datatype properties. For these datasets, we ignored the datatype properties that could contribute few examples. As a result, for each dataset, we tackled problems that consider a certain number of properties which span from 2 to 4 properties. Table 2 illustrates the properties, their range and the number of individuals employed as examples We used the *0.632 bootstrap* procedure as the design of the experiments to determine an estimate of the *Root Mean Squared Error* (RMSE) that is the root squared error computed on the test set and averaged w.r.t. the number of datatype properties.

Preparing Learning Problems on Linked Data Datasets. We considered datasets extracted from DBPEDIA by means of a crawler: the individuals and the considered target properties are retrieved by properly formulating SPARQL

Table 2. Target properties ranges, number of individuals employed in the learning problem

Ontology	Properties	Range	|Tr|
BCO	hasDuration	[1.0, 3.75]	63
	hasValue	[0.66, 1.2]	
GEOPOLITICAL	hasMaxLatitude	[12.2, 107.3]	222
	hasMinLatitude	[−22.34, 55.43]	
	hasMaxLongitude	[−87.83, 107.7]	
	hasMinLongitude	[−58.44, 176.13]	
MONETARY	ageOfMajorityFemale	[0, 21]	472
	ageOfMajorityMale	[0, 21]	
MUTAGENESIS	act	[−3.92, 2.62]	4794
	charge	[−0.398, 0.8608]	
	logp	[01.56, 07.13]	
	lumo	[−2.87, −0.937]	

Table 3. Datasets extracted from DBPEDIA

Datasets	Expr.	Axioms	#Classes	#Properties	#Ind.
FRAGM. #1	$\mathcal{ALCO}$	17222	990	255	12053
FRAGM. #2	$\mathcal{ALCO}$	20456	425	255	14400
FRAGM. #3	$\mathcal{ALCO}$	9070	370	106	4499

Table 4. Ranges, number of individuals employed in the learning problem on DBPEDIA

Datasets	Properties	Range	|Tr|
FRAGM. #1	elevation	$[-654.14, 19.00]$	10000
	populationTotal	$[0.0, 2255]$	
FRAGM. #2	areaTotal	$[0, 16980.1]$	10000
	areaUrban	$[0.0, 6740.74]$	
	areaMetro	$[0, 652874]$	
FRAGM. #3	height	$[0, 251.6]$	2256
	weight	$[-63.12, 304.25]$	

queries to the DBPEDIA endpoint. The resources resulting from the previous step were employed as seeds in order to start the crawling process: the RDF graph was traversed and the crawler retrieved all the immediate neighbors and the related schema information (direct classes and their super-classes). All the extracted knowledge have been used for building ontologies whose characteristics are summarized in Table 3. In the experiments various learning problems were considered. In the case of FRAGM. #1, the experiment aim at predicting the values for individuals that are instance of Geographical Information. The experiments with FRAGM. #2 aim at predicting areas for instances of the concept Populated Place. Finally, the third dataset was considered to determine the values for individuals that are instance of the class Person. Table 4 illustrates the properties, their range and the number of individuals employed as training/test examples We used the *10-fold cross validation* procedure as the design of the experiments to determine, unlike the experiments on small ontologies, an estimate of the *Relative Root Mean Squared Error* (RRMSE) averaged w.r.t. the number of datatype properties. This allows to evaluate the performance of the induced models without using normalization factors to consider the different ranges of the properties.

5.2 Outcomes and Discussion

Discussion About Small Ontologies. A first consideration regards the target properties: we note that the property values spanned wide ranges (see Table 2). Rare cases in which the filler is the maximum value observed in the knowledge base can be regarded as exceptions (a sort of outliers), especially for the datasets

Table 5. Outcomes: RMSE averaged over the number of the replications (and standard deviations)

Ontology	PCT	TRT	k-NN	LR
BCO	**0.0277** ± 0.01	0.0356 ± 0.01	0.0472 ± 0.01	0.0554 ± 0.01
GEOPOLITICAL	**0.0284** ± 0.01	0.03561 ± 0.03	0.057 ± 0.03	0.06 ± 0.02
MONETARY	**7.52** ± 0.15	8.46 ± 0.07	**7.53** ± 0.17	7.78 ± 0.34
MUTAGENESIS	**0.0445** ± 0.07	0.0637 ± 0.03	0.0547 ± 0.02	0.0647 ± 0.05

Table 6. Comparison in terms of RMSE between PCTs and other models for each property.

Ontology		PCT	TRT	k-NN	LR
BCO	hasDuration	**0.0291**	0.0302	0.0450	0.0308
	hasValue	**0.0263**	0.0400	0.0494	0.0800
GEOPOLITICAL	hasMaxLatitude	**0.0341**	0.054	0.054	0.06
	hasMinLatitude	**0.0425**	0.04312	0.053	0.061
	hasMaxLongitude	**0.0256**	0.03251	0.052	0.063
	hasMinLongitude	**0.0351**	0.01421	0.057	0.06
MONETARY	ageOfMajorityFemale	**8.64**	8.56	8.36	8.56
	ageOfMajorityMale	**6.36**	8.36	6.64	7
MUTAGENESIS	act	**0.0445**	0.07	0.0645	0.0647
	charge	**0.0446**	0.065	0.053	0.0652
	logp	**0.3445**	0.07	0.045	0.0654
	lumo	**0.5**	0.071	0.053	0.0642

with limited numbers of examples for the target properties. Table 5 reports the results of the experiments. In the experiments, we noticed that the RMSE for PCTs did not change significantly when the value of the parameter for controlling the growth was increased. Hence, in Table 5 reports results observed with the value set to 10.

As regards the comparison between PCTs and the other models, we noted that the former generally performed better than the latter, in terms of RMSE, particularly for the tests involving larger training sets like the case of MONE-TARY. Conversely, in the case of datasets with smaller training sets like BCO and GEOPOLITICAL, the performance is rather similar to the one observed using the model based on single TRTs. Throughout the empirical evaluation, PCTs outperformed k-NN and LR. In the case of k-NN, the poor performance might be due number of dimensions of the representation which may have affected the Euclidean distance that was employed. Conversely, the better performance of k-NN w.r.t. the LRs (exploiting the same feature set) was likely due to the locality principle characterizing instance-based methods: the choice of the neighborhood tended to exclude *spurious* individuals that were considered by the LRs. As a result, LR models were more error-prone than those induced by the k-NN. A Friedman statistical test with significance level of $\alpha = .05$ for each dataset showed a statistical significant difference with an average p-value of $p = 0.0432$ In the case of BCO and GEOPOLITICAL, we noticed that the refinement operator was crucial because it tended to generate candidates that differed from those obtained for the various single-target regression models. Finally, in the case of MONETARY and MUTAGENESIS datasets, we obtained again larger values through PCTs w.r.t. TRTs. Similarly to the experiments with BCO, this improvement is due to concept description installed as intermediate node of a PCT that led to minimize the error. On the other hand, the concepts employed in the single-target regression model w.r.t. a certain property led to predictive models whose intermediate concepts were different w.r.t. the ones installed in PCTs. In addition, we observed that the values for the individuals w.r.t. the properties considered in the experiments are quite similar. This suggests that no outliers were observed. In this way, the RMSE obtained through PCTs was low. For all datasets employed in the experiments, we noticed that the performance was stable in terms of standard deviation. This is due to the overlaps between training sets in many repeated runs of the bootstrap procedure. The effectiveness of the proposed approach is also confirmed by the results reported in Table 6. The table illustrates the RMSE separately for each target property: PCTs outperformed the other models employed in the experiments also when the results are not averaged w.r.t. the properties.

As regards the efficiency of the proposed solution, the time required for inducing and for exploiting PCTs (see Table 7) for solving the learning problem on a personal workstation was, in general, less than the total amount of time required to induce and to predict all the TRTs. On the other hand, the time for running k-NN and LR spans from few seconds to less than 10 min. Besides, Table 7 shows that the elapsed times tends to increase almost linearly w.r.t. the number of available individuals. Instead, the execution times of TRTs show a similar trend both w.r.t. the training set size and the number of target properties. For instance, if one considers GEOPOLITICAL,the training and test phases w.r.t. TRTs lasts almost six times more than the one required by PCTs. While these differences were rare in the experiments with BCO, which was a small dataset, the gap of efficiency was broader in the experiments with MONETARY, and MUTAGENESIS. The different performance was likely due to: the maximum depth of the PCTs, the number of instances, and the reasoning service adopted for splitting the instances that are sorted to the various recursive calls. Indeed, the time required for training the models increases more than proportionally to the number of levels (see Table 8). This was affected by the lengths of complex concept description installed in some nodes: the instance check test performed on each training individual w.r.t. more complex concepts requires more time than tests performed on simpler ones. As a result, the instance check test represented a bottleneck for the scalability of the algorithm. Obviously k-NN and LR required smaller amount of time because no specific reasoning services were necessary during the training phase.

Table 7. Elapsed times (secs) of the models in the first experiment. For TRTs, the times for each induced tree are also reported

Ontology		PCT	TRT	k-NN	LR
BCO	hasDuration		70.36		
	hasValue		73.14		
	Total	75.4	143.6	40.5	30.35
GEOPOLITICAL	hasMaxLatitude		183		
	hasMinLatitude		146		
	hasMaxLongitude		93		
	hasMinLongitude		95		
	Total	132	510	72	45
MONETARY	ageOfMajorityFemale		325		
	ageOfMajorityMale		430		
	Total	353	755	123	76
MUTAGENESIS	act		235		
	charge		126		
	logp		256		
	lumo		375		
	Total	540	978	231	124

Table 8. Training time for inducing the PCTs w.r.t. the maximal depth of the trees

Ontology	Depth			
	10	15	20	30
BCO	75.3	89	135	176
GEOPOLITICAL	132	265	272	350
MONETARY	353	460	729	824
MUTAGENESIS	540	752	1113	1650

A final remark to be discussed regards the quality of the models induced through the PCT learning algorithms. In the training phase, we noticed several cases in which most of the instances tended to be sorted down the right branch with fewer individuals routed to left one, which indicates that the selected test concepts where poor at capturing really divisive features on the data. This can be regarded as a sort of *small disjunct problem* occurring in classification problems. This issue was likely due the complexity of the concepts installed as a test. A more complex concept can be regarded as a constraint that can be hard to be satisfied by the individuals occurring in a knowledge base. For example, if a refinement is obtained by applying several existential restrictions, the new concept can be very specific so that no known individual can be assigned as its instance.

This means that the algorithm for growing PCTs routed the instances to the right branch.

Discussion About Linked Data Datasets. Also in these new experiments, we noted that the property values spanned wide ranges (see Table 4). For some properties, there were only few individuals with either the maximum or the minimum value for the target properties. Such individuals can be considered as spurious, likely due to the errors introduced in the process of knowledge extraction required to build DBPEDIA.

The evaluation confirmed the same results obtained on the small ontologies. Firstly, we noted a lower RRMSE of PCTs than the one obtained with other models, particularly for FRAGM. #1 and FRAGM. #2. Again, this difference was statistical significant according to a Friedman test with significance level of $\alpha = .05$ (average p-value of $p = 0.0341$). This may be likely due to the largest number of concepts available in the knowledge base that produced more discernible tests than the ones obtained in the experiments with FRAGM. #3. The comparison between the outcomes obtained through PCTs and TRTs suggests that the quality of the models proposed in this paper have some benefits from the employment of a multi-target selection criterion also in the case of Web of Data datasets. In particular, the standardization required for computing the heuristic played a crucial role: it mitigated the presence of aforementioned individuals having abnormal values that were routed to the node, which may represent one of the main cause for large errors. The standardization adjusted the error computation favoring the selection of concept descriptions producing even splits of the individuals sorted to the node. As in the previous experiments, PCTs outperformed k-NN and LR. In the case of k-NN, the largest number of concepts adopted as features have affected the Euclidean distance adopted in the experiments. In addition, the k-NN ruled out the *spurious* individuals. This explains the improved performance w.r.t. the LR model. (Table 9)

Table 9. RRMSE averaged on the number of runs

Datasets	PCT	TRT	k-NN	LR
FRAGM. #1	**0.42 ± 0.05**	**0.63 ± 0.05**	0.65 ± 0.02	0.73 ± 0.02
FRAGM. #2	**0.25 ± 0.001**	0.43 ± 0.02	0.53 ± 0.00	0.43 ± 0.02
FRAGM. #3	**0.24 ± 0.05**	0.36 ± 0.2	0.67 ± 0.10	0.73 ± 0.05

The time required for inducing and exploiting PCTs (see Table 10) was quite limited w.r.t. the number of individuals adopted as datasets. Also in this new evaluation, the time for training/test phase through PCTs was less than the total amount of time required to induce and to predict the fillers through all the TRTs. Particularly, the time required for solving each learning problem by inducing each TRT was approximately equal to the time needed for growing and exploiting a PCT. k-NN and LR were more efficient than the other models with execution times spanning from few seconds to less than 5 min. This result strictly depends on the different representation languages targeted by the algorithms: while PCTs and TRTs induction methods tries to learn models directly from DL knowledge bases exploiting the TBox as a background knowledge and the reasoning services instead of using propositionalization methods.

Qualitative Evaluation of the Rules. Due to the importance of DBPEDIA for the Web of Data, we discuss the quality of the rules derived from PCTs. In general, the fragment extracted with the crawling process DBPEDIA contains various concept names but not concept definitions. This affected the refinement process occurring during the learning of PCTs. The refinement operator adopted for generating the candidate concepts to use as tests tended to prefer specializations that introduces only concept names and its complement rather than existential and universal restrictions. In this case, the sparseness of the role assertions contained in the knowledge base did not allow adding useful existential/universal restrictions in order to split well the instances. Some interesting rules derived from the induction of PCTs are reported below. As an example, we reported some rules derived from the experiments with FRAGM #3:

Person(x) ∧ Athlete(x) ∧ AmericanFootballPlayer(x) → height(x, 195.4)
Person(x) ∧ Athlete(x) ∧ AmericanFootballPlayer(x) → weight(x, 113.5)
Person(x) ∧ Athlete(x) ∧ ¬(AmericanFootballPlayer)(x) ∧ SoccerPlayer(x) → height(x, 187)
Person(x) ∧ Athlete(x) ∧ ¬(AmericanFootballPlayer)(x) ∧ SoccerPlayer(x) → weight(x, 87.5)

The rules reported above can be used to determine that *American football players are (usually) taller* than *soccer players while American football players' weight is larger than soccer players' one* and, in general, there is a correlation between the kind of athlete and some physical features such as height and weight. Similar rules were obtained also in the experiments with FRAGM #1. In particular, in the

Table 10. Comparison in terms of elapsed times (secs) between PCTs and other models

Datasets		PCT	TRT	k-NN	LR
FRAGM #1	elevation		2454.3		
	populationTotal		2353.0		
	Total	2432	4807.3	547.6	234.5
FRAGM #2	areaTotal		2256.0		
	areaUrban		2345.0		
	areaMetro		2345.2		
	Total	2456	6946.2	546.2	235.7
FRAGM #3	height		743.5		
	weight		743.4		
	Total	743.3	1486.9	372.3	123.5

experiments with FRAGM #1 the rules showed negative correlations between the values of the properties elevation and populationTotal likely due to the fact that places with a high elevation (e.g. mountain villages) may be less populated than other places with a low elevation. In the perspective of an ontology engineer, the rules may be potentially evaluated and along the directions of *understandability* and *body coverage*. The first one can be evaluated in terms of length of the rules, i.e. the number of atoms that composes the rule (including also the head of the rule). The second one is the number of individuals in a knowledge base satisfying the antecedent of a SWRL rules. Such metrics can be integrated in an ontology repairing tool to provide a mechanism for ranking the induced rules.

6   Conclusion and Further Extensions

In this work, we tackled the problem of predicting numeric values for datatype properties through a solution based on the predictive clustering approach that descends from terminological regression trees [5]. Unlike other multi-relational data mining methods our approach target directly the representation languages that back the Web of Data. The outcomes of the experiments are promising: the models are effective and efficient and, in addition, useful rules can be derived from them. The method can be extended along various directions. Firstly, it may integrate to further refinement operators such as those available in other concept learning tools [18]. Then, we can extend the experiments by considering further datasets with a large number of numeric properties. A further extension concerns the employment of further criteria for selecting the best concept and the application of pruning strategies. Finally, we plan to consider linear functions [19] in place of constant vectors at leaf nodes, leading to *model trees*.

Acknowledgments. This work fulfills the objectives of the PON 02005633489339 project "Puglia@Service - Internet-based Service Engineering enabling Smart Territory structural development" funded by the Italian Ministry of University and Research (MIUR).

References

1. Heath, T., Bizer, C.: Linked Data: Evolving the Web into a Global Data Space. Synthesis Lectures on the Semantic Web. Morgan & Claypool, Palo Alto (2011)
2. Rettinger, A., Lösch, U., Tresp, V., d'Amato, C., Fanizzi, N.: Mining the semantic web. Data Min. Knowl. Discov. **24**, 613–662 (2012)
3. Baader, F., Calvanese, D., McGuinness, D., Nardi, D., Patel-Schneider, P. (eds.): The Description Logic Handbook, 2nd edn. Cambridge University Press, Cambridge (2007)
4. Badea, L., Nienhuys-Cheng, S.-H.: A refinement operator for description logics. In: Cussens, J., Frisch, A. (eds.) ILP 2000. LNCS (LNAI), vol. 1866, pp. 40–59. Springer, Heidelberg (2000). doi:10.1007/3-540-44960-4_3
5. Fanizzi, N., d'Amato, C., Esposito, F., Minervini, P.: Numeric prediction on OWL knowledge bases through terminological regression trees. Int. J. Semant. Comput. **6**, 429–446 (2012)

6. Breiman, L., Friedman, J.: Predicting multivariate responses in multiple linear regression. J. Roy. Stat. Soc. **59**, 3–54 (1997)

7. Ženko, B., Džeroski, S., Struyf, J.: Learning predictive clustering rules. In: Bonchi, F., Boulicaut, J.-F. (eds.) KDID 2005. LNCS, vol. 3933, pp. 234–250. Springer, Heidelberg (2006). doi:10.1007/11733492_14

8. Struyf, J., Džeroski, S., Blockeel, H., Clare, A.: Hierarchical multi-classification with predictive clustering trees in functional genomics. In: Bento, C., Cardoso, A., Dias, G. (eds.) EPIA 2005. LNCS (LNAI), vol. 3808, pp. 272–283. Springer, Heidelberg (2005). doi:10.1007/11595014_27

9. Fanizzi, N., d'Amato, C., Esposito, F.: Induction of concepts in web ontologies through terminological decision trees. In: Balcázar, J.L., Bonchi, F., Gionis, A., Sebag, M. (eds.) ECML PKDD 2010. LNCS (LNAI), vol. 6321, pp. 442–457. Springer, Heidelberg (2010). doi:10.1007/978-3-642-15880-3_34

10. Blockeel, H., De Raedt, L.: Top-down induction of first-order logical decision trees. Artif. Intell. **101**, 285–297 (1998)

11. Borchani, H., Varando, G., Bielza, C., Larrañaga, P.: A survey on multi-output regression. WIREs: Data Min. Knowl. Discov. **5**, 216–233 (2015)

12. Aho, T., Ženko, B., Džeroski, S., Elomaa, T.: Multi-target regression with rule ensembles. J. Mach. Learn. Res. **13**, 2367–2407 (2012)

13. Kocev, D., Vens, C., Struyf, J., Dzeroski, S.: Tree ensembles for predicting structured outputs. Pattern Recogn. **46**, 817–833 (2013)

14. Dimitrovski, I., Kocev, D., Loskovska, S., Džeroski, S.: Fast and scalable image retrieval using predictive clustering trees. In: Fürnkranz, J., Hüllermeier, E., Higuchi, T. (eds.) DS 2013. LNCS (LNAI), vol. 8140, pp. 33–48. Springer, Heidelberg (2013). doi:10.1007/978-3-642-40897-7_3

15. Horrocks, I., Patel-Schneider, P.F., Boley, H., Tabet, S., Grosof, B., Dean, M.: SWRL: a semantic web rule language combining OWL and RuleML. Technical report (2004). https://www.w3.org/Submission/SWRL/

16. Blockeel, H., DeRaedt, L., Ramon, J.: Top-down induction of clustering trees. In: Proceedings of ICML1998, pp. 55–63. Morgan Kaufmann (1998)

17. Lehmann, J., Hitzler, P.: A refinement operator based learning algorithm for the $\mathcal{ALC}$ description logic. In: Blockeel, H., Ramon, J., Shavlik, J., Tadepalli, P. (eds.) ILP 2007. LNCS (LNAI), vol. 4894, pp. 147–160. Springer, Heidelberg (2008). doi:10.1007/978-3-540-78469-2_17

18. Lehmann, J.: DL-learner: learning concepts in description logics. J. Mach. Learn. Res. **10**, 2639–2642 (2009)

19. Appice, A., Džeroski, S.: Stepwise induction of multi-target model trees. In: Kok, J.N., Koronacki, J., Mantaras, R.L., Matwin, S., Mladenič, D., Skowron, A. (eds.) ECML 2007. LNCS (LNAI), vol. 4701, pp. 502–509. Springer, Heidelberg (2007). doi:10.1007/978-3-540-74958-5_46

Option Predictive Clustering Trees
for Multi-target Regression

Aljaž Osojnik[1,2]([✉]), Sašo Džeroski[1,2], and Dragi Kocev[1,2]

[1] Department of Knowledge Technologies, Jožef Stefan Institute, Ljubljana, Slovenia
{aljaz.osojnik,saso.dzeroski,dragi.kocev}@ijs.si
[2] International Postgraduate School, Jožef Stefan Institute, Ljubljana, Slovenia

Abstract. Decision trees are one of the most widely used predictive modelling methods primarily because they are readily interpretable and fast to learn. These nice properties come at the price of predictive performance. Moreover, the standard induction of decision trees suffers from myopia: A single split is chosen in each internal node which is selected in a greedy manner; hence, the resulting tree may be sub-optimal. To address these issues, option trees have been proposed which can include several alternative splits in a new type of internal nodes called option nodes. Considering all of this, an option tree can be also regarded as a condensed representation of an ensemble. In this work, we propose to extend predictive clustering trees for multi-target regression by considering option nodes, i.e., learn option predictive clustering trees (OPCTs). Multi-target regression is concerned with learning predictive models for tasks with multiple continuous target variables. We evaluate the proposed OPCTs on 11 benchmark MTR datasets. The results reveal that OPCTs achieve statistically significantly better predictive performance than a single PCT. Next, the performance is competitive with that of bagging and random forests of PCTs. Finally, we demonstrate the potential of OPCTs for multifaceted interpretability and illustrate the potential of inclusion of domain knowledge in the tree learning process.

Keywords: Multi-target regression · Option trees · Interpretable models · Predictive clustering trees

1 Introduction

Supervised learning is one of the most widely researched and investigated areas of machine learning. The goal in supervised learning is to learn, from a set of examples with known class, a function that outputs a prediction for the (scalar-valued) class of a previously unseen example. However, in many real life problems of predictive modelling the output (i.e., the target) is structured, e.g., is a vector of class values of a tuple of target variables. There can be dependencies between the class values/targets (e.g., they can be organized into a tree-shaped hierarchy or a directed acyclic graph) or some internal relations between the class values (e.g., as in sequences).

© Springer International Publishing Switzerland 2016
T. Calders et al. (Eds.): DS 2016, LNAI 9956, pp. 118–133, 2016.
DOI: 10.1007/978-3-319-46307-0_8

In this work, we concentrate on the task of predicting multiple continuous variables. Examples thus take the form $(\mathbf{x_i}, \mathbf{y_i})$, where $\mathbf{x_i} = (x_{i1}, \ldots, x_{ik})$ is a vector of k input variables and $\mathbf{y_i} = (y_{i1}, \ldots, y_{it})$ is a vector of t target variables. This task is known under the name of *multi-target regression* (MTR) [1] (also known as multi-output or multivariate regression). MTR is a type of structured output prediction task which has applications in many real life problems, where we are interested in simultaneously predicting multiple continuous variables. Prominent examples come from ecology and include predicting the abundance of different species living in the same habitat [2] and predicting properties of forests [3,4]. Due to its applicability to a wide range of domains, this task is recently gaining increasing interest in the research community.

Several methods for addressing the task of MTR have been proposed [1,5]. These methods can be categorized into two groups of methods [6]: (1) local methods, that predict each of the target variable separately and then combine the individual model predictions to get the overall model prediction and (2) global methods, that predict all of the variables simultaneously (also known as 'big-bang' approaches). In the case of local models, for a domain with t target variables one needs to construct t predictive models – each predicting a single target. The prediction vector (that consists of t components) of an unseen example is then obtained by concatenating the predictions of the multiple single-target predictive models. Conversely, in the case of global models, for the same problem, one needs to construct only one model. In this case, the prediction vector of an unseen example is obtained by passing the example through the model and getting its (complete) prediction.

In the past, several researchers proposed methods for solving the task of MTR directly and demonstrated their effectiveness [1,4,7–9]. The global methods have several advantages over the local methods. First, they exploit and use the dependencies that exist between the components of the structured output in the model learning phase, which can result in better predictive performance. Next, they are typically more efficient: it can easily happen that the number of components in the output is very large (e.g., predicting the bioactivity profiles of compounds described with their quantitative structure activity relationships on a large set of proteins), in which case executing a basic method for each component is not feasible. Furthermore, they produce models that are typically smaller than the sum of the sizes of the models built for each of the components.

The state-of-the-art methods for MTR are based on tree and ensemble learning [1,5]. Trees for MTR (from the predictive clustering framework) inherit the properties of regression trees: they are interpretable models, but the learning them is greedy. The performance of the trees is significantly improved when they are used in an ensemble setting [1,7]. However, the myopia, i.e., greediness, of the tree construction process can lead to learning sub-optimal models. One way to alleviate this is to use a beam-search algorithm for tree induction [10], while another approach is to introduce option splits in the nodes [11,12].

In this work, we propose to extend predictive clustering trees (PCTs) for MTR towards option trees, hence we propose to learn option predictive clustering trees (OPCTs). An option tree can be seen as a condensed representation of an ensemble of trees which share a common substructure. More specifically, the heuristic function for split selection can return multiple values that are close to each other within a predefined range. These splits are then used to construct an option node. For illustration, see Fig. 1.

The remainder of this paper is organized as follows. Section 2 proposes the algorithm for learning option PCTs for MTR. Next, Sect. 3 outlines the design of the experimental evaluation. Section 4 continues with a discussion of the results. Finally, Sect. 5 concludes and provides directions for further work.

2   Option Predictive Clustering Trees

The predictive clustering trees framework views a decision tree as a hierarchy of clusters. The top-node corresponds to one cluster containing all data, which is recursively partitioned into smaller clusters while moving down the tree. The PCT framework is implemented in the CLUS system [13], which is available for download at http://clus.sourceforge.net.

Option predictive clustering trees (OPCT) extend the usual PCT framework, by introducing option nodes into the tree building procedure outlined in Algorithm 1. Option decision trees were first introduced as classification trees by Buntine [11] and then analyzed in more detail by Kohavi and Kunz [12]. Ikonomovska et al. [14] analyzed regression option trees in the context of data streams.

The major motivation for the introduction of option trees is to address the myopia of the *top-down induction of decision trees* (TDIDT) algorithm [15]. Viewed through the lens of the predictive clustering framework, a PCT is a nonoverlapping hierarchical clustering of the whole input space. Each node/subtree corresponds to a clustering go a subspace and prediction functions are placed in the leaves, i.e., lowest clusters in the hierarchy. An OPCT, however, allows the construction of an overlapping hierarchical clustering. This means that, at each node of the tree several alternative hierarchical clusterings of the subspace can appear instead of a single one.

When using an OPCT for prediction on a new example, we produce the prediction by aggregating over the predictions of the alternative subtrees (overlapping clusters) the example may encounter. However, as not all parts of the tree (hierarchical clustering) are necessarily overlapping, the example may encounter only nonoverlapping (sub)clusters. In that case, we produce the prediction as we would with a regular PCT.

When using TDIDT to construct a predictive clustering tree, and in particular when splitting a leaf, all possible splits are evaluated by using a heuristic and the best one is selected. However, other splits may have very similar heuristic values. The best split could be selected over another split as a consequence of noise or of the sampling that generated the data. In this case, selecting a different split could be optimal. To address this concern, the use of option nodes was proposed [12].

An option node is introduced into the tree when it would be hard to determine the best split, i.e., when the best splits have similar heuristic values. When this occurs, instead of selecting only the best split, we select several of them. Specifically, we select up to 5 splits s, called *options*, that satisfy the following

$$\frac{\text{Heur}(s)}{\text{Heur}(s_{best})} \geq 1 - \varepsilon \cdot d^{level},$$

where s_{best} is the best split, ε determines how similar the heuristics must be, $d \in [0, 1]$ is a decay factor and *level* is the level in the tree of the node we are attempting to split. After we have determined the candidate splits, we introduce an option node whose children are split nodes obtained by using the selected splits, i.e., an option node contains options as its children. Selecting more than 5 options is possible, vbut, uses exponentially more resources and is not advised [12].

As usual, we define the level of a node to be the number of its ancestor nodes, however, we do not count option nodes. This is motivated by the predictive clustering viewpoint, i.e., the option nodes only mark that there are overlapping clusters and not that there is an additional level of clustering.

The use of a decay factor makes the selection criterion more stringent in the lower nodes of the tree. The intuition behind this is that higher up, the split

---

Algorithm 1. The top-down induction algorithm for option PCTs.

Procedure OptionPCT
Input: A dataset E, parameter ε, decay factor d, current tree level l
Output: An option predictive clustering tree
    $candidates = \text{FindBestTests}(E, 5)$
    if $|candidates| > 0$ **then**
        if $|candidates| = 1$ **or** $l > 2$ **then**
            $(t^*, h^*, \mathcal{P}^*) = candidates[0]$
            for each $E_i \in \mathcal{P}^*$ **do**
                $tree_i = \text{OptionPCT}(E_i, \varepsilon, d, l+1)$
            return node($t^*, \bigcup_i \{tree_i\}$)
        else
            $(t_0^*, h_0^*, \mathcal{P}_0^*) = candidates[0]$
            $nodes = \{\}$
            for each $(t_i^*, h_i^*, \mathcal{P}_i^*) \in candidates$ **do**
                if $\frac{h_i^*}{h_0^*} \geq 1 - \varepsilon \cdot d^l$ **then**
                    for each $E_j \in \mathcal{P}_i^*$ **do**
                        $tree_j = \text{OptionPCT}(E_j, \varepsilon, d, l+1)$
                    $nodes = nodes \cup \{\text{node}(t^*, \bigcup_j \{tree_j\})\}$
            if $|nodes| > 1$ **then**
                return option_node($nodes$)
            else
                return $nodes[0]$
    else
        return leaf(Prototype(E))

selection is more important and a larger error would be inferred by introducing a non-optimal split. However, as we get deeper into the tree, the use of a non-optimal split would make decreasing impact. This intuition also allows us to prohibit the use of option nodes on levels 3 and greater, which severely mitigates the problem of combinatorial explosion.

When using a small ε, e.g., $\varepsilon = 0.1$, we are selecting only options whose heuristics are within 10 % of the best split. However, the use of larger ε, in the extreme case even $\varepsilon = 1$, can also be motivated through the success of methods such as random forests and ensembles of extremely randomized trees. Allowing the selection of options whose heuristics are considerably worse than the heuristic of the best split, might not necessarily reduce the performance of the tree, but actually increase it.

Once an OPCT is built, we want to use it for prediction. If we reiterate the methodology described above in a tree-prediction setting, we could say the following. In a regular PCT, it is simple to produce a prediction for a new example. It is sorted into a leaf (according to the splits of the tree) where a prediction is made by using a prototype function. When traversing an example through an OPCT, we behave the same when we encounter a split or leaf node. If we traverse an example to an option node, however, we clone the example for each of the options and traverse one of the copies down each of the options. This means that in an option node an example is (by proxy of its copies) traversed to multiple leaves, where multiple predictions are produced. To obtain a single prediction in an option node, we aggregate the obtained predictions. When addressing multi-target regression this is generally done by averaging all the predictions per target.

An option tree is usually seen a single tree, however, it can also be interpreted as a compact representation of an ensemble. To generate the ensemble of the *embedded trees*, we start recursively from the root node and move in a top-down fashion. Each time we encounter an option node we copy the tree above (and in "parallel") for each of the options and replace the option node with only the option, i.e., single split. This produces one tree for each option while removing the option node in question. We repeat this procedure on all of the generated trees until we are left with no option nodes. This is illustrated in Fig. 1. For this reason, we will sometimes refer to option trees as pseudo-ensembles.

On the other hand, a given OPCT is a direct extension of the PCT that would be learned on the same data. By definition, whenever we introduce an option node, we include the best split (in terms of the heuristic)[1]. In regular construction of PCTs, this is the only split we consider. Consequently, the PCT is embedded in the OPCT. Specifically, we can extract it if we, in a top-down fashion, only select the best option in each option node.

Let's consider a full option node, i.e., one where we have the full 5 options. Let's assume that there are no option nodes further down in the tree. Since we have 5 options and no options lower in the tree, we have a total of 5 embedded

---

[1] Not only is the best split included, other splits are compared to it to determine their inclusion in the option tree.

trees. Now, let's consider the size of the embedded ensemble, when we add two such nodes under a split node, i.e., each leaf of a split node was extended into a full option node. To construct an embedded tree we can now choose one of the 5 options when the test is satisfied and one of 5 options when it is not. This results in 25 different embedded trees. It can easily be inferred, that to calculate the number of embedded trees for an option node we need just to sum up the number of embedded trees for each of its options. In a (binary) split node, however, we multiply the numbers of embedded trees of the subtree that satisfies the split and of the subtree that doesn't, to obtain the total number of embedded trees.

Given the construction constraints described above, we know that option nodes with up to 5 options can appear only on the first three levels, i.e., levels 0, 1 and 2. If we now consider a full option tree, we calculate a maximum of $(5^2 \cdot 5)^2 \cdot 5 = 5^7 = 78125$ embedded trees. However, many of these trees overlap to a large extent.

Note that a given example will not traverse the entire tree. For example, in Fig. 1, if an example reaches S_1 and is traversed into the left child L_1, the same result would happen in both the first and second embedded tree. The example is "agnostic" of any option nodes in the right child of S_1. Therefore, in a general option tree, a given example will visit only up to $5^3 = 125$ leaves, as it will only traverse down one side of the tree in each split node.

In other words, there are a maximum of 125 different predictions that would be aggregated in order to obtain the final prediction of an option tree constructed this way. This can be compared to a single tree, where only 1 prediction will be made for each example, or to a tree ensemble, where 1 prediction would be made for each member of the ensemble and then aggregated to produce the final prediction.

3   Experimental Design

To evaluate the performance and efficiency of the OPCT method, we construct OPCTs by using different values for the algorithms' parameters, as well as standard PCTs and ensembles of PCTs. We first present the benchmark datasets used for evaluation of the methods and then give the specific experimental setup: parameter instantiations and evaluation measures.

3.1   Data Description

The 11 datasets with multiple continuous targets used in this study come mainly from the domain of ecological modelling. Table 1 outlines the properties of the datasets. This selection contains datasets with various number of examples described with different numbers of attributes. For more details on the datasets, we refer the reader to the referenced literature.

3.2   Experimental Setup

We use 10-fold cross-validation to estimate the predictive performance of the used methods. We assess the predictive performance of the algorithms using several evaluation measures. In particular, since the task we consider is that of MTR, we employed three well known measures: the correlation coefficient (CC), root mean squared error ($RMSE$) and relative root mean squared error ($RRMSE$). We present here only the results in terms of $RRMSE$, but similar conclusions hold for the other two measures. Finally, the efficiency of the proposed methods is measured with the time needed to construct a model and the size of the models (in terms of total number of leaf nodes available for making a prediction).

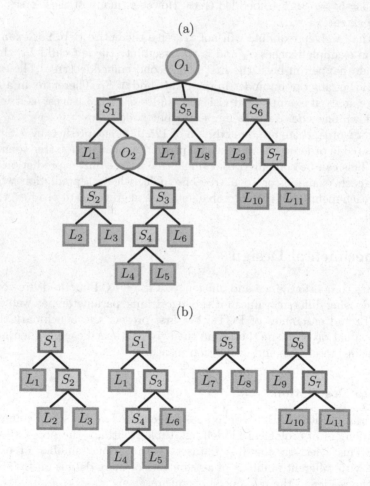

Fig. 1. An option tree (a) and the ensemble of its embedded trees (b). O_i are option nodes, S_j split nodes and L_k leaf nodes.

We parameterize OPCTs by selecting values for the parameters ε and d. When $\varepsilon = 1$, there are no constraints on the heuristic value of the selected splits with regards to the best test. However, since only the 5 best splits are selected, the risk that a split which would decrease the predictive performance would be selected is relatively low. This setting most resembles the ensemble setting, where greater variation is desired. If we were to select $\varepsilon = 0$, the resulting OPCT would almost always directly coincide with a regular PCT, as no split would likely reach exactly the same heuristic value as the best split. Hence, the only way an option node would be induced is if two splits had the exact same heuristic value. This configuration is not of interest, so for ε we consider the values $\{0.1, 0.2, 0.5, 1.0\}$, corresponding in order from the most stringent to the least stringent construction criterion.

As discussed above, the higher in the tree an option node is induced, the higher the variation in the learned subtrees. Induction of option nodes in lower levels of the tree not only contributes to the combinatorial explosion of the number of trees (and consequently the use of resources), but also generates less variation in the predictions, since the subtrees affected cover a smaller number of examples. Hence, we wish to curtail the number of options induced in option nodes lower in the tree. If we select a decay factor of $d = 1$, the depth of the option node induction will have no impact, while selecting a decay factor of 0.5 will effectively double the effective heuristic requirement $\varepsilon \cdot d^l$ at each level. For example, for $\varepsilon = 0.5$ and $d = 0.5$, the requirement would be 0.5 at the root level, 0.25 at the first level and 0.125 at the third level. We use the following values of the decay factor: $\{0.5, 0.9, 1\}$, these are the most to the least constrictive in terms of the construction of the OPCT.

Table 1. Properties of the datasets with multiple continuous targets (regression datasets): N is the number of instances, $|D|/|C|$ the number of descriptive attributes (discrete/continuous), and T the number of target attributes.

Name of dataset	N	$|D|/|C|$	T
Collembola [16]	393	8/39	3
EDM [17]	154	0/16	2
Forestry-Kras [3]	60607	0/160	11
Forestry-Slivnica-LandSat [18]	6218	0/150	2
Forestry-Slivnica-IRS [18]	2731	0/29	2
Forestry-Slivnica-SPOT [18]	2731	0/49	2
Sigmea real [19]	817	0/4	2
Soil quality [2]	1944	0/142	3
Vegetation clustering [20]	29679	0/65	11
Vegetation condition [4]	16967	1/39	7
Water quality [21]	1060	0/16	14

Note that, on a given dataset, all of the values of ε and d could produce the same OPCT, if the splits have very similar heuristic values, e.g., there could always be 5 splits that are within $10\% \cdot d^l$ of the heuristic value of the best split. Therefore, the evaluation of how ε and d affect both the predictive performance and efficiency must by design be evaluated on multiple datasets. The parameterized version of the OPCT method for a given ε and d is denoted OPCTeεdd, e.g., OPCTe0.5d0.9.

Next, we define the parameter values used in the algorithms for constructing single PCTs and ensembles of PCTs. The multi-target PCTs are obtained using F-test pruning. This pruning procedure uses the exact Fisher test to check whether a given split/test in an internal node of the tree results in a reduction in variance that is statistically significant at a given significance level. If there is no split/test that can satisfy this, then the node is converted to a leaf. An optimal significance level was selected by using internal 3-fold cross validation, from the following values: 0.125, 0.1, 0.05, 0.01, 0.005 and 0.001.

We consider two ensemble learning techniques: bagging [22] and random forests [23]. These are the most widely used tree-base ensemble learning methods. The construction of both ensemble methods takes as an input parameter the size of the ensemble, i.e., number of base predictive models to be constructed. We constructed ensembles with 100 base predictive models [1]. Furthermore, the random forests algorithm takes as input the size of the feature subset that is randomly selected at each node. For this purpose, we apply the logarithmic function of the number of descriptive attributes $\lfloor \log_2 |D| \rfloor + 1$, as recommended by Breiman [23].

In order to assess the statistical significance of the differences in performance of the studied algorithms, we adopt the recommendations by Demšar [24] for the statistical evaluation of the results. In particular, we use the Friedman test for statistical significance. Afterwards, to detect where statistically significant differences occur (i.e., between which algorithms), we use the Nemenyi post-hoc test to compare all the methods among each other.

We present the results from the statistical analysis with *average ranks diagrams*. The diagrams plot the average ranks of the algorithms and connect those whose average ranks differ by less than a given value, called *critical distance*. The critical distance depends on the level of the statistical significance (set in our case to 0.05). The difference in performance of the algorithms connected with a line is not statistically significant at the given significance level.

4   Results and Discussion

We discuss the results from the experimental evaluation along three major dimensions. First, we select the optimal parameter values for the construction of OPCTs. Second, we compare the performance of OPCTs with a single PCT and ensembles of PCTs. The comparison is performed on their predictive power, time consumption and size of the produced models. Third, we examine the interpretability of OPCTs in juxtaposition to PCTs.

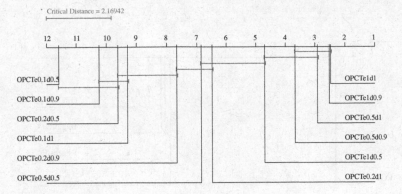

Fig. 2. Average rank diagram (in terms of predictive performance) for OPCTs obtained with different values of the parameters ε (e) and d.

4.1 Parametrization of OPCTs

In Fig. 2, we depict the performance of the different OPCTs obtained by using the experimental design outlined above. We can note that lower values for both the parameters leads to performance degradation. This is somewhat expected, because by imposing more stringent values, we are forcing the algorithm not to introduce option nodes, hence the obtained OPCTs are small. On the opposite side of the spectrum, we have the OPCTs obtained with larger values for the parameters. These achieve the best predictive performance and the corresponding OPCTs are large.

We can also observe that the ε parameter has a stronger influence on the performance than d. Namely, the OPCTs constructed with selecting 0.1 and 0.2 as values for ε (no matter the value of d) are the ones with the weakest predictive power. Furthermore, we can also note that the larger values for d also lead to better predictive performance. The best predictive performance is obtained when setting both parameters to 1. Such a setting produces large OPCTs and requires the most time to construct the OPCTs. All in all, we can recommend the use of two instantiations of the OPCT algorithm: OPCTe1d1 ($\varepsilon = 1$, $d = 1$) and OPCTe0.5d1 ($\varepsilon = 0.5$, $d = 1$). These two variants represent the parameters for the best performance and the trade-off between predictive power and efficiency, respectively.

4.2 Predictive Performance and Efficiency

Figure 3 shows the results of the statistical evaluation of the predictive performance of the proposed OPCT method in comparison to a single PCT and ensembles of PCTs. We can first note that both OPCTs (OPCTe1d1 and OPCTe0.5d1) achieve statistically significantly better predictive performance than the single PCT. Second, there is no statistically significant difference in the performance of the ensemble methods and the OPCTs. Furthermore, the two ensemble methods are bounded by the two OPCT variants: the best performing method is the

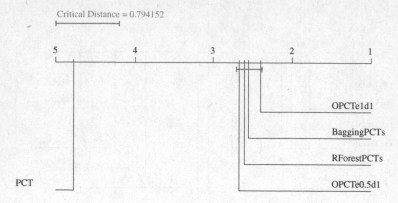

Fig. 3. Comparison of OPCTs predictive performance of OPCTs to the competing methods: average rank diagram in terms of predictive performance.

OPCTe1d1, followed by Bagging of PCTs, Random forests of PCTs, and the OPCTe0.5d1 method.

Next, we compare the methods in terms of their efficiency. First of all, learning PCTs is the most efficient method both in terms of time needed for model construction and model size. More specifically, single PCT models are statistically significantly smaller than all of the other models and are obtained statistically significantly faster than Bagging of PCTs and OPCTe1d1.

If we focus on the efficiency of the OPCT models, OPCTe1d1 is the least efficient: it produces models with largest numbers of leaves and takes the most time for model construction. Recall that this is due to the fact that, in the worst case, OPCTe1d1 is actually a condensed representation of what amounts to 125 PCTs. We also note that reducing the value of the ε parameter to 0.5 yields smaller models that are constructed faster than both bagging and random forests of PCTs. All in all, OPCTe0.5d1 offers the best trade-off between predictive performance and efficiency and should be considered for further use. However, if a given task requires the best predictive performance then one should use OPCTe1d1 (Fig. 4).

4.3   Interpretability of OPCTs

OPCTs, as well as option trees in general, offer a much higher degree of interpretability than ensemble methods. This is expressed through both the fact that the "ensemble" of an option tree is represented in a compact form, i.e., a single tree, as well as the fact that many of the embedded trees overlap. Additionally, the regular PCT that would be learned from the same data is always present in the OPCT, as described in Sect. 2. A PCT and an OPCT learned on the EDM dataset, and their relations are illustrated in Fig. 5.

Providing a domain expert with an option tree gives them a lot of choices with regards to the model. They can observe the selected options and attempt to determine which of the selected options were selected due to their actual

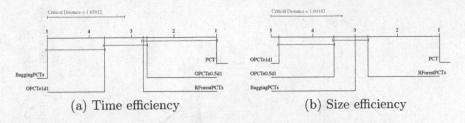

(a) Time efficiency   (b) Size efficiency

Fig. 4. Comparison of the efficiency of OPCTs to that of the competing methods as measured by the time needed to learn a model and the size of the models (number of leaf nodes).

importance as opposed to the sampling of the dataset or other artifacts of the data. If they are able to discard all but one of the options in each of the option nodes, we can collapse the OPCT into a single tree, specifically a PCT, which corresponds not only to the data, but also to the knowledge of the domain expert. In terms of the predictive clustering framework, this means selecting only one of the overlapping hierarchical clusters when multiple clusters are presented. This approach also has the advantage that the domain expert need not be available for interaction when the model is learned, but can assess the OPCT and chose the preferred options later on.

This process can also be looked at through a different lens. As we have introduced option trees (in part) to address myopia, by considering more options and later deciding on only one of them in each option node, we are essentially "looking ahead" of just the one split and utilizing, in the case of the domain expert, additional domain knowledge.

However, instead of using domain knowledge by proxy of interaction with a domain expert, we could also collapse the learned OPCT to a single PCT by using additional unseen data, i.e., by calculating an unbiased estimate of the predictive performance of the different options present in the OPCT. Since the collection and preparation of additional data examples could be expensive to the point of infeasibility, we could introduce a modified experimental setup. Part of the training data could be separated into a validation set which would not be used for the initial learning of the OPCT, but would be utilized to determine which of the selected options, and consequently embedded trees, has the best predictive performance. We would then collapse the OPCT into a PCT according to this validation set, after which we would test the obtained PCT on the test set. In this scenario, we could not only study the effect of myopia by comparing the collapsed OPCT to a PCT learned on the entire training dataset, but also observe the effect of averaging multiple predictions on the predictive performance by comparing the collapsed PCT and the original (pseudo-ensemble) OPCT.

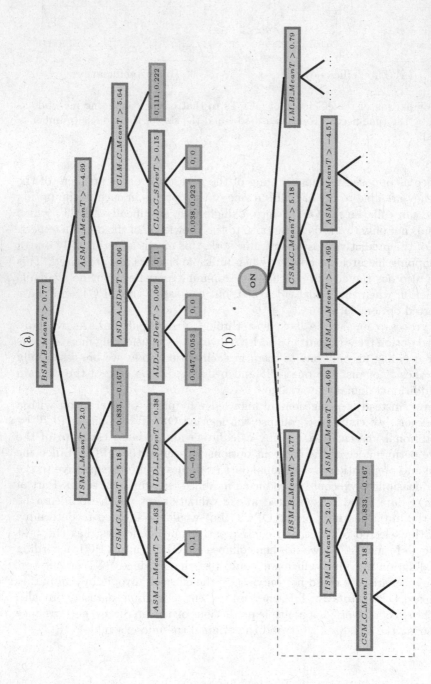

Fig. 5. A regular PCT (a) and an OPCT (b) learned on the EDM dataset. The left child of a split node corresponds to the subtree where the test is satisfied. Note that the regular PCT is included in the OPCT as the subtree enclosed in the dashed rectangle.

5 Conclusions

In this work, we propose an algorithm for learning option predictive clustering trees (OPCTs) for the task of multi-target regression (MTR). In contrast to standard regression, where the output is a single scalar value, in MTR the output is a data structure – a tuple/vector of continuous variable values. We consider learning of a global model, i.e., of a single model that predicts all of the target variables simultaneously.

More specifically, we propose OPCTs to address the myopia of the standard greedy PCT learning algorithm. OPCTs have the possibility to construct option nodes, i.e., nodes with a set of alternative sub-nodes, each containing a different split. These option nodes are constructed in the cases when the heuristic scores of the candidate splits are close to each other. Furthermore, OPCTs, and option trees in general, can be regarded as a condensed representation of an ensemble of trees.

The proposed method was experimentally evaluated on 11 benchmark MTR datasets. We first determined the optimal parameters of the algorithm. The results show that both parameters that control the number of option nodes (the range of heuristic scores considered ε and the decay factor d) need larger values to achieve better predictive performance. We then compared the performance of learning OPCTs (in terms of predictive power and efficiency) to the induction of standard PCTs and to two ensemble learning methods, i.e., bagging and random forests of PCTs.

The evaluation revealed that OPCTs yield statistically significantly better predictive performance than a single PCT. Next, the predictive performance of the OPCTs is not statistically significantly different than that of the other two ensemble methods, but OPCT with ε and d set to 1 achieves the best predictive performance on average. Moreover, in terms of efficiency, an OPCT with ε set to 0.5 and d set to 1 is faster to learn that the ensemble models and the OPCT with ε and d set to 1T. Finally, through an example, we illustrated the interpretability of the constructed OPCTs: they offer a multifaceted view on the data at hand.

We plan to extend this work along several directions. First of all, we will evaluate the OPCTs in the single tree context, i.e., we will use the induction of OPCTs as a beam-search algorithm for tree induction. Next, we will evaluate the influence of the two parameters at a more fine grained resolution. Finally, we will extend the algorithm towards other output types, i.e., machine learning tasks, such as multi-label classification, hierarchical multi-label classification and time series prediction.

Acknowledgments. We acknowledge the financial support of the European Commission through the grants ICT-2013-612944 MAESTRA and ICT-2013-604102 HBP, as well as the support of the Slovenian Research Agency through a young researcher grant and the program Knowledge Technologies (P2-0103).

References

1. Kocev, D., Vens, C., Struyf, J., Džeroski, S.: Tree ensembles for predicting structured outputs. Pattern Recogn. **46**(3), 817–833 (2013)
2. Demšar, D., Džeroski, S., Larsen, T., Struyf, J., Axelsen, J., Bruns-Pedersen, M., Krogh, P.H.: Using multi-objective classification to model communities of soil. Ecol. Model. **191**(1), 131–143 (2006)
3. Stojanova, D., Panov, P., Gjorgjioski, V., Kobler, A., Džeroski, S.: Estimating vegetation height and canopy cover from remotely sensed data with machine learning. Ecol. Inform. **5**(4), 256–266 (2010)
4. Kocev, D., Džeroski, S., White, M., Newell, G., Griffioen, P.: Using single- and multi-target regression trees and ensembles to model a compound index of vegetation condition. Ecol. Model. **220**(8), 1159–1168 (2009)
5. Tsoumakas, G., Spyromitros-Xioufis, E., Vrekou, A., Vlahavas, I.: Multi-target regression via random linear target combinations. In: Calders, T., Esposito, F., Hüllermeier, E., Meo, R. (eds.) ECML PKDD 2014. LNCS (LNAI), vol. 8726, pp. 225–240. Springer, Heidelberg (2014). doi:10.1007/978-3-662-44845-8_15
6. Bakır, G.H., Hofmann, T., Schölkopf, B., Smola, A.J., Taskar, B., Vishwanathan, S.V.N.: Predicting Structured Data. Neural Information Processing. The MIT Press, Cambridge (2007)
7. Kocev, D., Ceci, M.: Ensembles of extremely randomized trees for multi-target regression. In: Japkowicz, N., Matwin, S. (eds.) DS 2015. LNCS (LNAI), vol. 9356, pp. 86–100. Springer, Heidelberg (2015). doi:10.1007/978-3-319-24282-8_9
8. Struyf, J., Džeroski, S.: Constraint based induction of multi-objective regression trees. In: Džeroski, S., Struyf, J. (eds.) KDID 2006. LNCS, vol. 3933, pp. 222–233. Springer, Heidelberg (2006). doi:10.1007/11733492_13
9. Appice, A., Džeroski, S.: Stepwise induction of multi-target model trees. In: Kok, J.N., Koronacki, J., Mantaras, R.L., Matwin, S., Mladenič, D., Skowron, A. (eds.) ECML 2007. LNCS (LNAI), vol. 4701, pp. 502–509. Springer, Heidelberg (2007). doi:10.1007/978-3-540-74958-5_46
10. Kocev, D., Struyf, J., Džeroski, S.: Beam search induction and similarity constraints for predictive clustering trees. In: Džeroski, S., Struyf, J. (eds.) KDID 2006. LNCS, vol. 4747, pp. 134–151. Springer, Heidelberg (2007). doi:10.1007/978-3-540-75549-4_9
11. Buntine, W.: Learning classification trees. Stat. Comput. **2**(2), 63–73 (1992)
12. Kohavi, R., Kunz, C.: Option decision trees with majority votes. In: Proceedings of the 14th International Conference on Machine Learning, ICML 1997, pp. 161–169. Morgan Kaufmann Publishers Inc., San Francisco (1997)
13. Blockeel, H., Struyf, J.: Efficient algorithms for decision tree cross-validation. J. Mach. Learn. Res. **3**, 621–650 (2002)
14. Ikonomovska, E., Gama, J., Ženko, B., Džeroski, S.: Speeding-up hoeffding-based regression trees with options. In: Proceedings of the 28th International Conference on Machine Learning, ICML 2011, pp. 537–544 (2011)
15. Breiman, L., Friedman, J., Olshen, R., Stone, C.J.: Classification and Regression Trees. Chapman & Hall/CRC, Boca Raton (1984)
16. Kampichler, C., Džeroski, S., Wieland, R.: Application of machine learning techniques to the analysis of soil ecological data bases: relationships between habitat features and Collembolan community characteristics. Soil Biol. Biochem. **32**(2), 197–209 (2000)

17. Karalič, A.: First order regression. Ph.D. thesis, Faculty of Computer Science, University of Ljubljana, Ljubljana, Slovenia (1995)
18. Stojanova, D.: Estimating forest properties from remotely sensed data by using machine learning. Master's thesis, Jožef Stefan IPS, Ljubljana, Slovenia (2009)
19. Demšar, D., Debeljak, M., Džeroski, S., Lavigne, C.: Modelling pollen dispersal of genetically modified oilseed rape within the field. In: The Annual Meeting of the Ecological Society of America, p. 152 (2005)
20. Gjorgjioski, V., Džeroski, S., White, M.: Clustering analysis of vegetation data. Technical report 10065, Jožef Stefan Institute (2008)
21. Džeroski, S., Demšar, D., Grbović, J.: Predicting chemical parameters of river water quality from bioindicator data. Appl. Intell. **13**(1), 7–17 (2000)
22. Breiman, L.: Bagging predictors. Mach. Learn. **24**(2), 123–140 (1996)
23. Breiman, L.: Random forests. Mach. Learn. **45**(1), 5–32 (2001)
24. Demšar, J.: Statistical comparisons of classifiers over multiple data sets. J. Mach. Learn. Res. **7**, 1–30 (2006)

HSIM: A Supervised Imputation Method for Hierarchical Classification Scenario

Leandro R. Galvão$^{(\boxtimes)}$ and Luiz H.C. Merschmann

Computer Science Department, Federal University of Ouro Preto, Ouro Preto, Brazil
leandrodvmg@gmail.com, luizhenrique@iceb.ufop.br

Abstract. The missing value imputation process can be defined as a preprocessing step that fills missing values of attributes in incomplete datasets. Nowadays, the problem of incomplete datasets in the hierarchical classification scenario must be solved using unsupervised missing value imputation methods due to the lack of supervised methods to deal with the hierarchical context. Thus, in this work, we propose and evaluate a supervised missing value imputation method for datasets used in hierarchical classification problems in which the classes are organized into tree structure. Experiments were performed on incomplete datasets to evaluate the effect of the proposed missing value imputation method on classification performance when using a global hierarchical classifier. The results showed that, using the proposed method for dealing with missing attribute values, it provided higher classifier predictive performance than other unsupervised missing value imputation methods.

Keywords: Missing attribute value imputation · Hierarchical classification · Data mining

1 Introduction

The technological advances in the last decades have allowed the production and storage of a huge amount of data related to different types of applications. The transformation of this data in useful, valid and understandable information is essential. However, this is not an easy task, requiring automated strategies to analyse the data [1]. Thus, the Knowledge Discovery from Data (KDD) process adopted for this purpose is basically composed by three main steps: data preprocessing, data mining and results validation.

The missing value imputation process can be defined as a preprocessing step that fills missing values of attributes in incomplete datasets [2]. Attribute's missing value can occur in a dataset for several reasons, such as filling failure, omission of data by respondents in survey questions or even a failure on the sensor responsible to collect the data. Missing values can make the manipulation of datasets more complex and reduce the efficiency of data mining algorithms [3].

Classification is a data mining task that aims to identify an instance's class through its characteristics [1]. Different types of classification problems can be

© Springer International Publishing Switzerland 2016
T. Calders et al. (Eds.): DS 2016, LNAI 9956, pp. 134–148, 2016.
DOI: 10.1007/978-3-319-46307-0_9

found in the literature, each one with its own complexity level [4]. In flat classification problems each instance is assigned to a class, in which classes do not have relationships to each other. Nevertheless, there are more complex classification problems, known as hierarchical classification problems, in which the classes are hierarchically organized.

Several application domains such as text categorization [5,6], protein function prediction [7,8], music genre classification [9,10], image classification [11,12] and emotional speech classification [13] can benefit from hierarchical classification techniques, since the classes to be predicted are naturally organized into class hierarchies. Despite of some works have ignored the class hierarchy and performed predictions considering only leaf node classes (flat classification approach), hierarchical classification methods are overall better than flat classification methods when solving hierarchical classification problems [4]. Therefore, hierarchical classification is a research topic that certainly deserves attention.

In literature, it is possible to find several works that deal with missing attribute values. Expectation Maximization and KNNImpute are examples of popular methods often used to handle missing attribute values [14–16]. Since they are unsupervised methods (ignore the target class values), they can be applied to datasets used in flat or hierarchical classification scenario. Thus, in hierarchical classification context, due to the lack of suitable supervised missing value imputation methods (able to take into account the class relationships in the target problem), the researchers are limited to use unsupervised missing value imputation techniques. [17–19] are examples of works that have used an unsupervised missing value imputation method for hierarchical classification context.

Therefore, in this work, we fill this gap by presenting and evaluating a supervised missing value imputation method for datasets for hierarchical classification task. Experiments performed on incomplete datasets using a global hierarchical classifier showed that the method proposed to deal with missing attribute values provided higher classifier predictive performance than other popular unsupervised missing value imputation methods.

2   Background

2.1   Hierarchical Classification

Several classification problems are available in literature wherein the huge majority of them deals with the flat classification scenario. In flat classification problems there is no relationship among the classes. However, there are more complex classification problems where the classes are hierarchical organized as a tree or DAG (Direct Acyclic Graph). This group of problems is known as hierarchical classification problems [4].

Hierarchical classification methods can be analysed according to different aspects. The first aspect is related to the hierarchical structure the method is able to deal with. The classes can be hierarchically organized in a tree or DAG (Direct Acyclic Graph) structure. The main difference between these structures

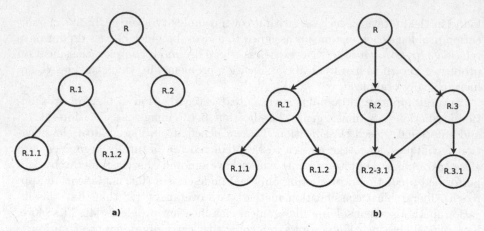

Fig. 1. (a) Class hierarchy structured as tree; (b) Class hierarchy structured as DAG.

is related to the number of parents of a class. The tree structure restricts each class to possess only one parent class. However, in the DAG structure a class (node) is allowed to have more than one parent class. Figure 1 presents a tree and a DAG example, where the nodes represent the classes and the edges indicate a relationship between them.

The second aspect is related to the prediction depth of the method. Thus, hierarchical classification problems can either be organized in mandatory leaf node prediction (MLNP) or non-mandatory leaf node prediction (NMLNP). In the mandatory leaf node prediction, the output of a classifier is always a class represented by a leaf node of the class hierarchy. For the non-mandatory leaf node prediction, the most specific class predicted by the classifier can be represented by a node at any level (internal or leaf) of the class hierarchy.

The third aspect refers to the number of different paths in the hierarchy a method can assign to a given instance. This aspect defines two different types of problems: single path of labels (SPL) and multiple path of labels (MPL). Single path problems restrict each instance to be assigned to at most one path of predicted labels whilst multiple path problems allow an instance to be assigned to multiple paths of predicted labels.

Finally, the fourth aspect concerns how the hierarchical structure is handled by the method. In [4], three approaches are listed: flat classification, which do not take into account the class hierarchy and performs predictions considering only the leaf node classes; local model approaches, when a group of flat classifiers are employed; and global model approaches, when a single classifier is built considering the class hierarchy as a whole.

Several works in literature consist of proposals to modify existing flat classifiers to cope with the entire class hierarchy in a single step and, therefore, creating global classifiers. Some examples of modifications of traditional flat classification algorithms are: HC4.5 [20] and HLC [21] (modified versions of the C4.5), Global Model Naive Bayes (modified version of the Naive Bayes) [22],

Clus-HMC (based on Predictive Cluster Trees) [23,24] and hAnt-Miner (adaptation of Ant-Miner algorithm) [25]. Given the relevance of global classifiers to the hierarchical classification scenario, one can see that it is important the development of preprocessing techniques to deal with the class hierarchy as a whole. Thus, in this work, we propose a supervised missing value imputation method for datasets used by global hierarchical classifiers.

2.2 Missing Value Imputation

Missing value imputation can be defined as the estimation of values based on the analysis of the known values of the attribute [26]. The missing values in data can be the result of failure during the data collection process, accidental removal of values or refusal to answer questions in surveys used to collect data. Whatever the reasons, missing attribute values can pose an obstacle for classification and other data mining tasks.

The missing value imputation methods can be categorized according to different criteria [2]. Methods that use the target class values during the imputation process are categorized as supervised, while methods that ignore the target class values are categorized as unsupervised. While univariate methods impute missing values on each attribute for each instance based on observed values for other instances on same attribute, multivariate methods impute missing values on each attribute for each instance based on observed values for same instance on other attributes.

The strategies used to impute missing attribute values vary from simple statistical algorithms like mean imputation to more complex approaches based on predictive models.

Mean imputation, one of the easiest ways to impute missing values, works by replacing each missing value with mean of observed values of that attribute for other instances. Despite its simplicity, this method changes the distribution of attribute, producing different kinds of bias.

Expectation Maximization [27] algorithm is a popular method to impute missing attribute values. It is an iterative refinement method that assumes the data are distributed based on a parametric model with unknown parameters. Basically, in each iteration, it estimates missing values and parameters model. After parameters model initialization, it has an Expectation step, where a missing attribute value for an instance i is substituted by its expected value computed from the estimates for parameters model and observed values for same instance i on other attributes. Then, in the Maximization step, parameters model are updated in order to maximize the complete data likelihood. These two steps are iteratively repeated until convergence is obtained. The Expectation Maximization method is categorized as unsupervised and multivariate.

KNNImpute [15] is another widely used method to impute missing attribute values. In short, it selects the k nearest instances (neighbours) from dataset instances with known value in the attribute to be imputed. Aiming at finding the nearest neighbours a distance measure (e.g., Euclidean distance) must be adopted. Then, the missing attribute value is replaced by a value calculated from the k selected neighbours. Usually, mode (for categorical data) and mean (for continuous data) are used to compute the replacement value. The KNNImpute method is categorized as unsupervised and multivariate.

3   Proposed Method

The missing value imputation method proposed in this work will be referenced as $\underline{H}ierarchical$ $\underline{S}upervised$ $\underline{I}mputation$ $\underline{M}ethod$ (HSIM). The HSIM is a supervised method able to deal with datasets containing classes hierarchically organized in a tree structure.

The motivation behind the HSIM is to take into account the class hierarchy to impute missing attribute values. In short, whenever there are no known attribute values for the instances associated to a particular class, the main idea of the proposed method is replacing each missing value with mean (for continuous data) or mode (for categorical data) of observed values of that attribute for other instances associated to a class descendant or ascendant of the class of the instance containing the missing value.

An example is now presented to illustrate the operation of the proposed method. Consider the incomplete dataset shown in Table 1, where the instances 2, 7, 9 and 13 contain missing value for attribute F. For each of these instances, a different strategy is adopted by HSIM to impute missing attribute value.

Note that instance 2 is associated to the class R.2, that is also the class associated to other instances (3, 6 and 10) containing observed values of the attribute F. In this case, the missing value is substituted by the average of the

Table 1. Incomplete dataset

ID	F	Class attribute
1	0.15	R.1.1
2	?	R.2
3	3.41	R.2
4	4.12	R.2.1
5	0.22	R.1.2
6	3.5	R.2
7	?	R.1
8	0.34	R.1.2
9	?	R.3.1.1
10	4.1	R.2
11	1.6	R.3.1
12	1.55	R.3
13	?	R.4
14	1.71	R.3.1

three observed values (3.41, 3.5 and 4.1) of the instances associated to the same class of the instance containing the missing value (R.2). Thus, the attribute value of instance 2 is imputed as 3.67.

In the case of instance 7, where the missing value is associated to the class R.1, since there are no other instances associated to that class, the missing value is imputed with the average of the observed values (0.15, 0.22 and 0.34) of instances associated to descendant classes of the class associated to the instance containing the missing value. Thus, the attribute value of the instance 7 is imputed as 0.24.

The previous strategies are not applicable to the case of instance 9, as the missing value is associated to the class R.3.1.1, which is not associated to any other dataset instance and neither is an ascendant class of any class associated to an instance with observed value of the attribute F. Then, the missing value is replaced with mean of the observed values (1.6, 1.55 and 1.71) of instances associated to ascendant classes of the class associated to the instance containing the missing value. Thus, the attribute value of the instance 9 is imputed as 1.62.

Finally, instance 13 has a missing value associated to the class R.4, which is not associated to any other dataset instance and neither is an ascendant or descendant class of any class associated to an instance with observed value of the attribute F. Then, the missing value is replaced with mean of observed values of attribute F for all other instances. In this example, the attribute value of the instance 13 is imputed as 2.07.

Table 2 shows the complete dataset achieved after application of the HSIM on the incomplete dataset presented in Table 1. In this table, the imputed attribute values are highlighted in bold. The steps of the proposed method described in the Algorithm 1 are detailed next.

Algorithm 1 describes the steps of the proposed method. First, HSIM receives as input a dataset represented by an $M \times N$ matrix, where M is the number of instances and N is the total number of attributes (predictive attributes and class attribute). In addition, we consider that the last matrix column is the class attribute. By scanning the data matrix (lines 1 and 2), whenever a missing value for attribute j in instance i is found (line 3), four empty vectors ($sameClass$, $descendantClass$, $ascendantClass$ and $differentClass$) are initialized (lines 4, 5, 6 and 7). Then, all known values for attribute j associated to the same class of the instance i (line 10) are stored in the vector $sameClass$ (line 11). Similarly, all known values for attribute j associated to a class descendant of the class of the instance i (line 12) are stored in the vector $descendantClass$ (line 13). In the same way, all known values for attribute j associated to a class ascendant of the class of the instance i (line 14) are stored in the vector $ascendantClass$ (line 15). The remaining known values for attribute j associated to a class non-descendant and non-ascendant of the class of the instance i are stored in the vector $differentClass$ (line 17). After, if the vector $sameClass$ has some element, the average or mode of the $sameClass$ elements is used to impute the missing value (lines 23 and 25). Otherwise, if the vector $descendantClass$ is not empty, the average or mode of the $descendantClass$ elements is used to impute the missing value (lines 29 and 31).

Table 2. Complete dataset

ID	F	Class attribute
1	0.15	R.1.1
2	**3.67**	R.2
3	3.41	R.2
4	4.12	R.2.1
5	0.22	R.1.2
6	3.5	R.2
7	**0.24**	R.1
8	0.34	R.1.2
9	**1.62**	R.3.1.1
10	4.1	R.2
11	1.6	R.3.1
12	1.55	R.3
13	**2.07**	R.4
14	1.71	R.3.1

Alternatively, when *sameClass* and *descendantClass* vectors are empty, if the vector *ascendantClass* is not empty, the average or mode of the *ascendantClass* elements is used to impute the missing value (lines 35 and 37). Finally, whenever *sameClass*, *descendantClass* and *ascendantClass* vectors are empty, the average or mode of the *differentClass* elements is used to impute the missing value (lines 41 and 43).

4   Computational Experiments

4.1   Datasets and Experimental Setup

Computational experiments were carried out on incomplete datasets to evaluate the effect of the proposed missing value imputation method on classification performance when using a global hierarchical classifier. Thus, the proposed method (HSIM) was compared against the following popular unsupervised missing value imputation methods: Mean Imputation (MI), Expectation Maximization (EM) and KNNImpute. Since there are no supervised missing value imputation methods proposed in literature for hierarchical classification context, we consider that it is fair to have a comparison of proposed method against unsupervised methods, given that they can be directly applied for datasets used in hierarchical classification scenario.

While KNNImpute algorithm was implemented in C++ programming language, for MI and EM, the experiments were executed using the WEKA [28]

Algorithm 1. HSIM

Input: DB matrix ; //Incomplete Dataset
Output: DB_full matrix ; //Complete Dataset
1: **for** $j = 1; j < N - 1; j + +$ **do**
2:    **for** $i = 1; i <= M; i + +$ **do**
3:       **if** $DB[i][j] ==$ "?" **then**
4:          sameClass $\leftarrow \emptyset$
5:          descendantClass $\leftarrow \emptyset$
6:          ascendantClass $\leftarrow \emptyset$
7:          differentClass $\leftarrow \emptyset$
8:          **for** $k = 1; k <= M; k + +$ **do**
9:             **if** $k \neq i$ and $DB[k][j] \neq$ "?" **then**
10:                **if** $DB[k][N] == DB[i][N]$ **then**
11:                   sameClass.insert($DB[k][j]$);
12:                **else if** $isDescendant(DB[k][N], DB[i][N])$ **then**
13:                   descendantClass.insert($DB[k][j]$);
14:                **else if** $isAscendant(DB[k][N], DB[i][N])$ **then**
15:                   ascendantClass.insert($DB[k][j]$);
16:                **else**
17:                   differentClass.insert($DB[k][j]$);
18:                **end if**
19:             **end if**
20:          **end for**
21:          **if** $size(sameClass) > 0$ **then**
22:             **if** $is\_continuous(j)$ **then**
23:                $DB\_full[i][j] =$ average(sameClass);
24:             **else**
25:                $DB\_full[i][j] =$ mode(sameClass);
26:             **end if**
27:          **else if** $size(descendantClass) > 0$ **then**
28:             **if** $is\_continuous(j)$ **then**
29:                $DB\_full[i][j] =$ average(descendantClass);
30:             **else**
31:                $DB\_full[i][j] =$ mode(descendantClass);
32:             **end if**
33:          **else if** $size(ascendantClass) > 0$ **then**
34:             **if** $is\_continuous(j)$ **then**
35:                $DB\_full[i][j] =$ average(ascendantClass);
36:             **else**
37:                $DB\_full[i][j] =$ mode(ascendantClass);
38:             **end if**
39:          **else**
40:             **if** $is\_continuous(j)$ **then**
41:                $DB\_full[i][j] =$ average(differentClass);
42:             **else**
43:                $DB\_full[i][j] =$ mode(differentClass);
44:             **end if**
45:          **end if**
46:       **else**
47:          $DB\_full[i][j] = DB[i][j]$
48:       **end if**
49:    **end for**
50: **end for**

---

implementations, named ReplaceMissingValues and EMImputation, respectively. For KNNImpute, the experiments were conducted by varying the parameter k (number of nearest instances considered to impute missing values) between 10 % and 50 % of the number of instances in the dataset, in increments of 10 %. Since the best results were obtained for $k = 10$ %, it was adopted to obtain the results presented here. For EM, WEKA's default parameters were used.

Experiments were conducted by running both the proposed and the baseline methods on 8 bioinformatics datasets related to gene functions of yeast. In these datasets, the predictor attributes include the following types of bioinformatics data: secondary structure, phenotype, homology, sequence statistics, and expression. In addition, the classes to be predicted are hierarchically organized in a tree structure. The datasets, initially presented in [20], were multi-label data. Since in this work we focus on single path label scenario, before running the missing value imputation algorithms, the datasets were converted into single label data by choosing one class for each instance. This process consisted of selecting, for each instance, the more frequent class in the original dataset. HSIM implementation and the single label datasets are available at https://github.com/leandrodvmg/HSIM. Table 3 provides the main characteristics of the datasets used in the experiments. This table shows, for each dataset, its number of predictive attributes, number of instances, number of classes at each hierarchy level (1st/2nd/3rd/4th/5th/6th levels), percentage of instances containing at least one missing attribute value and percentage of missing data in the $M \times P$ matrix, where M is the number of instances and P is the total number of predictive attributes.

After all datasets were processed by both the proposed and the baseline missing value imputation methods, the imputation quality was measured by running a global hierarchical classifier on these datasets. However, before running the classifier, as in [7,9,29], an unsupervised discretization algorithm based on equal-frequency binning (using 20 bins) was applied to continuous attributes.

Table 3. Characteristics of the datasets

Dataset	# Attributes Categorical / Continuous	# Instances	# Classes per level	% Incomplete instances	% Missing values
CellCycle	0 / 77	3758	8/37/73/46/25/2	93.45	5.57
Church	1 / 26	3756	8/37/72/47/25/2	61.76	9.65
Eisen	0 / 79	2425	5/26/55/34/22/2	75.45	1.93
Expr	4 / 547	3780	8/37/73/46/26/2	100.00	8.90
Gasch1	0 / 173	3765	8/37/73/46/26/2	86.34	2.27
Gasch2	0 / 52	3780	8/37/73/46/26/2	61.15	3.55
Sequence	5 / 473	3920	8/37/73/46/26/2	0.66	0.01
SPO	3 / 77	3704	8/37/73/46/26/2	99.43	2.28

The Global-Model Naive Bayes (GMNB) [22], an extension of the flat classifier Naive Bayes to deal with hierarchical classification problems, was the global hierarchical classifier adopted in these experiments. It makes possible predictions at any level of the class hierarchy. In order to evaluate the predictive performance of the hierarchical classifier GMNB, we used the 10-fold cross validation method [1] and the hierarchical F-measure, an adaptation of the flat F-measure customized for hierarchical classification scenario. For each dataset, the same ten folds were used in the evaluation of the GMNB classifier.

4.2   Computational Results

As mentioned earlier, the objective of experiments was to compare the hierarchical classifier performances when running on datasets preprocessed using different missing value imputation methods. More specifically, the HSIM method was compared against each one of the baseline methods (Mean Imputation, Expectation Maximization and KNNImpute). Therefore, for each dataset, in order to determine if there is a statistically significant difference between the F-measures of the GMNB classifier when running on the dataset preprocessed by HSIM and by other baseline method, we have used the Wilcoxon's Signed-Rank Test (two-sided test) with Bonferroni adjustment on the results as we are making many-to-one comparisons [30]. This statistical test was applied with 95 % of confidence level.

Experimental results are shown in Table 4 for each dataset listed in the first column. This table shows, from the second to fifth column, the average hierarchical F-measure (hF) achieved by GMNB classifier (with standard deviation in parentheses) when running on each dataset preprocessed by the missing value imputation methods Mean Imputation (MI), Expectation Maximization (EM), KNNImpute (KNN) and HSIM, respectively. In bold we mark the best result achieved for each dataset. In addition, the ♦ symbol after an hF value indicates that the difference between that baseline method and HSIM holds statistical significance. Finally, the last row of the table summarizes the results of statistical test, i.e., for each baseline method, it is presented the number of times the HSIM outperformed the baseline method by providing a better GMNB classifier performance.

From results presented in Table 4 it is possible to observe that for most of datasets the GMNB classifier achieved higher predictive performance when the dataset was preprocessed using the proposed HSIM. In 6 out of 8 datasets, HSIM obtained significantly better results than MI and, in the remaining two datasets there was no statistically significant difference between the two methods. EM is outperformed by HSIM, with statistical significance, in 5 out of 8 datasets and, in the remaining datasets, the difference between the methods was not statistically significant. Finally, HSIM outperformed KNNImpute in 7 out of 8 datasets with statistical significance and, in the remaining dataset there was no statistically significant difference between the methods. It is also interesting to note that, in 5 out of 8 datasets, the HSIM outperformed all baseline methods by providing significantly better GMNB predictive performance. For only one

Table 4. Experimental Results

Dataset	MI + GMNB hF (std. error)	EM + GMNB hF (std. error)	KNN + GMNB hF (std. error)	HSIM + GMNB hF (std. error)
CellCycle	15.57 (2.00) ♦	15.91 (1.13) ♦	16.00 (1.70) ♦	**27.17 (2.17)**
Church	8.42 (1.20) ♦	8.31 (1.17) ♦	8.26 (0.95) ♦	**13.10 (1.41)**
Eisen	20.59 (2.23)	20.56 (1.66)	20.06 (1.36) ♦	**21.88 (1.68)**
Expr	19.62 (1.84) ♦	20.27 (1.27) ♦	20.20 (1.51) ♦	**45.64 (2.32)**
Gasch1	18.29 (1.30) ♦	18.47 (2.06) ♦	18.22 (1.61) ♦	**22.97 (1.86)**
Gasch2	15.37 (1.30) ♦	15.44 (1.35) ♦	15.44 (1.65) ♦	**19.55 (1.78)**
Sequence	18.73 (1.13)	**18.95 (1.48)**	18.76 (1.15)	18.75 (1.15)
SPO	13.37 (1.03) ♦	13.22 (1.13)	13.37 (1.03) ♦	**14.36 (0.76)**
HSIM wins	6	5	7	

dataset (Sequence) HSIM was statistically equivalent to all baseline imputation methods.

When analysing the results showed in Table 4 and the percentage of missing values in datasets presented in Table 3, it is interesting to note that HSIM improves the most over the other methods on datasets with large percentage of missing values. Besides, the unique dataset (Sequence) where HSIM does not outperform any baseline method has very small percentage (0.01 %) of missing values.

In order to contribute to understand the results presented in Table 4, in the graphs of Fig. 2, it is presented the distribution of missing value per attribute as well as information on the predictive power of the attributes for classification purpose. The hierarchical Symmetrical Uncertainty (SU_H), originally proposed in [31] to deal with feature selection for hierarchical classification problems, was adopted as measure of predictive power of each attribute. A higher SU_H indicates better predictive power. In these graphs, the bars represent the percentage of missing value of each attribute while the solid and dashed lines correspond to SU_H of the attributes without missing value imputation and with missing value imputation using HSIM, respectively.

From the graphs presented in Fig. 2, we can observe that for most of datasets (6 out of 8) the predictive power (according to SU_H) of several attributes improved after missing value imputation process using HSIM. These graphs show that the missing value imputation does not necessarily imply higher predictive power of attributes, since it depends on the quality of imputation. Nevertheless, even for datasets where only a few attributes had their predictive power increased after missing value imputation process (e.g., Church dataset), it is possible to verify the improvement of the predictive performance of the GMNB classifier.

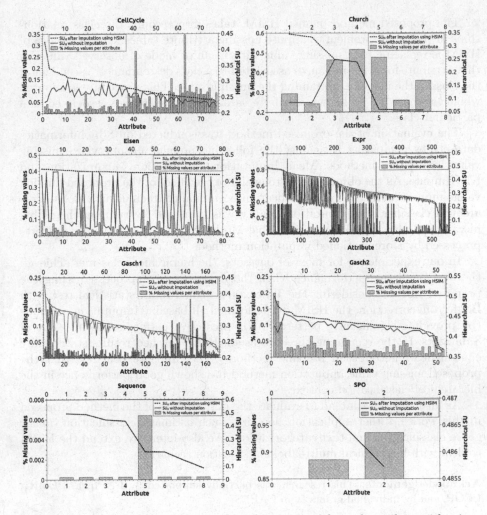

Fig. 2. Percentage of missing values and predictive power for each attribute with missing data.

5 Conclusion

In data mining applications, incomplete datasets is a very common situation. Since many classification algorithms are sensitive to missing attribute values, they can pose an obstacle for classification and other data mining tasks. Although several methods for substituting the missing values can be found in the literature, to the best of our knowledge, for hierarchical classification scenario, there are no supervised methods to deal with the hierarchical context. Therefore, in this work, we proposed and evaluated a supervised missing value imputation method for datasets used in the hierarchical classification problems.

The proposed method, named HSIM, takes into account the class relationships in the target problem to impute missing attribute values. The main idea of HSIM is replacing each missing value with mean or mode of observed values of that attribute for other instances associated to a class descendant or ascendant of the class of the instance containing the missing value. This procedure is adopted whenever there are no known attribute values for the instances associated to a particular class.

The evaluation of the proposed method was conducted on 8 bioinformatics datasets by comparing it against the following popular unsupervised missing value imputation methods: Mean Imputation, Expectation Maximization and KNNImpute. As the objective was to evaluate the effect of the proposed missing value imputation method on classification performance when using a global hierarchical classifier, the imputation quality was measured by running the global hierarchical classifier, known as Global-Model Naive Bayes, on datasets preprocessed by aforementioned imputation methods.

In our experiments, for most of datasets, the hierarchical classifier achieved the best predictive performance when the dataset was preprocessed using the proposed HSIM. Considering the Wilcoxon's Signed-Rank statistical test with Bonferroni correction, the HSIM outperformed all baseline imputation methods (by providing significantly better GMNB predictive performance) in 5 out of 8 datasets. In the remaining three datasets, HSIM reached results statistically equivalent or better than baseline methods. Therefore, we conclude that the proposed missing value imputation method has shown good performance in the hierarchical classification context.

As future work we intend to evaluate the performance of the method proposed in this work in other application domains, such as image classification, music genre classification and text categorization. We also intend to extend the HSIM to deal with hierarchical multi-label classification scenario.

Acknowledgements. This research was partially supported by CNPq, FAPEMIG, UFOP, and by individual grants from CAPES.

References

1. Han, J., Kamber, M.: Data Mining: Concepts and Techniques: Concepts and Techniques. Elsevier, Amsterdam (2011)
2. Little, R.J., Rubin, D.B.: Statistical Analysis with Missing Data. Probability and Statistics, vol. 1, 2nd edn. Wiley, New York (2002)
3. Schafer, J.L., Graham, J.W.: Missing data: our view of the state of the art. Psychol. Methods **7**(2), 147 (2002)
4. Silla Jr., C.N., Freitas, A.A.: A survey of hierarchical classification across different application domains. Data Min. Knowl. Disc. **22**(1–2), 31–72 (2011)
5. Qiu, X., Huang, X., Liu, Z., Zhou, J.: Hierarchical text classification with latent concepts. In: Proceedings of the 49th Annual Meeting of the Association for Computational Linguistics: Human Language Technologies: Short Papers, vol. 2, pp. 598–602. Association for Computational Linguistics (2011)

6. Dollah, R.B., Aono, M.: Classifying biomedical text abstracts based on hierarchical 'concept' structure. World Acad. Sci. Eng. Technol. Int. J. Comput. Electr. Autom. Control Inf. Eng. **5**(2), 178–183 (2011)
7. Campos Merschmann, L.H., Freitas, A.A.: An extended local hierarchical classifier for prediction of protein and gene functions. In: Bellatreche, L., Mohania, M.K. (eds.) DaWaK 2013. LNCS, vol. 8057, pp. 159–171. Springer, Heidelberg (2013). doi:10.1007/978-3-642-40131-2_14
8. Valentini, G.: Hierarchical ensemble methods for protein function prediction. ISRN Bioinf. **2014** (2014)
9. Silla, C.N., Freitas, A.A.: Novel top-down approaches for hierarchical classification and their application to automatic music genre classification. In: 2009 IEEE International Conference on Systems, Man and Cybernetics, SMC 2009, pp. 3499–3504. IEEE (2009)
10. Ariyaratne, H.B., Zhang, D.: A novel automatic hierachical approach to music genre classification. In: 2012 IEEE International Conference on Multimedia and Expo Workshops (ICMEW), pp. 564–569. IEEE (2012)
11. Binder, A., Kawanabe, M., Brefeld, U.: Efficient classification of images with taxonomies. In: Zha, H., Taniguchi, R., Maybank, S. (eds.) ACCV 2009. LNCS, vol. 5996, pp. 351–362. Springer, Heidelberg (2010). doi:10.1007/978-3-642-12297-2_34
12. Kramer, G., Bouma, G., Hendriksen, D., Homminga, M.: Classifying image galleries into a taxonomy using metadata and wikipedia. In: Bouma, G., Ittoo, A., Métais, E., Wortmann, H. (eds.) NLDB 2012. LNCS, vol. 7337, pp. 191–196. Springer, Heidelberg (2012). doi:10.1007/978-3-642-31178-9_20
13. Le, B.V., Bang, J.H., Lee, S.: Hierarchical emotion classification using genetic algorithms. In: Proceedings of the Fourth Symposium on Information and Communication Technology, pp. 158–163. ACM (2013)
14. Van Hulse, J., Khoshgoftaar, T.M.: Incomplete-case nearest neighbor imputation in software measurement data. Inf. Sci. **259**, 596–610 (2014)
15. Troyanskaya, O., Cantor, M., Sherlock, G., Brown, P., Hastie, T., Tibshirani, R., Botstein, D., Altman, R.B.: Missing value estimation methods for dna microarrays. Bioinformatics **17**(6), 520–525 (2001)
16. Rahman, M.G., Islam, M.Z.: IDMI: a novel technique for missing value imputation using a decision tree and expectation-maximization algorithm. In: 2013 16th International Conference on Computer and Information Technology (ICCIT), pp. 496–501. IEEE (2014)
17. Bi, W., Kwok, J.T.: Multi-label classification on tree-and dag-structured hierarchies. In: Proceedings of the 28th International Conference on Machine Learning (ICML 2011), pp. 17–24 (2011)
18. Sun, Z., Zhao, Y., Cao, D., Hao, H.: Hierarchical multilabel classification with optimal path prediction. Neural Process. Lett., 1–15 (2016)
19. Cerri, R., Barros, R.C., de Carvalho, A.: Hierarchical classification of gene ontology-based protein functions with neural networks. In: IEEE International Joint Conference on Neural Networks (IJCNN), pp. 1–8 (2015)
20. Clare, A., King, R.D.: Predicting gene function in saccharomyces cerevisiae. Bioinformatics **19**(suppl 2), ii42–ii49 (2003)
21. Chen, Y.L., Hu, H.W., Tang, K.: Constructing a decision tree from data with hierarchical class labels. Expert Syst. Appl. **36**(3), 4838–4847 (2009)
22. Silla, C.N., Freitas, A.A.: A global-model naive bayes approach to the hierarchical prediction of protein functions. In: 2009 Ninth IEEE International Conference on Data Mining, ICDM 2009, pp. 992–997. IEEE (2009)

23. Blockeel, H., Schietgat, L., Struyf, J., Džeroski, S., Clare, A.: Decision trees for hierarchical multilabel classification: a case study in functional genomics. In: Fürnkranz, J., Scheffer, T., Spiliopoulou, M. (eds.) PKDD 2006. LNCS (LNAI), vol. 4213, pp. 18–29. Springer, Heidelberg (2006). doi:10.1007/11871637_7

24. Vens, C., Struyf, J., Schietgat, L., Džeroski, S., Blockeel, H.: Decision trees for hierarchical multi-label classification. Mach. Learn. **73**(2), 185–214 (2008)

25. Otero, F.E.B., Freitas, A.A., Johnson, C.G.: A hierarchical classification ant colony algorithm for predicting gene ontology terms. In: Pizzuti, C., Ritchie, M.D., Giacobini, M. (eds.) EvoBIO 2009. LNCS, vol. 5483, pp. 68–79. Springer, Heidelberg (2009). doi:10.1007/978-3-642-01184-9_7

26. Brown, M.L., Kros, J.F.: Data mining and the impact of missing data. Ind. Manag. Data Syst. **103**(8), 611–621 (2003)

27. Dempster, A.P., Laird, N.M., Rubin, D.B.: Maximum likelihood from incomplete data via the EM algorithm. J. Roy. Stat. Soc.: Ser. B (Methodol.), 1–38 (1977)

28. Hall, M., Frank, E., Holmes, G., Pfahringer, B., Reutemann, P., Witten, I.H.: The weka data mining software: an update. ACM SIGKDD Explor. Newsl. **11**(1), 10–18 (2009)

29. Borges, H.B., Silla, C.N., Nievola, J.C.: An evaluation of global-model hierarchical classification algorithms for hierarchical classification problems with single path of labels. Comput. Math. Appl. **66**(10), 1991–2002 (2013)

30. Japkowicz, N., Shah, M.: Evaluating Learning Algorithms. Cambridge University Press, Cambridge (2011)

31. Dias, T.N., Merschmann, L.H.C.: Adaptação da medida incerteza simétrica para a seleção de atributos no contexto de classificação hierárquica monorrótulo. In: Anais do Encontro Nacional de Inteligência Artificial e Computacional, Natal, RN, Brazil, pp. 142–149 (2015)

Applications

Predicting Cargo Train Failures: A Machine Learning Approach for a Lightweight Prototype

Sebastian Kauschke[1](✉), Johannes Fürnkranz[2], and Frederik Janssen[1]

[1] Knowledge Engineering Group, Telecooperation Group,
TU Darmstadt, Darmstadt, Germany
{kauschke,janssen}@ke.tu-darmstadt.de
[2] Knowledge Engineering Group, TU Darmstadt, Darmstadt, Germany
fuernkranz@ke.tu-darmstadt.de

Abstract. In cargo transportation, reliability is a crucial issue. In the case of railway traffic, the consequences of locomotive failure are not limited to the affected machine, but are propagated through the railway network and may affect public transport as well. Therefore it is desirable to predict and avoid failures. In order to do this, constant monitoring of the trains' systems and measurement of the relevant variables is required, but often not implemented. In this paper we leverage the existing technology of the 185 locomotive series and build a layered model for power converter failure prediction that can be applied without additional technology. We train instance anomaly detectors based on the pattern structure of the locomotives' diagnostic messages from historical data records. For this purpose we selected rule and decision tree learning because they can be easily implemented in the existing software, whereas more complex classifiers would require costly software adaptations. In order to predict a time series of instances, we construct a meta classification layer. We then evaluate our model on the data of 180 locomotive tours by leave one out classification. The results show that the meta classifier improves classification accuracy, which will allow us to use this technology in a fielded prototype installation without disturbing daily operations.

1 Introduction

DB Schenker Rail started the *TechLok* project in 2011 with the goal to discover underlying processes of specific failures to implement counter-measures. One of the research directions is *Predictive Maintenance* (PM). PM targets the substitution of existing maintenance processes by adaptive maintenance intervals that respect the actual state of the equipment. In order to benefit from PM, a constant monitoring and recording of the machine status data is required, and prediction models for various systems have to be built. In [2] we evaluated the process of building such a model based on another failure type. In this paper we are focussing on the *power converter*, a unit converting high-voltage electricity from the power lines to be used to drive the electric train motors.

Usually in PM, historical data is used to train a model of either the standard behaviour of the machine, or—if enough example cases have been recorded—a

© Springer International Publishing Switzerland 2016
T. Calders et al. (Eds.): DS 2016, LNAI 9956, pp. 151–166, 2016.
DOI: 10.1007/978-3-319-46307-0_10

model of the deviant behaviour before the failure. These models are then used on new data to determine whether the machine is operating within standard parameters, or, in the second case, if the operating characteristics are similar to the failure scenario. If the model is trained properly, it will give an alarm in due time. An overview of various PM methods is given in [6]. In our use case, the machines are cargo trains. These trains are pulling up to 3000 tons of cargo, so a lot of parts are prone to deterioration effects.

In this paper we will present a method to predict *power converter failure* (PCF) based on 40 recorded historical events of this type. PCF causes the train to break down and is costly in repair. Previous attempts at predicting PCF showed that a prediction horizon of multiple days was not possible. By thorough analysis together with engineers and domain experts we concluded, that this type of failure has a short prediction horizon. Hence, we try predicting PCF in an online manner such that we use the historical data to simulate a live-environment and live-prediction. We analyse complete tours of locomotives containing the incidents (*incident* tours) as well as normal tours (*non-incident* tours) and build a two-level prediction method. The classifier is trained with a classical offline method. The classification phase takes place in an online setting (which we simulate here), where a steadily increasing incoming information stream about the tour has to be classified. This distinguishes our method from related methods (e.g. direct classification methods like [11] or multi-instance methods [8]) which deal with a complete set of information.

Caused by the limits of the existing data collection systems and our goal to create a prototype installation of our method, the resulting method has to be implemented in a script language. Therefore we chose to aim for easy-to-implement classifiers like decision trees or rule learners. The system will be implemented to generate trust in the method first, before making the effort to thoroughly integrate it into the existing database environment.

This paper is organized as follows. Section 2 gives an introduction to the data and the process of creating instances to train a classifier upon. In Sect. 3 we describe the data analysis and our conclusions thereof, which leads us to building the classification model in Sect. 4. Finally, we set up the experiment (Sect. 5), show the results in Sect. 6 and give a conclusion in Sect. 7.

2   Data Retrieval and Preparation

In this section we will give an insight to the locomotive data and show how we create instances for machine learning from them.

2.1   Diagnostic Data

Diagnostic data is collected on the locomotive in the form of a logfile, in which all events occurring in the various systems of the locomotive are recorded. In total, there are 6909 different event types, the so called *diagnostic messages* (DM). Diagnostic messages can contain status information, warnings or specific error messages.

Table 1. Examples of diagnostic messages in their raw format

train	code	from	to	environment
185004	4003	1394442136	1394443101	D3BA1C00500000000000000000B00000000000E8053
185004	4003	1393940175	1394792657	4FB11C00500002000000000BE00000005019005B

A set of system variables, the so called environment, which are encoded as strings (see Table 1), is attached to each DM. The variables are not monitored periodically, but only recorded when a DM occurs, since the whole system is event-based. Which variables are encoded depends on the diagnostic message that was recorded. This implies that some variables will be recorded rarely and sometimes not for hours or days. Overall, there are 2291 system variables available: Booleans, numeric values like temperature or pressure and system states of certain components (encoded as numeric values).

The diagnostic messages have two timestamps (Table 1), one for when the code occurred first (*from*), and one for when it disappeared (*to*). This was originally designed for status reporting messages that last a certain period. Most codes only occur once, so they do not have a timespan, others can last up to days. For example, code 8703 (Battery Voltage On) is always recorded when the train has been switched on, and lasts until the train is switched off again. We use it to determine the start and end of tours.

2.2   Converting Diagnostic Data to Instances

In our pursuit to classify tours in an online scenario with *live* data, we decide to keep the overhead for data transformation into our desired form as small as possible. Therefore we look at each DM as a boolean value, which is either *active* or *inactive* for any specific point in time. In the historical data this can be decided based on the *from* and *to* values that are set for each DM.

With this information, we create status vectors for each relevant point in time for each tour, where relevant is defined as a point where the status vector changes compared to the one before (see Table 2 as an example, C_i is the state of a DM). We could create such vectors in arbitrary or fixed intervals, at maximum once a second (that is the finest granularity in the data). Fixed intervals could generate multiple identical instances, because state changes occur sparsely, so we use arbitrary intervals and generate an instance on every status change. As environment variables we use temperature readings that are relevant to our prediction problem, e.g. temperature values that are dependent on the outside temperature. The outside temperature was considered as interesting and relevant by the engineering experts.

Preparing the data this way, we receive a set of instances per tour. The size of this set depends on the duration and the circumstances. A locomotive being taxied might generate a high number of messages in a short time, whereas a locomotive moving for hours in the same direction may not generate many messages.

Table 2. Example of status vectors as binary instances with relevant changes

Instance \ Code	C_1	C_2	C_3	C_4	...	C_{n-2}	C_{n-1}	C_n	$Temp_1$	$Temp_2$
Instance 1	1	0	0	1	...	0	1	0	?	?
Instance 2	1	1	0	1	...	0	1	0	17.5	18.3
Instance 3	0	1	0	1	...	0	1	0	?	?

As a result we see a high variance in temporal density of diagnostic messages. We will analyse the behaviour of the tour data we generated this way in Sect. 3.

2.3   The Learning Problem

Given by the historic data is a set of tours S, where each tour consists of data from a single locomotive. Tours of a locomotive are non-overlapping in time. The set of tours S consists of the set of *incident* tours F and the set of *non-incident* tours N. A tour s has N_s instances $I_{s,n}$ which depict the state of the locomotive at a certain timestamp $T_{s,n}$. Two instances from the same tour never have identical timestamps. $L_s \in \{normal, incident\}$ denotes the label of each tour. $LI_{s,n}$ denotes the label of each instance n in tour s, where $LI_{s,n} \in \{negative, positive\}$.

The *online* approach we are pursuing causes the situation to be as such: During the course of a tour, instances will be recorded and the instance set x for the tour will grow. The classifier CI predicts the instance classes $LI_{s,n}$ of the instances of x.

Based on the results of CI, a classifier CT has to make a decision if the tour is *normal* or a potential *incident*. Preferably, the classifier can decide if the tour is an *incident* before the actual failure occurs, so that a person can react accordingly. In this work, we pursue the following goals:

Learn an instance classifier function CI such that $CI(I_{s,n}) = LI_{s,n}$ from the instance data of all other tours $\{S \setminus s\}$.

Learn a tour classifier function CT such that $CT(x) = L_s$ from the instance data of all other tours $\{S \setminus s\}$.

Select an optimal CT so that the prediction takes place ahead of the actual failure.

In the following sections we will elaborate on the historic data and illustrate the problems that arise with it.

3   Tour Analysis

In our historic data we find a total of 90038 tours with a length of at least 100 km and a duration of 2 to 48 h. We assume tours with less duration/length are often not tours with actual payload and might show a different behaviour that we want to avoid learning. These tours contain 40 recorded cases of *power converter failure*, which is the failure type we want to predict. We refer to these tours as *incident* tours. All other tours without this specific failure occurrence are *non-incident* tours. In this section, we will give a short analysis over the instances' temporal behaviour.

3.1   Temporal Behaviour of Events in Tours

In Fig. 1 we show a 10 min excerpt from the diagnostic data that was recorded on a locomotive. Each mark represents the occurrence of a message. As stated before, a message only occurs when a change is detected or notification/warning issued in one of the locomotives' systems.

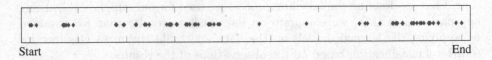

Start   End

Fig. 1. Visualised instance distribution of a 10 min tour excerpt

Unfortunately, the event-based character of the data poses a problem for our given task of online failure prediction. Since we are depending on information about the state of the machine—which in this case might change drastically before getting a new status reading in the form of a DM—we might not be able to give accurate predictions. The example (Fig. 1), shows gaps between the instances, although we selected a gratuitous example with 65 instances in 10 min. In less optimal cases there might be hours between two instances.

In Tables 3 and 4 we show the differences in behaviour of *incident tours* and *non-incident tours*. We can see, that the latter have significantly fewer instances per hour, which increases the average distance between instances. One intuitive explanation of this behaviour would be that locomotives that are about to fail issue more diagnostic messages. On the other hand it might be a random effect caused by the low number of *incident* tours these numbers are based on.

Table 3. Incident tour statistics (based on 40 tours longer than 2 h)

Incident tours	Instances/h	Duration	Average instance distance
Average	34	11.8 h	147 s
StdDev	22.3	16.2 h	87 s

The difference in average duration seems counter-intuitive. The *incident* tours should be shorter because they are aborted prematurely by the failure. The higher standard deviation shows that this value is misleading in this regard, so we can not make further assumptions based on these values.

Table 4. Non-incident tour statistics (based on 90038 tours from 2 to 48 h)

Non-incident tours	Instances/h	Duration	Average instance distance
Average	22.6	9.8 h	232 s
StdDev	10.9	5.1 h	114 s

4 Building a Classification Model

In this section we propose a two-level approach for classifying tours in an *online* manner. We will train a classifier based on the instances, and then introduce a meta classifier to combine individual predictions to an assessment of the entire tour. First we describe the labelling process we used for our supervised learning approach. After that, we elaborate on intermediate results that we obtained by applying the learned models on the data and build the meta classifier for tour-level classification based on the observations of the results.

4.1 Labelling with Variable Window Size

Since the data is unlabelled we only have information about when the incident happened. To train the classifier, we need to artificially generate positive instances to learn upon. This is a crucial aspect with influence on the quality of the resulting classifier. We use a modified version of the labelling method suggested by Létourneau et al. [4]. This method uses a windowed labelling approach: We label the instances such that they are labelled *positive* in a window before the failure occurs, and *negative* otherwise (cf. Sect. 2.3). Because we do not know which the indicative instances are or how they are different from non-indicative instances, we define an integer w as the duration of this window, the so called *warning epoch* (see Fig. 2). We will determine the value of w experimentally by applying the values shown in Table 5 (Window size).

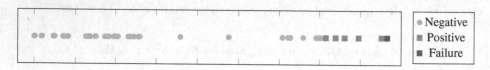

Fig. 2. The labels assigned to instances of an exemplary *incident* tour

As stated before, we are looking at a failure that has a very short prediction horizon, meaning that the selection of the *positive* instances is crucial to the success of the method. Otherwise our classifier will not be able to build an adequate model. The number of *positive* instances depends on the size of the labelling window w. We analysed the quantity of labelled instances in the *incident* tours by using window sizes from 60 to 9600 s and a setting with all instances of the tour

Table 5. Analysis of positive instances in 40 *incident* tours without failure instance

Window size	60 s	150 s	300 s	600 s	1200 s	2400 s	4800 s	9600 s	All
Number pos. instances	18	48	78	113	230	538	1298	2677	10903
Avg pos. inst. per tour	0.45	1.2	2	2.8	5.8	13.5	32.5	66.9	273

in order to determine plausible values for w. In this table, we excluded the actual instance of the moment the failure happened. By including this instance we gain one more positive instance per tour. In Sect. 5 we will use both variants to make sure we do not permit the classifiers to gain information from this instance.

We are dealing with a heavily skewed dataset in terms of *positive* and *negative* instance numbers since we have very few incidents but a large number of normal tours. As shown in Fig. 3 and Table 5, we see that for the lower end of the window size spectrum, e.g. 60 and 150 s, we receive low numbers of positive instances, so that some of the tours contribute no instances to train the classifier upon. The *all* setting enlarges the window to cover every instance in a tour. With this few instances to train upon, we expect poor classification performance from the models trained with these parametrisations in the evaluation. Only at the largest window size of 9600 s every tour contributes at least one instance to the training set.

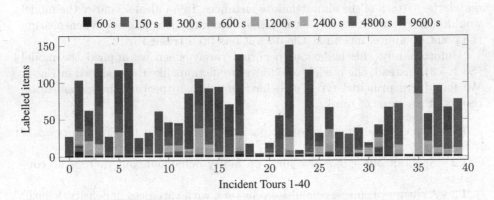

Fig. 3. Incident tours - labelled items w.r.t. window size (seconds)

By training the classifier this way, we do not expect to achieve a perfect classifier, but rather create an *anomaly detector* that can be deployed with little effort on the given systems. More complex anomaly detection methods are available in the form of e.g. One-class SVM [5], but they can not be implemented in this proof of concept prototype.

4.2   Instance Level Classification

After we constructed the models by the procedure given in Sect. 4.1, we apply them on the given tour data. Since the model is trained to recognize the instances

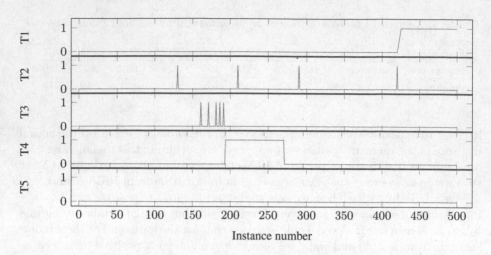

Fig. 4. Instance level classification behaviour in tours: principal types (Incident at 500)

right before the failure, and we assume those instances show a different behaviour than while operating normally, it will classify instances as *positive* when they match the pattern of the almost-failure instances. In an ideal scenario, the model would classify tour instances as *negative*, and switch to *positive* when nearing the point of failure, much like the data it has been trained with.

Unfortunately, this behaviour occurred rarely when we applied the model (Sect. 5.3). Instead, the model would give predictions like those shown in Fig. 4. We found five principal types of behaviour when inspecting the classification results of a sample of tours.

- **T1** - The classifier gives constant positive results when nearing the point of failure.
- **T2** - Spontaneous spikes give hints on faulty behaviour, but there is no consistency.
- **T3** - A cluster of spikes, or multiple clusters, with variances in density/length.
- **T4** - A positive phase is followed by a negative phase.
- **T5** - Purely negative, no indication is given.

If we want to make a prediction for a complete tour based on the results of the instance level classifier CI, we can do this in a naive way by assuming that $CT = CI$. Therefore if an instance is classified as *positive*, the tour will be classified as *incident*. If the classified tour is an actual *incident*, the types **T1-T4** would give true positive results, and a false negative in case of **T5**.

Besides the classification of *incident* tours, we also classified *non-incident* tours using the same model. Unfortunately, the behaviours shown in Fig. 4 could also be observed there, although **T1-T4** occurred less often. The naive method would create false positives in cases **T1-T4** when making predictions for the whole tour. To solve this on the instance level, we need more precisely trained models. This issue was anticipated: with only 40 examples of PCF, a well-performing generalised model was not to be expected.

Depending on the parametrisation of the model, types **T1** and **T4** did not occur at all in our sample set of *non-incident* tours. Behaviour type **T2** on the other hand occurred frequently. We will propose an advanced meta classifier to prevent these occurrences from creating *false positives* and improve tour-level classification in the next section.

4.3   Tour-Level Meta Classification

In this section we will propose two methods for the meta-layer of the prediction process. The baseline tour classifier *SimpleDirect*, and a method which will improve tour-classification performance beyond the limitations of *SimpleDirect*.

The *SimpleDirect* Tour Classifier. We use the *SimpleDirect* (SD) method as a baseline tour classification method. The classifier SD is defined such that it uses an instance classifier *CI*. If *CI* classifies an instance as *positive*, SD will classify the tour as *incident*.

The instance classifier gives us an indication based on a single instance as class: either *positive* (as in: failure is likely) or *negative*. It is un-intuitive to base the classification of a whole tour on a single instance. We would rather look at a certain number of instances and make the decision on multiple results of instance classifications. We present a naive approach to this problem which improves on the *SimpleDirect* schema.

Meta Classification by Percentage Threshold. This method requires a threshold $t = positive/total$ to be surpassed to predict the *incident* label for the whole tour. Instance level classification is applied separately to the instances in the order of their arrival time, and the result is used to calculate the threshold. The method requires a minimum number of seen instances w.r.t. the threshold ratio. This has the positive effect that it reduces false positives in the case of **T2**-type instance classifications, but it also limits the classifier to make a very early decision caused by the necessity of a minimum of seen instances. We will take this into account when evaluation the performance in Sect. 6.

In order to measure the performance of our tour meta classifiers, we will use the following metrics besides the usual candidates such as Hits (true positives), Misses (false negatives), true negatives and false positives:

- *Time to failure* (TTF): Remaining time delta (seconds) between the moment of the classifier declaring a tour as *incident* and the actual incident.
- *Aggregated time to failure* (aTTF): TTF averaged over all correctly classified *incident* tours.
- *Minimum time to failure* (mTTF): minimum TTF in all correctly classified *incident* tours.

These metrics have to be regarded in combination to sort out useful parametrisations and find the best-performing combinations of instance classifier and tour-level meta classifier. It is important to detect as many *incidents* in advance as possible, but it is also important to generate a low number of *false positives* in order to avoid unnecessary maintenance cost.

5   Experimental Setup

In the following, we will introduce the classifiers, elaborate on the methods we used to help improve the problem with imbalanced classes, explain our chosen evaluation method and state all variants of the tour-level meta classifier we used. Finally, we will explain the evaluation method.

5.1   Parametrisation of the Classifiers

Our goal is to create a lightweight prototype installation of the prediction system, which means we have to rely on classifiers that can be implemented on the given systems. We decided to go with two basic classifiers, JRIP (as the WEKA [10] implementation of the RIPPER rule learning algorithm [1]) and J48 (WEKA implementation of C4.5 decision tree learner [7]). A ruleset or a decision tree is easy to implement on a system that only allows a low-level scripting language, and therefore the right choice for our endeavours. Furthermore, they allow for easy interpretability. This is especially useful for the engineers to draw conclusions from the learned rules or trees and helps justify or discard a model.

5.2   Sampling and Balancing

As a result of the labelling process explained in Sect. 4.1 we have a set of instances with a highly skewed class imbalance. Training a classifier on this kind of data might give suboptimal results, because classifiers will focus on the majority class (and subsequently be correct more than 99 % of the time). We will apply the following techniques in order to improve classifier performance:

– Subsampling with the *SpreadSubsample*[1] filter from WEKA
– Class-weight balancing with the *ClassBalancer*[2] filter from WEKA.

Subsampling rebuilds the training set, such that a certain ratio between the rarest and the most frequent class is realised. For example we use this to create a 15:1 spread ratio of *negative* to *positive* class. Of course by this measure the number of *negative* instances in the training set will be reduced, which might have an influence on learning the class in some classifiers.

---

[1] http://weka.sourceforge.net/doc.stable/weka/filters/supervised/instance/SpreadSubsample. html.

[2] http://weka.sourceforge.net/doc.dev/weka/filters/supervised/instance/ClassBalancer.html.

Class-weight balancing adjusts the instance weights in the data set, so that all classes have the same total weight. Both, JRip and J48 support weighted instances. This is not the case for all classifiers implemented in WEKA.

We used both of the classifiers with their default WEKA configuration. The default configurations provide solid classification performance, which we can later optimise, but at the moment we will concentrate on the other parameters.

5.3   Leave One Tour Out Evaluation

Our data consists of all 40 *incident* tours, and we add a random sample of 140 *non-incident* tours from our pool of tours that have a duration of more than 2 h and 100 km.

For evaluation we use a leave-one-out cross-validation method, i.e. we train the model on 179 tours and apply it to the remaining one, for each of the 180 tours. This method is a version of the *leave one batch out* method described in [3], which leaves a batch of instances out of the training data. In our case, a batch consists of all instances from the left-out tour. We want the maximum possible number of *incident* tours in the training set, so that the classifier has the best chance to learn a proper model. By this we assure that at least 39 *incident* tours are in each training set, which gives the best possible conditions for the classifier.

At the instance level, the 40 *incident* tours contribute 10903 and the 140 *non-incident* tours 27512 instances to the train/test data. The number of positive instances depends on the w value.

5.4   Variants

We used the following parameters and values in combination to find the optimal result:

The labelled Data. We used the process described in Sect. 4.1 in two variants. The first variant excludes the instance that contained the actual failure, the second includes it.

Labelling Windows. We used 7 labelling window sizes as described in Sect. 4.1 from *300* s to *9600* s and a version that labels *all* instances, in order to empirically find the optimal window.

Class Balancing. For balancing we used Subsampling (see Sect. 5.2) in 3 variants with spread-factors of 15, 30 and 50, Class-weight balancing and a variant with no balancing.

Tour-level. In order to classify complete tours we applied the *percentage threshold* (see Sect. 4.3) in 9 variants (*2, 5, 7, 10, 15, 20, 30, 40* and *50 %* threshold) as well as *SimpleDirect* (*1 hit in total*). Since the result is determined by this parameter in combination with the instance level classifier we used a wide spread of values.

As stated before, we used both J48 and JRip as classifiers, resulting in a total of 1400 combinations of parameters. For each of these combinations we applied the leave-one-tour out evaluation method, resulting in around 10700 h computing time for the complete evaluation. Unfortunately, caused by time constraints, we were not able to compute multiple runs with different sample sets of *non-incident* tours.

6   Experiment Results

In this section, we will describe the baseline of our experiment, show the results we achieved, and compare the enhanced meta classifier to the *SimpleDirect* baseline.

6.1   The Accuracy Baseline

Since we are targeting the classification of tours and have no related method to use as a baseline, the ratio of the tour classes will be the starting point. We are using a dataset with 40 *incident* tours and 140 *non-incident* tours, so a classifier that always chooses the majority class would achieve an accuracy of 77.7 %, albeit with no *true positives*.

6.2   Relevant Results

Due to the amount of parameter combinations we will not be able to show all results here, instead we show a pre-filtered set based on the criterion, that the mTTF should never be zero, which would mean that a tour was classified based on the last instance in the tour, and be of no practical use. Furthermore, we sort the remaining results by the overall false positives and then the true positives. We want as few false positives as possible, but at the same time a maximum of true positives. After the filtering we are left with 824 resulting classification variants.

Caveat. In our evaluation we assume that *false positives* are costly. Unfortunately we do not know the real cost of PCF and false alarms, so we can not base our findings on it. With over 90000 tours in our dataset, even a 1 % *false positive* rate would lead to 900 locomotives having to be checked. Therefore we want to find a method that creates little to no false alarms. We also asked the people that will have to use this system, and they would rather work with a method that has a higher confidence and less hits than a classifier that creates many false alarms.

The results (Table 6) are ranked by the criteria mentioned above. The columns contain the following information:

1. *Rk*: Rank as determined by our sorting criteria
2. *f.I.*: Dataset contains the *failure instance* ($f.I. = 1$) or not ($f.I. = 0$)

3. *Window*: Size of the specific labelling window
4. *Classifier*: The classifier used
5. *Balancing*: Balancing method (Subsampling, Classweight balancing or None)
6. *Meta Classifier*: Parametrisation of the meta classifier, or *SimpleDirect* (SD)
7. *Accuracy*: Percent correctly classified instances
8. *aTTF*: Average TTF in correctly classified *incident* tours in hours
9. *mTTF*: Minimum TTF in correctly classified *incident* tours in minutes
10. *TP, TN, FP, FN*: True positive, true negative, false positive and false negative tours

Table 6 shows that we can achieve multiple combinations with no false positives. Only from rank 50 onwards some false positives occur. On the top ranks we have a maximum of 12.5 % (5 of 40) true positives. Although this does not seem impressive at first sight, with an expensive failure like the PCF this can be a valuable asset used for cost reduction. All the methods we selected show an *average time-to-failure* high enough for a person to react. Only in the worst case (mTTF of Rk. 12) the reaction time could be too short to react.

As far as the optimal labelling window is concerned, the best performing configurations were built on the 2400 s window size. Although we can not precisely determine why this works best, we assume it's a combination of the number of instances and the relevance of the instances they cover. Most of the top results

Table 6. Top results with $mTTF > 0$ ordered by FP asc, TP desc

Rk	f.I	Window	Classifier	Balanc.	Meta	Accur.	aTTF	mTTF	TP	TN	FP	FN
1	0	2400	J48	subs50	5perc	80.56	5.85 h	125.9 m	5	140	0	35
2	0	2400	J48	subs50	7perc	80.56	5.70 h	113.7 m	5	140	0	35
3	0	2400	J48		5perc	80.00	7.33 h	135.9 m	4	140	0	36
4	0	2400	J48		7perc	80.00	7.20 h	123.2 m	4	140	0	36
5	0	2400	J48	subs50	10perc	80.00	5.41 h	70.6 m	4	140	0	36
6	1	2400	J48	subs30	30perc	79.44	7.04 h	63.3 m	4	140	0	36
7	0	2400	JRip	subs50	2perc	79.44	8.24 h	142.1 m	3	140	0	37
8	0	2400	JRip		2perc	79.44	8.50 h	141.2 m	3	140	0	37
9	0	2400	J48		10perc	79.44	7.58 h	99.4 m	3	140	0	37
10	1	1200	J48	subs30	30perc	79.44	6.75 h	63.3 m	3	140	0	37
11	0	2400	J48		20perc	79.44	6.43 h	63.3 m	3	140	0	37
12	1	2400	JRip		20perc	78.89	5.79 h	1.0 m	3	140	0	38
13	1	300	J48	subs50	10perc	78.89	12.36 h	266.0 m	2	140	0	38
14	1	300	J48	subs50	5perc	78.89	12.35 h	264.1 m	2	140	0	38
15	0	600	JRip	subs50	10perc	78.89	8.61 h	140.6 m	2	140	0	38
...												
50	1	2400	JRip	subs30	7perc	80.00	16.39 h	138.1 m	6	139	1	35
51	1	2400	JRip	subs30	10perc	79.44	18.32 h	135.8 m	5	139	1	36
52	1	2400	J48	subs15	30perc	79.44	8.15 h	63.3 m	5	139	1	36
53	1	2400	JRip		2perc	79.44	5.89 h	37.2 m	5	139	1	36

Table 7. Top 5 results with $mTTF > 0$ and *SimpleDirect* meta classifier

Rk	f.I	Window	Classifier	Balanc.	Meta	Accur.	avgTTF	minTTF	TP	TN	FP	FN
228	0	600	J48	subs50	SD	78.33	10.76 h	28.8 m	5	136	4	35
505	0	1200	J48	subs15	SD	72.22	8.28 h	33.5 m	13	117	23	27
688	0	all	J48		SD	52.78	12.62 h	51.3 m	37	58	82	3
711	1	600	J48		SD	22.78	11.88 h	53.7 m	40	0	140	0
712	1	600	JRip		SD	22.78	11.88 h	53.7 m	40	0	140	0

also include some version of the SpreadSubsample filter, but the variants without balancing are not far behind (e.g. Rk.1 and Rk.3). All of the top results perform above baseline in terms of accuracy, although not by much.

We can also see that the *tour-level* meta classification increases the classification performance compared to *SimpleDirect* (Table 7). The first configuration that uses *SimpleDirect* is ranked on position 228 in the results. It performs marginally above baseline in terms of accuracy (78.33 %) but some false positives are present. The further *SimpleDirect* variants achieve even lower ranked positions, having even more false positives.

The percentage threshold method has the disadvantage that it has to gather a certain number of instances before it can make a decision. Naturally, we assume it would be slower in making decisions compared to *SimpleDirect*. As we see in Tables 6 and 7, this is not completely true. Based on the aTTF value, it seems that the meta classifier is slower. But when we compare the—more crucial— mTTF value, the meta classifier outperforms *SimpleDirect*. We attribute this to the fact that the *SimpleDirect* results are based on smaller labelling windows (600, 1200 s), which changes the behaviour of the instance classification.

6.3   Improvement by Percentage Threshold Meta Classifier

An example of how the tour-level classifier improves the performance can be seen in Table 8. Using *SimpleDirect* the instance classifier produces false alarms. By utilising the meta classifier with a 7 % positive instance threshold we are able to completely avoid false positives and still achieve 5 true positives.

Table 8. Example of improvements through percentage threshold metaclassification

	TP	TN	FP	FN	Accuracy
SimpleDirect	40	0	140	0	22.2 %
7 % threshold	5	140	0	35	80.56 %

7   Conclusions and Outlook

We have created the basis for an easy-to-implement prediction prototype that can be realised even on systems with limited capabilities. In our case, a decision tree with a meta classifier built on top shows good recognition performance, while creating no *false positives* on our dataset. This is especially useful to create trust in the method among the engineers and people that are supposed to use it on daily basis.

The next step will be the realisation of the prototype, but we also plan further enhancements that we could not yet incorporate into the process:

- Finetuning of the labelling windows
- Parameter optimisation of the classifier models
- Replacing the meta classifier with a learned model
- Using ensembles of instance level classifiers and tour level classifiers

At the moment we apply fixed labelling windows to all tours. This might not yield the optimal result for each of the tours. It is likely that we have to decide upon other criteria to improve the labelling, in order to train a better instance-level classifier. Also, the tour-classification is naive at the moment. We will try to learn another classification model on the output of the instance-level classifier w.r.t. the tour class, so that the results can be improved and we achieve more hits and less false alarms. Further improvements could be achieved by combining multiple classifiers and their decisions on the instance as well as the tour level. Finally, we will also apply more complex learning algorithms on the data. While the practical use may be limited at the moment, it is useful to see if the results can be improved.

Further approaches would include treating this problem as a multi-instance problem [8], especially regarding the question, if the *online* character of this situation can be handled in a multi-instance setting. Another way of approaching the topic would be via pattern learning [9], e.g. by discovering sequences of diagnostic messages that lead to failure.

Acknowledgments. This work has been co-funded by the DB Schenker Rail project "TechLok" and by the LOEWE initiative (Hessen, Germany) within the NICER project [III L 5-518/81.004].

References

1. Cohen, W.W.: Fast effective rule induction. In: Proceedings of the Twelfth International Conference on Machine Learning, pp. 115–123 (1995)
2. Kauschke, S., Schweizer, I., Janssen, F.: On the challenges of real world data in predictive maintenance scenarios: a railway application. In: Görg, S., Müller, G., Bergmann, R. (eds.) Proceedings of the LWA 2015 Workshops: KDML, FGWM, IR, and FGDB, pp. 121–132. CEUR Workshop Proceedings, October 2015
3. Kubat, M., Holte, R.C., Matwin, S.: Machine learning for the detection of oil spills in satellite radar images. Mach. Learn. **30**(2–3), 195–215 (1998)

4. Létourneau, S., Famili, F., Matwin, S.: Data mining for prediction of aircraft component replacement. IEEE Intell. Syst. Jr. - Special Issue on Data Mining **14**, 59–66 (1999)

5. Martinez-Rego, D., Fontenla-Romero, O., Alonso-Betanzos, A.: Power wind mill fault detection via one-class v-svm vibration signal analysis. In: Proceedings of International Joint Conference on Neural Networks (2011)

6. Peng, Y., Dong, M., Zuo, M.J.: Current status of machine prognostics in condition-based maintenance: a review. Int. J. Adv. Manufact. Technol. **50**(1–4), 297–313 (2010)

7. Ross Quinlan, J.: C4. 5: Programs for Machine Learning. Morgan Kaufman Publishers, Inc., San Francisco (1993)

8. Sipos, R., Fradkin, D., Moerchen, F., Wang, Z.: Log-based predictive maintenance. In: Proceedings of the 20th ACM SIGKDD International Conference on Knowledge Discovery and Data Mining, pp. 1867–1876. ACM (2014)

9. Vaarandi, R., et al.: A data clustering algorithm for mining patterns from event logs. In: Proceedings of the 2003 IEEE Workshop on IP Operations and Management (IPOM), pp. 119–126 (2003)

10. Witten, I.H., Frank, E.: Data Mining: Practical Machine Learning Tools and Techniques. Morgan Kaufmann, San Francisco (2005)

11. Zaluski, M., Létourneau, S., Bird, J., Yang, C.: Developing data mining-based prognostic models for cf-18 aircraft. J. Eng. Gas Turbines Power **133**(10), 101601 (2011)

Predicting Bug-Fix Time: Using Standard Versus Topic-Based Text Categorization Techniques

Pasquale Ardimento[1]($\boxtimes$), Massimo Bilancia[2], and Stefano Monopoli[3]

[1] Department of Informatics, University of Bari Aldo Moro,
Via Orabona, 4, 70125 Bari, Italy
pasquale.ardimento@uniba.it

[2] Ionian Department of Law, Economics and Environment,
University of Bari Aldo Moro, Via Lago Maggiore angolo Via Ancona,
74121 Taranto, Italy
massimo.bilancia@uniba.it

[3] Everis Italia S.p.A., Via Gustavo Fara, 26, 20124 Milano, Italy
stefano.monopoli@everis.com

Abstract. In modern software development, finding and fixing bugs is a vital part of software development and quality assurance. Once a bug is reported, it is typically recorded in the Bug Tracking System, and is assigned to a developer to resolve (bug triage). Current practice of bug triage is largely a manual collaborative process, which is often time-consuming and error-prone. Predicting on the basis of past data the time to fix a newly-reported bug has been shown to be an important target to support the whole triage process. Many researchers have, therefore, proposed methods for automated bug-fix time prediction, largely based on statistical prediction models exploiting the attributes of bug reports. However, existing algorithms often fail to validate on multiple large projects widely-used in bug studies, mostly as a consequence of inappropriate attribute selection [2]. In this paper, instead of focusing on attribute subset selection, we explore an alternative promising approach consisting of using all available textual information. The problem of bug-fix time estimation is then mapped to a text categorization problem. We consider a multi-topic Supervised Latent Dirichlet Allocation (SLDA) model, which adds to Latent Dirichlet Allocation a response variable consisting of an unordered binary target variable, denoting time to resolution discretized into FAST (negative class) and SLOW (positive class) labels. We have evaluated SLDA on four large-scale open source projects. We show that the proposed model greatly improves recall, when compared to standard single topic algorithms.

Keywords: Bug triage · Bug-fix time prediction · Text categorization · Supervised topic models · Supervised Latent Dirichlet Allocation (SLDA)

The authors equally contributed to this paper.

© Springer International Publishing Switzerland 2016
T. Calders et al. (Eds.): DS 2016, LNAI 9956, pp. 167–182, 2016.
DOI: 10.1007/978-3-319-46307-0_11

1   Introduction

In recent years, with the increasing complexity of software systems, the task of software quality assurance has become progressively more challenging. In modern software development, software repositories are specialized database storing the output of the development process. The large-scale and partially unstructured data stored in these facilities are often not fully suited to traditional analysis methods [27]. A major role is played by finding and fixing bugs, which is a vital part of software development and quality assurance to such an extent that it has been estimated that software companies spend over 45 percent of their costs in fixing bugs [21,27]. The largest and most complex software projects are most often supported by a specialized database known as Bug Tracking System (BTS), used by quality assurance personnel and programmers to keep track of software problems and resolutions. Once a bug is reported, it is typically recorded in the BTS, and is assigned to a developer to resolve (bug triage). Current practice of bug triage is largely a manual collaborative process, in that the triager first examines whether a bug report contains sufficient or duplicated informations, then she/he confirms the bug and sets severity and priority, and finally decides who has the expertise in resolving it. Any such process is expected to be costly and inaccurate when the number of bug reports is large. For example, [12] report that empirical studies on Eclipse and Mozilla show that 37 %–44 % of bugs have been re-assigned (tossed) at least once to another developer. As there are many bug reports requiring resolution and potentially many developers working on a large project, it is non-trivial to assign a bug report to the appropriate developers.

For the reasons above, predicting on the basis of past data the time to fix a newly-reported bug has been shown to be an important target to support the whole triage process (and, in particular, to make the assignment more effective), and hence to help project managers to better estimate software maintenance efforts and improve cost-effectiveness [29]. Broadly speaking, bug fix-time is defined as the calendar time from the triage of a bug to the time the bug is resolved and closed as fixed [19]. Many researchers have proposed methods for automated bug-fix time prediction, largely based on machine learning techniques. Most of existing approaches are building prediction models based on the attributes of bug reports. For example, [20] used a historical portion of the Eclipse Bugzilla database. The predictor variables consisted of selected fields included, at time of confirmation, in the textual description of bugs reports. Several data mining models were then built and tested, using a nominal target class based on discretized time to resolution on a logarithmic scale. A logistic regression classifier provided the best classification accuracy of 34.5 %. In a similar fashion, [11] combined the attributes of the initial bug report with post-submission information. Bugs reports in the training set were classified into fast and slowly fixed. Not surprisingly, they found that post-submission data of bug reports improved their prediction model based on decision tree analysis. See also [29], where further studies are reviewed in greater depth.

Despite these apparently positive findings, [2] showed how existing models fail to validate on multiple large projects widely-used in bug studies, indicating a poor predictive accuracy comprised between 30 % and 49 %. In addition, it was found that there was no correlation between the probability that a new bug will be fixed, bug-opener's reputation and the time it takes to fix a bug. These findings show that we must be able to identify attributes which are effective in predicting bug-fix time. On the other side, instead of focusing on attribute subset selection, an alternative promising approach consists of using all available textual information. The problem of bug-fix time estimation is then mapped to a text categorization problem. A new bug report is classified to a set of discretized time to resolution classes (discretized bug-fix time), based on a classifier which is trained using historical data.

Traditional text categorization techniques establish the relationship between a set of predefined categories and their respective documents by analyzing document contents. This class of models can be unified under the assumption that each textual bug report exhibits exactly one probability distribution over strings drawn from some vocabulary of terms [17]. This assumption is often too limiting to model a large collection of textual bug reports. In standard unsupervised Latent Dirichlet Allocation (LDA) each word in a document is generated from a Multinomial distribution conditioned on its own topic, which is a probability distribution over the terms in the vocabulary representing a particular underlying semantic theme [3]. However, as our main objective is to predict bug-fix time, we consider a supervised Latent Dirichlet Allocation (SLDA) model [4,28], which adds to LDA a response variable consisting of an unordered binary target variable, denoting time to resolution discretized into FAST (negative class) and SLOW (positive class) class labels. We have evaluated SLDA on four large-scale open source projects (see Sect. 4 for in-depth details). We show that the proposed model greatly improves recall, when compared to single topic algorithms.

The remainder of the paper is organized as follows. Section 2 describes how our model collects the data. Section 3 deals with the methods used in our prediction models. Section 4 presents the empirical study and the results. Section 5, finally, draws the conclusions.

2   Data Collection

Data collection is the first step of bug-fix time prediction process, whose overall conceptual design is shown in Fig. 1. Our design is largely application independent, albeit we will use the open source BTS Bugzilla for the actual implementation [7] (see also Sect. 4 for further details). First, data gathering consists of selecting only those historical bug records which are sensible to predict discretized time to resolution of previously unseen bugs. In other words, we select textual reports of resolved and closed bugs only, whose Status field has been assigned to VERIFIED, as well as Resolution field has been assigned to FIXED.

As we said before, our approach maps the prediction problem into a text categorization task. Hence, once the relevant textual bug reports have been selected, we consider the content of the following fields (which are first extracted and then re-collapsed into a single text identified by a unique Id, and used as input for

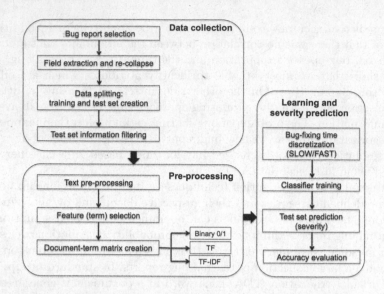

Fig. 1. Conceptual design of bug-fix time prediction process.

the subsequent phase described in Subsect. 3.1): **Product** (a real-world product, identified by a name and a description, having one or more bugs). **Component** (a given subsection of a Product, having one or more bugs). **Short_desc** (a one-sentence summary of the problem). **First_priority** (priority set by the user who created the report. The default values of priority are from P1, highest, to P5, lowest). **First_severity** (severity set by the user who created the report. This field indicates how severe the problem is, from blocker when the application is unusable, to trivial). **Reporter** (the account name of the user who created the report). **Assigned_to** (the account name of the developer to which the bug has been assigned to by the triager, and responsible for fixing the bug). **Days_resolution** (the calendar days needed to fix the bug). **Priority** (priority set either by the triager or a project manager). **Severity** (severity set either by the triager or a project manager). **First_comment** (the first comment posted by the user who created the report, which usually consists of a long description of the bug and its characteristics). **Comments** (subsequent comments posted by the **Reporter** and/or developers endowed with appropriate permissions, which can edit and change all bugs fields, and comment these activities accordingly).

We discarded a few fields, such as **Number_of_activities**, **CC_list**, **Status** and **Resolution**. For example, **Number_of_activities** is an integer value that would surely be removed during pre-processing steps, required to transform a raw text into a bag-of-word representation (see Sect. 3.1 and Fig. 2). Both **Status** and **Resolution** have been discarded because these fields have already been used for bug report selection, and they have always been assigned values VERIFIED and FIXED respectively. Finally, field **CC_list** contains a list of account names who get mail when the bug changes, and it is has been discarded because it has not been assigned any value in the great majority of cases. We also want to highlight that Days_resolution has been calculated as the number of days between the date the bug report has been assigned to a developer, and the date the Resolution field has been assigned to FIXED for the last time. This datum may be a very inaccurate estimate of the actual time spent on bug fixing, especially when developers of open source projects are considered (because of their discontinuous work patterns, for example during the weekend and/or their free time). Unfortunately, we cannot trace the actual time spent on a bug fixing, and provide a more accurate estimate expressed in person-hours.

When data collection and field extraction are complete, we randomly split the dataset into a training and a test dataset (given a fixed split percentage). This operation must necessarily take place after field extraction, because we have to filter out all post-submission information from test set. In fact, test instances simulate newly-opened and previously unseen bugs, and this makes compulsory to delete some of previously extracted fields that were not actually available before the bug was assigned. In particular, the deleted fields were Priority, Severity and Comments. On the contrary, fields First_priority, First_severity were not deleted, as they have been assigned a value by the user who created the report.

We want also to highlight that our design envisages that historical bug-fix times are discretized into two classes, conventionally labelled as SLOW and FAST. The SLOW labels indicates a discretized bug-fix time above some fixed threshold in right-tail of the empirical bug-fix time distribution. We also assume that SLOW indicates positive class, hence SLOW being the target class of our prediction exercise. In fact, we are interested in increasing the number of true positives for the positive class. In other words, over-estimation of bug-fix times can be considered as a less severe error than under-estimation.

3 Methods

3.1 Pre-processing Textual Description of Bug Reports

The forecasting models that we will use to predict bug-fix times are based on the representation of a document in terms of a bag-of-words. In this simplified representation, both the grammar and the order of occurrence of the words are not relevant. It is only relevant whether a term occur or not, as well as how many times it occurs in a textual description reproducing the bug.

A number of pre-processing steps are therefore necessary to convert the raw text into a bag-of-words representation. The steps followed in this paper, shown in Fig. 2, are quite common and well described in the literature on text categorization and Natural Language Processing (NLP; see, for example, [17]). The final goal is to precisely define a vocabulary of terms V, for example by eliminating those words that are very commonly used in a given language and focusing on the important words instead (stop word removal), or reducing inflectional forms to a common base form and then heuristically complete stemmed words by taking the most frequent match as completion (stemming and stem completion).

Two points appears to be particularly interesting and worthy of further eluci-
dation. The first is that probabilistic text modeling is often done ignoring multi-
word expressions that in specific contexts are given specific meanings. This issue
is particularly relevant for textual descriptions of software defects, and ignoring it can lead to sub-optimal outcomes for a numbers of reason that are well
described in [6]. The algorithm we have followed for detecting multi-words is a
modification of a very simple heuristic introduced by [14]. The detection consists
of the following steps:

1. Tokenize the text into bigrams (sequences of two adjacent words) and store
   candidate bigrams whose frequency of occurrence in the text is ≥ 3.
2. Pass the set of candidate bigrams through a part-of-speech (POS) filter.
3. Only let through the POS filter those part-of-speech patterns that are likely to
   be phrases, such as JJNN (Adjective+Noun, singular or mass, using the syn-
   tactic annotation scheme implemented by the Penn Treebank Project [18]).
   Ten predetermined patterns have been used to identify likely multi-words.

Surprisingly, the proposed algorithm has shown a reasonable accuracy in finding
multi-words of more than two words (results will be published elsewhere).

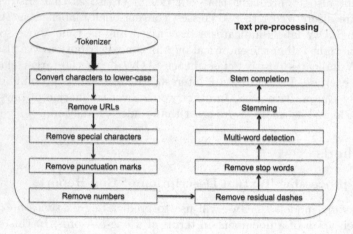

Fig. 2. The process of transforming a raw textual bug description into a bag-of-words.

Once text pre-processing has been completed, term selection is often necessary as it may even result in a moderate increase in predictive accuracy, depending on the classifier used and other factors [25]. In our approach, terms are sorted
(from best to worst) according to their estimated normalized expected mutual
information (NEMI, [17]) with the discretized time to resolution (SLOW/FAST).
We only include terms which have a NEMI greater than the average NEMI, ensuring that terms almost approximately independent with the target class label are
omitted.

The final pre-processing step consist of building, under different weighting schemes (see Fig. 1), the documents-terms arrays that will be used as inputs for the classification algorithms described in the following sections. Any global pre-processing parameters determined on the training data (such the number of documents that contain a given word) were subsequently applied to the test documents.

3.2 Prediction of Bug-Fix Times Based on Standard Text Categorization Models

As we said before, automated approaches to bug-fix time prediction, based on text-mining and machine-learning techniques, are not new [29]. One common generative model for a textual content is the Multivariate Bernoulli model (MB), which generates an indicator for each term of the vocabulary $|V|$, either 1 indicating presence of the term in the text or 0 indicating absence [17,26]. This amounts to assume that each document is represented as a binary vector $e_{1:|V|} = (e_1, \ldots, e_{|V|})$ of dimensionality $|V|$. For each bug report and independently of each other, the MB model assumes the following generative process, based on a Naïve Bayes conditional independence assumption:

1. Choose a discrete unordered label variable y (SLOW/FAST) from a discrete probability distribution.
2. For each word in V (i.e. for $t = 1, \ldots, |V|$) and independently of each other, choose a value of the indicator e_t from a Bernoulli distribution, $e_t|y \sim \mathsf{Bernouilli}(\pi_t)$, where $\pi_t = P(e_t = 1|y)$ is the probability that the word represented by e_t will occur at least once in any position, in a bug report labelled with severity y.

It is worth noting that we have suppressed writing the document index d. MB model implicitly relies on a bag-of-word assumption, and the natural input used by the model consists of a binary document-term incidence matrix (DTM). Hence, documents are modeled as a realization of a stochastic process, which is then reversed by standard machine learning techniques that return maximum-likelihood estimates of the posterior probabilities of a document (bug) d being in class y. Such estimates are in turn used for predicting discretized time to resolution y of newly-opened bugs. Despite its improved performance compared to the MB model, due to the incorporation of frequency information, we will not explicitly consider the Multinomial Naïve Bayes (MNB) model. Indeed, the supervised multi-topic model described in some detail in the next section, contains as a special case a supervised unigram model which is largely equivalent to the MNB model.

As an alternative way of considering the frequency of term occurrence in a document, we will use the Vector Space model (VS), with documents being represented as vectors in $\mathbb{R}^{|V|}$ [24]. The DTM can be weighted using the term frequency TF_{td} of word t in document d, as well as the term frequency–inverse document frequency $\mathsf{TF\text{-}IDF}_{td} = \mathsf{TF}_{td} \times \mathsf{IDF}_t$, which dampens the effects of local term-document counts. As usual, the inverse document frequency of word t is defined as $\mathsf{IDF}_t = \log\left(|\mathcal{D}|/\mathsf{DF}_t\right)$, where $|\mathcal{D}|$ is the number of documents in the training collection, and DF_t is the number of documents in $\mathcal{D}$ that contain the word t (document frequency). Using the vector-space approach, the documents of the training set correspond to a labeled set of points in an $|V|$-dimensional space. As state-of-the-art vector-based text classification algorithm [13,23], we will consider in Sect. 4 a non-linear Support Vector Machine (SVM) with soft margin classification.

3.3   Prediction Based on Supervised Latent Dirichlet Allocation

Models introduced so far limits each textual report to a single topic. This assumption may often be too limiting to model a large collection of textual bug reports, as any report typically concerns multiple topics and specific sub-issues in different proportions. As our objective is to discover this hidden thematic structure, and use it to predict the discretized time to resolution, we consider Supervised Latent Dirichlet Allocation (SLDA) as introduced in [4,28], which adds to standard Latent Dirichlet Allocation a target unordered binary variable.

We now briefly introduce the necessary notation to SLDA, as the representation of textual bug reports is different from that introduced in Sect. 3.2. As before, the target variabile is the unordered binary variable y denoting SLOW and FAST labels (also in this case, we suppress writing the document index d):

- A document d is a stream of N words, $d = (w_1, \ldots, w_N)$. Using superscripts to denote components, the νth word in $|V|$ is represented as a unit-basis vector w such that $w^\nu = 1$ and $w^u = 0$ for $u \neq \nu$.
- For each document we have K underlying semantic themes (topics), $\beta_{1:K} = (\beta_1, \ldots, \beta_K)$, where each β_k is a $|V|$-dimensional vector of probabilities over the elements of V, for $k = 1, \ldots, K$.
- $z_{1:N} = (z_1, \ldots, z_N)$ is a vector of K-dimensional vectors indicating, for $n = 1, \ldots, N$, the topic which has generated word w_n in document d. The indicator z of the k-th topic is represented as a K-dimensional unit-basis vector such that $z^k = 1$ and $z^j = 0$ for $j \neq k$.

The topic indicator uniquely selects a probability distribution in $\beta_{1:K}$, as $\beta_{z_n} \equiv \beta_k$ when $z_n^k = 1$. Independently of each other, the generative process of each pre-processed bug report d is the following:

1. Draw topic proportions from a symmetric Dirichlet distribution over the K-dimensional simplex, $\theta|\alpha \sim \mathsf{Dirichlet}_K(\alpha)$.
2. For each word w_n, $n = 1, \ldots, N$, and independently of each other:
   (a) Choose a topic from a Multinomial distribution with probabilities θ, $z_n|\theta \sim \mathsf{Multinomial}_K(\theta)$.
   (b) Choose a word from a Multinomial distribution with probabilities dependent on z_n and $\beta_{1:K}$, $w_n|z_n, \beta_{1:K} \sim \mathsf{Multinomial}_{|V|}(\beta_{z_n})$.

3. Draw response variabile y (SLOW/FAST, with SLOW $\equiv 1$) from a logistic Generalized Linear Model (GLM) of the form:

$$y|z_{1:N}, \eta \sim \mathsf{Bernouilli} \left(\frac{\exp(\eta^\top \overline{z})}{1 + \exp(\eta^\top \overline{z})} \right), \tag{1}$$

where $\overline{z} = (1/N) \sum_{n=1}^N z_n$ is the vector of empirical topic frequencies.

In this hierarchical specification, we model discretized bug-fix time as a nonlinear function of the empirical topic frequencies via the linear predictor $\eta^\top \overline{z}$ (readers unfamiliar with GLMs may profitably consult [9]). A graphical model representation of SLDA is depicted in Fig. 3, where observable stochastic nodes are represented as gray circles and latent stochastic variables are white circles. Parameters $\alpha, \beta_{1:K}$ and η (white boxes) are treated as unknown fixed hyperparameters rather than latent stochastic nodes.

It is apparent from Fig. 3 that the response y and the words share a common ancestor (the latent topic variables), and hence they are not conditionally independent. Documents are generated as a bag-of-words under full word exchangeability, and then topics are used to explain the response. Other specifications are indeed possibile, for example y can be regressed as a nonlinear function of topic proportions θ, but [4] claim that the predictive performance degrades as a consequence of the fact the topic probabilistic mass does not directly predicts discretized bug-fix times.

Posterior inference of latent model variables is not feasible, as the conditional posterior distribution $p(\theta, z_{1:N}|w_{1:N}, y, \alpha, \beta_{1:K}, \eta)$ has not a closed form. Consequently, we use a standard variational Bayes (VB) parameter estimation approach under a mean-field approximation, say $q(\theta, z_{1:N}|\phi)$, of the conditional posterior distribution [5,16]. In the variational E-step of the algorithm, the variational parameters ϕ are optimized to minimize the Kullback-Leibler divergence

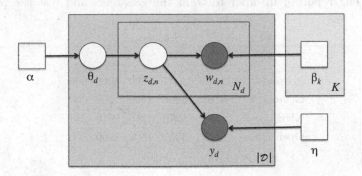

Fig. 3. A graphical model representation of Supervised Latent Dirichlet Allocation (SLDA). Parameters $\alpha, \beta_{1:K}$ and η are treated as unknown hyper-parameters to be estimated, rather than random variables.

between $q(\theta, z_{1:N}|\phi)$ and the conditional posterior distribution, given the hyper-parameters $\alpha, \beta_{1:K}$ and η. In the M-step, the hyper-parameters are optimized to maximize the evidence lower bound (ELBO) given the variational parameters.

Let $\{\tilde{\alpha}, \tilde{\beta}_{1:K}, \tilde{\eta}\}$ denote the hyper-parameters of a fitted model. As our main objective is to apply logistic SLDA to predict discretized bug-fix time of a newly-opened bug, whose words are $w_{1:N}^{\mathrm{new}}$, this amounts to approximating the marginal posterior expectation of the response variable:

$$E(y^{\mathrm{new}}|w_{1:N}^{\mathrm{new}}, \tilde{\alpha}, \tilde{\beta}_{1:K}, \tilde{\eta}) = E(\mu(\tilde{\eta}^\top \overline{z}^{\mathrm{new}})|w_{1:N}^{\mathrm{new}}, \tilde{\alpha}, \tilde{\beta}_{1:K}), \qquad (2)$$

with $\mu(\tilde{\eta}^\top \overline{z}^{\mathrm{new}}) = \exp(\tilde{\eta}^\top \overline{z}^{\mathrm{new}})/(1+\exp(\tilde{\eta}^\top \overline{z}^{\mathrm{new}}))$. The identity (2) easily follows from the law of iterated expectations. Using the multivariate Delta method and a suitable variational procedure, the RHS of (2) can be easily approximated (we defer specifics to [4,8]). Of course, if $E(y^{\mathrm{new}}|w_{1:N}^{\mathrm{new}}, \tilde{\alpha}, \tilde{\beta}_{1:K}, \tilde{\eta}) > 0.5$ then $y^{\mathrm{new}} \equiv 1 (\equiv \mathsf{SLOW})$.

4   Results

We have obtained bug report information from Bugzilla repositories of four large open source software projects: Eclipse, Gentoo, KDE and OpenOffice. Data were automatically extracted from Bugzilla data sources, using a suitable scraping routine written in PHP/JavaScript/Ajax. Raw textual reports were pre-processed and analyzed using the R software system [22]. The table below shows the total number of textual bug reports extracted for each project (n_1), having both Status field assigned to VERIFIED and $\mathsf{Resolution}$ field assigned to FIXED. Next, we deleted some textual reports because of corrupted and unre-coverable records, or missing XML report, or dimension being too large (the resulting sample sizes are in column n_2). If $n_2 > 1500$ (resp.: $n_2 \leq 1500$) we randomly selected $n_3 = 1500$ bug reports (resp.: we retained the whole set of $n_3 = n_2$ bug reports), in order to train the classifiers and test the proposed models.

	n_1	n_2	n_3	n_4	n_5
Eclipse	44435	44347	1500	1200	300
Gentoo	3704	2466	1500	1200	300
KDE	1275	1270	1270	1016	254
Open Office	3057	3057	1500	1200	300

These subsets were randomly divided into a training and a test part, using an 80:20 split ratio. The resulting sample sizes are indicated by n_4 (training set) and n_5 (test set). Before splitting, for each experiment we binned bug reports into FAST and SLOW using the third quartile $q_{0.75}$ of the empirical distribution of bug resolution times. The table below shows the resulting binned distribution.

We can see that the time needed to fix the bugs exhibits large variations, and that the underlying continuous-time effort distribution is long-tailed, with a large percentage of bugs being fixed within a relatively short time.

	Eclipse	Gentoo	KDE	Open Office
FAST	0–52	0–42	0–248	0–107
SLOW	53–3696	43–1854	249–2188	108–2147

As shown below, the SLOW/FAST ratio is approximately preserved after each dataset is split into a training and a test part. Hence, the two class are moderately imbalanced in both the training and the test set, with a low ratio of positive to negative class (as we said before, SLOW identifies the positive class).

	n_3	Training set			Test set		
		SLOW	FAST	ratio	SLOW	FAST	ratio
Eclipse	1500	299	901	25.00%	82	218	27.00%
Gentoo	1500	298	902	25.00%	79·	221	26.00%
KDE	1270	254	762	25.00%	67	187	26.00%
Open Office	1500	300	900	25.00%	65	235	22.00%

As we highlighted in Sect. 2, we are interested in increasing the number of true positives for the positive class, considering over-estimation of bug-fix times as a less severe error than under-estimation. We used accuracy, precision, recall and false positive rate (FPR) for measuring the performance of prediction models. Accuracy denotes the proportion of correctly predicted bugs: Accuracy = (TP+TN)/(TP+FP+TN+FN). Precision denotes the proportion of correctly predicted SLOW bugs: Precision = TP/(TP+FP). Recall denotes the proportion of true positives of all SLOW bugs: Recall = TP/(TP+FN). Finally, false positive rate measure the proportion of false positive of all FAST bugs: FPR = FP/(FP+TN). The following classification models were trained over the training set, and evaluated over the test set:

- Multivariate Bernoulli (MB), with either no posterior class probabilities smoothing, or Laplace λ smoothing parameter respectively set to $1, 2, 3$ [17].
- Support Vector Machines (SVM), with sigmoid kernel and soft-margin classification, cost parameter C respectively set to $0.1, 1, 10, 100$ and precision parameter γ respectively set to $0.001, 0.01, 0.1, 1$ [15].
- Supervised Latent Dirichlet Allocation (SLDA), with number of topics K being respectively set to $1, 2, 5, 10, 15, 20, 25, 30, 35, 40, 50, 100$. When $K = 1$, the SLDA model becomes essentially equivalent to the Multinomial Naïve Bayes (MNB) model.

In what follows, we optimize over the test set and show only the best models (along with the corresponding parameter settings), i.e. the model with the highest predictive accuracy among the models of the same class. The full result set is shown below. We look at the first table (Eclipse), as it exhibits a typical pattern. The highest achievable accuracy with SLDA ($K = 25$ topics) is lower than the best achievable accuracies with both MB and SVM. However, accuracy provides reliable comparisons only when the two target classes have equal importance. In our setting, we assume that the minority class (SLOW) is more important, because of its larger impact in terms of cost/effectiveness. If, consequently, we assume that the main goal is increasing the recall, the logistic SLDA model has the best performance. FPR increases too, as expected, because of the larger number of FAST bugs that are classified SLOW under the SLDA model. However, costs incurred in false positive are generally very low. On the contrary, both MB and SVM classifiers have high accuracy, but they failed to correctly predict most of SLOW bugs. On the whole, these result clearly show that the use of a supervised topic model greatly improves the recall of bug-fix time prediction.

Eclipse.

	Parameters	Accuracy	Precision	Recall	FPR
MB	$\lambda = 2$	0.73	0.60	0.04	0.01
SVM	$\gamma = 0.001$, $C = 10$	0.67	0.23	0.09	0.11
SLDA	$K = 25$	0.57	0.32	0.48	0.40

Gentoo.

	Parameters	Accuracy	Precision	Recall	FPR
MB	$\lambda = 2$	0.74	0.50	0.13	0.05
SVM	$\gamma = 0.001$, $C = 100$	0.74	0.67	0.03	0.00
SLDA	$K = 30$	0.43	0.27	0.70	0.67

KDE.

	Parameters	Accuracy	Precision	Recall	FPR
MB	$\lambda = 2$	0.83	0.64	0.79	0.02
SVM	$\gamma = 0.001$, $C = 100$	0.60	0.03	0.01	0.19
SLDA	$K = 40$	0.41	0.29	0.84	0.74

Open Office.

	Parameters	Accuracy	Precision	Recall	FPR
MB	$\lambda = 2$	0.78	0.00	0.00	0.00
SVM	$\gamma = 0.001, C = 100$	0.58	0.22	0.38	0.38
SLDA	$K = 10$	0.51	0.23	0.55	0.55

5 Discussion and Conclusion

Manual bug triage is expensive both in time and cost. But even more importantly, manual triage is error-prone due to the large number of daily newly-
opened bugs and the lack of knowledge about all bugs by the developers. We
have, therefore, proposed a novel model for automatic prediction of bug-fix time
of newly opened bugs, in order to support the whole triage process and, in particular, the assignment of any new bug to a developer who will try to fix it.
Our prediction models use text categorization techniques, mapping each textual
bug description into a bag-of-words after a suitable pre-processing stage. Each
bug in the training data set is classified as SLOW or FAST (discretized time to
resolution). The trained prediction model is then used to predict the discretized
time to resolution of each bug in the test set. Any post-submission information,
which was not actually available before the bug was assigned, has been removed
from the test set. We compared two single-topic supervised learning algorithms,
multivariate Bernoulli Model (MB) and Support Vector Machines (SVM), with
a multi-topic model known as Supervised Latent Dirichlet Allocation (SLDA),
recently introduced in [4].

To evaluate the forecasting accuracy of the proposed predictive model against
that of MB and SVM algorithms, we used bug datasets on bug repositories of
four large open source projects: Eclipse, Gentoo, KDE and OpenOffice. Results
show that the proposed model greatly improves recall, when compared to single topic algorithms. On the other hand, the loss of accuracy of our method is
quite significant. However, predictive accuracy provides meaningful and reliable
comparisons only when the two target classes have equal importance. In our
experimental setting the negative class (FAST) plays a minor role. Therefore, we
may assume that the main goal is increasing the recall, that is the true positives
for the positive class (SLOW). In this case, the number of false positives can also
be increased, even though costs incurred in false positives are generally very low.
Finally, a comparison with previously reported literature values shows a marked
improvement of the predictive accuracy of the two single topic algorithms. However, a direct comparison is not possible and further validation will be needed, as
different software projects were examined, and different attributes and/or parts
of textual bug reports were selected, as well as different pre-processing methods
were involved in extracting the bag-of-words representations.

In conclusion, the proposed method seems promising for implementing a large-scale bug-fix time prediction system. In the future, we plan to investigate the following threatens to internal validity:

- Using the quantile q_a with $a = 0.75$ to separate positive and negative instances is arbitrary. Letting a vary has an impact on both the imbalance ratio and the predictive accuracy. A sensitivity analysis is therefore needed.
- We need to assess the performance on a larger and independent validation set. Each presented method is indeed trained for a number of parameter settings and tested on the test set, but only the best results are presented. In the future, the methods will be optimized on a separate validation set, instead of the test set.
- Another issue consists of identifying the potential outliers of the distribution of bug-fix times, and using them to filter out the data sets. Many authors have demonstrated that filtering these outliers can improve the accuracy of the prediction models (see, for example, [1]).
- Most of defect tracking systems are just ticketing systems, that cannot keep track of actual person-hours spent to resolve a bug. This threat to internal validity of bug-fix time prediction models has not been investigated yet. In the same way, similarities and differences between open and non-open source software projects need to be investigated further.

Finally, some recent techniques tackle the bug-fix time prediction problem using a radically different (yet very promising) approach, based on predictive Process Mining, in which each ticket gives rise to a series of activities viewed as an instance of some ticket handling process, and which can be handled along with textual informations and descriptive fields, in order to assign a performance value to any partial process instance, and monitoring the resolution status during its enactment as well (see [10]). It is therefore highly desirable to compare (at least experimentally) our solution with this kind of approach. The future work will explore this interesting task.

References

1. AbdelMoez, W., Kholief, M., Elsalmy, F.M.: Improving bug fix-time prediction model by filtering out outliers. In: 2013 The International Conference on Technological Advances in Electrical, Electronics and Computer Engineering (TAEECE), pp. 359–364 (2013)
2. Bhattacharya, P., Neamtiu, I.: Bug-fix time prediction models: can we do better? In: Proceeding of the 8th Working Conference on Mining Software Repositories, MSR 2011, pp. 207–210. ACM Press, New York (2011)
3. Blei, D.M., Ng, A.Y., Jordan, M.I.: Latent Dirichlet allocation. J. Mach. Learn. Res. **3**, 993–1022 (2003)
4. Blei, D.M., McAuliffe, J.D.: Supervised topic models. In: NIPS (2007)
5. Blei, D.M., Kucukelbir, A., McAuliffe, J.D.: Variational inference: a review for statisticians, pp. 1–33 (2016). http://arxiv.org/abs/1601.00670

6. Boyd-Graber, J., Mimno, D., Newman, D.: Care and feeding of topic models: problems, diagnostics, and improvements. In: Airoldi, E.M., Blei, D., Erosheva, E.A., Fienberg, S.E. (eds.) Handbook of Mixed Membership Models and Their Applications. CRC Press, Boca Raton (2014)
7. The Bugzilla Team: Bugzilla Documentation 5.0.3+ (2016). https://www.bugzilla.org/docs/
8. Chang, J., Blei, D.M.: Hierarchical relational models for document networks. Ann. Appl. Stat. 4(1), 124–150 (2010)
9. Dobson, A.J., Barnett, A.: An Introduction to Generalized Linear Models: Chapman & Hall/CRC Texts in Statistical Science, 3rd edn. Taylor & Francis (2008)
10. Folino, F., Guarascio, M., Pontieri, L.: An approach to the discovery of accurate and expressive fix-time prediction models. In: Hammoudi, S., Maciaszek, L., Teniente, E., Camp, O., Cordeiro, J. (eds.) ICEIS 2015. LNBIP, vol. 241, pp. 108–128. Springer, Heidelberg (2015). doi:10.1007/978-3-319-22348-3_7
11. Giger, E., Pinzger, M., Gall, H.: Predicting the fix time of bugs. In: Proceedings of the 2nd International Workshop on Recommendation Systems for Software Engineering, RSSE 2010, pp. 52–56. ACM Press, New York (2010)
12. Hu, H., Zhang, H., Xuan, J., Sun, W.: Effective bug triage based on historical bug-fix information. In: 2014 IEEE 25th International Symposium on Software Reliability Engineering, pp. 122–132. IEEE (2014)
13. Joachims, T.: Text categorization with support vector machines: learning with many relevant features. In: Nédellec, C., Rouveirol, C. (eds.) ECML 1998. LNCS, vol. 1398, pp. 137–142. Springer, Heidelberg (1998). doi:10.1007/BFb0026683
14. Justeson, J.S., Katz, S.M.: Technical terminology: some linguistic properties and an algorithm for identification in text. Nat. Lang. Eng. 1(01), 9–27 (1995)
15. Karatzoglou, A., Meyer, D., Hornik, K.: Support vector machines in R. J. Stat. Softw. 15(1), 1–28 (2006)
16. Lakshminarayanan, B., Raich, R.: Inference in supervised latent Dirichlet allocation. In: 2011 IEEE International Workshop on Machine Learning for Signal Processing, pp. 1–6 (2011)
17. Manning, C.D., Raghavan, P., Schütze, H.: Introduction to Information Retrieval. Cambridge University Press, New York (2008)
18. Marcus, M., Kim, G., Marcinkiewicz, M.A., MacIntyre, R., Bies, A., Ferguson, M., Katz, K., Schasberger, B.: The Penn Treebank: annotating predicate argument structure. In: Proceedings of the Workshop on Human Language Technology, pp. 114–119. Association for Computational Linguistics, Stroudsburg (1995)
19. Marks, L., Zou, Y., Hassan, A.E.: Studying the fix-time for bugs in large open source projects. In: Proceedings of the 7th International Conference on Predictive Models in Software Engineering, Promise 2011, pp. 1–8. ACM Press, New York (2011)
20. Panjer, L.D.: Predicting eclipse bug lifetimes. In: Fourth International Workshop on Mining Software Repositories, MSR 2007: ICSE Workshops 2007, pp. 29–32. IEEE, Washington, DC (2007). doi:10.1109/MSR.2007.25
21. Pressman, R.S., Maxim, B.R.: Software Engineering: A Practitioner's Approach, 8th edn. McGraw-Hill Higher Education (2014)
22. Core Team, R.: R: A language and environment for statistical computing. R Foundation for Statistical Computing, Vienna, Austria (2016). https://www.R-project.org/
23. Rennie, J.D.M., Shih, L., Teevan, J., Karger, D.R.: Tackling the poor assumptions of Naïve Bayes text classifiers. In: Proceedings of the Twentieth International Conference on Machine Learning (ICML-2003), Washington DC, pp. 616–662 (2003)

24. Salton, G., Wong, A., Yang, C.S.: A vector space model for automatic indexing. Commun. ACM **18**(11), 613–620 (1975)
25. Sebastiani, F.: Machine learning in automated text categorization. ACM Comput. Surv. **34**(1), 1–47 (2002)
26. Wilbur, W.J., Kim, W.: The ineffectiveness of within-document term frequency in text classification. Inf. Retr. **12**(5), 509–525 (2009)
27. Xuan, J., Jiang, H., Hu, Y., Ren, Z., Zou, W., Luo, Z., Wu, X.: Towards effective bug triage with software data reduction techniques. IEEE Trans. Knowl. Data Eng. **27**(1), 264–280 (2015)
28. Zhang, C., Kjellström, H.: How to supervise topic models. In: Agapito, L., Bronstein, M.M., Rother, C. (eds.) ECCV 2014, Part II. LNCS, vol. 8926, pp. 500–515. Springer, Heidelberg (2015). doi:10.1007/978-3-319-16181-5_39
29. Zhang, J., Wang, X., Hao, D., Xie, B., Zhang, L., Mei, H.: A survey on bug-report analysis. Sci. China Inf. Sci. **58**(2), 1–24 (2015)

Predicting Wildfires

Propositional and Relational Spatio-Temporal Pre-processing Approaches

Mariana Oliveira[1,2(✉)], Luís Torgo[1,2], and Vítor Santos Costa[1,2]

[1] DCC – Faculdade de Ciências, Universidade Do Porto, Porto, Portugal
mariana.r.oliveira@inesctec.pt
[2] INESC TEC, Porto, Portugal

Abstract. We present and evaluate two different methods for building spatio-temporal features: a propositional method and a method based on propositionalisation of relational clauses. Our motivating application, a regression problem, requires the prediction of the fraction of each Portuguese parish burnt yearly by wildfires – a problem with a strong socio-economic and environmental impact in the country. We evaluate and compare how these methods perform individually and combined together. We successfully use under-sampling to deal with the high skew in the data set. We find that combining the approaches significantly improves the similar results obtained by each method individually.

1 Introduction

Wildfires are an environmental hazard that affects severely most southern European countries, and Portugal in particular. Although nature relies on fire to rejuvenate the forest, factors such as the introduction of non-indigenous species, the rise of industrial forestry, rural depopulation, and climate changes have compounded the problem [5], severely affecting the country's finances and environment, and sometimes even causing human losses. Given a limited amount of resources to address wildfires, a better understanding of the factors that lead to fire events, and namely to severe fire events, is needed.

Toward this goal, data on wildfires and corresponding geographical context has been continuously collected by several organisations, both at national and at European level. This is an example of the novel environmental and socio-economic databases that store data on entities and how these entities occupy and transform a space while interacting with each other. Besides background data on the entities or the location, such as a site's topology or a country's administrative units, most data will be about the events of interest.

Given the size and complexity of the data, it would be difficult even for a highly qualified expert to fully leverage it. Spatio-temporal data mining techniques offer the promise of finding human-interpretable patterns (e.g., automatically learning association rules), or of models that can be used to successfully predict unknown

---

We thank Dr. João Torres for providing the data we worked with.

T. Calders et al. (Eds.): DS 2016, LNAI 9956, pp. 183–197, 2016.
DOI: 10.1007/978-3-319-46307-0_12

or future values based on a set of explanatory variables (e.g., regression models to predict risk of fire). We focus on the latter. Dealing with both spatial and temporal dimensions with this goal in mind presents numerous challenges as: (i) the dimensions have different properties, (ii) relationships between spatio-temporal objects are often fuzzy or implicit [1], (iii) multiple levels of granularity and of abstraction of both dimensions impact results differently [27], and (iv) data is often voluminous making scalability a concern.

Propositional data mining methods work on a single table, often assuming that each instance in a data set has been independently sampled from the same underlying distribution. In contrast, multi-relational methods explicitly consider the complex nature of the data, often extending a corresponding propositional approach in order to work on multiple tables from a relational database [12]. In [15], Malerba argues that relational approaches are particularly suited to spatial data mining tasks since they can deal with heterogeneous spatial objects and naturally represent a wide variety of relationships between them. We believe the same argument can be made for spatio-temporal tasks.

Regarding wildfires, the goal will be to estimate the percentage of area burnt (or burn fraction) of a pre-defined unit (in this case, a Portuguese parish) over periods of one year. Moreover, we would like to differentiate major events, that are hard to control, from smallish fire events, that are more frequent but have little impact. To do so, we approach this problem as a regression task.

In order to construct the regression model, we follow the widely used approach of encoding the relevant spatio-temporal information in the form of propositional features through a pre-processing step. These features can be obtained by considering spatio and/or temporal properties in the data [6,18] or even learned through propositional and/or relational techniques. This approach makes it possible to benefit from standard (propositional) prediction models that are both efficient and easy to use.

In this work we present and evaluate two different methods for building spatio-temporal features: a propositional method and a relational method based on Inductive Logic Programming (ILP). We compare how these methods perform individually and combined together. We evaluate performance quantitatively and through the extra knowledge it provides. We also address the high skew in this data set, i.e., the fact that the most important cases of higher burn fraction are under-represented in the data, and demonstrate that under-sampling is quite effective in improving model performance under these conditions.

We proceed to mention some examples of relational and propositional approaches to spatio-temporal prediction that have already been proposed. Purely relational approaches include methods based on ILP [16] and the use of graphical models [4,23]. Propositional approaches include methods based on clustering [3], combinations of spatial and temporal methods [11], extensions of time series forecasting techniques such as ARIMA to account for spatial information [20] and of spatial techniques such as GWR to transfer across time [2]. Further, propositional and relational approaches have been contrasted before in a spatial associative classification setting [10].

In the following section we present the data set we worked with. In Sect. 3 we describe the pre-processing approaches we applied, comparing their results. Section 4 includes concluding remarks and future research directions.

2 Wildfires in Portugal

We proceed to describe the data set motivating our work. We also discuss the pre-computation of spatial relationships below.

2.1 Data Set

Our motivating application is the evolution of wildfires across mainland Portugal from 1991 to 2010.[1] Spatially, we work at the civil parish level (there are 2882 of them). Our target variable is the percentage of a parish's area burnt yearly by wildfires. The variable is non-cumulative, i.e., an area burning multiple times during the year is considered only once. We used 23 other numerical variables with different temporal granularity as background knowledge (variables measured only once are considered fixed; see Table 1).

Imbalanced Domain. The values taken by our target variable range from 0 % (when no wildfire occurred throughout the year) to 99.8 % (see Fig. 1). The distribution of values is imbalanced in a way that does not correspond to our preference bias, that is, while we are most interested in accurately predicting instances of high burn fraction, these cases are under-represented in the data set. In fact, only about a third of the 57 640 instances have non-zero burn fractions, while less than 9 % of cases present values of 5 % or above and only 0.5 % of cases have a burn fraction of 40 % or more.

Table 1. Explanatory variables used as background knowledge in our data set.

Land cover	Eucalyptus Tall scrubland Small scrubland Broad-leaved forest Pinewood Urban	(%) Fixed	**Road density**	All roads Roads > 6m wide Roads < 6m wide		Fixed
			Census data	Irrigable area Meadow area	(%)	Decennial (from 1989)
				Bovine population density Ovine population density Caprine population density	(ha^{-1})	
Terrain	Maximum altitude Mean altitude	(m)		Population density Population's mean age	(ha^{-1}) (years)	Decennial (from 1991)
	Maximum slope Mean slope			Population of age 65+ Housing density	(%) (ha^{-1})	Decennial (from 2001)

---

[1] Most data for this application (with the exception of census data downloaded from ine.pt) provided by Dr. João Torres, researcher at CIBIO. Details regarding data collection can be found in [26].

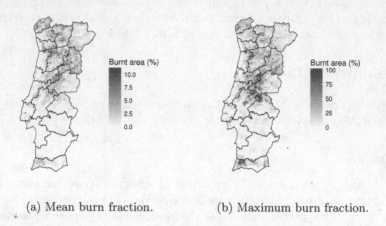

(a) Mean burn fraction.   (b) Maximum burn fraction.

Fig. 1. Mean and maximum percentage of area burnt yearly per parish. Note that they have different scales. Black lines delineate Portuguese districts.

2.2 Computing Spatial Relationships

Both pre-processing approaches we present require the computation of spatial relationships in the data set. We made use of the `PostGIS` spatial extension to a `PostgreSQL` database loaded with the shapefiles of the 2882 Portuguese civil parishes in order to determine spatial neighbourhoods and border parishes.

Defining Neighbourhoods. Neighbourhoods for each parish consist of all intersecting parishes, calculated using the `PostGIS` function `ST_Intersect` (prefix `ST_` identifies `PostGIS` functions).

Neighbour Direction. The relative direction of a neighbour in relation to a reference parish, O, is taken into consideration in an effort to capture effects of dominant winds affecting the spread direction of wildfires. This is not a straightforward problem, given the heterogeneous shapes presented by parishes. Our solution is meant to be fast and easily computable. We first compute cartographic azimuths using the parish of interest as reference and each neighbour's centroids (calculated with `ST_Centroid`). The azimuth is given clockwise relative to the north, resulting in the following definition:

$$\text{neighbour direction} = \begin{cases} \text{east} & \text{if } az \in [45, 135[\,° \\ \text{south} & \text{if } az \in [135, 225[\,° \\ \text{west} & \text{if } az \in [225, 315[\,° \\ \text{north} & \text{if } az \in ([315, 360] \cup [0, 45[)\,° \end{cases} \qquad (1)$$

where az is the result of applying the `PostGIS` function `ST_Azimuth` to the centroids of a reference parish and one of its neighbours.

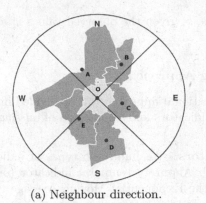

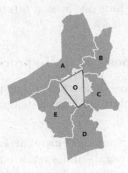

(a) Neighbour direction.   (b) Simplified borders.

Fig. 2. On the left, a parish's neighbourhood divided by cardinal directions. Red dots represent the parishes' centroids. The red lines centred at the reference parish, O, divide the neighbourhood in four directions. According to this division, O has no western neighbours. A and B are its northern neighbours; E and D, southern neighbours; and C, an eastern neighbour. On the right, coloured lines define simplified borders between O and each of its neighbours. (Color figure online)

Issue. The example in Fig. 2a illustrates a problem in our proposal. While it is clear O shares borders with neighbours A and E in the western direction, no western neighbour is found since their centroids fall under the northern and southern subspaces. Our propositional approach mitigates this by the way it fills in missing spatio-temporal indicators (Sect. 3.2), while the relational approach relies on the explicit neighbourhood relationship itself (Sect. 3.3).

3   Predicting Wildfires

In the following sections, we define our problem clearly and detail the different steps involved in each pre-processing methodology, as well as steps common to both approaches.

3.1   Problem Definition

Predictive data analysis tasks face the problem of approximating an unknown function $Y = f(X_1, X_2, \ldots, X_p)$ mapping values of a set of predictors or *explanatory* variables, $\mathbf{X}$, into the values of a *target* variable, Y, where the approximation is called the *model.*

In a spatio-temporal setting, the aim is to predict values at different times and locations. In this work, we aim at forecasting future values given past information from neighbouring locations in the past. Consider a data set $D = \{\{y_1^1, x_{a_1}^1, \ldots, x_{p_1}^1\}, \ldots, \{y_n^m, x_{a_n}^m, \ldots, x_{p_n}^m\}\}$ where y_t^l and $x_{i_t}^l$ correspond, respectively, to the values of the target variable Y and explanatory variables X_i at geographical location l and time t. The goal is to predict the value of Y at a location of interest, s, at a future time, k, given the observed values y_t^l and $\mathbf{x}_t^l$, such that $t < k$.

3.2   Propositional Pre-processing Approach

The pre-processing stage of our propositional approach can be divided into two steps: calculation of spatio-temporal indicators and imputation of missing data.

Building Spatio-Temporal Indicators. We build two types of indicators: purely temporal, and spatio-temporal. A purely temporal indicator (or, self-indicator) is obtained by calculating the Exponential Moving Average (EMA) of n past values of the target variable for the reference parish in the previous 9 years. We also build spatio-temporal indicators, inspired by the work of [18], considering historical values of the target variable for direct neighbours located at each cardinal direction in the previous 5 years. We compute the indicator for a particular direction in two steps. First, we calculate the EMA (with ratio $\frac{2}{n+1}$) of the target variable for each neighbour whose centroid falls in that direction. Then, if there is more than one, a weighted mean of these values is calculated.

Weighing Neighbours. The weights above are designed to roughly approximate the risk of exposure of a parish to wildfire spread from each neighbour. The strength of connection between neighbours could be directly measured by the fraction of the border shared with them. However, meandering borders can easily increase in length without proportionately increasing the degree of exposure of the reference parish to wildfires originating in that particular neighbour. Therefore, we define a simplified border as the maximum distance between any two points of the intersection using ST_MaxDistance (see Fig. 2b). The weight of a neighbour is the length of its simplified border divided by the sum of the lengths of the simplified borders of all neighbours in that direction.

Issues. Weighting a neighbour's EMA in this fashion raises a problem: the centroid of a neighbour may fall in a subspace of a certain direction while most of its border with the reference parish belongs to another. However, since we do not have information regarding which portion of a neighbour was burnt (whether it was close to the border or not), and the level of temporal and spatial granularity of our data is low, these approximations are still reasonable.

Filling in Missing Data. In order to use standard learning algorithms, we pre-selected reasonable procedures to fill in missing data. First, independent spatial-only Inverse Distance Weighting (IDW) (as implemented in [19]) is used to fill in values missing due to unavailability of spatial data (2.8 % of all cells). Next, missing values due to heterogeneous temporal granularity (20.6 % of all cells) are filled in with the latest measurement as there are not enough points to meaningfully smooth over values. Finally, spatio-temporal indicators missing due to no neighbour centroids falling within a certain direction (6.8 % of all cells) are filled in with zero if the parish borders with the sea/ocean or the average of the two contiguous directions, otherwise.

3.3   Relational Pre-processing Approach

The relational approach we propose follows three steps: first, we rely on the clause search mechanism implemented in the Aleph ILP system [22]; we then propositionalise by associating each clause to a different attribute; last, we construct the attribute examples table. In this table, given an example e and a clause (attribute) i, $e[i] = 1$ if the clause is true for the instance, and $e[i] = 0$ otherwise. This approach has been used before in diverse contexts, including spatial classification [10]. Although Aleph searches for clauses that are optimized to perform well for binary classification, we hypothesise that standard regression algorithms such as Support Vector Regression machines (SVRs) can successfully use the binary features representing interesting clauses to accurately approximate the numerical values of the target variable.

Background Knowledge and Examples

Explanatory Attributes. In order to use Aleph, each explanatory attribute was converted into a binary (if fixed) or ternary (if time-varying) Prolog predicate.

Spatial Relationships. Spatial relationships are expressed by the following predicates: neighbour(Parish, Neighbour) where recursion on Neighbour is avoided, neighbourDirection(Parish, Neighbour, Direction) where Direction is defined in Sect. 2.2, and border(Parish, Object) where Object can take the values sea or spain.

Temporal Relationships. We use the number of years past since a wildfire last affected a certain parish. This is represented by a pair of predicates: yearsSince FireLE(Parish, Year, TimeDist) and yearsSinceFireGE(Parish, Year, TimeDist), which are true if by Year the Parish has suffered a wildfire TimeDist or less years ago, or TimeDist or more years ago, respectively. We use two definitions, conditioned on TimeDist being a variable: the first holds when TimeDist in unbound, and is used in Aleph's saturation step; the second, is always called with the argument bound to one of the values found during saturation (TimeDist is a constant).

Additional Predicates. Since Aleph does not deal with numerical attributes directly, auxiliary predicates were designed. That is, each time-varying attribute had a corresponding attributeLE(Parish, Year, Value) and attributeGE (Parish, Year, Value) meaning that the value of attribute in Parish measured in or before Year is lesser or equal (or greater or equal) to Value. Similar predicates attributeLE(Parish, Value) and attributeGE(Parish, Value) were created for fixed attributes.

Examples. The predicate at the head of çlauses is `burnt(Parish, Year)`, where a positive example is a `Parish` burnt more than 5 % in a given `Year`.

Clause Search and Selection. The standard Aleph command `induce/0` is not appropriate in this context. Instead, we devise our own method of clause search and selection. We set clause cost (used for generalisation on the reduction step) to be the F_β-measure defined as

$$F_\beta\text{-measure} = (1 + \beta^2) \cdot \frac{precision \cdot recall}{(\beta^2 \cdot precision) + recall}. \tag{2}$$

First, we randomly chose an example as the seed for search. We then generate clauses until a certain threshold is reached. Instead of trying to find a theory covering all examples, we store each and every clause that has been the best so far for each saturated example according to our chosen metric. In the style of Gleaner [13], we experimented with different values of β for the F_β-measure, trying 60 random seed examples for each $\beta \in \{0.75, 0.9, 1.0, 1.1, 1.25\}$. Note that this requires that the clause found to be the best so far be reset every time we change the value of β. By varying β, we hope to add some diversity to our discovered clauses, while keeping it around 1.0 assigns similar importance to their precision and recall. We set the Aleph parameters controlling the maximum number of layers of new variables and nodes to 3 and 7500, respectively.

Propositionalisation. Having found the clauses, a Prolog program converts the stored clauses into a CSV file with rows corresponding to instances and columns to clauses. This program is capable of filtering out clauses that are exact repetitions of others, but cannot filter clauses that are even extremely similar except for some minor change in à constant numeric literal, for example.

3.4   Common Steps

After the pre-processing methods described in Sects. 3.2 and 3.3, we apply a re-sampling technique that aims to deal with the imbalanced target domain as an extra pre-processing step. The learning algorithms used and post-processing steps are also shared by both approaches.

Re-sampling. Several pre-processing techniques exist to tackle the problem of imbalanced domains that do not correspond to the user's preference bias. Re-sampling techniques are quite effective and have the advantage of having already been proposed for both classification and regression [9], working equally well with numerical and categorical attributes. Besides balancing the domain, under-sampling reduces the dimensionality of the data set by removing instances of the most common class (or range of values), mitigating scalability issues in the process. We pre-selected the re-sampling technique for regression proposed by [25] as implemented in [8]. This method automatically calculates the amount

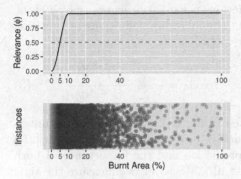

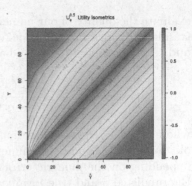

(a) Relevance function (full black line) used for re-sampling technique and performance metrics adapted to regression under imbalanced domains. Note that the threshold of relevance (dashed red line) coincides with 5% of burnt area. Below, a visual representation of instances across the domain (most are concentrated at 0%).

(b) Contour map of regression utility. The x-axis shows predicted values ($\hat{Y}$) for true values (Y) in the y-axis. Colouring and contour lines map the utility of each prediction as defined by the relevance function ϕ pictured on the left.

Fig. 3. Relevance function and resulting regression utility map. (Color figure online)

of under-sampling needed to balance the domain, given a user-specified relevance function for the target's domain and a threshold of relevance below which instances can be removed. We have settled on the function shown in Fig. 3a with a relevance threshold of 0.5, corresponding to 5 % of burnt area. This relevance function is also used for performance evaluation in Sect. 3.5.

Modelling and Post-processing We applied the Random Forest (RF) and SVR algorithms, as implemented in [14] and [17], respectively, to our transformed data sets. The predictions were then forced into the range of our target variable.

3.5   Experimental Analysis

The main goal of our experimental analysis is to compare the results obtained by standard regression models on our data set after a propositional versus a relational pre-processing approach. We also test if a combination of the two pre-processing approaches results in improved predictions of burn fraction. We try to provide insight into the effect of different variables in the results. Further, we assess the impact of the re-sampling step mentioned in Sect. 3.4.

Experimental Setup Standard metrics such as Mean Squared Error (MSE) are not well equipped to deal with domains where the user preference bias does not correspond to the target domain distribution as in our case [9]. Therefore, we evaluate the quality of our numeric predictions using F_1-measure$_R$, which is the standard F_1-measure (2) calculated using the following definitions of precision and recall[2], adapted to utility-based regression [7],

$$\text{precision}_R = \frac{\sum\limits_{\phi(\hat{y}_i)>t_R} (1 + u_i)}{\sum\limits_{\phi(\hat{y}_i)>t_R} (1 + \phi(\hat{y}_i))}, \qquad \text{recall}_R = \frac{\sum\limits_{\phi(y_i)>t_R} (1 + u_i)}{\sum\limits_{\phi(y_i)>t_R} (1 + \phi(y_i))} \qquad (3)$$

where ϕ is the relevance function (depicted in Fig. 3a), t_R is a relevance threshold (set to 0.5), and u is a function of utility of a prediction (defined in [21]) depending on the numeric error of the prediction and the importance of both the predicted $\hat{y}$ and true y values (see Fig. 3b). Moreover, we measure the time spent pre-processing data, training and testing models.

In order to obtain reliable estimates of these metrics, we divide the data set into 10 pairs of sliding training and test sets, and calculate statistics of the results over them. We train our methodologies on data for a stretch of 8 years (23056 instances), and test them in the following 3 years (8646 instances). The first training set starts in 1991 and the last in 2000.

We repeat the experiments for each pre-processing approach (and their combination) with and without the under-sampling step[3]. We test for statistical significance in difference of performance using the Wilcoxon signed-rank test.

Results and Discussion. Tables 2 and 3 summarise the results obtained. For each setup described above, we only show the results achieving the highest F_1-measure$_R$ after grid search parameter optimisation. The propositional and the relational pre-processing approach both obtain very similar results, behaving well when the under-sampling step is performed. However, the best results are obtained by the combination of the two methods, with statistical significance in terms of F_1-measure$_R$ (as determined by pairwise comparisons with this approach as baseline). This seems to validate the incorporation of both approaches when dealing with this kind of data sets, although there is an obvious trade-off on pre-processing and training time when adopting a relational approach. Note also, that in terms of recall, this combination of approaches does not work significantly better than simply applying either approach individually; and in terms of precision, it is also not significantly better than using a relational approach only. The fact that the relational approach works so well, with results competitive with or even better than the propositional approach, confirms our hypothesis that it is possible to build a good regression model by applying a standard algorithm to a table with a numerical target variable and Boolean features optimised for classification (of the categorised target variable).

Moreover, it is clear that under-sampling not only greatly improves the predictive ability of the models, but also decreases the training time needed to build the model. On average, the under-sampling methodology used reduces the train-

---

[2] Implemented in R package `uba` (http://www.dcc.fc.up.pt/~rpribeiro/uba/).

[3] Using R package `performanceEstimation` [24].

Table 2. Median (med) and interquartile range (IQR) of results obtained with each methodology. The best results for each pre-processing method are in italic, while the best results overall are in bold. The Wilcoxon signed rank test was used to obtain p-values of the differences in performance with the best approach as baseline (bold p-values mean that the difference is statistically significant).

Method	Re-sample	Model	$Precision_R$		$Recall_R$		$F1\text{-measure}_R$	
			med±IQR	p-val.	med±IQR	p-val.	med±IQR	p-val.
Propositional	None	RF	0.70 ± 0.13	(0.002)	0.22 ± 0.13	(0.002)	0.33 ± 0.13	(0.002)
		SVR	0.68 ± 0.10	(0.002)	0.49 ± 0.10	(0.002)	0.56 ± 0.10	(0.002)
	Under	RF	0.81 ± 0.13	(0.002)	0.67 ± 0.13	(0.002)	0.72 ± 0.13	(0.002)
		SVR	*0.84 ± 0.07*	(0.002)	*0.76 ± 0.07*	(0.01)	*0.80 ± 0.07*	(0.002)
Relational	None	RF	0.71 ± 0.12	(0.002)	0.18 ± 0.12	(0.002)	0.29 ± 0.12	(0.002)
		SVR	0.68 ± 0.09	(0.002)	0.50 ± 0.09	(0.002)	0.57 ± 0.09	(0.002)
	Under	RF	0.80 ± 0.09	(0.002)	0.58 ± 0.09	(0.002)	0.66 ± 0.09	(0.002)
		SVR	*0.85 ± 0.06*	(0.02)	*0.76 ± 0.06*	(0.04)	*0.80 ± 0.06*	(0.002)
Propositional + Relational	None	RF	0.72 ± 0.11	(0.002)	0.22 ± 0.11	(0.002)	0.33 ± 0.11	(0.002)
		SVR	0.70 ± 0.10	(0.002)	0.52 ± 0.10	(0.002)	0.59 ± 0.10	(0.002)
	Under	RF	0.80 ± 0.12	(0.002)	0.65 ± 0.12	(0.002)	0.70 ± 0.12	(0.002)
		SVR	**0.85 ± 0.06**	–	**0.77 ± 0.06**	–	**0.81 ± 0.06**	–

Table 3. Median and IQR of time taken per observation, in seconds, by various methodologies. The pre-processing time shown for propositional approaches includes time spent calculating spatio-temporal indicators and imputing missing data; for relational approaches, it includes time spent finding clauses using the Aleph system and converting them to propositional form (but not time spent encoding auxiliary predicates). Both exclude time spent computing spatial relationships, but include re-sampling time when appropriate. Since part of the pre-processing was performed on the data set as a whole, we omit its IQR.

Method	Re-sample	Model	Pre-proc.	Training	Testing	Total time
Propositional	None	RF	2.2e-3	5.8e-1 ± 6e-2	8.0e-4 ± 4e-4	5.8e-1
		SVR	1.7e-3	8.5e-3 ± 5e-4	6.7e-4 ± 7e-5	1.1e-2
	Under	RF	6.8e-3	2.6e-2 ± 6e-3	3.3e-4 ± 6e-5	3.3e-2
		SVR	3.1e-3	1.8e-4 ± 6e-5	2.1e-4 ± 4e-5	**3.5e-3**
Relational	None	RF	1.7	2.1e-1 ± 7e-2	3.6e-4 ± 7e-5	1.9
		SVR	1.7	2.0e-2 ± 1e-2	2.7e-3 ± 6e-4	1.7
	Under	RF	1.7	2.2e-2 ± 6e-3	5.0e-4 ± 4e-4	1.7
		SVR	1.7	6.0e-4 ± 1e-4	7.0e-4 ± 2e-4	*1.7*
Propositional + Relational	None	RF	1.7	1.5e-1 ± 2e-2	2.8e-4 ± 5e-5	1.9
		SVR	1.7	7.0e-2 ± 1e-2	6.0e-3 ± 2e-3	1.8
	Under	RF	1.7	1.9e-2 ± 7e-3	3.2e-4 ± 8e-5	1.7
		SVR	1.7	1.0e-3 ± 3e-4	1.0e-3 ± 1e-3	*1.7*

ing sets to 20 % of their original size, but increases the F_1-measure$_R$ obtained by RFs and SVRs by 118 % and 42 %, respectively.

Furthermore, SVRs consistently (and significantly) outperformed RFs. SVR presented higher susceptibility to parameter tuning, as evidenced by the fact that, when using default parameters, RF routinely outperformed SVR (these results are omitted in favour of those obtained after parameter optimisation).

Figure 4 allows us to examine the spatial distribution of results. We should remark that some parishes have zero or very low numbers of wildfires that exceed the minimum threshold in the considered time period. In this case, one error may have a disproportionate impact on our measure, as it can be observed in areas such as the center-south Alentejo region and highly urbanised areas such as Lisbon metro area. The higher average F_1-measure$_R$ per parish, depicted in Fig. 4a, is strongly (and positively) correlated with the average historic and neighbourhood values of the target variable itself, i.e., with our spatio-temporal indicators. This is not too surprising considering their higher level of temporal granularity but, coupled with the fact that they also achieve the highest RF importance (computed from permuting OOB data), still validates our propositional approach. This strong correlation is closely followed by positive correlations with mean altitude and slope, percentage of area covered by scrubland and caprine population density. Negative correlations were topped by mean percentage of urbanised area, housing, population and road density. We believe the negative correlations are explained by the very few cases of wildfires in urban regions (our data set does not include house fires) as discussed above. By examining Fig. 4b, we can see that although parishes benefiting from the use of each pre-processing approach on its own differ, they are similarly distributed across regions.

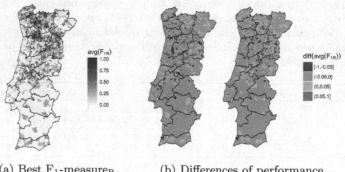

(a) Best F_1-measure$_R$.   (b) Differences of performance.

Fig. 4. On the left, best average F_1-measure$_R$ overall per parish (obtained by the combination of propositional and relational pre-processing approaches). To the right, the categorised difference of average F_1-measure$_R$ per parish obtained by the best propositional only (in the middle) and relational only (on the right) approaches in relation to the combination of approaches on the left. A negative difference means that the combination of approaches, on average, performs better in that parish. Note that these were obtained using under-sampling and SVR.

Note that the propositional feature space is the same for every training set (29 explanatory variables), while the number of features used by the relational approach depends on the number of clauses found, ranging from 39 to 89 binary features, with a median of 67 features – more than double the amount for the propositional approach. Below, we present two interesting examples of clauses found on the first and last training sets, respectively:

```
burnt ( ParishA , Year )  :−
    yearsSinceFireGE ( ParishA , Year , 2 ) ,
    neighbourDirection ( ParishA , ParishB , west ) ,
    yearsSinceFireGE ( ParishB , 2 ) .

burnt ( ParishA ,  Year )  :−
    maxAltitudeGE ( ParishA ,  507 ) ,
    neihgbourdirection ( ParishA ,  ParishB ,  south ) ,
    yearsSinceFireLE ( ParishB ,  Year ,  5 ) .
```

They can be written as "A parish burned if it has a western neighbour and both of them hadn't burnt for at least a year" and "A parish burned if its maximum altitude was higher than (or equal to) 507m and it has a southern neighbour that burnt at least once in the last five years". Both clauses include temporal predicates as well as information on a parish's neighbours. For a notion of the coverage of these clauses, see Fig. 5. The clauses appear, respectively, on the top 30 and top 50 most important features (according to RF importance) when using under-sampling and a combination of propositional and relational approaches (top 15 and top 10 if using relational only).

Overall, the propositional and relational approaches obtain very good (and similar) results. Although the relational approach performs slightly better in some cases, it requires longer processing times. The results are significantly improved by combining both strategies, which is interesting.

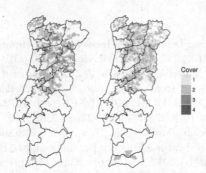

Fig. 5. Spatial coverage of example relational clauses ordered from left to right. Coverage is defined as the number of years for which only the body of the clause is true subtracted from the total number of years that the whole clause is true for each parish. If the body of the clause is always false, the parish is left white.

4   Conclusion

In order to predict the annual burn fraction of Portuguese parishes, we compared two approaches that encode spatio-temporal information in propositional form, each using different pre-processing methods, so that standard regression algorithms can be used. The fully propositional approach builds spatio-temporal indicators considering simplified borders. The relational one, uses an ILP system to find relational clauses that can be transformed into binary features. Both used a notion of spatio-temporal neighbourhood including spatial direction and an utility-based re-sampling technique to deal with this imbalanced domain. Further, we compared each method with a combination of both.

In spite of the features produced by the relational approach having been optimised for classification, the results obtained by the former method are still competitive with (and sometimes slightly better than) the propositional approach in this regression task. Propositional features are, however, much faster to compute. Despite both strategies behaving well after under-sampling, they still perform significantly worse than their combination.

Future work includes exploration of other propositional clustering-based approaches (such as [3]) and graphical modelling techniques (such as Markov Logic Networks), and their application to different data sets. We also plan on investigating whether our results transfer between different countries.

Acknowledgements. We would like to thank Dr. João Torres for providing most of the data we worked with. This work is financed by the ERDF European Regional Development Fund through the Operational Programme for Competitiveness and Internationalisation - COMPETE 2020 Programme within project POCI-01-0145-FEDER-006961, and by National Funds through the FCT Fundação para a Ciência e a Tecnologia (Portuguese Foundation for Science and Technology) as part of project UID/EEA/50014/2013.

References

1. Andrienko, G., Malerba, D., May, M., Teisseire, M.: Mining spatio-temporal data. J. Intell. Inf. Syst. **27**(3), 187–190 (2006). doi:10.1007/s10844-006-9949-3
2. Appice, A., Ceci, M., Malerba, D., Lanza, A.: Learning and transferring geographically weighted regression trees across time. In: Atzmueller, M., Chin, A., Helic, D., Hotho, A. (eds.) MSM/MUSE 2011. LNCS, vol. 7472, pp. 97–117. Springer, Heidelberg (2012). doi:10.1007/978-3-642-33684-3_6
3. Appice, A., Pravilovic, S., Malerba, D., Lanza, A.: Enhancing regression models with spatio-temporal indicator additions. In: Baldoni, M., Baroglio, C., Boella, G., Micalizio, R. (eds.) AI*IA 2013. LNCS, pp. 433–444. Springer, Heidelberg (2013). doi:10.1007/978-3-319-03524-6_37
4. Barber, C., Bockhorst, J., Roebber, P.: Auto-regressive HMM inference with incomplete data for short-horizon wind forecasting. In: NIPS, pp. 136–144 (2010)
5. Bassi, S., Kettunen, M., Kampa, E., Cavalieri, S.: Forest Fires: Causes and Contributing Factors in Europe. European Parliament, Brussels (2008)

6. Bilgili, M., Sahin, B., Yasar, A.: Application of artificial neural networks for the wind speed prediction of target station using reference stations data. Renew. Energ. **32**(14), 2350–2360 (2007). doi:10.1016/j.renene.2006.12.001

7. Branco, P.: Re-sampling approaches for regression tasks under imbalanced domains. Master's thesis, University of Porto (2014)

8. Branco, P., Ribeiro, R.P., Torgo, L.: UBL: utility-based learning (2014). R package version 0.0.1

9. Branco, P., Torgo, L., Ribeiro, R.P.: A survey of predictive modelling under imbalanced distributions. CoRR abs/1505.01658 (2015)

10. Ceci, M., Appice, A.: Spatial associative classification: propositional vs structural approach. J. Intell. Inf. Syst. **27**(3), 191–213 (2006). doi:10.1007/s10844-006-9950-x

11. Cheng, T., Wang, J.: Integrated spatio-temporal data mining for forest fire prediction. Trans. GIS **12**(5), 591–611 (2008). doi:10.1111/j.1467-9671.2008.01117.x

12. Dzeroski, S.: Multi-relational data mining: an introduction. SIGKDD Explor. **5**(1), 1–16 (2003). doi:10.1145/959242.959245

13. Goadrich, M., Oliphant, L., Shavlik, J.W.: Gleaner: creating ensembles of first-order clauses to improve recall-precision curves. Mach. Learn. **64**(1–3), 231–261 (2006). doi:10.1007/s10994-006-8958-3

14. Liaw, A., Wiener, M.: Classification and regression by randomforest. R News **2**(3), 18–22 (2002)

15. Malerba, D.: A relational perspective on spatial data mining. IJDMMM **1**(1), 103–118 (2008). doi:10.1504/IJDMMM.2008.022540

16. McGovern, A., Gagne, D.J., Williams, J.K., Brown, R.A., Basara, J.B.: Enhancing understanding and improving prediction of severe weather through spatiotemporal relational learning. Mach. Learn. **95**(1), 27–50 (2014). doi:10.1007/s10994-013-5343-x

17. Meyer, D., Dimitriadou, E., Hornik, K., Weingessel, A., Leisch, F.: Misc Functions of the Department of Statistics, Probability Theory Group (Formerly: E1071), TU Wien (2014). R package version 1.6-4

18. Ohashi, O., Torgo, L.: Wind speed forecasting using spatio-temporal indicators. In: ECAI, pp. 975–980 (2012). doi:10.3233/978-1-61499-098-7-975

19. Pebesma, E.J.: Multivariable geostatistics in S: the gstat package. Comput. Geosci. **30**(7), 683–691 (2004). doi:10.1016/j.cageo.2004.03.012

20. Pravilovic, S., Appice, A., Malerba, D.: An intelligent technique for forecasting spatially correlated time series. In: Baldoni, M., Baroglio, C., Boella, G., Micalizio, R. (eds.) AI*IA 2013. LNCS, pp. 457–468. Springer, Heidelberg (2013). doi:10.1007/978-3-319-03524-6_39

21. Ribeiro, R.P.: Utility-based regression. Ph.D. thesis, University of Porto (2011)

22. Srinivasan, A.: The Aleph Manual. University of Oxford, Oxford (2007)

23. Thompson, C.S., Thomson, P.J., Zheng, X.: Fitting a multisite daily rainfall model to New Zealand data. J. Hydrol. **340**(1), 25–39 (2007). doi:10.1016/j.jhydrol.2007.03.020

24. Torgo, L.: An infra-structure for performance estimation and experimental comparison of predictive models in R. CoRR abs/1505.01658 (2014)

25. Torgo, L., Ribeiro, R.P., Pfahringer, B., Branco, P.: SMOTE for regression. In: Pereira, F., Machado, P., Costa, E., Cardoso, A. (eds.) EPIA 2015. LNCS, vol. 9273, pp. 378–389. Springer, Heidelberg (2013). doi:10.1007/978-3-642-40669-0_33

26. Torres, J.: Patterns and drivers of wildfire occurrence and post-fire vegetation resilience across scales in Portugal. Ph.D. thesis, University of Porto (2014)

27. Yao, X.: Research issues in spatio-temporal data mining. In: Workshop on Geospatial Visualization and Knowledge Discovery, UCGIS, pp. 18–20 (2003)

Recognizing Family, Genus, and Species of Anuran Using a Hierarchical Classification Approach

Juan G. Colonna[1(✉)], João Gama[2], and Eduardo F. Nakamura[1]

[1] Institute of Computing (Icomp), Federal University of Amazonas (UFAM),
Avenida General Rodrigo Octávio 6200, Manaus, AM 69077-000, Brazil
{juancolonna,nakamura}@icomp.ufam.edu.br
[2] Laboratory of Artificial Intelligence and Decision Support (LIAAD), INESC Tec,
Campus da FEUP, Rua Dr. Roberto Frias, 4200-465 Porto, Portugal
jgama@fep.up.pt

Abstract. In bioacoustic recognition approaches, a "flat" classifier is usually trained to recognize several species of anuran, where the number of classes is equal to the number of species. Consequently, the complexity of the classification function increases proportionally to the amount of species. To avoid this issue we propose a "hierarchical" approach that decomposes the problem into three taxonomic levels: the family, the genus, and the species level. To accomplish this, we transform the original single-label problem into a multi-dimensional problem (multi-label and multi-class) considering the Linnaeus taxonomy. Then, we develop a top-down method using a set of classifiers organized as a hierarchical tree. Thus, it is possible to predict the same set of species as a flat classifier, and additionally obtain new information about the samples and their taxonomic relationship. This helps us to understand the problem better and achieve additional conclusions by the inspection of the confusion matrices at the three levels of classification. In addition, we carry out our experiments using a Cross-Validation performed by individuals. This form of CV avoids mixing syllables that belong to the same specimens in the testing and training sets, preventing an overestimate of the accuracy and generalizing the predictive capabilities of the system. We tested our system in a dataset with sixty individual frogs, from ten different species, eight genus, and four families, achieving a final Micro- and Average-accuracy equal to 86 % and 62 % respectively.

1 Introduction

Amphibians are directly affected by environmental changes [3,4]. This observation has motivated many researchers to monitor the decline of amphibian populations through time and use it as an indicator of environmental problems [1,13]. Among all amphibian species that may be monitored anuran (frogs and toads) are preferred, because these have a semi-permeable skin which makes them sensitive to aquatic and terrestrial conditions [19]. Nowadays, the most widely used

T. Calders et al. (Eds.): DS 2016, LNAI 9956, pp. 198–212, 2016.
DOI: 10.1007/978-3-319-46307-0_13

method to monitor frog populations takes advantage of the vocalization capability to apply acoustics surveys [20, 25]. However, manual application of these acoustic surveys requires many human and economic resources, as well as expert knowledge, being difficult to apply in remote tropical areas of the Amazon rainforest. Therefore, our goal is to develop an Automatic Calls Recognition (ACR) system to monitor frog populations automatically and in a less invasive manner using acoustic sensors. The general idea consists of treating the challenge of anuran monitoring as a species recognition task using their calls and Machine Learning (ML) techniques [6, 7, 14, 28].

In bioacoustics most of the related works deal with the species recognition problem using "flat" classifiers, where each instance belongs to one class (or species name in this case), and there is no hierarchical relationship between the classes [7, 12, 18, 22, 27, 28]. This work we addressed the problem of anuran species recognition through their calls using a "hierarchical" classifier that considers its family and genus taxonomy. For this purpose, the family and genus information of each species was added as new labels, transforming the original problem with a single label into a multi-dimensional approach, i.e., a problem where the outputs are multi-class and multi-label.

The hierarchical approach allows us to test three hypothesis:

(i) the decomposition of the main problem into three levels of small hierarchically related problems, in which the results may improve compared to a normal flat classifier;

(ii) this configuration allows us to understand the relationship between the misclassifications and the acoustic proximity of the species, and their taxonomy, into feature space; and

(iii) the method has better predictive capabilities for new individuals that were not present in the original training set.

Thus, the two main contributions in this work are: a customization of the existing hierarchical models, specially adapted to the anuran species taxonomy; and the advantages of this model in our bioacoustics application context.

In order to test the first hypothesis we introduce our hierarchical system in Sect. 5. To accomplish this, we give a detailed explanation about how this approach can reduce the complexity of the model from a feature space point of view and, consequently, simplify the decision function. To test the second hypothesis we compare the confusion matrix of each classification level, i.e., family, genus, and species levels in Sect. 7. To test the third hypothesis we carry all of our experiments using Cross-Validation (CV) by individuals (or specimens) as explained in Sect. 6.3. The results and conclusion are supported by the calculation of the Micro- and Average-accuracy by level (see Sects. 6.4, 7 and 8).

In addition (Sect. 4) we also discuss how hierarchical models were applied to the anuran recognition task (particularly the prediction of frogs and toads species) and, in general, to the bioacoustics problems, in which a hierarchical relationship between the labels could be modeled. Finally, we would like to emphasize that our work is the first one regarding the combination of a hierarchical approach together with a CV procedure by individuals using the Linnaeus taxonomy.

2   Motivation for Using a Hierarchical Approach

Anura is the name of an order in the Amphibia class of animals that includes frogs and toads. According to recent reports there are more than 6600 different species of anuran in the world, classified into 56 families and several genus [9]. The anuran diversity in the tropical areas of South America is the greatest, concentrating approximately 70 % on the global biodiversity of amphibians [16]. In order to develop a flat classifier we need to train it with the number of classes equal to the number of species that we intend to recognize. Therefore, the complexity of the decision function increases with the number of classes, becoming an intractable problem in certain scenarios. A hierarchical approach can alleviate this problem by decomposing the classification function in several levels, similarly to a decision tree. Thus, we use the well known Linnaeus taxonomy to construct a system with three levels: family, genus, and species (see Sect. 5). With this, every time we go down through the tree to another level, the output space of possible solutions is simplified.

3   Fundamentals

In order to understand the methodology adopted in this work, two concepts are described in this section: how a bioacoustic recognition framework works, and how to create a hierarchical classification approach.

3.1   Bioacoustics Systems

Anuran call classification systems are traditionally composed of three main steps with different purposes (see Fig. 1). Formally, the input bioacoustic signal $X = \{x_1, x_2, \cdots, x_N\}$ is a time series of length N, in which its values represent the acoustics pressure levels (or amplitude). A syllable $\mathbf{x}_k = \{x_t, x_{t+1}, \cdots, x_{t+n}\}$ is a subset of n consecutive signal values. Thus, the pre-processing step segments the signal X by identifying the beginning and the endpoints of $\mathbf{x}_k$ (Fig. 2(a)) [6].

After the syllable extraction we need to represent each $\mathbf{x}_k$ by a set of features, commonly called Low Level Descriptors (LLDs). The most frequent LLDs are the Mel-Frequency Spectral Coefficients (MFCCs). The MFCCs perform a spectral analysis based on a triangular filter-bank logarithmically spaced in the frequency

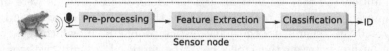

Fig. 1. An automatic call recognition system (ACR).

domain (Fig. 2(b)) [7,21]. The feature extraction using the MFCCs allows to represent any syllable by a set of coefficients (MFCC($\mathbf{x}_k$) $\rightarrow$ $\mathbf{c}_k$), i.e., $X \rightarrow$ $\{(\mathbf{c}_1, s_i), (\mathbf{c}_2, s_i), \ldots, (\mathbf{c}_k, s_i)\}$, where each $\mathbf{c}_k = [c_1, c_2, \ldots, c_l]$ is a feature vector with l coefficients, and s_i is the species name (or label). The representation of $\mathbf{x}_k$ through $\mathbf{c}_k$ is more robust, more compact, and easier to recognize, compared to use raw data.

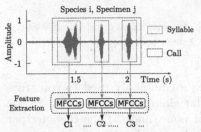 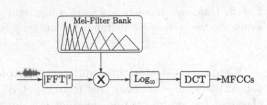

(a) An audio record of the species Hyla minuta.

(b) MFCCs steps. Here, FFT stands for Fast Fourier Transform and DCT for Discrete Cosine Transform.

Fig. 2. A framework for automatic frog's calls recognition.

Finally, the challenge is how to assign the species name to a new syllable by using the MFCC values. This is a supervised classification task and is performed by the last step of the system. For this purpose several ML algorithms could be applied to create and train a model $f(\cdot)$ with capabilities to predict new incoming samples, i.e., given an unknown $\mathbf{c}$ estimates the most probable label by evaluating $f(\mathbf{c}) \rightarrow s_i$, where $S = \{s_1, s_2, \ldots, s_i\}$ is the set of species names.

3.2 Review of Hierarchical Classification Approaches

Hierarchical methods are widely used to solve multi-label problems in which the classes have an inherent taxonomy structure, i.e., an instance that belongs to a subclass, naturally belongs to its higher level classes. These methods help to simplify complex multi-class problems transforming these into a multi-label approach by considering the hierarchical relationship between the labels. For instance, every time we go down in a level of the hierarchy, the number of possible solutions is reduced, simplifying the decision function, as showed by Fig. 5. There are two common models to describe the hierarchical relationships between the classes: (a) trees, and (b) Direct Acyclic Graphs (DAG). A tree structure connects a set of leaf nodes to a single parent node forming several subtrees not interconnected on the same level. A DAG is a more flexible structure allowing the leafs to have more than one parent node [8]. In our approach we adopted the tree structure, due to the taxonomic constraints of our problem, in which every species can belong to just one genus class and one family class at the same time.

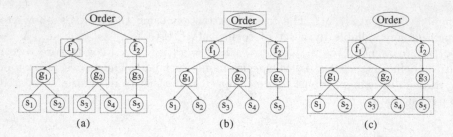

Fig. 3. Different manners to create a hierarchical classifier combining flat classifiers. From top-to-down levels: **f** stands for family, **g** for genus, and **s** for species.

Figure 3 illustrates three different approaches commonly used to construct a multi-label hierarchical tree from a set of flat classifiers. These are: (1) one classifier per node, (2) one classifier per parent node, and (3) one classifier per level [24]. These trees may be imbalanced depending on the taxonomic structure of the problem. The classifiers inside the nodes should be trained separately and assembled after that. During the prediction phase the strategy adopted to determine the class of a new sample is top-down. This strategy starts from the top nodes performing the corresponding predictions and goes down until it reaches a leaf node in the last hierarchical level. Thus, the decision results in a unique relationship between the set of predicted labels. An obvious disadvantage associated with this top-down approach is the error propagation from the higher levels of the tree. However, this approach is well suited for the context of species recognition where the number of classes is too high to train a flat classifier. Moreover, the configuration shown in Fig. 3(b) fits better the characteristics of our problem presented in Fig. 4.

4   Related Works

Several authors have already studied the problem of recognition and classification of anuran species through their calls. Among these Huang *et al.* [14] and Colonna *et al.* [7] studied the best acoustics features to recognize different species. Jaafar *et al.* [17] and Colonna *et al.* [6] focused on comparing some syllable extraction procedures as a pre-processing step. Finally, Ribas *et al.* [22] and Colonna *et al.* [5] evaluated the possibility of embedding a classifier into the nodes of a wireless sensor network (WSN). However, little effort was made to link the hierarchical taxonomy of the species with an automatic classification system. The hierarchical taxonomic organization of the species is a standard approach in ecology since it was defined in 1935 by Carl Linnaeus.

Gingras et al. [11] formulated the hypothesis that anuran species which are phylogenetically or taxonomically close have more similar calls. To test this hypothesis the authors developed a three-parameter model using the mean values of dominant frequency, the variation coefficient of root-mean square energy, and spectral flux of the signals. Calls from 142 species belonging to four genera were analyzed and classified applying a logistic regression model, a Support Vector Machine (SVM), a k-Nearest Neighbors (kNN), and a Gaussian Mixture Model (GMM), achieving an accuracy of approximately 70 %. During the test different specimens (or individuals) were used for training and testing in order to prove the generalization capabilities of the model.

An acoustic feature extraction and a comparative analysis of these features, for developing a hierarchical classification technique of Australian frog calls, was proposed by Xie *et al.* [29]. This work studies which acoustics features should be used in each classification level, considering the taxonomy information separated in three levels: family, genus and species. The contribution was a correlation method, able to select the better features for each level, but the final classification was addressed as three separate problems using SVM. The levels were not integrated into one single approach leaving two open questions: (1) how to integrate these classifiers in one single method capable of reducing the complexity by taking advantage of the hierarchical taxonomy, and (2) how to handle the disagreement between the levels.

The technique called Balance-Guaranteed Optimized Tree with Reject option (BGOTR) is a hierarchical classification system including the reject option. This was developed by Phoenix *et al.* [15] for fish image recognition using underwater cameras. In this system a multi-class classifier and a feature selection are built together into a hierarchical tree, and this is optimized to maximize the classification accuracy by grouping the classes based on their inter-class similarities. The rejection option is performed after the hierarchical classification by applying a Gaussian Mixture Model (GMM) to fit the distribution of the features in the images. Despite the interesting results the authors highlight that this approach does not consider the taxonomy of the problem. Indeed, this method was not developed for a multi-label purpose, and therefore it is not possible to evaluate the similarities between family, genus and species.

An evaluation on different hierarchical approaches applied to the bird species recognition was performed by Sillas et al. [23]. The authors compared three different approaches: a flat classification where the class hierarchy is disregarded, one classifier per parent node (see Fig. 3), and one global approach where a single algorithm is used to predict classes at any level of the hierarchy based on Global-Model Hierarchical Classification Naive Bayes (GMNB). Moreover, an extension of the metrics Precision, Recall, and F-measure was introduced, tailored to the hierarchical classification scenario. The results show that the hierarchical approaches outperform flat classifiers when the number of species is large, and that the labels can be organized in an adequate hierarchy.

To the best of our knowledge, no study has yet been published integrating the family, the genus, and the species labels of anuran in one unique hierarchical approach, to be solved as a multi-dimensional problem and, at the same time, performing a CV by individuals (or specimens) to test the model generalization capabilities.

5   Proposed Approach

The phylogenetic taxonomy aims to organize animals into hierarchical categories. Using this pre-defined organization for anuran, we can build our hierarchical classification system adding two extra labels to the original dataset (g and f):

$$\text{Dataset} = \begin{bmatrix} \mathbf{c}_1 = [c_1, c_2, \ldots, c_l], & s, & g, & f \\ \mathbf{c}_2 = [c_1, c_2, \ldots, c_l], & s, & g, & f \\ \vdots & & \vdots & \vdots & \vdots \\ \mathbf{c}_k = [c_1, c_2, \ldots, c_l], & s_j, & g_i, & f_m \end{bmatrix}$$

with these new labels we have turned our multiclass problem with a single label into a multi-label and multi-class problem (MM). This MM is a generalization of the common multi-label problems, where the classes are binary in each column. This MM problems are also called Multi-dimensional problems because the output is composed by a tuple of labels [2], which are three in our case.

This is possible because there is a unequivocal relationship between the species names and its genus and family names. That is, a subset $S^0 = \{s_1, \ldots s_p\}$ of species belongs to a singular genus ($S \subseteq g_m$), while a subset of genus $G^0 = \{g_1, \ldots, g_m, \ldots, g_p\}$ also belongs to a particular family ($G \subseteq f_m$) such that $f_m \subseteq F^0$. Therefore, any s_j is from G^0 and F^0 without ambiguity. Thus, if a flat classifier correctly predicts a particular species, the system is effectively predicting not only the species at the last level, but also the genus and the family classes at the first two levels together.

With this concept we can apply reverse engineering and develop a hierarchical top-down approach as shown in Fig. 3(b). Our hierarchical tree is represented in Fig. 4. An example of problem simplification by the hierarchical decomposition using an example with two attributes is shown in Fig. 5. As we can note, in the beginning all the samples belong to two families (or classes). After the family classification, the problem is reduced and consequently simplified by the simple decomposition of the feature space, in which only the samples of the first family remain. This process is repeated until the last classification level is reached (the species label). Thus, the class of a leaf node is used to estimate the label of new samples.

A remarkable advantage of this approach is that we do not have to perform every classification for some branches in all levels. This is the main advantage of the customization based on Linnaeus taxonomy and the reason why we chose the approach described in Fig. 3(b). For instance, if the first classifier assigns the Bufonidae label to a new sample at the top level, it is not necessary to continue classifying the remaining levels, because there are no more splits for this branch. Therefore, the genus label Rhinella and the species label Rhinella granulosa are assigned automatically. The remaining settings of our approach are detailed in Sect. 6.

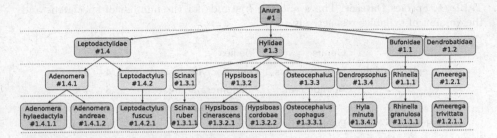

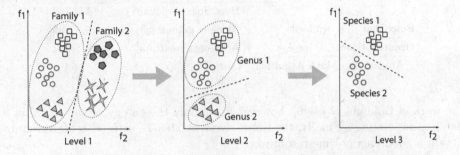

Fig. 4. Species tree. From Top-to-Down levels: Order, Family, Genus and Species. The # stands for node ID.

Fig. 5. Problem decomposition stages when performing the hierarchical classification from top to down described by an example of prune training data.

6   Methodology Description

In order to develop our hierarchical method, the first step is to obtain the family and genus labels for each sample of our dataset. For this, we used the taxonomy information available at [9]. The dataset description, the classifier setting, the validation procedure and the metrics used are described in the following subsections.

6.1   Dataset Description

The dataset used in our experiments is summarized in Table 1. It has 10 different species, 60 specimens and 5998 syllables[1]. These records were collected *in situ* under real noise conditions. Some species are from the Federal University of Amazonas, others from Mata Atlântica, Brazil and the last from Córdoba, Argentina. These recordings were stored in *wav* format with 44.1 kHz of sampling frequency and 32 bit, which allows us to analyze signals up to 22.05 kHz. From each extracted syllable, 24 MFCCs were calculated by using 44 triangular filters and these coefficients were normalized between $-1 \leq c_l \leq 1$ (see Sect. 3.1). For the segmentation and syllable extraction tasks we based our approach on

---

[1] Available at https://goo.gl/61IoXc.

Table 1. Species Dataset. The *s* and the *k* stands for the number of specimens and the amount of syllables respectively.

Family	Genus	Species	s	k
Leptodactylidae	Leptodactylus	Leptodactylus fuscus[a]	4	222
	Adenomera	Adenomera andreae[a]	8	496
		Adenomera hylaedactyla[b]	11	3049
Hylidae	Dendropsophus	Hyla minuta[b]	11	229
	Scinax	Scinax ruber[b]	5	96
	Osteocephalus	Osteocephalus oophagus[a]	3	96
	Hypsiboas	Hypsiboas cinerascens[a]	4	429
		Hypsiboas cordobae[c]	4	702
Bufonidae	Rhinella	Rhinella granulosa[a]	5	135
Dendrobatidae	Ameerega	Ameerega trivittata[b]	5	544

[a]Amazonas, [b]Mata Atlântica, [c]Córdoba.

the work of Colonna *et al.* [6], however using only the energy of the signal in a batch mode setting[2]. Finally, the frame size was 0.0464 s with 66 % of overlap to obtain a good energy-time resolution.

6.2   Node Classifier Description

In our experiments we chose kNN with k = 3 as the base classifier for the parent nodes. As kNN is considered a subspace technique, then the predicted result is the similarity between the samples in the feature space and consequently, the acoustics similarity of the syllable's frequencies. Besides that, in all parent nodes we decomposed the multiclass model $f(\cdot)$ into a combination of smaller binary models $f'(\cdot)$ applying the One-against-One (1A1) procedure [10]. After that, the result of each binary model was combined by using the majority voting rule. This decomposition technique reduces the complexity of each sub-problem compared to the multiclass approach.

6.3   Special Type of Cross-Validation

Because we are dealing with a supervised problem, and we want to consider the generalization capabilities of the system, we need to apply a cross-validation (CV) procedure to estimate the expected error in a real situation. With *k*-CV the original dataset is split into *k* disjoint folds, and for each one the conditional error (e_k) is estimated training the model $f(\cdot)$ with *k*-1 folds. Thus, this procedure is repeated *k* times and the expected generalized error can be obtained by averaging e_k. When the information of the individuals (or specimens) is omitted, we may fall into a situation in which the split could leave syllables of the

---

[2] The segmentation code is available at http://goo.gl/vjVQ2c.

same individuals in the testing and training sets. This causes an overestimate on the accuracy. To overcome this problem, we consider the specimen information during the k-CV fold splitting, i.e., we leave all the syllables that belong to the same specimen together, avoiding mixing them in the testing and training sets. To accomplish this we introduce an extra label with the record ID that will only be considered during the k-CV split. Thus, we assume that the generalization error will be more realistic, because we are training with one specimen to predict a different one.

6.4   Performance Measure (Average-Accuracy)

Diverse species of anuran have different syllable rates (amount of syllables per unit time) in their calls. This is a particular vocalization characteristic of each anuran species. Therefore, an unequal number of samples could be retrieved from each record producing an unbalanced dataset [6]. This is a secondary problem that affects the classical accuracy measure. Thus, a classification model that always predicts the species with the higher number of samples might have a high accuracy, even in the extreme case of losing all syllables from the other classes. To overcome this matter we suggest to use the average-accuracy instead of the traditional micro-accuracy [6, 26]. It means, the final accuracy value is calculated as the average accuracy of each species individually as:

$$\text{Average-Acc} = \frac{1}{m} \sum_{i=1}^{m} \text{Acc}_i = \frac{1}{m} \sum_{i=1}^{m} \frac{tp_i}{k_i}, \tag{1}$$

where Acc_i is the accuracy per row i of confusion matrix, m the total number of rows, tp_i are the true positives, and k_i the total number of syllables per row.

7   Experiments and Results

The structure of our hierarchical approach was introduced in Fig. 4. The first parent node corresponds to the order (Anura) and is responsible for the classification of the samples into four family classes (column 1 in Table 1). In the second level (the family level) the parent nodes are trained with the genus labels that correspond to each particular family. Thus, the family branches are able to predict their owns genus labels. The last prediction takes place at the genus level being responsible for predicting their owns species names, as shown by their leaf nodes. With this configuration we can obtain a confusion matrix per level, i.e.: one matrix for the family labels (Table 2), one for the genus labels (Table 3), and one for the species labels (Table 4).

The rows of the Tables 2, 3 and 4 are the Ground Truth (GT) labels and the columns indicate the predicted labels. The main diagonal corresponds to the number of hits. From these matrix we can obtain the micro- and average-accuracy by level. The last column of each matrix (Acc) is the accuracy by class, from which we can get the average-accuracy averaging the values

Table 2. Confusion matrix of family level with kNN (k = 3). Last column (Acc) is the accuracy of each column.

	Bufonidae	Dendrobatidae	Hylidae	Leptodactylidae	Acc_i
Bufonidae	**43**	0	21	71	0.31
Dendrobatidae	27	**488**	0	29	0.89
Hylidae	3	0	**1465**	84	0.94
Leptodactylidae	16	36	322	**3393**	0.90

Table 3. Confusion matrix of genus level with kNN (k = 3). Legend: (a) Adenomera, (b) Ameerega, (c) Dendropsophus, (d) Hypsiboas, (e) Leptodactylus, (f) Osteocephalus, (g) Rhinella, and (h) Scinax. Last column (Acc) is the accuracy of each column.

	a	b	c	d	e	f	g	h	Acc_i
a	**3186**	36	18	61	58	184	0	2	0.90
b	23	**488**	0	0	6	0	27	0	0.90
c	51	0	**123**	35	0	0	0	20	0.54
d	7	0	0	**1117**	0	6	0	1	0.99
e	15	0	20	14	**134**	21	16	2	0.60
f	7	0	0	48	4	**34**	3	0	0.35
g	8	0	0	9	63	0	**43**	12	0.32
h	15	0	50	11	0	0	0	**20**	0.21

Table 4. Confusion matrix of species level with kNN (k = 3). Legend: (a) Adenomera andreae, (b) Adenomera hylaedactyla, (c) Ameerega trivittata, (d) Hyla minuta, (e) Hypsiboas cinerascens, (f) Hypsiboas cordobae, (g) Leptodactylus fuscus, (h) Osteocephalus oophagus, (i) Rhinella granulosa, and (j) Scinax ruber. Last column (Acc) is the accuracy of each column.

	a	b	c	d	e	f	g	h	i	j	Acc_i
a	**156**	0	35	2	61	0	58	184	0	0	0.31
b	0	**3030**	1	16	0	0	0	0	0	2	0.99
c	23	0	**488**	0	0	0	6	0	27	0	0.90
d	3	48	0	**123**	3	32	0	0	0	20	0.54
e	1	6	0	0	**415**	0	0	6	0	1	0.97
f	0	0	0	0	0	**702**	0	0	0	0	1.00
g	1	14	0	20	0	14	**134**	21	16	2	0.60
h	7	0	0	0	48	0	4	**34**	3	0	0.35
i	8	0	0	0	6	3	63	0	**43**	12	0.32
j	0	15	0	50	9	2	0	0	0	**20**	0.21

Table 5. Results summary. G stands for the accuracy gains compared to the naive baseline approach for each case.

	Level					
	Family	G (%)	Genus	G (%)	Species	G (%)
Micro-Baseline	0.63		0.59		0.51	
Micro-Acc	0.90	+0.27	0.86	+0.27	0.86	+0.35
Average-Baseline	0.25		0.13		0.10	
Average-Acc	0.76	+0.51	0.60	+0.48	0.62	+0.52

of this column. A summary of the results is presented in Table 5. For the micro-accuracy case (Micro-Acc) the baseline values are given by a naive classifier which always chooses the most numerous classes (Micro-Baseline), and for the average-accuracy (Average-Acc) the baseline values are given by a naive classifier that always chooses a label randomly (Average-Baseline).

Just analyzing the confusion matrix at the family level, we can note that Bufonidae family lost about 70 % of the samples, in which almost 50 % fall in the class Leptodactylidae. That means, the Bufonidae family seems to find it hard to recognize in the presence of Leptodactylidae. However, the opposite case is not equally true. This means that the samples of the Bufonidae family are probably surrounded by samples of the Leptodactylidae in the feature space. A similar conclusion can be achieved analyzing the genus and the species levels. For instance, several samples of the Scinax were confused with Adenomera, Dendropsophus and Hypsiboas. Inside the Adenomera genus, hylaedactyla was the only confused species with the Scinax ruber.

Previously, we highlighted that we are carrying out a cross-validation by specimens, therefore we can infer that different individuals of the Leptodactylidae family share high similarities or regularities between them in the feature space. The same is valid for the Hylidae family. Therefore, we are able to recognize better the specimens that belong to these species and, consequently, they are good candidates to use them in a monitoring acoustic project.

In addition, we performed a similar test by species using only one flat classifier with the same configurations used for the nodes of the tree (that is, 1A1 and kNN with k = 3). In this test the Micro-accuracy was 0.85 and the Average-accuracy was 0.61 achieving results comparable to our approach. However, with our approach we can obtain several complementary information related to the taxonomy. Unfortunately, our dataset is not big enough to be able to observe the gains of the hierarchical classification at the last level.

8   Conclusion

We presented a hierarchical classification approach for frog species recognition using their calls and the biological taxonomy information. The main algorithmic contribution is how to prune the training data using a tree customized structure, i.e., after the high level class is decided, the number of class options in the lower levels are reduced.

First, we transform the original multiclass problem with a single label into a multidimensional problem adding the genus and family labels. It allows us to understand and investigate with greater depth the relationship between the samples and their taxonomy. Our hierarchical system is able to decompose and simplify this multidimensional problem into smaller subproblems avoiding the disadvantages of flat classifiers in this application context. In addition we present the confusion matrices and the Micro-accuracy and Average-accuracy at the three levels of the decomposition, useful to understand the nature of our problem and the relationship between the samples.

The combination of the phylogenetic taxonomy together with the cepstral frequency coefficients and the proximity obtained through the kNN classifier, enables us to notice the bioacoustics similarities between different species from a classification point of view. We can conclude that the species Adenomera hylaedactyla and Hypsiboas cinerascens are clearly recognizable in the presence of other species, and therefore are good candidates for an automatic acoustic monitoring program. We would like to emphasize that these two species belong to different families and genus confirming that our hierarchical strategy is indeed advantageous for this type of application. Another interesting fact is that the species Hypsiboas cordobae, which belongs to another country, far away from the tropical area, is easy to recognize.

However, one major drawback in most of the hierarchical classification approaches is the error propagation. Unfortunately, each level of the hierarchical tree could have some misclassifications that will compound the final error when we go down through the tree. As a result, practical applications usually require corrections to eliminate the confusing cases, especially when the database is imbalanced or when the hierarchy is deeper and composed from many levels, i.e., the accuracy decreases when the number of levels increases. As future work, in order to handle this problem, we propose the development of a hierarchical tree with a soft decision strategy based on the posterior probabilities of each level. With this, we intend to correct the misclassifications of the highest levels using the confidence of the lower levels.

Acknowledgments. Juan G. Colonna gratefully acknowledge to National Council of Technological and Scientific Development (CNPq, Brazil) by the PhD fellowship. Eduardo F. Nakamura acknowledge to FAPEAM by the support granted through the Anura Project (FAPEAM/CNPq PRONEX 023/2009). We also thank to professors Marcelo Gordo and the biologist Celeste Salineros for the help with the recordings.

This work was supported by the research project "TEC4Growth-Pervasive Intelligence, Enhancers and Proofs of Concept with Industrial Impact/NORTE-01-0145-FEDER-000020", financed by the North Portugal Regional Operational Programme (NORTE 2020), under the PORTUGAL 2020 Partnership Agreement, and through the European Regional Development Fund (ERDF) and by European Commission through the project MAESTRA (Grant number ICT-2013-612944).

References

1. Adams, M.J., Miller, D.A.W., Muths, E., Corn, P.S., Grant, E.H.C., Bailey, L.L., Fellers, G.M., Fisher, R.N., Sadinski, W.J., Waddle, H., Walls, S.C.: Trends in amphibian occupancy in the United States. PLoS One **8**(5), 1–5 (2013)
2. Borchani, H., Larrañaga, P., Gama, J., Bielza, C.: Mining multi-dimensional concept-drifting data streams using Bayesian network classifiers. Intell. Data Anal. **20**(2), 257–280 (2016)
3. Carey, C., Alexander, M.A.: Climate change and amphibian declines: is there a link? Divers. Distrib. **9**(2), 111–121 (2003)
4. Cole, E.M., Bustamante, M.R., Reinoso, D.A., Funk, W.C.: Spatial and temporal variation in population dynamics of andean frogs: effects of forest disturbance and evidence for declines. Glob. Ecol. Conserv. **1**, 60–70 (2014)
5. Colonna, J.G., Cristo, M.A.P., Nakamura, E.F.: A distribute approach for classifying anuran species based on their calls. In: 22nd International Conference on Pattern Recognition (2014)
6. Colonna, J.G., Cristo, M.A.P., Salvatierra, M., Nakamura, E.F.: An incremental technique for real-time bioacoustic signal segmentation. Expert Syst. Appl. **42**(21), 7367–7374 (2015)
7. Colonna, J.G., Ribas, A.D., dos Santos, E.M., Nakamura, E.F.: Feature subset selection for automatically classifying anuran calls using sensor networks. In: International Joint Conference on Neural Networks (IJCNN), pp. 1–8. IEEE, June 2012
8. Freitas, A.A., Carvalho, A.C.P.L.F.: A tutorial on hierarchical classification with applications in bioinformatics. In: Taniar, D. (ed.) Research and Trends in Data Mining Technologies and Applications, Chap. 7, pp. 175–208. Idea Group Publishing (2007)
9. Frost, D.R.: Amphibian species of the world: an online reference. Electronic Database accessible at http://goo.gl/3WRZhx, April 2016. American Museum of Natural History, New York, USA
10. Fürnkranz, J.: Round Robin rule learning. In: Proceedings of the Eighteenth International Conference on Machine Learning, ICML 2001, pp. 146–153 (2001)
11. Gingras, B., Fitch, W.T.: A three-parameter model for classifying anurans into four genera based on advertisement calls. J. Acoust. Soc. Am. **133**(1), 547–559 (2013)
12. Han, N.C., Muniandy, S.V., Dayou, J.: Acoustic classification of Australian anurans based on hybrid spectral-entropy approach. Appl. Acoust. **72**(9), 639–645 (2011)
13. Houlahan, J.E., Findlay, C.S., Schmidt, B.R., Meyer, A.H., Kuzmin, S.L.: Quantitative evidence for global amphibian population declines. Nature **404**(6779), 752–755 (2000)
14. Huang, C.J., Yang, Y.J., Yang, D.X., Chen, Y.J.: Frog classification using machine learning techniques. Expert Syst. Appl. **36**(2), 3737–3743 (2009)
15. Huang, P.X.: Hierarchical classification system with reject option for live fish recognition. In: Fisher, R.B., et al. (eds.) Fish4Knowledge: Collecting and Analyzing Massive Coral Reef Fish Video Data. Intelligent Systems Reference Library, vol. 104, pp. 141–159. Springer, Switzerland (2016)
16. IUCN. Geographic patterns, April 2016. http://goo.gl/nq2qt7. The IUCN Red List of Threatened Species
17. Jaafar, H., Ramli, D.A.: Automatic syllables segmentation for frog identification system. In: 9th International Colloquium on Signal Processing and Its Applications (CSPA), pp. 224–228. IEEE (2013)

18. Jaafar, H., Ramli, D.A., Rosdi, B.A.: Comparative study on different classifiers for frog identification system based on bioacoustic signal analysis. In: Proceedings of the 2014 International Conference on Communications, Signal Processing and Computers (2014)
19. King, V.: A study of the mechanism of water transfer across frog skin by a comparison of the permeability of the skin to deuterated and tritiated water. J. Physiol. **200**(2), 529–538 (1969)
20. Marques, T.A., Thomas, L., Martin, S.W., Mellinger, D.K., Ward, J.A., Moretti, D.J., Harris, D., Tyack, P.L.: Estimating animal population density using passive acoustics. Bio. Rev. **88**(2), 287–309 (2013)
21. Rabiner, L., Schafer, R.: Introduction to digital speech processing. Found. Trends Sign. Process. **1**, 1–194 (2007)
22. Ribas, A.D., Colonna, J.G., Figueiredo, C.M.S., Nakamura, E.F.: Similarity clustering for data fusion in wireless sensor networks using k-means. In: International Joint Conference on Neural Networks (IJCNN), pp. 1–7. IEEE, June 2012
23. Silla, C.N., Kaestner, C.A.A.: Hierarchical classification of bird species using their audio recorded songs. In: International Conference on Systems, Man, and Cybernetics (SMC), pp. 1895–1900. IEEE, October 2013
24. Silla Jr., C.N., Freitas, A.A.: A survey of hierarchical classification across different application domains. Data Min. Knowl. Discov. **22**(22), 31–72 (2011)
25. Da Silva, F.R.: Evaluation of survey methods for sampling anuran species richness in the neotropics. S. Am. J. Herpetology **5**(3), 212–220 (2010)
26. Sokolova, M., Lapalme, G.: A systematic analysis of performance measures for classification tasks. Inf. Process. Manag. **45**(4), 427–437 (2009)
27. Vaca-Castaño, G., Rodriguez, D.: Using syllabic mel cepstrum features and k-nearest neighbors to identify anurans and birds species. In: 2010 IEEE Workshop on Signal Processing Systems (SIPS), pp. 466–471 (2010)
28. Xie, J., Towsey, M., Truskinger, A., Eichinski, P., Zhang, J., Roe, P.: Acoustic classification of australian anurans using syllable features. In: IEEE Tenth International Conference on Intelligent Sensors, Sensor Networks and Information Processing (ISSNIP 2015). IEEE (2015)
29. Xie, J., Zhang, J., Roe, P.: Acoustic features for hierarchical classification of australian frog calls. In: 10th International Conference on Information, Communications and Signal Processing (2015)

Evolution Analysis of Call Ego-Networks

Shazia Tabassum[✉] and João Gama

LIAAD, Inesctec, University of Porto, Porto, Portugal
up201402360@fe.up.pt, jgama@fep.up.pt

Abstract. With the realization of networks in many of the real world domains, research work in network science has gained much attention now-a-days. The real world interaction networks are exploited to gain insights into real world connections. One of the notion is to analyze how these networks grow and evolve. Most of the works rely upon the socio centric networks. The socio centric network comprises of several ego networks. How these ego networks evolve greatly influences the structure of network. In this work, we have analyzed the evolution of ego networks from a massive call network stream by using an extensive list of graph metrics. By doing this, we studied the evolution of structural properties of graph and related them with the real world user behaviors. We also proved the densification power law over the temporal call ego networks. Many of the evolving networks obey the densification power law and the number of edges increase as a function of time. Therefore, we discuss a sequential sampling method with forgetting factor to sample the evolving ego network stream. This method captures the most active and recent nodes from the network while preserving the tie strengths between them and maintaining the density of graph and decreasing redundancy.

Keywords: Evolving networks · Evolution analysis · Call networks · Densification · Ego-networks · Forgetting factor

1 Introduction

Enormous streams of graphs are generated by some of the real-time applications, at a speed of millions of nodes and billions of edges per day. Such social streams provide an abstraction of interactions between real world social entities or individuals. Studying the structural properties of these streams enables powerful insights and extrapolations of real world. Space and time complexity is one of the challenging issues related to analyzing these streams. Networks representing real world social structures are usually temporal and evolving. The rapidly changing and evolving structure of these graphs, calls for an exigency of latest and up to date results. Processing the real-time network stream as it arrives, is one of the best solutions for the above problem. Therefore, we employ the stream processing approach to process enormous data. Some of the social network analysis methods that can be applied over streams of graphs are given in [12]. Furthermore, we use a streaming ego network approach over a telecommunications' call graph stream of temporal edge/calls' as in [14].

© Springer International Publishing Switzerland 2016
T. Calders et al. (Eds.): DS 2016, LNAI 9956, pp. 213–225, 2016.
DOI: 10.1007/978-3-319-46307-0_14

An ego network is based on the relationships of a single node called "ego" with the other nodes in a social network. An ego can represent an individual, entity, object or organization. All the other nodes related to ego in the network are called alters. An ego network maps the relationships of an ego with alters and also between themselves. In the recent work [4] by Google.com, the authors argue that it is possible to address important graph mining tasks by analyzing the ego-nets of a social network and performing independent computations on them. The studies made by Everett and Borgatti [5] indicate that the local ego betweenness is highly correlated with the betweenness of the actor in the complete network. In [17] Wellman describes an ego network as a personal network. The author explains that the importance of local ties becomes apparent by redefining the composition of personal community networks in terms of the number of contacts (interactions) that egos have with the active members of the networks instead of the traditional procedure of counting the number of ties (relationships). In this work, we analyze the evolution of ego networks by using a bunch of social network analysis metrics.

We also discuss the growth pattern of our ego networks. In [8] the authors discussed how large graphs evolve over time. They stated a densification power law which is followed by these networks. In our work, we test the densification power law over the temporal ego networks of call graph stream and observe that it obeys the densification power law and follows the similar properties of large graphs. We also consider the properties of real world graphs depicted by [1, 2, 10, 16] such as diameter, path length etc.

As we observe the call ego networks satisfy the densification power law and the number of edges grow superlinearly to the number of nodes, the evolving graphs can get humongous in no time. There are a few sampling strategies discussed in [14] for sampling real time streaming graphs, but none of them preserve the tie strengths between nodes in the network. Nevertheless, there are no sampling techniques designed for ego networks to preserve the tie strengths, while maintaining the active and most recent nodes. Now the obvious question is, how do we capture the ego network of an evolving multi-graph stream over time with least possible edges, while preserving the structure, properties and efficiency of an ego network? For which, we proposed a streaming ego network sampling method using a forgetting factor [13]. The proposed method is suitable for dynamically evolving multi-graphs. We use this method over a real world temporal stream of edges/calls to generate a sample stream in real time. Our results show that the proposed method preserves tie strengths in the networks. We also show that our method decreases redundancy in the network while preserving the importance of ego. We measure the importance and efficiency of network using some socio-metrics. We evaluate our method by comparing the samples generated by varying parametric values, with the original ego network. The proposed method can also be implemented over a socio-centric network.

The following paper is organized as follows: In Sect. 2 we discuss some related works. In Sect. 3, we described our call network data and the metrics we used in our experiments in Sect. 4. We proved the densification power law for our evolving ego networks in Sect. 5. In Sect. 6, we analyzed the properties and structure of evolving call ego network. Further in Sect. 7, we proposed a sampling method for ego network multi graph streams with forgetting factor. Sections 8 and 9, we evaluated the above method by comparing the samples with the original network.

2 Related Work

The concept of ego networks was discussed by L.C. Freeman in [6], where he described an ego network as a social network, built around a particular social unit called ego. In [17] Wellman discusses the importance of local ties in personal networks. In [3] Burt studied the affects, gaps and relationships between the neighborhood of a node, referring them as structural holes. He also introduced metrics to evaluate an efficient-effective network which strives to optimize structural holes in order to maximize information benefits.

Most of the research works in this field are carried out by analyzing the structure and growth pattern of evolving socio centric networks and evolutionary nature of socio centric graphs [1,2,8,10,11,16]. Nevertheless, there are few works which studied the structure of ego networks [3,6,7,13,15]. To the best of our knowledge, this is the first work about analyzing the evolution of ego networks. We would analyze the evolution of ego network for 31 days using an extensive list of graph level and node level metrics.

[9] proposed an ego-centric network sampling approach for viral marketing applications. The authors employed a variation of forest fire algorithm for sampling ego network. They compared the degree and clustering coefficient distributions of sampled ego networks with the original ego network. In this work, we discuss an edge based sampling method with forgetting factor over an evolving ego network stream of temporal edges.

3 Description of Call Network Data

Telecommunications' call graphs are one of the massive streams of calls generated in real-time. We made use of such anonymised temporal call stream of 31 days available from a service provider. The network data stream is generated a speed of 10 to 280 calls per second around mid-night and mid-day. On an average we have 12.4 million calls made by 4 million subscribers per day. Streaming approach is highly feasible for this kind of rapidly evolving data.

From the above massive stream of calls for 31 days, we built the ego networks by selecting egos with five different properties. The first ego network $egonet_1$ is built by selecting an ego with a degree equal to average degree of network. The $egonet_2$ with an ego of highest in-degree centrality of graph and is also the node with highest eigen vector centrality of graph. The $egonet_3$ and $egonet_4$ with highest betweenness centrality and lowest out-degree centrality respectively for enhancing the diversity of ego networks. We built these ego networks by accumulating all the adjacent edges of ego and their adjacent edges i.e. a network of radius 2. We generated the ego network streams from a call network stream of 400 million calls made by 12 million subscribers on an aggregated scale. In order to avoid duplicated number of edges as it is a multi-graph, we maintained unique edges between any pair of nodes in the network and map them onto a weighted graph.

4   Metrics for Evaluating Ego Networks

In this section we discuss an extensive list of graph metrics we would use in the later sections to analyze the densification of call ego-network, to analyze the evolution of structural, topological and behavioral properties of call ego network and to evaluate the proposed sampling method of forgetting factor. We exploit these properties at graph level and node level.

4.1   Graph Level Metrics

We studied the properties of ego network graphs using average degree, average weighted degree, density, diameter and average path length.

Additionally for evaluating evolving samples using our proposed method, we compared the degree distributions of the samples at the end of 31 days with the original network using kolmogorov-Smirnov test. We use the D-statistics from the test and also p-values to evaluate our null hypothesis (H_0) that our sampled ego networks follow the same distribution as the original ego network. The degree distributions of the networks is obtained by counting the frequency of each degree d in the network. The frequency of each degree d is given by the number of nodes with degree d in the network snapshots at the end of 31 days.

We compared the effective size and efficiency of samples with that of ego network using ego metrics introduced by Burt in [3]. Effective size of the ego network (ES) is the number of alters that an ego has, minus the average number of ties that each alter has to other alters. In the simplest form, for an undirected ego network of radius 1, the effective size can be given with the Eq. 1. Efficiency (EF) of an ego network is the proportion of ego's ties to its neighborhood that are "non-redundant." Efficiency is the normalized form of effective network size (Eq. 2). Therefore, it is a good measure for comparing ego networks of different sizes.

$$ES = n_a - \frac{\sum_{a=1}^{n_a}(d_a - 1)}{n_a} \tag{1}$$

$$EF = \frac{ES}{n_a} \tag{2}$$

where n_a is the number of alters in the ego network and d_a is the degree of an alter a.

4.2   Node Level Metrics

The node level centrality metrics discussed in the later sections are Degree, Weighted Degree, Closeness (CC), and Eigen Vector Centralities (EVC). We also explored the Eccentricity and Clustering Coefficient of the ego.

5   Densification Law for Evolving Call Ego-Network

In [8] the authors studied the temporal evolution of, number of nodes vs number of edges. Besides, the authors employed the measures of average out degree to ascertain the densification law proposed by them. They validated that, most of these graphs densify over time, with the number of edges growing super-linearly to the number of nodes and their average degree increases. They investigated the above properties in an evolving citation graph, autonomous systems graph and affiliation graph. The authors stated that as the graphs evolve over time, they follow the relation given by the Eq. 3.

$$e(t) \propto n(t)^a \tag{3}$$

where e(t) and n(t) denote the number of edges and nodes of the graph at time t, and a is an exponent that generally lies strictly between 1 and 2. The authors refer to such a relation as a densification power law, or growth power law where the number of edges grow super-linearly to the number of nodes. The authors also show that the average degree of these graphs gradually increases. With this justification the authors prove that the graphs densify over time. In this section we investigate the densification power law (DPL) over the temporal stream of call ego network by depicting a densification power law plot (DPL plot) for the number of nodes n(t) and the number of edges e(t) at each timestamp t. In our experiments, we used a time stamp of one day.

We used the four temporal evolving ego networks from the call/edge stream as described in Sect. 3. We grabbed the snapshots of ego networks at the end of each day and calculated the number of nodes and edges. Figure 1 shows the DPL plots for the call ego networks. As the slope of the line in a log-log plot gives the exponent in a power law relation, in the figure discussed above, we derived the lines obeying power relation with the best fits of 0.99 and 1.0 with their respective points. Therefore, the slope of these lines gives the densification exponents as a = 1.03, 1.1, 1.08, and 1.05 (in Fig. 1a, b, c and d respectively) which shows a super-linear growth of edges over nodes. Hence, we deduce that the ego networks of a call network also follow the densification power law as many other socio centric networks, with the number of edges growing super linearly to the number of nodes with their respective exponents a.

We consider the average degree of ego networks per time stamp, which is plotted in Fig. 2. We see that the average degree of graphs for Fig. 2b and d (ego network of highest in-degree centrality node and highest eigen vector centrality of graph) is gradually increasing. Average degree of graphs in Figs. 2a and c are is slightly increasing with the evolution. From the above experiments with the densification power law and the average degree, we see that the graphs are densifying. Hence we require sampling techniques in real-time to analyze such enormous evolving data.

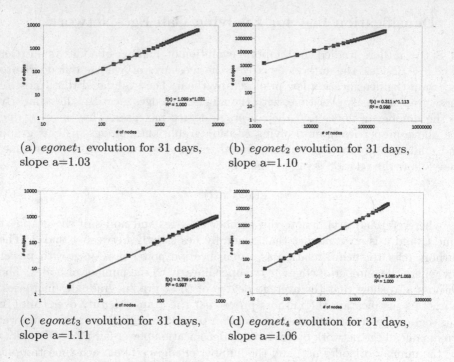

(a) *egonet₁* evolution for 31 days, slope a=1.03

(b) *egonet₂* evolution for 31 days, slope a=1.10

(c) *egonet₃* evolution for 31 days, slope a=1.11

(d) *egonet₄* evolution for 31 days, slope a=1.06

Fig. 1. DPL plot for temporal call ego networks

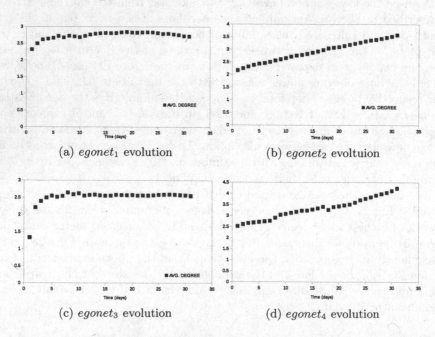

(a) *egonet₁* evolution

(b) *egonet₂* evoltuion

(c) *egonet₃* evolution

(d) *egonet₄* evolution

Fig. 2. Average degree evolution in temporal call ego networks for 31 days

6   Evolution Analysis of a Temporal Ego-Network

In this section, we have analyzed the evolution of the ego network ($egonet_1$) over a period of one month using the metrics mentioned in Sect. 4. As described in earlier section, we have constituted the adjacent nodes of an ego and their adjacent nodes in the ego network as the stream progresses for a month. Then we took snapshots of the ego network per equal intervals of time stamp i.e. one day in our case. To investigate the evolving structure and properties of a call ego network, we undertook a piecemeal structural analysis of network by employing the following metrics per day i.e. average degree, average weighted degree, Density, Diameter and Average Path Length and derive some empirical observations. We also made use of a bunch of centrality metrics to study the importance of ego in the network and compare the position of ego during evolution, they are Degree, Weighted degree, Closeness (CC), and Eigen vector centralities (EVC). We also explored the Eccentricity and Clustering coefficient of the ego.

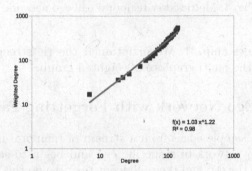

Fig. 3. Degree vs weighted degree of ego network

Figure 3 plots the degree vs weighted degree of the ego in the ego network over a log log plot. The equation of the line that best fits our temporal data points is given in the figure. The slope of the line is given as 1.22 which shows a power relation between degree and weighted degree of a node. Therefore, we can say that the weighted degree of the node is growing superlinearly over its degree. The above analysis demonstrates a social behavior, that the people are more interested in maintaining their old relationships or friends than making new friends. However they also show interest in making new pals.

Figure 4 depicts the graph metrics and node metrics over the evolving ego network. When considering node metrics we see that the CC and eccentricity of the ego increases with the evolution but the EVC remains constant, as the ego remains the important person in the network with highest betweenness centrality. The betweenness centrality of the ego also increases with the function of network size.

Figure 5 displays the call ego network of a particular ego for day 1 and final accumulated network on day 31. The ego is represented with the red dot in the center. The figure illustrates the evolution of network for one month from

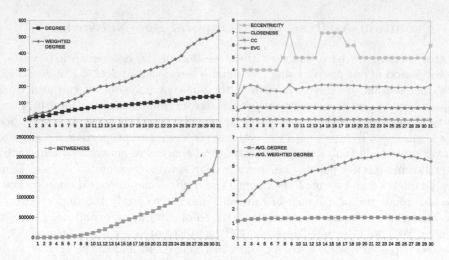

Fig. 4. Metrics over temporal call ego network

timestamp 1 to timestamp 31. We maintained the tie strengths between the nodes by mapping the multi graph to a weighted graph.

7   Sampling Ego Network with Forgetting Factor (SEFF)

In this method, we sample edges from a stream of temporal network. We start by building the ego network of a specific ego and begin to scrape together all the adjacent ties to the ego and their adjacent ties. We do this by using a set for storing adjacent nodes. For every recurring edge, we increment the edge weight of the corresponding edge by maintaining a hash table. We impose a forgetting factor over edges, following successive grace periods. In our experiments, we use a grace period of 1 day. This means we apply the forgetting factor over the ego network as soon as the stream enters a new day, i.e. we forget the old edges each of a kind (i.e. edges between a pair of nodes), by some fixed percentage defined by the forgetting factor. The forgetting factor is given by two parameters, an attenuation factor α and a threshold θ. Where $0 < \alpha < 1$ and also $0 < \theta < 1$. After every grace period or update time t the tie strength between two nodes is given by the Eq. 4.

$$w_t = w_t + (1 - \alpha)w_{t-1} \tag{4}$$

where w_t is the tie strength between any two nodes in the ego network at time t. After every successive grace period, we decrease the edge weight by α and consequently remove the alter/alters adjacent to the corresponding edge, as the edge weight decreases than the threshold value θ. When $\alpha = 1$ we have a maximum forgetting i.e. we forget the whole network except the network of current day. When $\alpha = 0$ we get the original network. If the removed edge corresponds to an alter adjacent to the ego, we remove the adjacent edge and the alter, and all the second level alters adjacent to the alter itself, if the above condition is

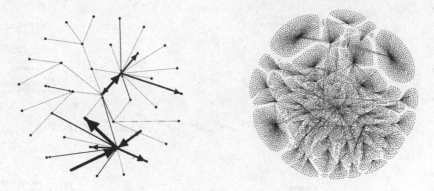

Fig. 5. Evolution of a call ego network

satisfied. If we forget a second level edge, not having a direct connection to ego then we only forget the corresponding node. Following this strategy, we can have most active alters in the ego network at the end of each day.

8   Evaluation Methodology

In order to evaluate our method SEFF discussed in Sect. 7, we applied it over a real world streaming call Graph G of 31 days by randomly choosing an ego e and generating a sample stream of depth $d = 2$ at any point of flow. This was done by generating six real time sample streams, where each sample stream S_i is generated by different combinations of $\alpha \in \{0.9, 0.8, 0.7, 0.5\}$ and $\theta \in \{0.1, 0.2, 0.3, 0.4, 0.5\}$ discussed in Sect. 7. For investigating the above sample streams, we captured their snapshots of sample streams at the end of 31 days each. Each snapshot $S_i^{31} \subset G$. Beforehand, we took a snapshot G_e^{31} of original ego network stream G_e of e (where $d(G_e) = 2$) at the end of 31 days from the socio-centric call graph G. Each sample graph $S_i^{31} \subset G_e^{31}$. We then compared the conclusive sample snapshots S_i^{31} where $1 \leq i \leq 6$, with the original ego network snapshot G_e^{31} by employing metrics discussed in Sect. 4. We use Kalmogorov-Smirnoff test to compare the degree distributions of the original network with that of samples. Conclusively, we derive some conclusions about the properties preserved by the sample networks.

9   Experimental Evaluation

The call networks are the special application scenario for employing our method as these networks are multi-graphs with more than one edge between two users, representing the strength of their relationship unlike a social network based on friendship and, follower and followee relations, where there is a single binomial relation between two nodes. However, the proposed method can be applied to

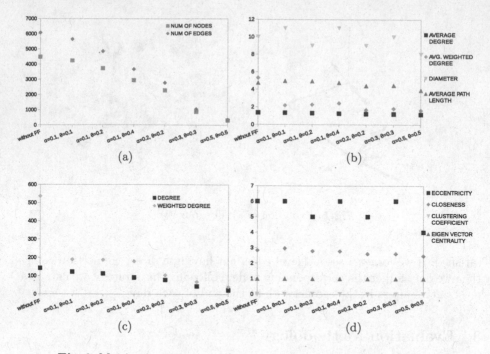

Fig. 6. Metrics over ego networks with and without forgetting factor

networks with binomial relationships as it forgets edges and eventually forgets nodes. SEFF method is also appropriate for sampling weighted networks.

We selected an arbitrary user "ego" from the real world call/edge stream described in Sect. 3 and start building the ego network of ego with a two step neighborhood, i.e. by acquiring the neighbors of ego and the neighbor of neighbors of ego. We take a snapshot of the ego network at the end of 31 days stream. Using the same ego we start constructing the sample ego networks (using SEFF) gradually as the stream flows for 31 days. For which, we have used six different combinations of α and θ corresponding to six different samples depicted in Fig. 6. The figure also plots the values of computed metrics discussed in Sect. 4 over the conclusive sampled ego networks and the original ego network.

Figure 6(a) shows the number of nodes and the number of edges in the above described ego networks. We observe that the number of nodes gradually decrease with the increasing forgetting factor. For an attenuation value of 0.5 and threshold value of 0.5 we forget 50 % of the edges per day, between two adjacent nodes. This shows we always have the most active nodes with the increased forgetting factor. We also observed that, the number of edges decrease in greater proportion than the number of nodes, Almost reaching equal for the highest forgetting factor in the illustration. This exhibits that the proposed SEFF method decreases redundant edges.

We also compare the degree distributions of the original ego network with the samples generated by using SEFF method at the end of 31 days. We applied

Table 1. Comparison of degree distributions using KS-Test

Samples	$\alpha = 0.1,$ $\theta = 0.1$	$\alpha = 0.1,$ $\theta = 0.2$	$\alpha = 0.1,$ $\theta = 0.4$	$\alpha = 0.2,$ $\theta = 0.2$	$\alpha = 0.3,$ $\theta = 0.3$	$\alpha = 0.5,$ $\theta = 0.5$
D-stat	0.146	0.138	0.173	0.146	0.191	0.096
p-value	0.114	0.124	0.065	0.182	0.105	0.724

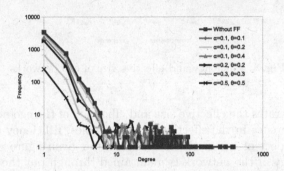

Fig. 7. Degree distributions of ego networks at the end of 31 days with and without forgetting factor

Kolmogorov–Smirnov test to compare the degree distributions of the samples with the original network. The D-statistics and P-values of tests are given in the Table 1. The p-values are computed using exact method. The significance level used for the comparisons is 5 %, i.e. $\alpha = 0.5$. The results show that all the sampled distributions follow the distribution of original graph. We also observe that the value of θ has a greater impact on the similarity of distributions, than α in the SEFF method. We can see the pictorial representation of the degree distributions of the original graph and sample graphs in Fig. 7.

Figure 6(b) depicts metrics over the ego networks. The diameter of the graphs varies with the inclusion and removal of the connecting nodes from the ego network. It depends on the network of ego selected. Average degree and the average path length decreases with the increasing forgetting, this shows that the networks shrink with increased forgetting. The SEFF method has a noticeable effect over the weighted degree of graphs.

The degree and weighted degree of the ego are plotted in Fig. 6(c). Both the values decreased with the increased forgetting, while the drop in weighted degree is higher, this suggests that when we increased forgetting we decreased the tie strengths but relatively maintained the ties. In Fig. 6(d) we see that the eccentricity has a similar effect of diameter in the ego network graphs. This corresponds to the conceptual relation between diameter and eccentricity. Closeness of the ego with alters also decreased gradually with the increased forgetting factor. The clustering coefficient of ego is too low to compare. The eigen vector centrality portrays the important node in the network. SEFF preserves the importance of ego along side forgetting.

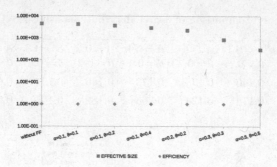

Fig. 8. Efficiency and effective size of ego networks

Figure 8 illustrates the effective size and efficiency of the ego networks. There is negligible difference in the effective size of samples. Efficiency of the network indicates the impact of ego in the network. In the given figure we can observe that the efficiency of the network is maintained through out the samples using SEFF. The measure of effective size of the network is not normalized with the size of network, therefore it decreases with the average number of ties that each alter has to other alters.

10   Conclusions

In this work, we analyzed the evolution of ego network for a period of one month. We exploited the structural properties of network and related them with the natural behavior of users. We also proved the densification law over the ego networks of call graphs for a period of one month and found that the graphs are densifying along time.

As we observed the properties of evolving ego network, we proposed a sampling method with forgetting factor for streaming multi-graph networks which preserves the density of graph and retains the tie strengths between nodes. We evaluated our method by exploiting the ground truth of original graph vs samples generated by varying parameter values. Based on the empirical experiments we prove that our method maintains the importance and efficiency of the network and decreases the redundancy while preserving most active and recent nodes from the network.

Acknowledgments. This work was partly supported by the European Commission through MAESTRA (ICT-2013-612944) and the Project TEC4Growth - Pervasive Intelligence, Enhancers and Proofs of Concept with Industrial Impact/NORTE-01-0145-FEDER-000020 is financed by the North Portugal Regional Operational Programme (NORTE 2020), under the PORTUGAL 2020 Partnership Agreement. Shazia Tabassum is financed by the ERDF – European Regional Development Fund through the Operational Programme for Competitiveness and Internationalisation - COMPETE 2020 Programme within project (POCI-01-0145-FEDER-006961), and by National

Funds through the FCT – Fundação para a Ciência e a Tecnologia (Portuguese Foundation for Science and Technology) as part of project UID/EEA/50014/2013. The authors also thank WeDo Business for providing the data.

References

1. Albert, R., Jeong, H., Barabási, A.-L.: Internet: diameter of the world-wide web. Nature **401**(6749), 130–131 (1999)
2. Broder, A., Kumar, R., Maghoul, F., Raghavan, P., Rajagopalan, S., Stata, R., Tomkins, A., Wiener, J.: Graph structure in the web. Comput. Netw. **33**(1), 309–320 (2000)
3. Burt, R.S.: Structural Holes: The Social Structure of Competition. Harvard University Press, Cambridge (2009)
4. Epasto, A., Lattanzi, S., Mirrokni, V., Sebe, I.O., Taei, A., Verma, S.: Ego-net community mining applied to friend suggestion. Proc. VLDB Endowment **9**(4), 324–335 (2015)
5. Everett, M., Borgatti, S.P.: Ego network betweenness. Soc. Netw. **27**(1), 31–38 (2005)
6. Freeman, L.C.: Centered graphs and the structure of ego networks. Math. Soc. Sci. **3**(3), 291–304 (1982)
7. Hanneman, R.A., Riddle, M.: Introduction to social network methods (2005)
8. Leskovec, J., Kleinberg, J., Faloutsos, C.: Graphs over time: densification laws, shrinking diameters and possible explanations. In: Proceedings of the Eleventh ACM SIGKDD International Conference on Knowledge Discovery in Data Mining, pp. 177–187. ACM (2005)
9. Ma, H.H., Gustafson, S., Moitra, A., Bracewell, D.: Ego-centric network sampling in viral marketing applications. In: Ting, I.-H., Wu, H.-J., Ho, T.-H. (eds.) Mining and Analyzing Social Networks. SCI, vol. 288, pp. 35–51. Springer, Heidelberg (2010)
10. Milgram, S.: The small world problem. Psychol. Today **2**(1), 60–67 (1967)
11. Newman, M.E.: The structure and function of complex networks. SIAM Rev. **45**(2), 167–256 (2003)
12. Sarmento, R., Oliveira, M., Cordeiro, M., Tabassum, S., Gama, J.: Social network analysis of streaming call graphs. In: Japkowicz, N., Stefanowski, J. (eds.) Big Data Analysis: New Algorithms for a New Society. Studies in Big Data, vol. 16, pp. 239–261. Springer, Switzerland (2016)
13. Tabassum, S., Gama, J.: Sampling ego-networks with forgetting factor. In: IEEE Workshop on High Velocity Mobile Data Mining (2016, in press)
14. Tabassum, S., Gama, J.: Sampling massive streaming call graphs. In: ACM Symposium on Advanced Computing, pp. 923–928 (2016)
15. Tabassum, S., Gama, J.: Social network analysis of mobile streaming networks. In: IEEE Conference on Mobile Data Mining, Ph.d. Forum (2016, in press)
16. Watts, D.J., Strogatz, S.H.: Collective dynamics of 'small-world' networks. Nature **393**(6684), 440–442 (1998)
17. Wellman, B.: Are personal communities local? A dumptarian reconsideration. Soc. Netw. **18**(4), 347–354 (1996)

Ensemble Learning

Ensemble Diversity in Evolving Data Streams

Dariusz Brzezinski[✉] and Jerzy Stefanowski

Institute of Computing Science, Poznan University of Technology,
ul. Piotrowo 2, 60–965 Poznan, Poland
{dariusz.brzezinski,jerzy.stefanowski}@cs.put.poznan.pl

Abstract. While diversity of ensembles has been studied in the context
of static data, it has not still received such research interest for evolving data streams. This paper aims at analyzing the impact of concept
drift on diversity measures calculated for streaming ensembles. We consider six popular diversity measures and adapt their calculations to data
stream requirements. A comprehensive series of experiments reveals the
potential of each measure for visualizing ensemble performance over time.
Measures highlighted as capable of depicting sudden and virtual drifts
over time are used as basis for detecting changes with the Page-Hinkley
test. Experimental results demonstrate that the κ interrater agreement,
disagreement, and double fault measures, although designed to quantify
diversity, provide a means of detecting changes competitive to that using
classification accuracy.

Keywords: Classifier ensemble · Diversity measure · Data stream ·
Concept drift · Drift detection

1 Introduction

Recent decades have increased interest in collecting big data, which resulted
in new challenges for data storage and processing. Apart from their massive
volumes, these demanding data sources are also characterized by the speed at
which data is passed to analytical systems. These properties are especially relevant when data are continuously generated in the form of *data streams*.

Compared to static data, classification in streams implies new requirements
for algorithms, such as constraints on memory usage, restricted processing time,
and one scan of incoming examples [5,6]. An even more challenging aspect of
analyzing streaming data is that learning algorithms often act in dynamic, non-
stationary environments, where the data and target concepts change over time.
This phenomenon, called *concept drift*, deteriorates the predictive accuracy of
classifiers, as the instances the models were trained on differ from the current
data. Examples of real-life concept drifts include spam categorization, weather
predictions, monitoring systems, financial fraud detection, and evolving customer
preferences; for their review see, e.g. [8,11,20].

Since typical batch learning algorithms for supervised classification are not
capable of fulfilling the aforementioned data stream requirements, several new

© Springer International Publishing Switzerland 2016
T. Calders et al. (Eds.): DS 2016, LNAI 9956, pp. 229–244, 2016.
DOI: 10.1007/978-3-319-46307-0_15

learning algorithms have been introduced [5,6]. They are based on using sliding windows to manage memory and provide a forgetting mechanism, sampling techniques, drift detectors, and new online algorithms. Out of several proposals, *ensemble* methods play an important role. Ensembles of classifiers are quite naturally adapted to non-stationary data streams, as they are capable of incorporating new data by either introducing a new component or updating existing components. Forgetting of outdated knowledge can be implemented by removing components that perform poorly at a given moment or by continuously adapting component weights accordingly to performance on recent data. Classifier ensembles for streaming data are typically divided into block-based (batch-incremental) and online (instance-incremental) approaches, depending on the way they process incoming examples.

Most of the existing experimental studies on stream classifiers focus on predictive abilities and computational costs of ensembles in several scenarios of concept drifts [3]. However, in earlier research on batch ensembles for static data, several researchers were also interested in the *diversity* of ensembles, which is usually calculated as the degree in which component classifiers make different decisions for a single case [12]. Some authors hypothesize that high predictive accuracy and diversity among component classifiers should be related. As a result, many researchers considered special techniques for: visualizing diversity [13], selecting the most diverse ensemble [10], or using diversity measures to prune a large pool of component classifiers [1,9,13].

On the other hand, such interest in diversity measures is not so visible in research on data stream ensembles. As ensemble components are typically learned form different parts of the data stream, potentially referring to different concept distributions, most researchers claim that they are diversified but do not measure it directly [18]. There have been rare attempts at directly promoting diversity during classifier training [14,16,19], yet once again diversity over time was not reported in these studies. Notably, Minku et al. [14] discuss the impact of diversity on online ensemble learning and reactions to drift by modifying the Poisson distribution used in Online Bagging. However, doing so they only measure accuracy of the modified ensemble, not its diversity.

In this paper, we analyze the more general problem of measuring ensemble diversity in evolving data streams. More precisely, we are interested in answering the following research questions, which are not answered by previous works:

1. Which commonly used diversity measures can be calculated for streams processed: in blocks, incrementally, incrementally with forgetting?
2. How is ensemble diversity affected by concept drifts? Does diversity change over time?
3. Do incremental component classifiers enhance or degrade diversity, compared to batch component classifiers?
4. Can diversity measures be used as additional information during classifier training or drift detection?

To answer the above questions, in the following sections we perform a review of the most popular diversity measures known from static learning, analyze the possibility of calculating these measures online, and perform a comprehensive series of experiments to evaluate the use of each measure for visualizing and detecting various types of concept-drift.

2   Related Work

2.1   Ensemble Diversity Measures

To the best of our knowledge, there have been no proposals of specialized ensemble diversity measures for changing data streams. Therefore, we will analyze the use of diversity measures known from static learning in streaming scenarios. For this purpose, we selected six popular definitions of diversity based on the comprehensive review done by Ludmila Kuncheva [12].

To illustrate the calculation of each measure, we will consider the joined outputs of two component classifiers C_i and C_j shown in Table 1. The table presents proportions of correct/incorrect answers of one of or both components, thus, the total of all the cell values $a + b + c + d = 1$. An ensemble of L classifiers will produce $L(L-1)/2$ pairwise diversity values based on such tables. To get a single value we average across all pairs.

Table 1. The 2×2 ensemble component relationship table with probabilities [12]

	C_i correct	C_i wrong
C_j correct	a	b
C_j wrong	c	d

The six analyzed diversity measures are: disagreement (D), Kohavi-Wolpert variance (KW), double fault (DF), interrater agreement (κ), Yule's Q statistic (Q), and coincident failure diversity (CFD). The definitions of all the measures, using values from Table 1, are presented in Eqs. 1–6. Note that we use shorter equivalents of definitions given by primary authors, to make measure descriptions shorter. For a broader discussion on ways of computing each measure, please review [12].

$$D_{i,j} = b + c \tag{1}$$

$$KW = \frac{L-1}{2L} D_{av} \tag{2}$$

$$DF_{i,j} = d \tag{3}$$

$$\kappa = 1 - \frac{1}{2\bar{p}(1-\bar{p})} D_{av} \tag{4}$$

$$Q_{i,j} = \frac{ad - bc}{ad + bc} \tag{5}$$

$$CFD = \begin{cases} 0, & p_0 = 1; \\ \frac{1}{1-p_0} \sum_{i=1}^{L} \frac{L-i}{L-1} p_i, & p_0 < 1. \end{cases} \tag{6}$$

The disagreement measure D (1) is equal to the probability that two classifiers will disagree on their decision. It is worth noting that for binary classification the true label of an example is not needed to determine if components disagree. The Kovavi-Wolpert variance KW (2) is inspired by the variance of the predicted class label across different training sets that were used to build the classifier. However, here we use the property that KW differs from the averaged disagreement D_{av} by a coefficient. Double fault DF (3) counts the number of times both classifiers make mistakes, whereas κ (4) measures the level of agreement between classifiers, where $\bar{p}$ is the arithmetic mean of the components' classification accuracy. The Q statistic (5) varies between -1 and 1, where components that tend to recognize the same objects correctly will have positive values and components which tend to classify different examples incorrectly will have negative values. Finally, CFD (6) is a measure that originates from software reliability, and achieves its best value of 1 when all misclassifications are unique. In Eq. (6), p_i denotes the probability that exactly i out of L components fail on a randomly chosen input.

2.2   Stream Classifiers and Drift Detectors

As an increasingly important data mining technique, data stream classification has been widely studied by different communities; a detailed survey can be found in [5,6]. In our study, we focus on representatives of block-based and online ensembles. As an example of that first category, we will use the *Accuracy Updated Ensemble* (AUE) [3], which creates a new component with each block of examples and adds it to the ensemble, incrementally trains previously created components, and weights (evaluates) components according to their performance on the newest data block. As an example of online ensemble learning, we will use *Online Bagging* [15], which incrementally updates components with each incoming example and makes a final prediction with simple majority voting. The sampling, crucial to batch bagging, is performed incrementally by presenting each example to a component k times, where k is defined by the Poisson distribution.

In this paper, we investigate ensemble diversity measures not only as a means of visualizing ensemble and stream characteristics, but also as a basis for drift detection. For this purpose, we modify the Page-Hinkley (PH) test [7], however, generally other drift detection methods could also have been adapted [8]. The PH test considers a variable m^t, which measures the accumulated difference between observed values e (originally error estimates) and their mean till the current moment $\bar{e}^t$, decreased by a user-defined magnitude of allowed changes δ: $m^t = \sum_{i=1}^{t} (e^i - \bar{e}^t - \delta)$. After each observation e^t, the test checks whether the difference between the current m^t and the smallest value up to this moment $\min(m^i, i = 1, \ldots, t)$ is greater than a given threshold λ. If the difference exceeds λ, a drift is signaled. In this paper, we propose to use the studied diversity measures as the observed value.

3   Calculating Diversity Measures for Streaming Data

We will discuss the possibility of calculating ensemble diversity measures in three basic stream processing scenarios: in blocks, incrementally, and prequentially [6].

Block-based processing is the most natural framework for calculating diversity measures, as examples arrive in portions (chunks) of sufficient size. Thus, one can recalculate ensemble diversity on each incoming data block in a similar way as for static data. We note that each of the presented measures is based on individual component predictions and their summary in the form of pairwise component relationships presented in Table 1. Therefore, for a data block of d examples and L ensemble components, measures 1–6 can be computed in $O(d \cdot L^2)$ time and $O(d \cdot L)$ memory. Since d and L are user-defined constants, this resolves to constant time and memory per block, therefore, the analyzed measures can be successfully used in block processing.

Incremental calculation assumes that a measure can be computed based only on a summary of all previous examples and a single new example. This is slightly less trivial, however, if we monitor the number of processed examples and update a, b, c, d counts with each instance, each of the measures considered in this study can be calculated for a new example in $O(L^2)$ time and $O(L^2)$ memory.

Finally, if a stream is subject to changes, one may be interested in calculating diversity measures prequentially, that is, incrementally with forgetting [4,7]. Two basic approaches to calculating values with forgetting are used: *sliding windows* and *fading factors* [7]. Sliding windows provide a way of limiting the amount of analyzed examples by retaining a set of only d most recent examples at each time point. Fading factors, on the other hand, discount older information across time by multiplying the previous summary by a factor and adding a new value computed on the incoming example. Sliding windows resemble data blocks updated after each example, and can be similarly used to calculate diversity measures with forgetting. Furthermore, due to the fact that all of the analyzed measures are based on counts, all of the measures can be also computed using fading factors. For this purpose, it suffices to calculate the *fading sum* $S_{x,\alpha}(t)$ and *fading increment* $N_\alpha(t)$ from a stream of objects x at time t [7]:

$$S_{x,\alpha}(t) = x^t + \alpha \times S_{x,\alpha}(t-1)$$
$$N_\alpha(t) = 1 + \alpha \times N_\alpha(t-1)$$

where x can be counts of any of the values a, b, c, d from Table 1. For example, if d^t is 1 when both components misclassify an example, then double fault can be calculated as $DF_\alpha(t) = S_{d,\alpha}(t)/N_\alpha(t)$. As with incremental computation, the prequential calculation of any of the analyzed diversity measures for a new example requires $O(L^2)$ time and $O(L^2)$ memory.

To sum up, all the considered diversity measures can be computed on blocks, incrementally, and prequentially, while fulfilling limited time and memory requirements of stream processing. As we are interested in using these diversity measures on concept-drifting data, in the following sections we will visualize and analyze diversity calculated prequentially. To the best of our knowledge, this is the first study of diversity measures from this perspective.

4   Experimental Study

We performed two basic groups of experiments, one visualizing and comparing diversity measures over time, and another assessing the possibility of using them as a basis for drift detection. In the first group, we tested two different ensemble classifiers: Online Bagging (Bag) and Accuracy Updated Ensemble (AUE). Bag was chosen as an online approach, whereas AUE represents block-based ensembles. As component classifiers we compared: Naive Bayes (NB), Linear Perceptron (P), Decision trees (J48), and Hoeffding Trees (HT). For the second group of experiments, we compared drift detectors using Online Bagging with HT components.

All the algorithms and evaluation methods were implemented in Java as part of the MOA framework [2]. The experiments were conducted on a machine equipped with a dual-core Intel i7-2640M CPU, 2.8 Ghz processor and 16 GB of RAM. For all the experiments, base learners where parametrized with default values proposed in MOA.

4.1   Datasets

In experiments showcasing visualizations of diversity measures over time, we used 2 real and 10 synthetic datasets[1]. For the real-world datasets it is difficult to precisely state when drifts occur. In particular, Airlines (Air) is a large, balanced dataset, which encapsulates the task of predicting whether a given flight will be delayed and no information about drifts is available. However, the second real dataset (PAKDD) was intentionally gathered to evaluate model robustness against performance degradation caused by market gradual changes and was studied by many research teams [17].

Additionally, we used the MOA framework [2] to generate 10 artificial datasets with different types of concept drift. The SEA generator [16] was used to create a stream without drifts (SEA_{ND}), as well as three streams with sudden changes and constant 1:1 (SEA_1), 1:10 (SEA_{10}), 1:100 (SEA_{100}) class imbalance ratios. Similarly, the Hyperplane generator [18] was used to simulate three streams with different class ratios, 1:1 (Hyp_1), 1:10 (Hyp_{10}), 1:100 (Hyp_{100}), but with a continuous incremental drift rather than sudden changes. Streams with subscripts $_{10}$ and $_{100}$ were created to assess measures in the presence of class imbalance, which usually remains undetected by classification accuracy [4]. We also tested the performance of the analyzed measures in the presence of very short, temporary changes in a stream (RBF) created using the RBF generator [2].

---

[1] Source code, test scripts, and generator parameters available at:
http://www.cs.put.poznan.pl/dbrzezinski/software.php.

Apart from data containing real drifts, we additionally created four streams with virtual drifts, i.e., class distribution changes over time. SEA_{RC} contains three sudden class ratio changes (1:1/1:100/1:10/1:1), whereas Hyp_{RC} simulates a continuous ratio change from 1:1 to 1:100 throughout the stream. All the synthetic datasets, apart from RBF, contained 5–10 % examples with class noise.

For experiments assessing diversity measures as potential drift detectors, we created 7 synthetic datasets using the SEA (SEA), RBF (RBF), Random Tree (RT), and Agrawal (Agr) generators [2]. Each dataset tested for a single reaction (or lack of one) to a sudden change. $SEA_{NoDrift}$ contained no changes, and should not trigger any drift detector, while RT involved a single sudden change after 30 k examples. The Agr_1, Agr_{10}, Agr_{100} datasets also contained a single sudden change after 30 k examples, but had a 1:1, 1:10, 1:100 class imbalance ratio, respectively. Finally, SEA_{Ratio} included a sudden 1:1/1:100 ratio change after 10 k examples and RBF_{Blips} contained two short temporary changes, which should not trigger the detector. The main characteristics of all the datasets are given in Table 2.

Table 2. Characteristic of datasets

Dataset	#Inst	#Attrs	Class ratio	Noise	#Drifts	Drift type
SEA_{ND}	100 k	3	1:1	10 %	0	None
SEA_1	1 M	3	1:1	10 %	3	Sudden
SEA_{10}	1 M	3	1:10	10 %	3	Sudden
SEA_{100}	1 M	3	1:100	10 %	3	Sudden
Hyp_1	500 k	5	1:1	5 %	1	Incremental
Hyp_{10}	500 k	5	1:10	5 %	1	Incremental
Hyp_{100}	500 k	5	1:100	5 %	1	Incremental
RBF	1 M	20	1:1	0 %	2	Blips
SEA_{RC}	1 M	3	1:1/1:100/1:10/1:1	10 %	3	Virtual
Hyp_{RC}	500 k	3	1:1 → 1:100	5 %	1	Virtual
Air	539 k	7	1:1	-	-	Unknown
PAKDD	50 k	30	1:4	-	-	Unknown
Elec	45 k	8	1:1	-	-	Unknown
KDDCup	494 k	41	1:4	-	-	Unknown
$SEA_{NoDrift}$	20 k	3	1:1	10 %	0	None
Agr_1	40 k	9	1:1	1 %	1	Sudden
Agr_{10}	40 k	9	1:10	1 %	1	Sudden
Agr_{100}	40 k	9	1:100	1 %	1	Sudden
RT	40 k	10	1:1	0 %	1	Sudden
SEA_{Ratio}	40 k	3	1:1/1:100	10 %	1	Virtual
RBF_{Blips}	40 k	20	1:1	0 %	0	Blips

4.2   Diversity Analysis over Time

In our first group of experiments, we plotted diversity measures (1–6) over 2 real and 10 synthetic datasets with various types of drift. The measures where prequentially calculated on a sliding window of $d = 1000$ examples for Bag and AUE. Both ensemble classifiers where tested with NB, P, J48, and HT component classifiers. For subsequent plots, we also changed the number of component classifiers $k \in \{2, 3, 4, 5, 7, 10, 15, 25, 50\}$. By changing the mentioned parameters, we are interested in assessing the influence of:

– the type of visualized diversity measure,
– type of drift occurring in the stream,
– number of component classifiers,
– type of component base learner,
– ensemble adaptation procedure.

A set of plots depicting disagreement (D) measured on Bag with different base learners and varying number of components is presented in Fig. 1. Due to the overwhelming number of subplots, we only present a full figure for D, however, a report containing plots of all the analyzed measures, for both Bag and AUE, is available online.[2]

Looking at Fig. 1, one can notice that D changes over time, and does so differently for each dataset (grid column). Moreover, subplots within one column are similar to each other. As grid rows represent the number of ensemble components, this shows that, for a given dataset, changes in diversity are not very sensitive to the ensemble size. This pattern was true for all the analyzed diversity measures.

Since the shape of each single diversity plot was very similar for varying ensemble sizes, in Fig. 2 we visually compare all the measures for Bag with fixed $k = 10$ components. The first two rows in Fig. 2 present prequentially calculated accuracy [7] and the area under the ROC curve (AUC) [4] as reference metrics, and measures CFD, D, DF, κ, KW, Q, in consecutive rows. The plot clearly showcases that the analyzed diversity measures differ from each other. For example, CFD is very sensitive to ensemble changes, whereas D, DF, and KW have relatively smooth plots. It is also worth noticing that D and KW have plots of identical shape, yet on different y-axis scales. This is expected as looking at Eqs. (1) and (2), one can notice that KW is a scaled version of D. Additionally, it is worth pointing out that some of the measures seem to depict sudden (D, DF, κ) and class ratio changes (κ, CFD) over time. This suggests, that some of the analyzed measures could be monitored over time to signal drifts or problems with the performance of an ensemble.

Figure 3 presents disagreement D of Bag and AUE with $k = 10$ components on a dataset with sudden changes (SEA_1). This pair of plots shows that AUE, which periodically replaces existing components with new classifiers, showcases high variability over time. Furthermore, Fig. 3 gives a closer look at the impact of

---

[2] http://www.cs.put.poznan.pl/dbrzezinski/software/DiversityInStream.html.

using different component base learners. The Naive Bayes (NB) algorithm does not promote diversity among components and does not depict diversity changes over time. The Hoeffding Tree (HT) and Perceptron (P) are much better at depicting changes over time due to their incremental nature. Finally, batch decision trees (J48) are only applicable to block-based ensembles. These properties were shared by all the analyzed plots.

In the following section, we will take a closer look at the possibility of detecting drifts using ensemble diversity measures.

4.3   Drift Detection Using Diversity Measures

The second group of experiments involved using the PH test to detect drifts based on changes in prequential accuracy and diversity measures. To compare all the analyzed metrics, we used sliding window sizes (1000–5000) and PH test parameters ($\lambda = 100$, $\delta = 0.1$) as proposed in [7]. Table 3 presents the number of missed versus false detection counts, with average delay time for correct detections; subscripts in column names indicate the PH test window size. The results refer to total counts and means over 10 runs of streams generated with different random seeds.

First, we note that two diversity measures, Q and KW, are missing from Table 3. We omitted these two measures from the presentation, because the drift detector never triggered for these measures. That means that for datasets with drifts Q and KW always had 10 missed detections and 0 false alarms. Thus, our first observation is that Q and KW are not good candidates for drift monitors, at least when using the PH Test. One explanation of this fact may be that the Kohavi-Wolpert variance KW is a measure with a small range of values, which most probably makes it difficult for the detector to trigger. The Q statistic, on the other hand, puts common misclassifications and correct answers on two ends of its scale, this way introducing difficulties for the used PH Test.

Another outlying measure is CFD. The Coincident Failure Diversity is very susceptible to small changes in the ensemble, causing a very large number of false alarms. Therefore, just as Q and KW, CFD is not a good choice of monitored value when detecting drifts using the PH Test.

The remaining three measures (κ, DF, D) showcase good drift detection properties. Particularly, κ offers detection rates comparable to those of prequential accuracy with smaller delay. Additionally, κ successfully detected 9 out of 10 class ratio changes, whereas accuracy did not detect any of them. DF and D have slightly more missed detections and are slower at signaling changes. However, it is worth noting that for binary classification problems, D has the potential of working in unlabeled or partially labeled stream settings. This opens an interesting option for future research, and might mean that if predictions of components start to disagree in an unusual way, we may be able to observe sudden changes even without true labels of incoming examples.

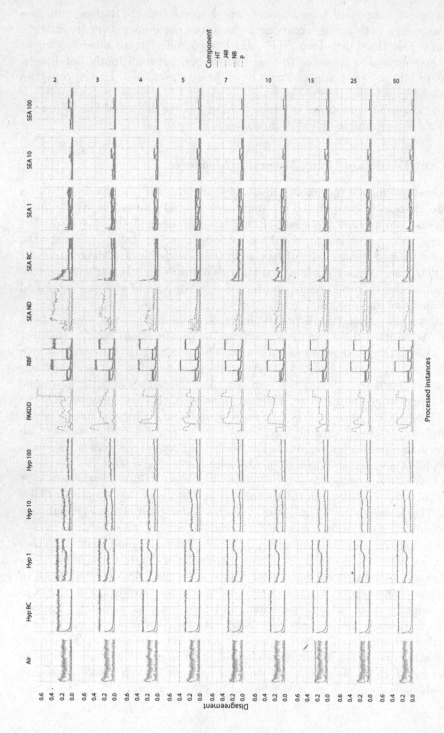

Fig. 1. Disagreement visualizations on all the analyzed datasets for Bag with $k \in \{2, 3, 4, 5, 7, 10, 15, 25, 50\}$ components

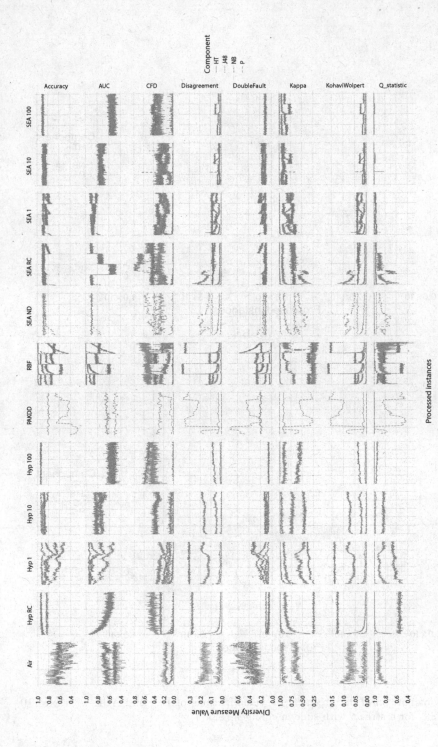

Fig. 2. Performance and diversity measure visualizations on all the analyzed datasets for Bag with $k = 10$ components

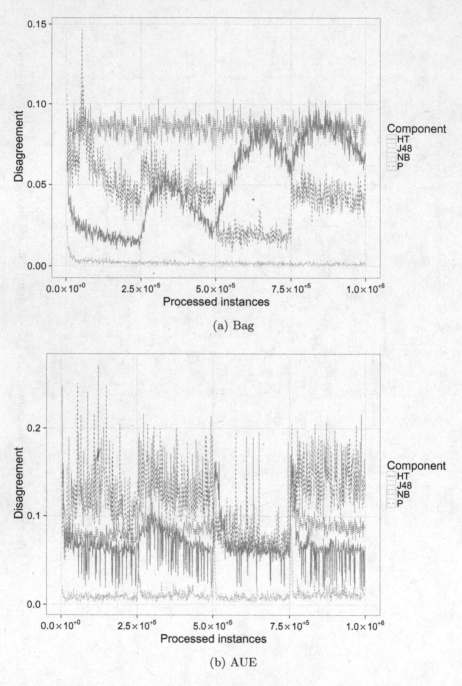

(a) Bag

(b) AUE

Fig. 3. Comparison of disagreement D visualizations of Bag and AUE with $k = 10$ components for a stream with sudden changes (SEA_1)

Table 3. Number of missed and false detections (in the format missed:false) obtained using the PH test with prequential accuracy and diversity measures. Mean delays of correct detections are given in parenthesis, where (-) means that the detector was not triggered or the dataset did not contain any change.

	Acc_{1k}	Acc_{2k}	Acc_{3k}	Acc_{4k}	Acc_{5k}
$SEA_{NoDrift}$	0:0 (-)	0:0 (-)	0:0 (-)	0:0 (-)	0:0 (-)
Agr_1	0:0 (946)	0:0 (1614)	0:0 (2265)	0:0 (2920)	0:0 (3582)
Agr_{10}	0:5 (805)	1:6 (1287)	0:1 (1685)	0:1 (2197)	0:1 (2909)
Agr_{100}	4:13 (1416)	4:11 (1706)	5:13 (2637)	4:10 (3035)	4:9 (3748)
RT	6:0 (1851)	7:0 (2414)	7:0 (3428)	8:0 (3656)	8:0 (4514)
SEA_{Ratio}	10:0 (-)	10:0 (-)	10:0 (-)	10:0 (-)	10:0 (-)
RBF_{Blips}	0:0 (-)	0:0 (-)	0:0 (-)	0:0 (-)	0:0 (-)
	κ_{1k}	κ_{2k}	κ_{3k}	κ_{4k}	κ_{5k}
$SEA_{NoDrift}$	0:0 (-)	0:0 (-)	0:0 (-)	0:0 (-)	0:0 (-)
Agr_1	0:0 (608)	0:0 (986)	0:0 (1352)	0:0 (1719)	0:0 (2082)
Agr_{10}	0:6 (453)	1:8 (648)	5:8 (757)	1:7 (1115)	1:8 (1810)
Agr_{100}	10:10 (-)	8:3 (596)	9:3 (1945)	9:3 (2769)	9:1 (3558)
RT	5:0 (1456)	6:0 (2057)	6:0 (2890)	6:0 (3809)	6:0 (4851)
SEA_{Ratio}	1:0 (1073)	1:0 (1976)	1:0 (2874)	1:0 (3755)	1:0 (4635)
RBF_{Blips}	0:0 (-)	0:0 (-)	0:0 (-)	0:0 (-)	0:0 (-)
	DF_{1k}	DF_{2k}	DF_{3k}	DF_{4k}	DF_{5k}
$SEA_{NoDrift}$	0:0 (-)	0:0 (-)	0:0 (-)	0:0 (-)	0:0 (-)
Agr_1	0:0 (1200)	0:0 (2112)	0:0 (3051)	0:0 (4019)	0:0 (5027)
Agr_{10}	0:3 (881)	0:1 (1387)	0:1 (1938)	0:1 (2727)	0:1 (3817)
Agr_{100}	10:10 (-)	10:5 (-)	10:5 (-)	10:5 (-)	10:5 (-)
RT	6:0 (2125)	8:0 (2092)	8:0 (2881)	8:0 (3688)	8:0 (4561)
SEA_{Ratio}	10:0 (-)	10:0 (-)	10:0 (-)	10:0 (-)	10:0 (-)
RBF_{Blips}	0:0 (-)	0:0 (-)	0:0 (-)	0:0 (-)	0:0 (-)
	D_{1k}	D_{2k}	D_{3k}	D_{4k}	D_{5k}
$SEA_{NoDrift}$	0:0 (-)	0:0 (-)	0:0 (-)	0:0 (-)	0:0 (-)
Agr_1	1:0 (1582)	1:0 (2342)	1:0 (3154)	1:0 (4008)	1:0 (4909)
Agr_{10}	9:1 (3120)	9:1 (3885)	9:0 (4704)	9:0 (5580)	9:0 (6464)
Agr_{100}	7:12 (3693)	8:8 (2337)	8:4 (4818)	9:3 (2991)	9:3 (4272)
RT	10:0 (-)	10:0 (-)	10:0 (-)	10:0 (-)	10:0 (-)
SEA_{Ratio}	10:0 (-)	10:0 (-)	10:0 (-)	10:0 (-)	10:0 (-)
RBF_{Blips}	0:0 (-)	0:0 (-)	0:0 (-)	0:0 (-)	0:0 (-)
	CFD_{1k}	CFD_{2k}	CFD_{3k}	CFD_{4k}	CFD_{5k}
$SEA_{NoDrift}$	0:176 (-)	0:120 (-)	0:80 (-)	0:59 (-)	0:41 (-)
Agr_1	0:580 (281)	0:284 (558)	0:176 (1449)	0:121 (1903)	0:93 (1386)
Agr_{10}	0:516 (506)	0:267 (908)	0:181 (741)	0:131 (1919)	0:102 (1673)
Agr_{100}	0:190 (1242)	0:187 (1085)	0:148 (1294)	0:113 (2463)	0:95 (2073)
RT	0:676 (218)	0:330 (531)	0:213 (810)	0:152 (1117)	0:117 (1972)
SEA_{Ratio}	1:218 (2501)	0:158 (1483)	0:114 (3152)	0:100 (2402)	0:80 (2067)
RBF_{Blips}	0:533 (-)	0:260 (-)	0:170 (-)	0:120 (-)	0:100 (-)

5   Conclusions and Outlook

Diversity is often perceived as one of the most important characteristics of ensemble classifiers. However, even though ensembles are among the most often proposed approaches for concept-drifting streams, up till now ensemble diversity measures have not been thoroughly studied in the context of time-evolving data. In this paper, we reviewed diversity measures known from static data, and analyzed the possibility of calculating them on blocks, incrementally, and prequentially. Additionally, a comprehensive series of experiments was performed to evaluate the use of each measure for visualizing and detecting various types of concept-drift.

Regarding the first question posed in the introduction, we can state that all six of the analyzed measures can be adapted to data stream requirements and computed according to three basic processing paradigms: on consecutive blocks, incrementally, and prequentially. We find the answer to the second question much more interesting. Visualizations of diversity measures calculated on streams with various types of drifts have shown that ensemble diversity visibly changes over time. In particular, we were able to highlight κ interrater agreement, double fault, disagreement, and coincident failure diversity, as measures that were able to depict sudden changes. Additionally, it is worth noting that diversity of the tested ensembles was generally low in terms of absolute values, which might signal that there is still pending research in the field of adaptive ensembles.

The third research question raised the problem of using incremental versus batch classifiers as ensemble components. Our results show that incremental base learners have greater potential for depicting diversity over time. In particular, Hoeffding trees and linear perceptrons were better at visualizing changes over time than batch decision trees and the Naive Bayes algorithm. We also noticed differences between using an adaptive ensemble that only updates existing components and one that periodically creates new components. The latter, represented in our experiment by the Accuracy Updated Ensemble exhibits slightly higher but also much more variable diversity. Surprisingly, one of the most commonly tuned ensemble parameters, the number of components, showcased little impact on diversity plots over time.

Finally, we were interested whether diversity measures can be used to detect concept drifts. A separate set of experiments employing the Page-Hinkley test showed that κ interrater agreement, double fault, and disagreement are capable of detecting sudden changes. In particular, κ was capable of detecting changes equally effectively as accuracy, with smaller delays. Moreover, contrary to accuracy, κ was capable of detecting class ratio changes.

Observations made in this study open several lines of future research. Drawing parallels from static ensembles, diversity measures could be used to prune large ensembles online. Moreover, the complementary nature of various diversity and performance measures suggests it might be worth investigating ideas of combining multiple detectors, which would monitor more than one metric. It is also worth recalling that for binary classification the true label of an example is not needed to calculate disagreement. Thus, there might be a possibility of using disagreement to detect sudden drifts in partially labeled streams, where supervised detectors cannot be applied. Finally, data stream characteristics call for new specialized diversity measures and visualizations. For example, one could take into account differences in component age when calculating pairwise diversity measures or visualize the variability of disagreement among component pairs by using whiskers or box plots.

Acknowledgments. The authors' research was funded by the Polish National Science Center under Grant No. DEC-2013/11/B/ST6/00963. Dariusz Brzezinski acknowledges the support of an FNP START scholarship.

References

1. Banfield, R.E., Hall, L.O., Bowyer, K.W., Kegelmeyer, W.P.: A new ensemble diversity measure applied to thinning ensembles. In: Windeatt, T., Roli, F. (eds.) MCS 2003. LNCS, vol. 2709, pp. 306–316. Springer, Heidelberg (2003). doi:10. 1007/3-540-44938-8_31
2. Bifet, A., Holmes, G., Kirkby, R., Pfahringer, B.: MOA: Massive online analysis. J. Mach. Learn. Res. **11**, 1601–1604 (2010)
3. Brzezinski, D., Stefanowski, J.: Reacting to different types of concept drift: the accuracy updated ensemble algorithm. IEEE Trans. Neural Netw. Learn. Syst. **25**(1), 81–94 (2014)
4. Brzezinski, D., Stefanowski, J.: Prequential AUC for classifier evaluation and drift detection in evolving data streams. In: Appice, A., Ceci, M., Loglisci, C., Manco, G., Masciari, E., Ras, Z.W. (eds.) NFMCP 2014. LNCS (LNAI), vol. 8983, pp. 87–101. Springer, Heidelberg (2015). doi:10.1007/978-3-319-17876-9_6
5. Ditzler, G., Roveri, M., Alippi, C., Polikar, R.: Learning in nonstationary environments: a survey. IEEE Comp. Intell. Mag. **10**(4), 12–25 (2015)
6. Gama, J.: Knowledge Discovery from Data Streams. Chapman and Hall, Boca Raton (2010)
7. Gama, J., Sebastião, R., Rodrigues, P.P.: On evaluating stream learning algorithms. Mach. Learn. **90**(3), 317–346 (2013)
8. Gama, J., Zliobaite, I., Bifet, A., Pechenizkiy, M., Bouchachia, A.: A surveyon concept drift adaptation. ACM Comput. Surv. **46**(4), 44:1–44:37 (2014)
9. Giacinto, G., Roli, F.: An approach to the automatic design of multiple classifier systems. Pattern Recogn. Lett. **22**(1), 25–33 (2001)
10. Giacinto, G., Roli, F.: Design of effective neural network ensembles for image classification purposes. Image Vis. Comput. **19**(9–10), 699–707 (2001)
11. Krempl, G., Zliobaite, I., Brzezinski, D., Hüllermeier, E., Last, M., Lemaire, V., Noack, T., Shaker, A., Sievi, S., Spiliopoulou, M., Stefanowski, J.: Open challenges for data stream mining research. SIGKDD Explorations **16**(1), 1–10 (2014)
12. Kuncheva, L.I.: Combining Pattern Classifiers: Methods and Algorithms. Wiley-Interscience, Hoboken (2004)
13. Margineantu, D.D., Dietterich, T.G.: Pruning adaptive boosting. In: Proceedings of 14th International Conference on Machine Learning, pp. 211–218 (1997)
14. Minku, L.L., White, A.P., Yao, X.: The impact of diversity on online ensemble learning in the presence of concept drift. IEEE Trans. Knowl. Data Eng. **22**(5), 730–742 (2010)

15. Oza, N.C., Russell, S.J.: Experimental comparisons of online and batch versions of bagging and boosting. In: Proceedings of 7th ACM SIGKDD International Conference on Knowledge Discovery Data Mining, pp. 359–364 (2001)
16. Street, W.N., Kim, Y.: A streaming ensemble algorithm (SEA) for large-scale classification. In: Proceedings of 7th ACM SIGKDD International Conference on Knowledge Discovery Data Mining, pp. 377–382 (2001)
17. Theeramunkong, T., Kijsirikul, B., Cercone, N., Ho, T.B.: PAKDD data mining competition (2009)
18. Wang, H., Fan, W., Yu, P.S., Han, J.: Mining concept-drifting data streams using ensemble classifiers. In: Proceedings of 9th ACM SIGKDD International Conference on Knowledge Discovery Data Mining, pp. 226–235 (2003)
19. Woźniak, M.: Application of combined classifiers to data stream classification. In: Saeed, K., Chaki, R., Cortesi, A., Wierzchoń, S. (eds.) CISIM 2013. LNCS, vol. 8104, pp. 13–23. Springer, Heidelberg (2013). doi:10.1007/978-3-642-40925-7_2
20. Zliobaite, I., Pechenizkiy, M., Gama, J.: An overview of concept drift applications. In: Japkowicz, N., Stefanowski, J. (eds.) Big Data Analysis: New Algorithms for a New Society, Studies in Big Data, vol. 16, pp. 91–114. Springer, Heidelberg (2016)

Learning Ensembles of Process-Based Models by Bagging of Random Library Samples

Nikola Simidjievski[1,2]([✉]), Ljupčo Todorovski[3], and Sašo Džeroski[1,2]

[1] Department of Knowledge Technologies, Jožef Stefan Institute, Ljubljana, Slovenia
{nikola.simidjievski,saso.dzeroski}@ijs.si
[2] Jožef Stefan International Postgraduate School, Ljubljana, Slovenia
[3] Faculty of Administration, University of Ljubljana, Ljubljana, Slovenia
ljupco.todorovski@fu.uni-lj.si

Abstract. We propose a new method for learning ensembles of process-based models for predictive modeling of dynamic systems from data and knowledge. Previous work has shown that ensembles based on sampling data (i.e., bagging), significantly improve predictive performance of process-based models. However, this improvement comes at the cost of a substantial computational overhead needed for learning. On the other hand, methods for constructing ensembles based on sampling knowledge (i.e., random library samples, RLS) allow for efficient learning ensembles of process-based models, while maintaining their long-term predictive performance. This paper aims at checking the conjecture whether the combination of these methods has a potential for further performance improvements. The proposed method, *bagging of random library samples* for learning ensembles of process-based models combines the aforementioned approaches in terms of sampling both data and knowledge. We apply the method to and evaluate its performance on a set of automated predictive modeling tasks in two lake ecosystems from data and library of knowledge for modeling population dynamics. The experimental results serve both to identify the optimal design decisions regarding the proposed method as well as to asses its predictive ability. The results show that such ensembles outperform single process-based model, but also outperform each of the two methods for learning ensembles from data samples (bagging) and knowledge samples (RLS).

1 Introduction

Mathematical models of dynamic systems are constructed and employed to recreate, simulate and/or predict the behavior of dynamic systems under various conditions [12]. The twin pillars of every approach to modeling dynamics are (1) structure identification and (2) parameter estimation. The former addresses the task of establishing a suitable model structure in terms of defining the components that are involved in the system and how they interact (commonly formalized as ordinary differential equations, ODEs). Parameter estimation, given a model structure, deals with approximating the constant parameters and the initial values of the variables in the model. Frequently used approach to modeling

© Springer International Publishing Switzerland 2016
T. Calders et al. (Eds.): DS 2016, LNAI 9956, pp. 245–260, 2016.
DOI: 10.1007/978-3-319-46307-0_16

dynamic systems is theoretical (knowledge-driven) modeling. A domain expert derives a proper structure of the model based on an extensive knowledge about the system at hand. Measured data are then used to estimate the (constant) parameters in the model. The alternative approach, i.e. empirical (data-driven) modeling, uses measured data in a trial-error process to search for such a combination of model structure/parameters that best fit the measurements.

Equation discovery, a sub-field of machine learning, joins the previous two approaches of modeling, by studying methods for learning both model structure and parameter values from observations [11,19]. The most recent approach, referred to as process-based modeling [6,18], combines heuristic search methods with parameter estimation techniques to simultaneously tackle the tasks of structure identification and parameter estimation. While the search methods explore the space of candidate model structures defined by a library of domain knowledge, the parameter estimation techniques aim at finding optimal values of the constant parameters for each candidate structure and evaluate its fit against the measured data. The utility of process-based modeling has been demonstrated in a variety of tasks ranging from modeling population dynamics in aquatic systems [3,20] to systems biology [9,16]. However, these applications focus primarily on establishing explanatory models of the system at hand. Typically, the obtained models are simulated and analyzed on the same data used for learning them, thus their ability to predict the systems' behavior is limited, i.e., evaluating their generalization error has not even been considered nor evaluated.

The main focus of this study is improving the predictive performance of process-based models (PBMs) by constructing ensembles. Recent studies [13,14] have introduced ensembles in the context of PBMs. The former study addresses the task of learning ensembles of PBMs by sampling data instances. The proposed method for bagging of PBMs yields a significant gain in predictive performance over a single PBMs. However, such ensembles suffer from substantial computational overhead when learning them [14]. In contrast, the latter study focuses on efficient learning of ensembles of PBMs by sampling data features. Here, the ensemble constituents are learned from the whole data set using samples of the knowledge library. This is in contrast to bagging, where ensemble constituents are being learned on data samples using the same knowledge library (or feature space). The results of the performed experiments show that these efficient ensembles also significantly outperform single PBMs in accurately predicting long-term behavior of dynamic systems, however they can not match the performance of bagging.

In this paper we propose a novel method for learning ensembles of PBMs which combines the afore-mentioned approaches to constructing ensembles in terms of sampling both data instances and domain knowledge. We conjecture that the novel method for learning ensembles besides outperforming single PBMs in terms of long-term predictive performance, will also perform better than both individual ensemble methods of bagging and by sampling the library of domain knowledge. To validate this conjecture we perform an empirical evaluation of the implemented method on the task of modeling and predicting population dynamics in two aquatic ecosystems. This empirical evaluation will allow us to identify the most appropriate design decisions within the algorithm for learning such ensembles. It will also allow us to compare the predictive performance of the new method to the performance of the existing methods for learning ensembles of PBMs by bagging and sampling the library of domain knowledge.

This work is closely related to that of [5], where an ensembles of PBMs is constructed by fusing the most frequent structure fragments of the individual base models learned from different data samples. The results show that the resulting ensemble still provides a process-based explanation of the observed system structure, while being more robust in terms of over-fitting. Note, however, that authors estimate the out-of-sample error of the models, by taking random subsamples of the observed time-series data and removing them from the training data. Therefore, the ability of such ensemble to generalize outside the time span of the training data has not been considered. In a similar context, the work of [1] also is related since it tackles the tasks of predictive modeling of (discrete) nonlinear dynamic systems with machine learning approaches. Here, the modeling problem is first transformed into a non-linear regression approximation problem which is subsequently tackled by learning fuzzy linear model trees and ensembles of fuzzy linear model trees. The results show that the ensembles improve the performance over the single fuzzy linear trees. Note that, this study focuses on short-term (one-step ahead) prediction of discrete-time dynamic systems, where the value of the time series in the next time-point is predicted, as opposed to long-term prediction of continuous non-linear dynamic systems.

The remainder of the paper is organized as follows. In the next section, we provide overview of the process-based modeling paradigm. Section 3 introduces the novel method for learning ensembles of process-based models. Next, in Sect. 4 we present the experimental setup for evaluating the proposed method and present the results in Sect. 5. Finally, the last section concludes the paper and suggests directions for further work.

2   Process-Based Modeling

The process-based modeling paradigm aims at constructing models which contain a high-level conceptual structure and a low-level mathematical formulation which allows for making predictions. Process-based models integrate domain-specific modeling knowledge and data into explanatory models of the observed systems. A process-based model consists of two types of components: entities and processes. Entities represent the state of the system. They incorporate the variables and the constants related to the components of the modeled system. The entities are involved in complex interactions represented by the processes. Each process includes specifications of how entities interact in terms of processes alternatives and sub-processes.

The entities and processes represent specific components and interactions observed in the particular system at hand. To elucidate such specific entities and processes, the process-based modeling approach allows for a higher-level representation of knowledge in terms of entity and process templates.

These serve as placeholders for general properties and definitions of the modeling components. The template entities and processes are hierarchically organized together into a library of components that defines the space of modeling alternatives for modeling a specific dynamic system.

More specifically, the entity templates define the constant parameters and incorporate the general definitions of variables in terms of their roles, initial values, ranges and aggregation function. The role specifies whether a particular variable is modeled (endogenous) or used as a forcing term in the system (exogenous). The initial value of a variable denotes the state of the system in the initial time point, which is a required input for a model simulation. The aggregation function specifies how the influence of multiple processes is combined for a particular entity. On the other hand, the processes templates include specifications of the entity templates that interact, in terms of (process) constants and algebraic/ordinary differential equations. In sum, the templates provide general modeling recipes for any instantiation to specific components or interactions, which in turn allow for a high-level qualitative conceptualization of a model to be translated into a low-level quantitative mathematical formalization which can be simulated.

The task of learning PBMs is comprised of two sub-tasks: (i) instantiating the library of entity and process templates and (ii) estimating the parameters in the resulting model structures to fit the measured data. The algorithm for learning PBMs (presented in Algortihm 1) takes three inputs: (1) a library of domain knowledge, (2) an incomplete model and (3) measured data. Given the library of model components the former sub-task is formulated as a combinatorial search problem. Taking the incomplete model into account, one can instantiate the template entities and processes from the library into a set of specific components (entities and processes) to be included in the PBM. The incomplete model represents modeling assumptions in terms of expected logical structure of the model, which limits the search space of the possible and plausible model structures, i.e., combinations of model structures. Some of the combinations can be rejected as implausible, due to their inconsistency with the incomplete model in terms of presence or absence of certain processes.

Algorithm 1. Outline of the generic algorithm for learning process-based models from knowledge and data.

Input: $library, data, incompleteModel$
Output: $modelList$
1   $components \leftarrow$ instantiate$(library, incompleteModel)$
2 **foreach** $structure \in$ enumerate$(components, incompleteModel)$ **do**
3      $modelEq \leftarrow$ compileToEquation$(structure)$
4      $\{model, error\} \leftarrow$ parameterEstimation$(modelEq, data)$
5      $modelList \leftarrow modelList \bigcup \{model, error\}$
   end
6   $modelList \leftarrow$ rank$(modelList, error)$

The latter sub-task, i.e., estimating the model's parameters is formulated as an optimization task. Each of the candidate model structures considered during the search task is compiled into a system of equations, for which a parameter estimation task is solved to obtain values of the model parameters that best fit the measured data. The parameter estimation process is based on the meta-heuristic optimization framework that implements a number of global optimization algorithms. The objective function usually considered for such problems is minimizing the discrepancy between the model simulation and the observed system behavior. Finally, the output of the algorithm is a set of complete models sorted according to their performance, i.e., the difference error between the simulation and the measured data, where the highest ranked model is considered as an output.

Algortihm 1 presents an outline of a generic algortihm for learning PBMs. In this study, we use ProBMoT [20], a recent process-based modeling tool which allows for complete modeling, parameter estimation, and simulation of PBMs. ProBMoT is employed as the learning algortihm for inducing base PBMs, i.e., learning the constituents of the process-based ensemble models.

Figure 1 depicts an exemplary output of a process-based modeling algortihm addressing the task of modeling population dynamics (of a phytoplankton concentration) in a simple aquatic ecosystem. More specifically the toy system represents a cyclic relationship involving a primary producer (phytoplankton, abbrev. *phyto*) that grows by feeding on nutrients (nitrogen and phosphorous), the concentrations of which are influenced by the environment (temperature) and the process of respiration. Figure 1 (top) presents a model of such a system in a qualitative process-based modeling formalism. Given that only the concentration of the phytoplankton is modeled, the respective entity *phyto* is defined as an endogenous variable whereas the rest of the variables (corresponding to the other three entities of *nitro*, *phos* and *env*) are exogenous. The interactions of these entities are then specified by four processes. For example, the process GROWTH defines the interaction between the phytoplankton and the two nutrients of nitrogen (*nitro*) and phosphorous (*phos*). In terms of mathematical formulation such a high-level representation of a dynamic system is compiled into a system of algebraic and ordinary differential equations adequate for simulation. Figure 1 (middle) provides the quantitative formulation of the PBM presented above, where $p_c(t)$, $ph_c(t)$ and $n_c(t)$ correspond to the concentration variables of the entities of *phyto*, *phos* and *nitro*, respectively. The environmental temperature is denoted with T, and the growth rate of the phytoplankton with $gRate$.

These equations can then be fully simulated, which results in a trajectory that allows for further analyses of the modeled system. Figure 1 (bottom) presents the simulation of phytoplankton concentration obtained from the PBM. Here we used real measurements from [3] for the exogenous variables involved in the system (nutrients and temperature) and for estimating the parameters of the system. The resulting simulation spans the time period of approximately one year, excluding the period when real measurements were not available. The *DATA* trajectory (represented by a dashed line) represents real measurements that can be used for a visual assessment of the PBM performance.

```
//Entities
entity phyto : PrimaryProducer {
    vars: conc { role: endogenous; initial: 1.665; aggr:sum};
    consts: maxGrowthRate = 0.88; }
entity phos : Nutrient { vars: conc { role: exogenous;};}
entity nitro : Nutrient { vars: conc { role: exogenous;};}
entity env : Environment { vars: temp { role: exogenous;};}

//Processes
process GROWTH(phyto, [phos, nitro]): GROWTH
    { processes: ExpSaturatedGrowthRate;}
process NITROGENLIM(phyto, nitro): ExpSaturatedGrowthRate
    { consts: saturationRate =14.9;}
process PHOSOPHOROUSLIM(phyto, phos): ExpSaturatedGrowthRate
    { consts: saturationRate =8.08;}
process RESPIRATION(phyto, [phos, nitro], env): Respiration
    { consts: respRate=0.036, minTemp=0.542, refTemp=17.4;}
```

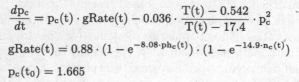

$$\frac{dp_c}{dt} = p_c(t) \cdot gRate(t) - 0.036 \cdot \frac{T(t) - 0.542}{T(t) - 17.4} \cdot p_c^2$$

$$gRate(t) = 0.88 \cdot (1 - e^{-8.08 \cdot ph_c(t)}) \cdot (1 - e^{-14.9 \cdot n_c(t)})$$

$$p_c(t_0) = 1.665$$

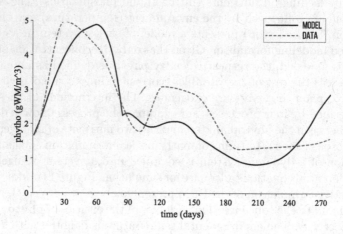

Fig. 1. A process-based model (top) of population dynamics compiled to ODE (middle) and its simulation (bottom) in a simple lake ecosystem.

3   Bagging of Random Library Samples

The algorithm for learning ensembles of PBMs with the bagging of random library samples (BRLS) method is presented in Algorithm 2. It takes four inputs: a library of domain knowledge (*lib*), a dataset consisting of training data (D_T)

and validation data (D_V), an incomplete model (*incompleteModel*) and an integer k denoting how many ensemble constituents are to be generated.

At each iteration (out of k), the algorithm learns a PBM from different samples of the training data and the library. For the task of randomly sampling data instances, BRLS borrows some ideas from bagging for regression tasks, however it distinctively differs in one important aspect. The difference relates to the temporal ordering of the training data, which here has to be retained in each data sample. To achieve this, we implement sampling by retaining the order of the instances by introducing a weight for each instance (time-point) that is provided as part of the data (line 4 in Algorithm 2). The weight corresponds to the number of times the instance has been selected in the process of sampling with replacement. Instances that have not been selected (the ones with weight 0) are simply omitted from the sample.

To take into account the weights when learning a model from the sample, we implement the weighted root mean squared error (WRMSE) as an objective function in the process of parameter estimation in ProBMoT:

$$WRMSE(m) = \sqrt{\frac{\sum_{t=0}^{N} \omega_t (y_t - \hat{y}_t)^2}{\sum_{t=0}^{N} \omega_t}}, \tag{1}$$

where y_t and $\hat{y}_t$ correspond to the *measured* and *simulated* values (simulating the base model m) of the system variable y at time point t, N denotes the number of instances in the data sample, and ω_t denote the weight of the data instance at time point t.

On the other hand, for the task of sampling the library (line 3 in Algorithm 2) we relate to the well-known ensemble method for sampling data features, i.e., the Random Subspace Method (RSM) [10]. In the realm of process based modeling, we can think of the feature space as being defined by the model components instantiated from the process templates. This space of components is determined by the number of process alternatives defined in the library of

---

Algorithm 2. Constructing ensemble of PBMs by bagging of random library samples

---

Input: $lib, \{D_T, D_V\}, incompleteModel, k$
Output: *Ensemble*
1 $Ensemble \leftarrow \varnothing$ /* set of base models */
2 **for** $i = 1$ *to* k **do**
3      $lib_S \leftarrow$ sampleLib(lib)   /* randomly sample the library lib */
4      $D_S \leftarrow$ sampleData(D_T)   /* randomly sample the training set D_T */
5      $modelList_i \leftarrow$ probmot($lib_S, D_S, incompleteModel$)
6      $bestModel_i \leftarrow$ rank($modelList_i, D_V$)
7      $\beta_i \leftarrow$ confidence($bestModel_i, D_V$)
8      $Ensemble \leftarrow Ensemble \bigcup \{bestModel_i, \beta_i\}$
   end

---

domain knowledge. Therefore, randomly sampling data features in the context of the traditional RSM, is analogous to generating random samples of the library of domain knowledge by randomly sampling (excluding) process alternatives from the original library. The sampling algorithm takes as input the complete library and enumerates all the process templates defining more than one modeling choice. In turn, for each process template on the list, it takes a random sample of the available modeling choices to be included in the sampled library. Note that the library sampling does not assume a uniform distribution of samples: the probability of a library sample is proportional to the size of the induced space of candidate models. In particular, the probability of a library sample lib_S of the whole library lib equals:

$$P(lib_S) = \frac{|L_S|}{\sum\limits_{L_i \in \mathbb{P}(L)} |L_i|}, \tag{2}$$

where L and $L_S \subseteq L$ correspond to the sets of candidate models induced by lib and lib_S (for a given incomplete model specification), respectively. Moreover, $|\cdot|$ denotes set cardinality and $\mathbb{P}(L)$ denotes the powerset of L, i.e., the set of all the possible subsets of L^1.

The output of a modeling task, when using ProBMoT, is a list of PBMs, which is a posteriori sorted according to their performance (line 5 in Algorithm 2). Depending on the input in the method, this ranking can be based either on the performance on a separate validation data set D_V, or on the training sample (if $D_V == D_T$). The highest ranked model from each modeling task i (out of k) denoted as $bestModel_i$, is selected to be an ensemble constituent in the output $Ensemble$.

Once the ensemble constituent set has been populated, the question then arises as to how the individual predictions are to be combined into a single prediction. In this context, for obtaining the output of an ensemble of PBMs the different model simulations from each base model are combined time-point-wise: The combination is performed by well known schemes usually applied in the case of aggregating numeric values, such as averaging, median and their weighted variants [14]. Note that, each ensemble constituent is paired with its own confidence β - an indicator of the performance of the base models, used for the weighted combination schemes.

4   Experimental Design

4.1   Data

The data originates from two real-world aquatic ecosystems: Lake Bled in Slovenia and Lake Zurich in Switzerland. The measurements for Lake Bled, performed by the Slovenian Environment Agency, consist of physical, chemical and

---

[1] Note that the sampling procedure does not require for generation of the power set $\mathbb{P}(L)$.

biological data for the period from 1996 to 2002. All the measurements were performed once a month and depth-averaged for the upper 10 m of the lake. To obtain daily approximations, the data was interpolated with a cubic spline algorithm and daily samples were taken from the interpolation [3]. Lake Zurich data comprises measurements performed by the Water Supply Authority of Zurich in the period from 1996 to 2002. The measurements, taken once a month, include profiles of physical, chemical and biological variables from 19 different sites. They were weight averaged to the respective epilimnion (upper ten meters) and hypilimnion (bottom ten meters) depths. The data was interpolated with a cubic spline algorithm and daily samples were taken from the interpolation [8].

We use the same structure of population dynamics model in both aquatic ecosystems. It includes a single equation (ODE) for a system variable representing the phytoplankton biomass. The exogenous variables include the concentration of zooplankton *Daphnia hyalina*, dissolved inorganic nutrients of nitrogen, phosphorus, and silica, as well as two input variables representing the environmental influence of water temperature and global solar radiation (light).

For each aquatic ecosystem, we used seven data sets corresponding to the last seven years of available measurements. Five of these were used (one at a time) for training the base-models, one was used for validating the models in the process of selecting the ensemble constituents, and one was used to measure the predictive performance of the learned models and ensembles. In each learning experiment, we take a single (year) training data set, learn a single model or an ensemble using the training and the validation data set, and test the predictive performance of the learned models on the test data set. We therefore perform 10 experiments. In the tables which report the results, the cases are labeled with the labels 1–5 corresponding to the training data set using in the experiments, where, e.g., Bled 3 denotes the Lake Bled data set for the third year (i.e., 1998).

4.2   Modeling Knowledge

In the experiments we use the library of domain knowledge for modeling population dynamics in aquatic ecosystems, presented by [20]. This library is based on the previous work of [2]. The library of domain knowledge, combined with the modeling assumptions (incomplete model), results in 27216 candidates for both of the lakes.

Note, however, an important difference between the setup of the bagging method and the experiments performed with the rest two methods of RLS and BRLS. In the last two, we use the whole library of domain knowledge as described previously. The use of such library is prohibitive for bagging, due to the high computational overhead considering the large space of candidate models in each learning iteration. To address this issue, we use a simplified version of the original library that results in 128 candidates for both lakes. We carefully prepared the simplified library, omitting only modeling alternatives (process templates) that are rarely observed to be among the top-ranked models in the single-model experiments with ProBMoT.

4.3   Experimental Setup

ProBMoT implements the Differential Evolution (DE) [15] method for parameter estimation. In the experiments the DE parameters were set as follows:
a population size of 50, strategy *rand/1/bin*, differential weight (F) and the
crossover probability (Cr) of 0.6. The limit on the number of evaluations of the
objective function is 1000 per parameter. The particular choice of parameters
setting of DE is based on previous studies of the sensitivity of DE for estimating
parameters of PBMs [17].

For simulating the ODEs, we used the CVODE simulator [7] with absolute
and relative tolerances set to 10^{-3}. Full simulation in our experimental evaluation
is performed with the CVODE package, a general-purpose ODE solver that
uses linear multistep variable-coefficient methods for integration. Here, we use
the backward differentiation method combined with Newton iteration and a
preconditioned Krylov method.

To evaluate the predictive performance of a given model m, we use the measure of relative root mean squared error (*ReRMSE*) [4], defined as:

$$ReRMSE(m) = \sqrt{\frac{\sum_{t=0}^{N}(y_t - \hat{y}_t)^2}{\sum_{t=0}^{N}(\bar{y} - \hat{y}_t)^2}},$$

where N denotes the number of measurements in the test data set, y_t and $\hat{y}_t$
correspond to the *measured* and *predicted*[2] value of the system variable y at
time point t, and $\bar{y}$ denotes the mean value of y in the test data set. Note
that the usual root mean squared error is observed here relative to the standard
deviation of the system variable in the test data, thus allowing us to compare the
errors of models for different system variables with measured values. Given the
performance metrics we also compute average ranks (averaged over all test data
sets) of model's performance, however we do not perform statistical analyses of
the results due to small number of test data sets.

To properly assess the predictive performance of the proposed method for
learning ensembles, in our last set of experiments we compare it to two other
methods for constructing ensembles of PBMs. These methods include learning
ensembles by sampling data instances and sampling domain knowledge, i.e.,
bagging and random library samples (RLS). Following the findings of [13,14]
the ensembles learned with bagging and RLS include 25 and 10 constituents,
respectively. The constituents are chosen based on their performance on a separate validation set and combined by averaging their predictions. These settings
were chosen by following a similar experimental setup that we use in this paper
to select the appropriate design choices for learning ensembles with bagging of
random library samples.

---

[2] Predictions are obtained by simulating the model m on the test (and not the validation) set.

5   Results

First, we explore the effects of the decisions related to the design of the proposed ensemble learning method. In particular, we aim at deciding upon an appropriate approach to selecting ensemble constituents, the optimal number thereof and the optimal way of combining their predictions. Second, we compare the predictive performance of the bagging-of-random-library-samples ensembles with the performance of single models, bagging ensembles and ensembles based on random library samples.

The first design choice in an algorithm for learning ensembles of PBMs is related to the way of selecting the ensemble constituents. Recall that in each iteration, the highest ranked model is selected to be an ensemble constituent. ProBMoT ranks candidate models with respect to their performance on the training set. Alternatively, to avoid overfitting, models can be ranked according to their performance on a separate validation data set. We compare these two in a set of experiments using ensembles with 5, 10, 25 and 50 constituents whose predictions are averaged.

Table 1 shows that using a validation set to select ensemble constituents is better. More specifically, the upper part of the table presents the results of the comparison of the predictive performance (ReRMSE on the testing set), where as the lower part presents the comparisons of the descriptive performance (ReRMSE on the training set) of the different ensembles. The average ranks are reported at the bottom of each subtable. The results in the lower subtable confirms our conjecture that this is due to overfitting. Ensembles comprising of base models selected based on the performance on the training data exhibit substantially better descriptive/training performance than the ones selected based on their performance on a separate validation dataset.

Another important decision regarding the optimal number of ensemble constituents is also presented in the table. Moreover, when comparing the predictive performance of the ensembles with different sizes (i.e., 5, 10, 25, 50), the ensemble comprised with 25 constituents selected on a separate validation set lead to the best performance (avg. rank 3.85). Finally, we focus on the design decision concerning the most appropriate scheme for combining the simulations of the base models in the ensemble. We compare the performance of four methods commonly used in learning ensembles of regression models: average, median, weighted average, and weighted median. Table 2 reports on the comparison of the predictive performances of the four methods for combining 25 base-model simulations. It can be observed that on average the median combining scheme for aggregation performs best.

In sum, based on the performed experiments we make the following design decisions. First, the ensemble constituents are chosen based on their performance on a separate validation set, as they exhibit better performance than the ones chosen based on their performance on the original training data set. Second, the ensembles should consist of relatively small number (25) of base models whose predictions should be combined using median. In the last set of experiments,

Table 1. Comparison of the predictive performance(top)(ReRMSE on the testing set) and descriptive performance (bottom)(ReRMSE on the training set) of the two methods for choosing ensemble constituents (V=validation, T=training). Bold typeface is used to mark the best-performing method for a given case (table row).

Case	Ens.5-V	Ens.10-V	Ens.25-V	Ens.50-V	Ens.5-T	Ens.10-T	Ens.25-T	Ens.50-T
Bled1	1.142	1.113	1.101	1.126	1.122	**1.045**	1.071	1.077
Bled2	1.248	1.238	1.237	**1.214**	1.304	1.248	1.268	1.259
Bled3	0.967	0.912	0.946	0.898	0.972	**0.706**	0.750	0.766
Bled4	0.805	0.811	0.767	0.804	**0.718**	0.809	0.737	0.757
Bled5	0.585	0.628	0.627	0.667	**0.561**	0.579	0.601	0.618
Zurich1	**0.925**	1.004	1.461	1.218	1.282	1.130	1.087	1.133
Zurich2	**0.794**	0.825	0.844	0.885	1.285	1.074	0.946	1.234
Zurich3	**0.877**	0.934	0.901	0.897	2.511	1.611	1.183	1.073
Zurich4	1.035	1.034	**1.005**	1.007	1.028	1.054	1.074	1.058
Zurich5	1.017	1.009	0.999	**0.999**	1.590	1.389	1.271	1.216
Avg. Ranks	4.05	4.30	3.85*	4.05	5.80	4.55	4.50	4.90
Case	Ens.5-V	Ens.10-V	Ens.25-V	Ens.50-V	Ens.5-T	Ens.10-T	Ens.25-T	Ens.50-T
Bled1	0.288	0.257	0.263	0.264	0.178	0.177	**0.174**	0.177
Bled2	0.242	0.302	0.268	0.279	**0.148**	0.153	0.150	0.156
Bled3	0.382	0.374	0.362	0.371	0.260	0.260	**0.258**	0.274
Bled4	0.408	0.402	0.359	0.374	0.268	0.273	**0.256**	0.263
Bled5	0.344	0.341	0.321	0.314	0.151	0.166	**0.150**	0.152
Zurich1	0.492	0.502	0.562	0.543	0.429	0.373	**0.343**	0.345
Zurich2	0.625	0.581	0.594	0.587	0.461	0.468	**0.448**	0.454
Zurich3	0.796	0.790	0.806	0.807	0.716	0.672	**0.636**	0.645
Zurich4	0.792	0.776	0.777	0.779	0.723	0.718	**0.702**	0.710
Zurich5	0.773	0.763	0.759	0.758	0.512	0.501	**0.491**	0.493
Avg. Ranks	7.2	6.2	6.2	6.4	3.15	3.2	1.1*	2.55

we use these algorithm settings for learning ensembles with bagging of random library samples.

Having made the design decisions for learning ensembles of PBMs, we here focus on testing our central hypothesis that ensembles improve the predictive performance of PBMs. To this end, we compare the predictive performance of the proposed method for construing ensembles with the one of single PBMs and ensembles constructed with bagging and RLS, individually.

Table 3 depicts the comparison of predictive performance and the average ranks, averaged over ten data sets. The results show that the proposed ensembles substantially outperform single PBMs. More specifically, the BRLS method outperforms the single model in 8 out of 10 cases. Moreover, in 4 cases out of these 10, BRLS is the best performing algorithm, overall. This result confirms the central hypothesis of this paper that ensembles of PBMs, more specifically ensembles learned with BRLS, improve the predictive performance over single

Table 2. Comparison of the predictive performance of the BRLS method with 25 constituents combined by four different methods. Bold typeface is used to mark the best-performing mehod for a given case (table row).

Case	Average	Weighted Average	Median	Weighted Median
Bled1	1.101	1.095	1.091	**1.089**
Bled2	**1.237**	1.241	1.262	1.262
Bled3	0.946	**0.901**	0.995	0.985
Bled4	0.767	0.746	**0.728**	0.731
Bled5	0.627	0.621	**0.604**	0.606
Zurich1	1.461	1.422	**0.889**	0.898
Zurich2	0.844	0.833	**0.8**	0.792
Zurich3	0.901	**0.898**	0.913	0.922
Zurich4	**1.005**	1.005	1.01	1.01
Zurich5	**0.999**	1.011	1.004	1.006
Avg. ranks	2.75	2.45	2.3*	2.5

Table 3. Comparison of the predictive performance of a single model with that of ensembles of process-based models learned with Bagging, Random Library Samples and Bagging of Random Library Samples. Bold typeface is used to mark the best-performing mehod for a given case (table row).

Case	Single Model	Bagging	Random Library Samples	Bagging of Random Library Samples
Bled1	5.184	**1.068**	1.103	1.091
Bled2	**0.938**	1.11	1.143	1.262
Bled3	1.042	0.968	**0.794**	0.995
Bled4	2.84	0.76	0.75	**0.728**
Bled5	0.842	0.859	0.858	**0.604**
Zurich1	**0.744**	0.78	0.797	0.889
Zurich2	1.323	0.881	0.948	**0.8**
Zurich3	29.463	0.939	**0.881**	0.913
Zurich4	27.593	**0.96**	1.011	1.01
Zurich5	1.489	1.37	1.13	**1.004**
Avg. Ranks	3.2	2.2	2.5	2.1*

PBMs. Note, however, that their predictive performance is also slightly better than the performance of the other two methods for learning ensembles of PBMs, which on average ranks them at the top, followed by bagging and RLS.

6 Conclusion

In this paper, we address the task of learning ensembles of process-based models via bagging of random library samples. We design, implement and evaluate a new method for learning ensembles by sampling both the data and the library of knowledge. This method for learning ensembles is the major contribution of our paper, since it improves the predictive performance of the state-of-the-art methods for learning PBMs and ensembles, thereof.

First, it improves the state-of-the art of learning ensembles of PBMs with a new, robust and accurate method based on sampling both data and domain knowledge. The proposed method most often leads to models with better predictive performance as compared to the models obtained with the state-of-the-art ensemble methods of bagging and RLS. We conjecture that this is due to the ability of the BRLS method to constraint the estimation bias (as an artifact from RLS) as well as to reduce the variance (as an artifact from bagging). Second, it mainly contributes to the area of ecological modeling. The results of the experimental evaluation confirm our central hypothesis that ensembles constructed with BRLS provide more accurate predictions of concentrations of species in an aquatic ecosystems than a single PBM. This is a significant improvement of predictive performance over the state-of-the-art models of population dynamics, which, while focusing on providing an accurate explanation of the behavior of the observed system, struggle to achieve a satisfactory performance at predicting population dynamics over a long periods [3].

While our empirical study is limited to the domain of aquatic ecosystems, i.e. modeling population dynamics in two lake ecosystems, the proposed approach to learning ensembles of PBMs is general enough to be applied to any other domain and to any other type of models of dynamic systems. An immediate continuation of the work presented here is to investigate the generality of the proposed methodology and extend the scope of learning ensembles of PBMs of population dynamics to other aquatic domains. Next, we plan to investigate whether the findings of this paper are also consistent when applied in an artificial setting, for predictive tasks of synthetic dynamic systems. Finally, we also intend to analyze the applicability of proposed methodology to different scientific domains, such as systems biology and systems neuroscience. Here, however, an additional challenge lies in developing a suitable and extensive library of domain knowledge and obtaining valid measurements of the systems.

When learning ensembles, there is a trade-off between the predictive performance of the ensemble and its interpretability. Considering this issue, we plan to follow ideas from [5] and improve our methodology by incorporating understandable structure into the resulting ensemble. A similar direction involves improving the ensembles comprehensibility by computing and visualizing each feature's (processes) contribution (in terms of information difference, log-odds ratio or probabilities difference) in the resulting ensemble [21].

References

1. Aleksovski, D., Kocijan, J., Džeroski, S.: Ensembles of fuzzy linear model trees for the identification of multi-output systems. IEEE Trans. Fuzzy Syst. **24**(4), 916–929 (2015)
2. Atanasova, N., Todorovski, L., Džeroski, S., Kompare, B.: Constructing a library of domain knowledge for automated modelling of aquatic ecosystems. Ecol. Model. **194**(1–3), 14–36 (2006)
3. Atanasova, N., Todorovski, L., Džeroski, S., Remec, R., Recknagel, F., Kompare, B.: Automated modelling of a food web in Lake Bled using measured data and a library of domain knowledge. Ecol. Model. **194**(1–3), 37–48 (2006)
4. Breiman, L., Friedman, J.H., Stone, C.J., Olshen, R.A.: Classification and Regression Trees. Chapman & Hall, London (1984)
5. Bridewell, W., Asadi, N.B., Langley, P.W., Todorovski, L.: Reducing overfitting in process model induction. In: Proceedings of the 22nd International Conference on Machine Learning, ICML 2005, pp. 81–88. ACM, New York (2005)
6. Bridewell, W., Langley, P.W., Todorovski, L., Džeroski, S.: Inductive process modeling. Mach. Learn. **71**, 1–32 (2008)
7. Cohen, S.D., Hindmarsh, A.C.: CVODE, a stiff/nonstiff ODE solver in C. J. Comput. Phys. **10**(2), 138–143 (1996)
8. Dietzel, A., Mieleitner, J., Kardaetz, S., Reichert, P.: Effects of changes in the driving forces on water quality and plankton dynamics in three swiss lakes – long-term simulations with BELAMO. Freshw. Biol. **58**(1), 10–35 (2013)
9. Džeroski, S., Todorovski, L.: Modeling the dynamics of biological networks from time course data. In: Choi, S. (ed.) Systems Biology of Signaling Networks. Systems Biology, pp. 275–295. Springer, New York (2010)
10. Ho, T.K.: The random subspace method for constructing decision forests. IEEE Trans. Pattern Anal. Mach. Intell. **20**(8), 832–844 (1998)
11. Langley, P.W., Simon, H.A., Bradshaw, G., Zytkow, J.M.: Scientific Discovery: Computational Explorations of the Creative Processes. The MIT Press, MA (1987)
12. Ljung, L.: System identification - Theory for the User. Prentice-Hall, Upper Saddle River (1999)
13. Simidjievski, N., Todorovski, L., Džeroski, S.: Predicting long-term population dynamics with bagging and boosting of process-based models. Expert Syst. Appl. **42**(22), 8484–8496 (2015)
14. Simidjievski, N., Todorovski, L., Džeroski, S.: Modeling dynamic systems with efficient ensembles of process-based models. PLoS ONE **11**(4), 1–27 (2016)
15. Storn, R., Price, K.: Differential evolution – a simple and efficient heuristic for global optimization over continuous spaces. J. Glob. Optim. **11**(4), 341–359 (1997)
16. Tanevski, J., Todorovski, L., Džeroski, S.: Learning stochastic process-based models of dynamical systems from knowledge and data. BMC Syst. Biol. **10**(1), 30 (2016)
17. Taškova, K., Šilc, J., Atanasova, N., Džeroski, S.: Parameter estimation in a nonlinear dynamic model of an aquatic ecosystem with meta-heuristic optimization. Ecol. Model. **226**, 36–61 (2012)
18. Todorovski, L., Bridewell, W., Shiran, O., Langley, P.W.: Inducing hierarchical process models in dynamic domains. In: Veloso, M.M., Kambhampati, S. (eds.) Proceedings of the 20th National Conference on Artificial Intelligence, NCAI 2005, pp. 892–897. AAAI Press, Pittsburgh (2005)

19. Todorovski, L., Džeroski, S.: Integrating domain knowledge in equation discovery. In: Džeroski, S., Todorovski, L. (eds.) Computational Discovery of Scientific Knowledge. LNCS (LNAI), vol. 4660, pp. 69–97. Springer, Heidelberg (2007). doi:10.1007/978-3-540-73920-3_4
20. Čerepnalkoski, D., Taškova, K., Todorovski, L., Atanasova, N., Džeroski, S.: The influence of parameter fitting methods on model structure selection in automated modeling of aquatic ecosystems. Ecol. Model. **245**(0), 136–165 (2012)
21. Štrumbelj, E., Kononenko, I.: An efficient explanation of individual classifications using game theory. J. Mach. Learn. Res. **11**, 1–18 (2010)

Early Random Shapelet Forest

Isak Karlsson[✉], Panagiotis Papapetrou, and Henrik Boström

Department of Computer and Systems Sciences, Stockholm University,
Stockholm, Sweden
{isak-kar,panagiotis,henrik.bostrom}@dsv.su.se

Abstract. Early classification of time series has emerged as an increasingly important and challenging problem within signal processing, especially in domains where timely decisions are critical, such as medical diagnosis in health-care. Shapelets, i.e., discriminative sub-sequences, have been proposed for time series classification as a means to capture local and phase independent information. Recently, forests of randomized shapelet trees have been shown to produce state-of-the-art predictive performance at a low computational cost. In this work, they are extended to allow for early classification of time series. An extensive empirical investigation is presented, showing that the proposed algorithm is superior to alternative state-of-the-art approaches, in case predictive performance is considered to be more important than earliness. The algorithm allows for tuning the trade-off between accuracy and earliness, thereby supporting the generation of early classifiers that can be dynamically adapted to specific needs at low computational cost.

1 Introduction

Early classification of time series has emerged as an increasingly important task [19], especially in domains where timely decisions are critical, such as medical diagnoses in health-care. Consider, for example, the task of screening patients for heart-related diseases, where physicians are commonly investigating the electrocardiogram (ECG) of patients. In the context of early prediction, the goal here is to as early as possible detect whether the condition of a patient requires special attention or not. In this case there is a trade-off between *earliness*, i.e., how early the prediction can be made, and *accuracy*, i.e., the rate of correct assessments. For medical diagnosis, both earliness and accuracy is important, however, false positives, i.e., the classifier incorrectly predicts that the patient is ill, are less dangerous than false negatives, i.e., the classifier incorrectly predicts that the patient is healthy. It should, however, be note that the former errors still incur a cost, e.g., by requiring unnecessary additional investigations. To address these types of classification problems, a plethora of early time series methods have been proposed in the literature, e.g., based on boosting [11], early distinctive shapelets [25], nearest neighbour classification [23], myoptic decision making [5] and reliable early classification [15].

To the best of our knowledge, the earliest mentioning of early time series classification is given by [19], where they describe time-series segments with

© Springer International Publishing Switzerland 2016
T. Calders et al. (Eds.): DS 2016, LNAI 9956, pp. 261–276, 2016.
DOI: 10.1007/978-3-319-46307-0_17

relative statements, such as increasing or decreasing. These statements are subsequently fed to an ensemble learner for processing. By considering unavailable segments as missing data, the proposed method is able to make predictions early [19]. Furthermore, one of the first definitions of early classification of temporal sequences appears in [23], where it is defined as making predictions as early as possible while maintaining an expected accuracy [23]. In that work, a set of frequent features which are both *frequent* and *early* are selected. From the selected features, a decision tree (or rule set) with a user specified accuracy in each branch or rule is constructed. The trees are subsequently used to make predictions such that an (incomplete) input sequence is matched with all branches (or rules) and a classification is given once the user supplied accuracy level is achieved [23].

Since time-series are numerical and *not* discrete, as assumed by [23], another approach based on early k-nearest neighbor under any suitable distance measure is proposed by [24,26]. In that work, the k-nearest neighbor classifier is extended with a training phase to determine how early an accurate classification can be made using the concept of *minimum prediction length* (MPL). It is noted, however, that this method has two major drawbacks: (1) the method may overfit the training set if the set contains few and non-uniformly sampled instances; and (2) the requirement of stability in MPL may be too restrictive, which, in some cases, inhibits early classification. To overcome these limitations, [24] investigates various clustering based approaches, concluding that these methods improve the earliness without sacrificing accuracy. Another, approach for early classification based on probabilistic classifiers [15], which, similar to the MPL-based approaches, estimates the earliest reliable time point for prediction, outperform the nearest neighbour approaches by utilizing a Gaussian process classifier.

Shapelets [27] have, recently, emerged as important primitives for time series classification [9,10]. A shapelet is a phase-independent, i.e., location invariant, sub-sequence of a longer time-series used as a local primitive for classification. In this setting, the idea is to find the closest matching position within each time-series to a shapelet and use the distance between them as a discriminatory feature. For traditional time series classification, several approaches have been proposed, e.g., shapelet-based decision trees [27], logical combinations of shapelets [16] and shapelet-based transformations [10,14]. For early classification of time series, [8] presents an approach based on early distinctive shapelets [25], which is able to provide the uncertainty of each class at each time step and issue a prediction once the uncertainty drops below a user specified threshold.

One limiting factor of shapelet-based classifiers is their relatively high computational cost [27] and comparatively low classification accuracy compared to, e.g., learning shapelets [9] or similarity-based approaches, such as dynamic time warping [1]. To overcome some of these limitations, forests of shapelet trees (RSF) [12] have been put forward as competitive and accurate alternative. Although the RSF algorithm is unable to provide early classification, random shapelets, i.e., short discriminatory subsequences, are by nature well-suited for early classification, since the full time series is not required at prediction time.

In this work, two *novel* approaches are presented that extend the random shapelet forest algorithm to support early classification of time series. In addition, a novel approach for triggering early classification is presented, which is based on out-of-bag (OOB) performance estimates, providing a computationally less costly alternative to the cross-validated triggering function outlined in [15]. The purpose of this study is, hence, to empirically evaluate how these approaches affect the predictive performance of the forest when performing early classification. More specifically, we explore the trade-off between accuracy, i.e., the fraction of correctly classified instances, and the earliness, i.e., the fraction of time-points required before making a prediction.

In summary, the **main contributions** of this paper include: (1) the introduction of a novel method for early classification using decision forests based on shapelets; (2) an optimization of the cross-validated triggering function based on out-of-bag estimates of reliability; and (3) an extensive empirical evaluation of the proposed approach compared to several state-of-the-art algorithms. In the next section, we formalize early classification and present related work. In Sect. 3, we present the proposed approach and in Sect. 4, we present the results from an empirical investigation of the proposed approach compared to several state-of-the-art methods. Finally, in Sect. 5, we summarize the main conclusions an outline directions for future work.

2   Background

The problem studied in this paper is early time series classification, and our focus is on the Random Shapelet Forests (RSF) [12] in this context. Specifically, the task is, given a set of time-evolving variables ordered by time and sampled at fixed and regular intervals of length m, we want to infer a single model from a training set consisting of completely evolved variables that is able to predict the (correct) class labels of previously unseen time-evolving time series examples as early and accurately as possible.

A time series $\mathcal{T} = \{T_1, \ldots, T_m\}$ of length m is a time-evolving variable measured at m regular time points, such that $T_i \in \mathbb{R}$, $\forall i \in \{1, \ldots, m\}$. Given a time series $\mathcal{T}$ of length m and a time point $t \leq m$, we can define the corresponding *incomplete time series* $\mathcal{T}^t = \{T_1, \ldots, T_t\}$ that contains all measurements from the first time point up until t. In time series classification, we assume that we have a collection of n time series $\mathbf{X} = \{(\mathcal{T}_1, y_1), \ldots, (\mathcal{T}_n, y_n)\}$ which defines a *training set*, where each time series is labeled with a label $y_i \in \mathbf{Y}$ and $\mathcal{Y}$ is a finite set of class labels. For notational simplicity, we here assume that every time series in the dataset has the same length and the same number of dimensions. In the framework, we assume that the training set $\mathbf{X}$ has been used to train a classification function $F(\mathcal{T})$ which is able to output a class label for an incomplete time series and a *trigger* function $R(\mathcal{T}^t)$ which indicates whether or not to allow an early classification. To illustrate, the early classification framework is outlined in Algorithm 1.

Algorithm 1. The early classification framework, consisting of an incoming time series $\mathcal{T}$, a classification function F and a trigger function R.

```
1: procedure PREDICT-EARLY(T, F, R)
2:     while true do
3:         assign the most recent time series data to T^t
4:         if R(T^t) is true then
5:             return F(T^t)
6:         end if
7:         t ← t + 1
8:     end while
9: end procedure
```

2.1 Related Work

In general, most time-series classifiers relies on instance based classification methods such as, e.g., nearest neighbours under various distance measures of which the most common and simplest is the Euclidean norm [7]. To improve accuracy, elastic distance measures, such as Dynamic Time Warping (DTW) [1] or its constrained version (cDTW) [22] and longest common sub-sequences (LCSS), that allow for localized distortion, have been proposed. Dynamic time warping, for example, finds the optimal match between two sequences by allowing non-linearity in the distance calculation. By regularization using e.g., a band [18] the search performance and generalization behavior of k-NN can be improved [7].

Currently, most available early classifiers follow the framework outlined in Algorithm 1, the major differences being the selection of the trigger function, i.e., R and the choice of algorithm for implementing the incomplete hypothesis function F. For example, in [24,26] the trigger function R relies on estimating the MPL, i.e., the earliest time for which a reliable prediction can be made. The algorithm proposed in [24,26] is based on the l-length prefix-space of the time series, where, for a given time series, the algorithm finds the smallest l such that for any length $k < l \leq m$, the nearest neighbors in the $k, \ldots, l$ prefix is not different from the nearest neighbors in the L-spaces. The authors define this as the *minimum prediction length* (MPL). Hence, the nearest neighbor algorithm in the l:th prefix-space is used to make a classification if and only if the nearest neighbor has a MPL of at most l, otherwise the prediction is postponed until more data has arrived [24]. To improve predictive performance and avoid overfitting, [24] proposed an alternative method in which a clustering approach is utilized. This method is known as *early classification on time series* (ECTS). Early classification based on probabilistic classification, which uses the accuracy for each class to estimate a point in time when an early classification can be made, is proposed by [15]. In this framework, the minimum prediction length is estimated separately for each class using the cross-validation error for Gaussian process models built using separate time slices, providing one of the strongest early classification algorithms [15]. Both the ECTS and ECDIRE algorithms require several models be constructed (one for each time step). For the nearest neighbor based model, the computational cost is negligible [24], for ECDIRE, however, this imposes an additional cost, which is also amplified by the cost of repeated cross-validation to estimate the threshold [15]. This contrasts to the method proposed in this work, by which a single model is built and evaluated using the OOB instances, hence, decreasing this cost significantly.

Since similarity-based early time-series classifiers cannot be interpreted by domain practitioners, an approach based on *early distinctive shapelets* (EDSC) was proposed in [25]. In this approach, shapelets that are both distinctive, i.e., separate the classes, and appear early in the time-series are iteratively selected until the union of shapelets cover at least a predefined fraction of the time series, or until the number of shapelets exceeds a user defined threshold [25]. To issue a classification, the shapelets are scanned in order of importance and a classification is made once a shapelet is matched. Somewhat similarly, [8] defines an effective method for providing uncertainty measures for EDSC [25], providing the certainty of each class at each time step, triggering a prediction once a predefined certainty threshold is met. Recently it has also been noted that most early time-series classifiers are *myopic*, i.e., the prediction time is greedily chosen. To overcome this, a non-myopic early sequence classifier is proposed in [5], which balances the expected gain in accuracy with the cost of delaying the prediction. This process, however, require that we are given a misclassification cost function associated with each time point and a monotonically increasing cost function associated with each time-point [5].

2.2   Random Shapelet Forests

As briefly introduced earlier, shapelets are local, phase independent sub-sequences of time series. Originally, shapelets have been introduced as local primitives for classification, where the idea is to find the closest matching position within each data series to a particular shapelet and use this similarity as a discriminatory feature [27]. The first shapelet classifier [27] embedded the extraction algorithm in a decision tree learning algorithm, achieving competitive classification accuracy. To improve the classification accuracy further and reduce learning time, an algorithm for generating ensembles of randomized shapelet trees was presented in [12]. In general, ensemble methods rely on the combined voting of several, relatively weak, models which all performs better than random guessing and makes somewhat independent errors [3]. In the random shapelet forest, randomization in the weak models is introduced both in the selection of training instances and in the selection of shapelet to use at each node. The first is achieved by generating a bootstrap sample, i.e., bagging [2], of instances. Bagging works by randomly selecting instances with replacement, duplicating some and excluding others. The process results in two subsets, where each shapelet tree is built using the larger in-bag instances and an unbiased estimate of the running performance is given by the OOB instances [3]. The second randomization works by evaluating a random sample of shapelets at each node [12].

The random shapelet forest construction algorithm consists of two parts: ensemble creation and tree generation, where the former algorithm requires one parameter; the number of shapelet trees to generate (p), and the latter requires the number of shapelets to examine at each split (r). In the ensemble part, p trees are constructed from an in-bag sample drawn with replacement from the training set, $\mathbf{X}$. The shapelet tree construction algorithm, which can be run in parallel for each bootstrap replicate, starts by selecting r random shapelets and, using their distances, computes an impurity measure[1] which ranks the shapelets in decreasing order of importance. Finally, the shapelet that reduces error the most if split upon is selected. The instances are subsequently partitioned into two subsets according to the selected shapelet and distance threshold, i.e., one subset for those instances with a distance less than the found threshold and one for those with a distance greater than the threshold. The tree generation continues to recursively build sub-trees until each node is pure, i.e., instances of only one class remains. The complete algorithm is outlined in [12].

3   Early Random Shapelet Forests

One approach to support early classification of a random shapelet forest (RSF) is to build one forest per accumulated time instant t and assess the earliest time where a classification can be made using cross-validation, similar to the approach presented by [15]. This approach is, however, computationally costly and requires a large number of models to be built and evaluated. To overcome this limitation, the main idea here is that since many of the randomly sampled shapelets of each tree will be shorter than the incomplete time series only the prediction function must be modified. Hence, there is no modifications to the training procedure of the complete forest. Instead, the classification function, F, is modified to support incomplete time series classification, i.e., allow $F(\mathcal{T}^t)$ where $t \leq 1$. This works since many of the randomly selected shapelets in the construction phase will have a length $l < m$, and be usable for classification of time series of length $tm < l$. It is, however, unclear what action should be taken if $tm > l$, i.e., what action to take if the shapelet is longer than the current incomplete time series. In this study, two approaches are investigated to allow the RSF prediction function to support incomplete time-series classification using weight propagation (WERSF) and class proximities (PERSF). Additionally, to avoid costly operations such as cross-validation [15] to identify the time instant, $t*$, after which a particular class can reliably be predicted, we here present a class conditional trigger function (Algorithm 4) based on out-of-bag estimated precision.

Weight-propagated eRSF. The first investigated strategy for performing incomplete time series classification using a RSF is based on the idea that when making predictions with a random shapelet tree and a shapelet is encountered that is longer than the current incomplete series for which a prediction is to be made, then a similar procedure to how missing values are handled in standard decision trees [17] should be employed. More concretely, given an incomplete time series $\mathcal{T}^t$, and a random shapelet tree ST_i, let $\mathcal{T}^t$ traverse the tree according to

---

[1] In this study the information gain [17,27] is used,

Algorithm 2. Predict the probability of each class label for an incomplete time series using weight propagation (WERSF).

1: **procedure** WERSF(e, $\mathcal{T}^t$, ω)
2:      **if** e is a leaf node **then**
3:          **return** the leaf class probability distribution weighted by ω
4:      **end if**
5:      assign the shapelet and distance threshold of node e to $\mathcal{S}$ and τ respectively
6:      **if** $\mathcal{S}$ is shorter than $\mathcal{T}^t$ **then**
7:          recursively visit a branch depending on $\text{dist}(\mathcal{T}^t, \mathcal{S})$ propagating ω
8:      **end if**
9:      find the relative frequency of instances in each branch as ω_{left} and ω_{right}
10:     $p_{left} \leftarrow$ WERSF(e_{left}, $\mathcal{T}^t$, ω_{left})
11:     $p_{right} \leftarrow$ WERSF(e_{right}, $\mathcal{T}^t$, ω_{right})
12:     **return** $p_{left} + p_{right}$
13: **end procedure**

Algorithm 2 initialized with $\omega = 1$ and the root node e_i of tree. In the algorithm, a prediction is made once a leaf node is reached, in which case the local class distribution, weighted by ω, is returned. If the current shapelet, $\mathcal{S}$, is longer than $\mathcal{T}^t$ both branches are inspected, recursively propagating the relative frequencies of instances in the left, ω_{left}, and right, ω_{right}, node as new weights. The recursively weighted relative frequencies are returned as a sum of the right and left probability distributions. If the current shapelet is shorter than the incomplete time series, the algorithm continues by inspecting either the left or right branch depending on the minimum distance between $\mathcal{S}$ and $\mathcal{T}^t$ compared the nodes distance threshold.

Hence, Algorithm 2 returns a probability distribution over the possible class labels using the local class distributions in the reached leaf nodes, each weighted by the probability of a particular path. In the extreme case that all tree nodes consists of shapelets longer than the given incomplete time series, then the *a priori* class distribution is assigned. In the other extreme case that all tree nodes consists of shapelets shorter than the given time series, then the class distribution in the leafs are returned. Otherwise, the algorithm returns a probability distribution weighted by the nodes used for prediction. We call this algorithm WERSF.

Proximity eRSF. Instead of discarding shapelets longer than the incomplete time series, the second investigated approach utilizes the available information, i.e., the overlap between the current shapelet and the incomplete time series, by finding the class centroid with the shortest mean distance to the shapelet and select the most probable path based on the prevalence of the found class in the left or right child nodes. More specifically, given an incomplete time series $\mathcal{T}^t$ and a random shapelet tree ST_i and its root node e_i, Algorithm 3 returns the local class distribution if the current node is a leaf node. Otherwise, if the shapelet $\mathcal{S}$ associated with the current node is longer than the time series $\mathcal{T}^t$, the overlapping Euclidean distance between the shapelet and the time series is

Algorithm 3. Predict the probability of each class label for an incomplete time series using class proximites (PERSF).

1: **procedure** PERSF(e, $\mathcal{T}^t$)
2:     **if** e is a leaf node **then**
3:         **return** the leaf class probability distribution
4:     **end if**
5:     assign the shapelet and distance threshold of node e to $\mathcal{S}$ and τ respectively
6:     **if** $\mathcal{S}$ is shorter than $\mathcal{T}^t$ **then**
7:         recursively visit a branch depending on dist($\mathcal{T}^t, \mathcal{S}$)
8:     **end if**
9:     compute the overlapping distance between $\mathcal{T}^t$ and $\mathcal{S}$ as d
10:     find the mean distance between $\mathcal{S}$ and the instances of each class $c \in \mathcal{Y}$ as $\mathbf{D}_c$
11:     find the class c with the smallest absolute difference, $\min\limits_{c \in \mathcal{Y}}(\|d - \mathbf{D}_c\|)$
12:     find the branch with highest relative frequency of class c as e_{best}
13:     **return** PERSF(e_{best}, $\mathcal{T}^t$)
14: **end procedure**

---

computed and compared to the mean distance of each time series of each class to the threshold. The algorithm continues by finding the class that minimizes this distance and subsequently selects the branch in which this class has highest prevalence among the subset of training examples reaching the node. If the time series is longer than the shapelet, the algorithm continues by inspecting the right or left branch depending on the sub-sequence distance. The main rational for this approach is, similar to how missing values are treated using surrogate splits [21], that the most probable branch is the one where time series are maximally proximate. We call this approach PERSF.

Finally, the prediction of an incomplete time series using RSF is given by the maximum sum of probabilities from the individual trees, i.e., using distributional summation [20]. More specifically, given the trees $ST_1, \ldots, ST_p$ and an incomplete time series $\mathcal{T}^t$, the summed probabilities is given by $\frac{1}{p} \sum_{i=1}^{p} F(ST_i, \mathcal{T}^t)$, where F is determined by either WERSF or PERSF and the final prediction $\hat{y}$ is given by the label with the highest probability.

Out-of-bag estimated early classification. To determine the point in time when the final prediction should be made, we modify the algorithm proposed by [15] to instead of utilizing a cross-validation estimated early classification threshold, utilize the less costly out-of-bag estimate provided by the RSF. More specifically, given a set of predetermined prediction time stamps, $\mathbf{E}$, and a collection of random shapelet trees, $ST_1, \ldots, ST_p$ and a prediction function F (here either WERSF or PERSF), the point in time, $t*$, when to make the final prediction is determined using the out-of-bag precision for each time step and class. Algorithm 4, starts by computing the OOB precision and minimum probability margin for each class and time step and subsequently finds the earliest time step for a particular class where all subsequent time steps has a higher precision than the precision of the full time series weighted by a specified constant w.

Algorithm 4. Algorithm for finding the earliest reliable time point for each class using the out-of-bag instances.

1: **procedure** ESTIMATE-RELIABILITY-THRESHOLD($\{ST_1, \ldots, ST_p\}$, **X**, w)
2:     **for** each class $c \in \mathcal{Y}$ **do**
3:         **for** each time step $t \in \mathbf{E}$ **do**
4:             estimate the OOB precision of class c for time step t as $\mathbf{P}_c^t$
5:             estimate the minimum probability margin of class c for time step t as
$\mathbf{M}_c^t = \min\limits_{y^c \in \mathbf{Y}} P(y^c | \mathcal{T}^t) - \max\limits_{y \neq y^c} P(y | \mathcal{T}^t)$
6:         **end for**
7:     **end for**
8:     **for** each class $c \in \mathcal{Y}$ **do**
9:         find the earliest time point $t*_c$ where each $\mathbf{P}_c^{t'}, \ldots, \mathbf{P}_c^1$ are above $w\mathbf{P}_c^1$
10:     **end for**
11:     **return** **P** and **M**
12: **end procedure**

At prediction time of an incomplete time series, of size $\hat{t}$, a prediction is only made if $t*_{\hat{y}} \leq \hat{t}$ and $M_{\hat{y}}^{\hat{t}} \leq P(\hat{y} | \mathcal{T}^{\hat{t}}) - \max\limits_{y \neq \hat{y}} P(y | \mathcal{T}^{\hat{t}})$, where $\hat{y}$ is the prediction [15].

4   Experiments

In this section, the general empirical methodology is outlined together with the experimental design protocol for evaluating the predictive performance and earliness of RSF using the proposed prediction functions, WERSF and PERSF, compared to state-of-the-art approaches based on nearest neighbors [24] and probabilistic classifiers [15].

4.1   Experimental Setup

In supervised time series classification, the most common measure of classifier performance is the accuracy (or error rate), i.e., the fraction of (in)correctly labeled instances, cf. [14,27]. For early time series classification, we are not only interested in the accuracy of different methods, but also in how early a classification can be made. The most frequent measure for assessing this is the *earliness*, i.e., the fraction of time-points required before classification, cf. [5,25,26]. Although these measurements can provide insights in isolation, one should, however, take into account the trade-off between accuracy and earliness, which is not captured when employing the measures separately. To overcome this limitation, [4] proposes a way of evaluating the two conflicting properties: prediction quality Q and prediction earliness E. The authors propose that a significance test is performed over multiple datasets, independently for both measures. One can, thus, conclude if there is any statistically significant difference between the methods under Q or E. In this framework, a classifier is considered better if it improves one objective without degrading the other. More specifically, a classifier is considered superior to another classifier if it is significantly better with respect to one of the criteria and *not* significantly worse with respect to the other [4]. In this definition, Q can be any suitable measure of predictive performance, including e.g., area under the ROC curve or accuracy; similarly E can be any measure of how early a prediction is made, where the simplest is the average prediction time. In this work, we opt for using accuracy as a measure of predictive performance and average prediction time as a measure of earliness.

Both the RSF algorithm and the early prediction framework requires a few parameters to be set. First, RSF requires: p the number of random shapelet trees; r the number of shapelets; and second, the trigger function requires: w the fraction of the full precision to be used as an early classification threshold. Since the purpose of this paper is not to evaluate the predictive performance of different parameter configurations, we here avoid searching for the optimal parameter configuration by setting: $p = 100$ trees, $r = 100$ random shapelets and weighting the full precision by $w = 0.9$. To evaluate different values for w, cross-validation accuracy on the training set is investigated in Fig. 2.

A widely accepted procedure for testing for significant differences with respect to a single measure, such as classification accuracy, when comparing a number of classifiers over several datasets, is the non-parametric Friedman test based on the ranks of each method per datasets [6]. To detect if there are any pair-wise differences between classifiers, we employ a Nemenyi post-hoc test. In the case of early classification, the null hypothesis of no difference can be stated as: there is no significant difference in predictive performance nor in average prediction time. Furthermore, for the purpose of the empirical comparison, the complete set of available datasets are selected from the UCR time series repository [13], with the main motivation being that these datasets are frequently used in studies involving state-of-the-art approaches to which the proposed algorithms is compared [15,24]. To improve reproducability, we utilize the already provided training and testing split when computing the evaluation metrics. Note, however, that the results for different precision multipliers (w) are computed using 10-fold cross validation on the training split. The included datasets represents a wide range of tasks such as: image outline classification (e.g., OSULeaf), motion classification (e.g., Gun Point) and sensor classification (e.g., NIFECG_Thorax1) with different characteristics (from 16 to 1,800 training instances) and different lengths (from 24 to 1,882 measurements). Finally, the proposed algorithms are compared to two state-of-the-art early classifiers: the reliable early classification time series (ECDIRE [15]) algorithm and the early classification of time series (ECTS [24]) algorithm.

4.2   Empirical Evaluation

A summary of the results for all datasets and algorithms is given in Table 1. Statistical tests for determining if the predictive performance and average prediction time differ significantly from what can be expected under the null hypothesis are undertaken by comparing the performance ranks of the algorithms. As can be seen in Table 1, one of the proposed methods, WERSF, has the highest average accuracy of 0.81, compared to 0.71 and 0.73 for ECTS and ECDIRE, respectively, which is also confirmed by the average rank. By inspecting the ranks for accuracy, one can see that WERSF and PERSF are ranked the highest (with an approximate average rank of 2), while ECTS and ECDIRE are ranked lowest (with an approximate average rank of 3). According to the Friedman test, the accuracy ranks deviates, at $p < 0.01$, significantly from what can be expected under the null hypothesis of no difference for these algorithms and datasets. To investigate if there are any significant differences between pairs of algorithms when considering classification accuracy, a post-hoc Nemenyi test is performed (see Fig. 1 (left)). The post-hoc test reveals that at the $p = 0.05$ level, the ECTS and ECDIRE algorithms are significantly less accurate than both the PERSF and WERSF algorithms. For ECTS and ECDIRE, no significant differences can be identified for these datasets. If we partition the algorithms into groups for $p < 0.05$, we can, hence, identify two groups: PERSF and WERSF are significantly more accurate than ECDIRE and ECTS.

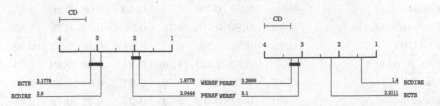

Fig. 1. Comparison of pair-wise Nemenyi test for the accuracy (left) and earliness (right) of all classifiers over all datasets in Table 1. Classifiers that are not significantly different at $p = 0.05$ are connected.

Considering the second performance metric, average prediction time, the proposed methods perform worse than the alternative methods. As can be seen in Table 1, ECDIRE can provide predictions after, on average, inspecting only 50 % of the measurements, while both PERSF and WERSF require roughly 80 % of measurements and ECTS require approximately 70 %. This is also confirmed by the average ranks, where ECDIRE is ranked highest (with an average rank of 1.4), ECTS ranked second and PERSF and WERSF ranked third. A Friedman test, again, reveals that the differences are significantly deviating from what can be expected under the null hypothesis of no difference in earliness. The Nemenyi post-hoc test reveals that there is a significant difference between ECDIRE and ECTS, WERSF and PERSF. Again, if we partition the algorithms into groups based on earliness, we see that there are two main groups: ECDIRE is significantly earlier than alternative methods. In terms of both performance metrics, and the Pareto optimality, we can see that the only significant difference we can detect is that between ECDIRE and ECTS, where the former is able to provide predictions significantly more early while not being significantly less accurate. For the other methods, the number of observations again do not allow for reaching any conclusions regarding the relative performance. We can, however, in Fig. 2 see that for alternative

Table 1. Predictive performance, as measured by accuracy and earliness, for all compared approaches. The highest accuracy and lowest earliness are highlighted.

Dataset	Accuracy				Earliness			
	ECDIRE	ECTS	WERSF	PERSF	ECDIRE	ECTS	WERSF	PERSF
50words	0.53	0.57	0.67	**0.69**	**0.40**	0.73	0.94	0.93
Adiac	0.55	0.40	0.71	**0.73**	**0.38**	0.59	0.95	0.96
Beef	0.50	0.50	**0.70**	0.60	**0.68**	0.77	0.98	0.92
CBF	0.89	0.85	0.97	**0.99**	**0.29**	0.77	0.51	0.67
ChlorineConcentration	0.56	0.62	**0.65**	**0.65**	**0.14**	0.66	1.00	1.00
CinC_ECG_torso	0.81	**0.87**	0.81	0.83	**0.50**	0.58	0.96	0.99
Coffee	**0.96**	0.75	**0.96**	0.89	**0.82**	0.84	0.85	0.92
Cricket_X	0.57	0.56	**0.77**	0.76	**0.48**	0.72	0.87	0.91
Cricket_Y	0.63	0.63	**0.76**	0.74	**0.36**	0.66	0.92	0.94
Cricket_Z	0.60	0.59	**0.76**	0.75	**0.46**	0.68	0.86	0.88
DiatomSizeReduction	**0.80**	**0.80**	0.76	0.79	0.24	**0.15**	0.79	0.87
ECG200	**0.91**	0.89	0.83	0.84	0.90	**0.60**	0.81	0.92
ECGFiveDays	0.60	0.62	**1.00**	**1.00**	**0.21**	0.64	0.89	0.85
FaceAll	**0.87**	0.76	0.75	0.76	**0.57**	0.64	0.85	0.89
FaceFour	0.61	0.82	**0.99**	0.74	**0.22**	0.72	0.79	0.86
FacesUCR	0.74	0.71	**0.87**	0.86	**0.59**	0.87	0.90	0.91
fish	0.81	0.75	**0.93**	0.90	**0.55**	0.61	0.74	0.81
Gun_Point	0.91	0.87	**1.00**	0.99	**0.32**	0.47	0.82	0.79
Haptics	0.44	0.37	0.45	**0.48**	0.87	0.94	1.00	0.94
InlineSkate	0.26	0.33	0.36	**0.37**	**0.34**	0.85	0.92	0.96
ItalyPowerDemand	0.93	0.94	**0.95**	**0.95**	0.72	0.79	0.95	0.97
Lighting2	0.54	**0.70**	0.54	0.67	0.09	0.89	**0.05**	0.83
Lighting7	0.48	0.58	**0.71**	**0.71**	0.20	0.87	0.93	0.94
MALLAT	0.78	0.85	**0.92**	**0.92**	0.45	0.69	0.79	0.86
MedicalImages	0.74	0.68	**0.69**	**0.69**	0.21	0.54	0.99	0.99
MoteStrain	0.80	0.88	**0.94**	**0.94**	0.12	0.79	0.85	0.77
NIFECG_Thorax1	0.89	0.81	**0.90**	0.89	0.64	0.78	0.89	0.94
NIFECG_Thorax2	0.93	0.88	0.92	**0.93**	0.57	0.77	0.89	0.93
OliveOil	0.40	**0.90**	**0.90**	**0.90**	0.30	0.87	0.92	0.91
OSULeaf	0.52	0.49	**0.82**	0.81	0.48	0.77	0.87	0.83
SonyAIBORobotSurface	0.83	0.69	**0.88**	0.82	0.62	0.68	0.72	**0.54**
SonyAIBORobotSurfaceII	0.74	0.85	**0.90**	0.85	0.18	0.55	0.54	0.57
StarLightCurves	0.95	0.15	**0.96**	0.95	0.53	0.82	0.63	0.54
SwedishLeaf	0.87	0.78	0.90	**0.91**	0.46	0.76	0.89	0.93
Symbols	**0.81**	**0.81**	0.71	0.18	0.45	0.51	0.80	**0.05**

(Continued)

Table 1. *(Continued)*

Dataset	Accuracy				Earliness			
	ECDIRE	ECTS	WERSF	PERSF	ECDIRE	ECTS	WERSF	PERSF
synthetic_control	0.96	0.88	**0.98**	0.97	**0.62**	0.88	0.76	0.81
Trace	0.77	0.74	**1.00**	**1.00**	**0.42**	0.51	0.78	0.79
Two_Patterns	0.87	0.86	**0.95**	**0.95**	0.99	0.87	**0.86**	0.88
TwoLeadECG	0.81	0.73	**0.99**	**0.99**	**0.69**	0.64	0.78	0.78
uWaveGestureLibrary_X	0.77	0.73	**0.78**	0.76	**0.74**	0.86	0.93	0.92
uWaveGestureLibrary_Y	0.70	0.63	**0.71**	**0.71**	0.97	**0.86**	0.95	0.91
uWaveGestureLibrary_Z	0.71	0.65	**0.72**	**0.72**	0.76	**0.85**	0.92	0.92
wafer	0.97	**0.99**	0.89	0.94	0.11	0.44	**0.05**	**0.05**
WordsSynonyms	0.52	**0.59**	0.54	0.52	**0.66**	0.83	0.94	0.92
yoga	**0.85**	0.81	0.79	0.80	1.00	**0.69**	0.85	0.80
Rank	2.90	3.18	**1.88**	2.04	**1.40**	2.20	3.10	3.30
Average	0.73	0.71	**0.81**	0.80	**0.50**	0.71	0.82	0.83

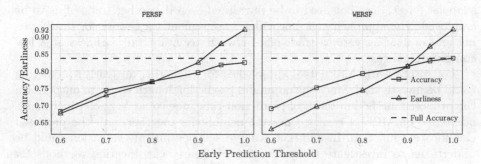

Fig. 2. Evolution of the earliness and accuracy (10-fold cross-validation on the training set) for different multipliers (w) of the full precision.

precision multipliers, the average accuracy for WERSF can exceed that of ECDIRE without having significantly lower early classification time.

To investigate the effect of the early prediction multiplier w, the earliness and accuracy for the proposed methods are computed using cross-validation on the training set. In Fig. 2, we see the evolution of early classification accuracy and earliness (solid lines) compared to the full time series accuracy (dashed line) for both PERSF (left) and WERSF (right). As seen in Fig. 2, the prediction time and accuracy increases, as expected, with an increasing precision multiplier. For WERSF, we can also see that when setting the w parameter to be maximally conservative ($w = 1$), the accuracy approaches the accuracy of the complete time series classifier for both methods. Furthermore, we can also see that the precision threshold seems to, on average, empirically lower bound the accuracy of the forest on the full time series, indicating that the threshold indeed functions as expected and provide means for trading accuracy for earliness or vice versa.

5   Concluding Remarks

In this paper, we have proposed and investigated a novel approach for early time series classification using the random shapelet forest together with two prediction functions and an out-of-bag estimated reliablity threshold. The proposed approach utilizes less computational resources than alternative methods, since a single classifier is built utilizing the full time series, with a modified prediction function that allow for early classification. Depending on how the weighting parameter (w) is set, accuracy vs. earliness can be balanced in different ways. An extensive empirical evaluation, in which w was optimized towards accuracy, showed that the proposed approach (in two different instantiations) could lead to significantly higher predictive performance. On the other hand, for this optimization criterion, the improved accuracy came with an associated cost of reduced earliness. By tuning the parameter differently, earliness can be improved (at the cost of accuracy). It should be noted that the underlying model, i.e., the random shapelet forest, does not need to be re-trained even if another trade-off is to be employed. The approach hence provides a convenient framework for investigating and evaluating various trade-offs between accuracy and earliness at a low computational cost.

For future work, alternative approaches for triggering an (early) prediction could be investigated. Since the minimum prediction length seems to outperform the proposed method in terms of prediction time, one could, e.g., consider methods based on MPL of reversed nearest neighbours together with the proposed classification function. In addition, a large scale empirical evaluation could be undertaken to investigate the performance of early classification methods for both univariate and multivariate time series. Finally, future work could also include investigating alternative approaches for early classification in which the process is able to at any time-point guarantee a user supplied confidence in the classification, e.g., employing confidence predictions.

Source code. The source code for replicating the experiments is available at the supporting website (https://people.dsv.su.se/~isak-kar/ersf/).

Acknowledgments. This work was partly supported by project High-Performance Data Mining for Drug Effect Detection at Stockholm University, funded by the Swedish Foundation for Strategic Research (IIS11-0053).

References

1. Berndt, D.J., Clifford, J.: Using dynamic time warping to find patterns in time series. In: KDD Workshop, vol. 10, pp. 359–370. Seattle, WA (1994)
2. Breiman, L.: Bagging predictors. Mach. Learn. **24**(2), 123–140 (1996)
3. Breiman, L.: Random forests. Mach. Learn. **45**(1), 5–32 (2001)

4. Dachraoui, A., Bondu, A., Cornuéjols, A.: Evaluation protocol of early classifiers over multiple data sets. In: Loo, C.K., Yap, K.S., Wong, K.W., Teoh, A., Huang, K. (eds.) ICONIP 2014. LNCS, vol. 8835, pp. 548–555. Springer, Heidelberg (2014). doi:10.1007/978-3-319-12640-1_66

5. Dachraoui, A., Bondu, A., Cornuéjols, A.: Early classification of time series as a non myopic sequential decision making problem. In: Appice, A., Rodrigues, P.P., Costa, V.S., Soares, C., Gama, J., Jorge, A. (eds.) ECML PKDD 2015. LNCS (LNAI), vol. 9284, pp. 433–447. Springer, Heidelberg (2015). doi:10.1007/978-3-319-23528-8_27

6. Demšar, J.: Statistical comparisons of classifiers over multiple data sets. J. Mach. Learn. Res. **7**, 1–30 (2006)

7. Ding, H., Trajcevski, G., Scheuermann, P., Wang, X., Keogh, E.: Querying and mining of time series data: experimental comparison of representations and distance measures. Proc. VLDB Endowment **1**(2), 1542–1552 (2008)

8. Ghalwash, M.F., Radosavljevic, V., Obradovic, Z.: Utilizing temporal patterns for estimating uncertainty in interpretable early decision making. In: Proceedings of the 20th ACM SIGKDD International Conference on Knowledge Discovery and Data Mining, pp. 402–411. ACM (2014)

9. Grabocka, J., Schilling, N., Wistuba, M., Schmidt-Thieme, L.: Learning time-series shapelets. In: Proceedings of the 20th ACM SIGKDD International Conference on Knowledge Discovery and Data Mining, pp. 392–401. ACM (2014)

10. Hills, J., Lines, J., Baranauskas, E., Mapp, J., Bagnall, A.: Classification of time series by shapelet transformation. Data Min. Know. Discovery **28**(4), 851–881 (2014)

11. Ishiguro, K., Sawada, H., Sakano, H.: Multi-class boosting for early classification of sequences. In: Proceedings of the British Machine Vision Conference, pp. 24.1–24.10. BMVA Press (2010). doi:10.5244/C.24.24

12. Karlsson, I., Papapetrou, P., Boström, H.: Forests of randomized shapelet trees. In: Gammerman, A., Vovk, V., Papadopoulos, H. (eds.) SLDS 2015. LNCS (LNAI), vol. 9047, pp. 126–136. Springer, Heidelberg (2015). doi:10.1007/978-3-319-17091-6_8

13. Keogh, E., Zhu, Q., Hu, B., Hao, Y., Xi, X., Wei, L., Ratanamahatana, C.A.: The UCR time series classification/clustering homepage (2015). www.cs.ucr.edu/~eamonn/time_series_data/

14. Lines, J., Davis, L.M., Hills, J., Bagnall, A.: A shapelet transform for time series classification. In: Proceedings of the 18th ACM SIGKDD International Conference on Knowledge Discovery and Data Mining, pp. 289–297. ACM (2012)

15. Mori, U., Mendiburu, A., Keogh, E., Lozano, J.A.: Reliable early classification of time series based on discriminating the classes over time. Data Min. Knowl. Discovery, 1–31 (2016)

16. Mueen, A., Keogh, E., Young, N.: Logical-shapelets: an expressive primitive for time series classification. In: Proceedings of the 17th ACM SIGKDD International Conference on Knowledge Discovery and Data Mining, pp. 1154–1162. ACM (2011)

17. Quinlan, J.R.: C4.5: Programs for Machine Learning. Elsevier (1993)

18. Ratanamahatana, C.A., Keogh, E.: Everything you know about dynamic time warping is wrong. In: 3rd Workshop on Mining Temporal and Sequential Data, pp. 22–25 (2004)

19. Rodrıguez, J.J., Alonso, C.J., Boström, H.: Boosting interval based literals. Intell. Data Anal. **5**, 245–262 (2001)

20. Rokach, L.: Ensemble-based classifiers. Artif. Intell. Rev. **33**(1–2), 1–39 (2010)

21. Rokach, L., Maimon, O.: Top-down induction of decision trees classifiers-a survey. IEEE Trans. Syst. Man Cybern. Part C: Appl. Rev. **35**(4), 476–487 (2005)

22. Sakoe, H., Chiba, S.: Dynamic programming algorithm optimization for spoken word recognition. Trans. ASSP **26**, 43–49 (1978)
23. Xing, Z., Pei, J., Dong, G., Yu, P.S.: Mining Sequence Classifiers for Early Prediction, pp. 644–655 (2008)
24. Xing, Z., Pei, J., Philip, S.Y.: Early classification on time series. Knowl. Inf. Syst. **31**(1), 105–127 (2012)
25. Xing, Z., Pei, J., Philip, S.Y., Wang, K.: Extracting interpretable features for early classification on time series. In: SDM, vol. 11, pp. 247–258. SIAM (2011)
26. Xing, Z., Pei, J., Yu, P.S.: Early prediction on time series: a nearest neighbor approach. In: Proceedings of the 21st International Jont Conference on Artifical Intelligence, pp. 1297–1302. Morgan Kaufmann Publishers Inc. (2009)
27. Ye, L., Keogh, E.: Time series shapelets: a new primitive for data mining. In: Proceedings of the 15th ACM SIGKDD. ACM (2009)

Classification

Shorter Rules Are Better, Aren't They?

Julius Stecher, Frederik Janssen, and Johannes Fürnkranz[✉]

Technische Universität Darmstadt, Knowledge Engineering, Darmstadt, Germany
jlstecher@gmail.com, {janssen,juffi}@ke.tu-darmstadt.de

Abstract. It is conventional wisdom in inductive rule learning that shorter rules should be preferred over longer rules, a principle also known as Occam's Razor. This is typically justified with the fact that longer rules tend to be more specific and are therefore also more likely to overfit the data. In this position paper, we would like to challenge this assumption by demonstrating that variants of conventional rule learning heuristics, so-called inverted heuristics, learn longer rules that are not more specific than the shorter rules learned by conventional heuristics. Moreover, we will argue with some examples that such longer rules may in many cases be more understandable than shorter rules, again contradicting a widely held view. This is not only relevant for subgroup discovery but also for related concepts like characteristic rules, formal concept analysis, or closed itemsets.

1 Introduction

It is conventional wisdom in machine learning that shorter explanations are better. Occam's Razor, *"Entia non sunt multiplicanda sine necessitate"*[1], is often cited as support for this principle. Typically, it is understood as *"given two explanations of the data, all other things being equal, the simpler explanation is preferable"* [2].

There are many plausible reasons why shorter explanations should be preferred, among them that simpler theories are easier to falsify, that there are fewer simpler theories than complex theories, so the a priori chances that a simple theory fits the data are lower, or that simpler rules tend to be more general, cover more examples and their quality estimates are therefore statistically more reliable. While it is well-known that striving for simplicity often yields better predictive results—mostly because pruning or regularization techniques help to avoid overfitting—the exact formulation of the principle is still subject to debate [3], and several cases have been observed where more complex theories perform better [1,17,22].

Much of this debate focuses on the aspect of predictive accuracy. When it comes to understandability, the idea that simpler rules are more comprehensible is typically unchallenged. In this paper, we will argue that this is not necessarily the case and motivate this argument with several examples of well-known concepts in inductive rule learning, such as characteristic rules, formal concepts and

---

[1] Entities should not be multiplied beyond necessity.

© Springer International Publishing Switzerland 2016
T. Calders et al. (Eds.): DS 2016, LNAI 9956, pp. 279–294, 2016.
DOI: 10.1007/978-3-319-46307-0_18

closed itemsets (Sect. 2). We will then present a subgroup discovery algorithm that is based on a classification learner using inverted heuristics (Sect. 3) and show quantitative and qualitative results that aim at demonstrating that a rule learner with a bias towards learning longer rules does not necessarily produce worse rules (Sect. 4).

2   Understandability and Rule Length

In this section, we try to relate the length of the learned rules to their *under-standability*—a very elusive concept which we use in an intuitive sense, without being able to give a precise definition, let alone a quantitative characterization.

Inductive rule learning is typically concerned with learning a set of rules that discriminate positive from negative examples [6]. For this task, a bias towards simplicity is necessary because for a contradiction-free training set, it is trivial to find a rule set that perfectly explains the training data, simply by converting each example to a maximally specific rule that covers only this example. Obviously, although the resulting rule set is clearly within the hypothesis space, it is not useful because it essentially corresponds to rote learning and does not generalize to unseen examples. Essentially for this reason, Mitchell [16] has noted that learning and generalization need a bias in order to avoid such elements of the version space.

Despite this necessity of a bias for simplicity, we argue that simpler rules are not necessarily more understandable, not even if all other things (such as coverage) are equal. Already Michalski [15] has noted that there are two different kinds of rules, discriminative and characteristic. *Discriminative rules* can quickly discriminate an object of one category from objects of other categories. A simple example is the rule

```
elephant :- trunk.
```

which states that an animal with a trunk is an elephant. The rule provides a simple but effective rule for recognizing the target class among all animals. However, it does not provide a very clear picture on properties of the elements of the target class. For example, from the above rule, we do not understand that elephants are also very large and heavy animals with a thick grey skin, tusks and big ears.

Characteristic rules, on the other hand, try to capture all properties that are common to the objects of the target class. A rule for characterizing elephants could be

```
heavy, large, grey, bigEars, tusks, trunk :- elephant.
```

Note that here the implication sign is reversed: we list all properties that are implied by the target class, i.e., by an animal being an elephant. From the point of understandability, characteristic rules are often preferable to deterministic rules. For example, in a customer profiling application, we would prefer to not only list a few characteristics that discriminate one customer group from the other, but would be interested in all characteristics of each customer group.

Characteristic rules are very much related to *formal concept analysis* [8,23]. Informally, a concept is defined by its intent (the description of the concept, i.e., the conditions of its defining rule) and its extent (the instances that are covered by these conditions). A *formal concept* is then a concept where both the extension and the intension are maximal, i.e., a concept where no conditions can be added without reducing the number of covered examples. In Michalski's terminology, a formal concept is both discriminative and characteristic, i.e., a rule where the head is equivalent to the body.

It is well-known that formal concepts correspond to *closed itemsets* in association rule mining, i.e., to maximally specific itemsets [21]. Closed itemsets have been mined primarily because they are a unique and compact representative of equivalence classes of itemsets, which all cover the same instances [24]. However, while all itemsets in such an equivalence class are equivalent with respect to their support, they may not be equivalent with respect to their understandability or interestingness.

Consider, e.g., the infamous {diapers,beer} itemset that is commonly used as an example for a surprising finding in market based analysis. A possible explanation for this finding is due to young family fathers who are sent to shop for their youngster and have to reward themselves with a six-pack. However, if we consider that a young family may not only need beer and diapers, the closed itemset of this particular combination may also include baby lotion, milk, porridge, bread, fruits, vegetables, cheese, sausages, soda, etc. In this combination, diapers and beer appear to be considerably less surprising, although the same set of customers may buy these sets of items. Conversely, an association rule beer :- diapers with an assumed confidence of 80 % appears considerably less interesting if we learn that 80 % of all customers buy beer, i.e., if its lift is equal to one.

These examples should help to motivate that the complexity of rules may have an effect on the understandability and plausibility of a rule, even in cases where a simpler and a more complex rule cover the same number of examples, but that it is not necessarily the case that shorter rules are more understandable. In the remainder of the paper, we will look at examples of rules that are learned from a pair of rule learning algorithms, one using a conventional heuristic that aims at finding maximally discriminative conditions, and one using inverted heuristics, which aim at finding longer rules that resemble characteristic rules.

3   Inverted Heuristics for Supervised Descriptive Rule Induction

In this section, we will explain the subgroup discovery algorithm that we used, and highlight its differences to the covering algorithm used in previous experiments with inverted heuristics [20]. Section 3.1 will briefly recapitulate the underlying base concepts before we explain the modifications for subgroup discovery in Sect. 3.2.

3.1   Inverted Heuristics

In [20], we have proposed to differentiate between *rule selection* and *rule refinement* heuristics in top-down separate-and-conquer rule learning, as opposed to using the same heuristic for both phases of the algorithm as conventional rule learning algorithms do [4,6]. While conventional heuristics match the bottom-up procedure of rule *selection*, we have argued that it can be beneficial for the top-down specialization of the rule *refinement* process to use heuristics reflecting this difference.

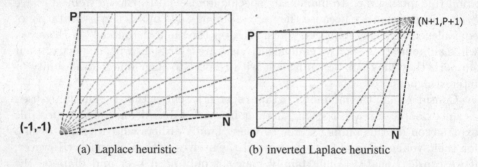

(a) Laplace heuristic   (b) inverted Laplace heuristic

Fig. 1. Visualization of the coverage space isometrics of conventional and inverted Laplace.

More specifically, we have taken a look at the *Laplace-* family of heuristics (namely *precision, Laplace* and the *m-Estimate*), all of which operate from a bottom-up perspective, differing only in the origin of their coverage space *isometrics* [5]. By creating an inverted counterpart for each of these heuristics, we were able to obtain new heuristics that better reflect the top-down nature of the rule refinement process in theory, as well as exhibit some interesting properties in practice. This is illustrated in Fig. 1, which shows that the isometrics of conventional Laplace heuristic h_{Lap} rotate around the point $(-1, -1)$ below the origin, whereas their counter-parts for the inverted Laplace η_{Lap} rotate around the point $(N + 1, P + 1)$ above the upper right corner. This point represents the empty, unconstrained rule, which has no conditions and covers all examples, and is therefore the starting point for a top-down refinement process.

While conventional heuristics of the *Laplace*-type tend to focus on consistency in rule refinement, mainly rewarding the early exclusion of false positives, their top-down counterparts instead try to maintain completeness by preserving coverage of true positives. One of the properties we deem relevant to the field of *subgroup discovery* is the refinement process taking smaller steps in coverage space without critically impacting the readability of the resulting rules. Consequently, we had observed that this results in rules that resemble characteristic rules instead of the conventionally learned discriminative rules. For example, Fig. 2 shows the two best rules that have been found for the mushroom dataset with the conventional Laplace heuristic h_{Lap} (top) and the inverted Laplace

```
[2160|0]  p :- odor = f.
```

(a) using h_{Lap} for refinement

```
[2192|0]  p :- veil-color = w, gill-spacing = c, bruises? = f,
              ring-number = o, stalk-surface-above-ring = k.
```

(b) using q_{Lap} for refinement

Fig. 2. The two best rules learned for the class poisonous in the *mushroom* dataset.

heuristic q_{Lap} (bottom). Although both rules cover approximately the same number of examples and would thus be about equal according to most quality measures, the former consists only of a single condition, whereas the latter has five conditions, which clearly makes a difference in the understandability of the rule.

Note that not all heuristics can be inverted. For example, the isometric lines of weighted relative accuracy (WRA), which is frequently used in subgroup discovery, are parallel to the diagonal of the coverage space, so that their inverted version coincides with the original heuristic.

In [20], we dealt with a rule selection and refinement selection in a conventional covering algorithm, and showed that their use tends to increase predictive accuracy, even though longer rules are learned. In this paper, we will focus on the use of inverted heuristics in a simple subgroup discovery algorithm based on Michalski's AQ family [14]. After introducing this algorithm in the next section, we will proceed by showing some of the properties of the learned rules and take a look at a couple of exemplary rules that highlight the characteristics and potential use of inverted heuristics in descriptive rule discovery.

3.2 Algorithm

The rules found with conventional covering algorithms are not ideal results for subgroup discovery, because rules are learned in order, and all examples that are covered by a rule are removed before subsequent rules are learned. Thus, each found rule (except the first one) has been learned and evaluated on a subset of the data. Subgroup discovery algorithms such as CN2-SD [12,13] try to alleviate this problem by resorting to weighted covering, where instead of removing covered examples their weight is reduced so that they have less influence on the learning of subsequent rules.

In this work, we adopt a similar but somewhat different strategy, which is based on Michalski's AQ family [14]. (cf. Algorithm 1). Unlike traditional covering algorithms, we do not remove covered instances from the data but instead keep an internal list of instances that have not yet been covered. This list is initialized with all positives instances.

A rule is then learned by selecting a random uncovered example, and by restricting the possible attribute values in the rule refinement step to those present in this instance. By doing so, coverage of this instance is guaranteed

---

Algorithm 1. Procedure SeCo-AQR

Data: Instances (labeled examples), classLabel
Result: unordered ruleset $\mathcal{R}$

1 $\mathcal{R}$ = empty unordered set of rules
2 C = Concept (list of instances with class = classLabel)
3 c = RANDOMSELECTION(C)
4 **while** positive examples left in C **do**
5   Rule **r** = FINDBESTRULE(Instances, c, classLabel)
6   **forall** *instance in C* **do**
7   **if** COVERS*(r, instance)* **then**
8   remove instance from C

9   c = RANDOMSELECTION(C)
10    $\mathcal{R} = \mathcal{R} \cup \mathbf{r}$

11 **return** $\mathcal{R}$

(while the rule can still cover other instances, both of the same as well as of other classes). Single rules are learned from the restricted set of conditions in essentially the same way as in [20] (Algorithm 2). The algorithm uses two heuristic functions, one for evaluating *rule refinements* and one for evaluating *rule selection*. These heuristics may be identical, but we advocate the use of inverted heuristics for rule refinement. The key difference to the covering and weighted covering approaches is that in this way, each rule is evaluated independently on the entire set of examples and not in the context of previously learned rules.

After a rule has been learned (see Algorithm 2), all covered examples are marked as such, and a new instance is selected from the remaining uncovered examples until none remain (all instances are covered by at least one rule). In addition, note that no stopping criteria or pruning methods were used at all, limiting the learning process only to the basic underlying algorithm and most importantly, the heuristics being used in both the selection and refinement steps.

4   Experiments

In this section, we compare statistics of rules learned with conventional and inverted heuristics with respect to rule length, number of rules and conditions and coverage information. The used algorithms will be denoted by the pair of heuristic functions ($h_{selection}$, $h_{refinement}$) they use. For example, (h_{Lap}, $ꓱ_{Lap}$) uses conventional Laplace for rule selection, and inverted Laplace for rule refinement, whereas (h_m, h_m) uses the conventional m-estimate in both cases. We always set the parameter $m = 22.446$ as has been empirically determined in [10].

4.1   Data

Table 1 shows the datasets that we use in our study. They consist of two medical datasets that have been previously used in subgroup discovery [7,11], and a

---

Algorithm 2. Procedure FINDBESTRULE

Data: Instances (labeled examples), c (random positive Instance), classLabel
Result: best rule r_{best} covering at least c

1 $r_{best} = r_{ref} = \emptyset$
2 *bestValue* = SELECTIONHEURISTIC(r_{best})
3 **repeat**
4   refinements = REFINERULE(Instances, r_{ref}, c, classLabel)
5   **forall** refinements r' **do**
6   evaluation = REFINEMENTHEURISTIC(r')
7   r_{ref} = best refinement w.r.t. evaluation
8   **if** SELECTIONHEURISTIC(r_{ref}) $\geq$ *bestValue* **then**
9           $r_{best} = r_{ref}$
10   *bestValue* = SELECTIONHEURISTIC(r_{best})
11 **until** no refinements left;
12 **return** r_{best}

Table 1. Number of classes (C), examples (E), and attributes(A) of the used datasets.

Dataset	C	E	A	Dataset	C	E	A
coronary heart disease	5	238	39	brain ischemia	2	300	28
deptmts-dbpedia-type	3	100	82	regions-dbpedia-type	3	26	40
deptmts-dbpedia-data	3	100	189	regions-dbpedia-data	3	26	149
deptmts-dbpedia-qrel	3	100	10041	regions-dbpedia-qrel	3	26	4608
regionWithLGD	3	27	334	regions-eurostat	3	26	16

number of datasets that have been generated from the Semantic Web with the Explain-A-LOD tool [18]. The latter group has been chosen because some of them contain large numbers of sparsely populated attributes, which, as we will discuss in Sect. 4.6, poses interesting problems for the use of inverted heuristics.

The *coronary heart disease* dataset [7] consists of 238 patients that have five different heart-disease-related diagnoses.

The *brain ischemia* dataset [11] contains the records of 300 patients of the Intensive Care Unit of the Department of Neurology in the University Hospital Center Zagreb, Croatia, where the task is to discriminate between patients with a positive computer tomography test for brain stroke and patients with a normal CT test.

The remaining datasets try to explain unemployment rates in France, on the level of *regions* or *départements* [19]. Each dataset has the same examples, with class values discretized into *high*, *medium*, and *low* using equal frequency discretization. They differ with respect to the information that has been extracted from the Semantic Web to be used as features in the prediction task. For *data*, all direct data values have been extracted from DBpedia for each region, for *type* all types for particular regions, and for *qrel* all qualified relations where the region

Table 2. Average rule length and (in parantheses) total number of rules and total number of conditions of the learned rules.

Dataset	Standard (., .)			$(h_{Lap}, .)$	
	h_{Lap}	h_m	h_{WRA}	q_{Lap}	q_{Prec}
coronary heart disease	2.73 (64/175)	5.03 (31/156)	3.24 (17/55)	9.78 (54/528)	10.26 (61/626)
brain ischemia	2,76 (46/127)	3.33 (33/110)	2.00 (13/26)	7.46 (56/418)	7.09 (55/390)
average (medical data)	**2.75**	**4.18**	**2.62**	**8.62**	**8.68**
deptmts-dbpedia-data	1.96 (45/88)	2.72 (29/79)	2.94 (16/47)	6.36 (42/267)	6.77 (43/291)
deptmts-dbpedia-qrel	1.13 (53/60)	2.85 (22/77)	10.80 (15/162)	21.75 (4/87)	23.33 (3/70)
deptmts-dbpedia-type	2.08 (12/25)	2.20 (10/22)	2.00 (7/14)	3.25 (12/39)	3.75 (12/45)
regions-dbpedia-data	1.54 (13/20)	1.89 (9/17)	2.13 (8/17)	3.00 (12/36)	3.42 (12/41)
regions-dbpedia-qrel	1.08 (13/14)	2.00 (7/14)	2.50 (6/15)	3.75 (4/15)	5.00 (3/15)
regions-dbpedia-type	2.36 (14/33)	2.71 (7/19)	2.71 (7/19)	3.92 (12/47)	4.17 (12/50)
regions-eurostat	1.46 (13/19)	1.67 (6/10)	1.67 (6/10)	3.36 (11/37)	3.58 (12/43)
regionWithLGD	1.06 (16/17)	2.30 (10/23)	4.33 (6/26)	6.60 (5/33)	7.75 (4/31)
average (region data)	**1.58**	**2.29**	**3.64**	**6.50**	**7.22**

occurs as an object. The datasets *withLGD* include information extracted from Linked Geo Data, and *eurostat* all data values from Eurostat. The latter two are only available for regions, whereas the first three are available for both regions and départements. All features generally have an open world semantics, i.e., the two values of a binary feature should be interpreted as *true* and *unknown*.

4.2   Rule Complexities

Table 2 shows the average rule lengths learned by three conventional heuristics, as well as two that use the Laplace-heuristic for selection, and an inverted heuristic for refinement. Our expectation is confirmed considering the two algorithms using inverted heuristics, (h_{Lap}, q_{Prec}) and (h_{Lap}, q_{Lap}), learn the longest rules by a fair margin. Generally, use of (h_{Lap}, h_{Lap}) and (h_{WRA}, h_{WRA}) will result in more general rules, where the former will learn many very short rules, and the latter will learn a considerably lower number of rules that exhibit similar length characteristics. Interestingly, the number of *rules* learned by the standard algorithm (h_m, h_m) is usually found between the two aforementioned algorithms, while the number of *conditions* is considerably higher, leading to a greater rule length average that still does not match the rule lengths produced by the two inverted variants.

4.3   Rule Coverage

Of course, learning longer rules by itself is not a noteworthy result. What makes it interesting is that these longer rules are nevertheless not more specific in the sense that they cover fewer examples. Table 3 shows the average coverage (number of examples explained by a single rule), including both false and true positives. Again, (h_{Lap}, h_{Lap}) is surpassed by both modified algorithms,

Table 3. Average number of examples (both true and false positives) covered by a rule.

Dataset	Standard (.,.)			$(h_{Lap},.)$	
	h_{Lap}	h_m	h_{WRA}	q_{Lap}	q_{Prec}
coronary heart disease	10.53	32.87	68.00	18.52	19.33
brain ischemia	37.11	82.09	155.85	44.14	46.71
average (medical data)	**23.82**	**57.48**	**111.93**	**31.33**	**33.02**
deptmts-dbpedia-data	5.09	13.41	32.38	6.62	6.79
deptmts-dbpedia-qrel	3.47	8.22	20.40	33.00	33.30
deptmts-dbpedia-type	42.25	54.80	72.71	42.42	42.42
regions-dbpedia-data	2.69	6.22	9.88	3.42	3.33
regions-dbpedia-qrel	2.61	6.00	7.17	8.00	8.67
regions-dbpedia-type	4.14	11.00	11.00	5.00	5.00
regions-eurostat	3.46	8.00	8.00	4.72	4.50
regionWithLGD	2.38	5.10	7.50	8.40	9.00
average (region data)	**8.26**	**14.09**	**21.13**	**13.95**	**14.13**

which use the same selection heuristic but an inverted refinement heuristic, by a fair margin w.r.t. average coverage. This behaviour was what we expected based on our previous results [20].

Weighted relative accuracy (WRA) is a heuristic that is commonly used in subgroup discovery but is known to overgeneralize in a predictive setting [10]. Consequently, (h_{WRA}, h_{WRA}) exhibits very large coverage values, with a high amount of both true and false positives, corresponding to its very short average rule length. A possible explanation is that especially in multiclass datasets, where usually $P << N$ for any given class, WRA tends to penalize the exclusion of true positives relatively early in the learning process, thus sticking with a relatively short best rule found w.r.t. selection heuristic. Algorithms with separated heuristics have the advantage of not evaluating entire rules with the same heuristic function used to evaluate single conditions, thus working bottom-up in the selection step, and top-down in the refinement step.

The m-estimate (h_m, h_m) is an interesting case, because w.r.t. to rule lengths, it lies between the traditional Laplace heuristics and the modified algorithms using split heuristics, while at the same time maintaining a high coverage without a large fraction of false positives. From the rule statistics alone, the m-Estimate seems to be a strong contender for the best rule *quality* (meaning high coverage *combined* with long descriptive rules) produced for subgroup discovery; it is thus necessary to take an in-depth look at the actual subgroups described by the rules of both (h_m, h_m) as well as the two split-heuristics variants (h_{Lap}, q_{Prec}) or (h_{Lap}, q_{Lap}) to determine the potential utility of inverted heuristics in subgroup discovery. We will highlight some of the differences further below in Sect. 4.5.

Table 4. Average number of rules covering a single example. Since every example must be covered by at least one rule, this value is always greater than or equal to one.

Dataset	Standard (.,.)			$(h_{Lap},.)$	
	h_{Lap}	h_m	h_{WRA}	$ч_{Lap}$	$ч_{Prec}$
coronary heart disease	2.84	4.28	4.86	4.2	4.95
brain ischemia	5.69	9.03	6.75	8.24	8.56
average (medical data)	**4.27**	**6.66**	**5.81**	**6.22**	**6.76**
deptmts-dbpedia-data	2.29	3.89	5.18	2.78	2.92
deptmts-dbpedia-qrel	1.84	2.22	3.06	1.32	1.00
deptmts-dbpedia-type	5.07	5.48	5.09	5.09	5.09
regions-dbpedia-data	1.35	2.15	3.04	1.57	1.54
regions-dbpedia-qrel	1.31	1.62	1.65	1.23	1.00
regions-dbpedia-type	2.23	2.96	2.96	2.31	2.31
regions-eurostat	1.73	1.85	1.85	2.00	2.08
regionWithLGD	1.41	1.89	1.67	1.56	1.33
average (region data)	**2.15**	**2.76**	**3.06**	**2.23**	**2.16**

4.4 Rule Overlap

We also evaluated the rule overlap, i.e., how many rules cover any given example on average (cf. Table 4). Every example is covered by at least one rule—either because the example was explicitly selected as a random seed example, or because one of the rules learned from a different seed example covered this example.

In direct comparison to their counterpart using the same *selection* heuristic, rules learned by $(h_{Lap}, ч_{Prec})$ or $(h_{Lap}, ч_{Lap})$ seem to overlap more (a characteristic also exhibited by (h_m, h_m) and (h_{WRA}, h_{WRA})), resulting in two or more rules covering the same example. This is especially the case with the medical datasets (*brain ischemia* and *coronary heart disease*). These datasets cover both a binary-class as well as a multiclass case and provide a sufficient number of samples, which is not the case with some of the region datasets, most notably *deptmts-dbpedia-qrel* and *regions-dbpedia-qrel*. The algorithms making use of inverted heuristics for the most part learned just a single rule for each class on these two datasets. This is due to the sparse encoding of these data, so that in many cases, an entire class can be covered with just one rule consisting of a conjunction of many zero values from such sparse attributes.

4.5 Example Rules

In this section we will compare some of the rules learned with conventional and with inverted heuristics. In order to reduce influence from a possibly bad random pick for the seed examples, we will always list the three best rules learned for a class as estimated by the rule selection heuristic. The conditions of the rules are shown in the order in which they have been learned, which also highlights some differences between the algorithms.

We will primarily focus on a comparison of (h_{Lap}, q_{Lap}) with (h_{WRA}, h_{WRA}) and (h_m, h_m) because the latter two seem to produce the best results among the conventional heuristics, and (h_{Lap}, q_{Lap}) and (h_{Lap}, q_{Prec}) behave very similarly, so that it is usually sufficient to look at the output of one of these two. A detailed semantic interpretation is beyond the scope of the paper and the expertise of its author, so we will focus on rule properties like rule length and coverage. With each rule, we will also show the number of positive and negative examples that are covered by this rule.

```
[17| 0] class 1 :- holst < 0.0001, vkgq = 1, ergfr = 1, ergrt = 1.
[17| 0] class 1 :- ergg < 180.0001, ergst < 0.2001, vkgq = 1,
                   ehoef >= 68, ergfr = 1.
[14| 0] class 1 :- holst < 0.2001, vkgq = 1, ecgfr >= 70, ergd >= 100.
```

(a) using (h_{Lap}, h_{Lap})

```
[40| 1] class 1 :- ergst < 0.2001, vkgq = 1, ergrt = 1, ergny = 1,
                   ecgpr = 1, ehoef >= 63, holst < 0.4001, ergfr = 1.
[36| 1] class 1 :- holst < 0.2001, vkgq = 1, ergrt = 1, ergny = 1,
                   ecgpr = 1, hight >= 163.
[35| 1] class 1 :- ergst < 0.2001, vkgq = 1, ergrt = 1, ergg < 200.000,
                   ergny = 1, ergfr = 1, ehoef >= 63, ecgpr = 1.
```

(b) using (h_m, h_m)

```
[45| 9] class 1 :- ergst < 0.2001, vkgq = 1, ergrt = 1, ergny = 1, ergfr = 1.
[43|11] class 1 :- ergst < 0.2001, ecgst = 1, ergrt = 1, ergny = 1,
                   ergfr = 1, holpr = 1, holfr = 1.
[42| 9] class 1 :- holst < 0.3001, vkgq = 1, ergny = 1, ergfr = 1, ergrt = 1.
```

(c) using (h_{WRA}, h_{WRA})

```
[32|0] class 1 :- vkgq = 1, ergkp = 1, ergny = 1, ergrt = 1,
                  hight >= 154, ergfr = 1, holst < 0.3001, ecgpr = 1,
                  holfr = 1, ehoef >= 65.
[28|0] class 1 :- ergst < 0.3001, vkgq = 1, ergny = 1, hdl >= 0.72,
                  ergfr = 1, ecgrt = 1, ecgpr = 1, fib < 4.5001,
                  vkghl = 1, holst < 0.2001, ecgst = 1, holrt = 1, ldl < 4.7601.
[25|0] class 1 :- ergst < 0.2001, vkgq = 1, ergny = 1, ergrt = 1,
                  ergfr = 1, ecgpr = 1, ergkp = 1, ecgrt = 1, holfr = 1,
                  ehoef >= 64, ua < 308.0001.
```

(d) using (h_{Lap}, q_{Lap})

Fig. 3. Three best rules learned for **class 1** in the *coronary heart disease* dataset.

Figure 3 shows the three best rules learned for a specific class in *coronary heart disease*. Not unexpectedly, it can be seen that the conventional heuristics position learn rules with different characteristics. Laplace has a preference for more specific, pure rules, whereas WRA has a preference for more general but possibly impure rules. This is consistent with previous results which showed that Laplace has a tendency to overfit, whereas WRA tends to over-generalize [10]. As the m-estimate may be viewed as a parametrized trade-off between these two extremes [5], it not surprisingly learns rules that have a somewhat higher coverage than those learned by Laplace, but less coverage than those learned with WRA, and, conversely, purer rules than those learned by WRA, but less pure than those of Laplace. The use of inverted heuristics (Fig. 3(d)) results in rules that have about the same coverage as those learned with the m-estimate, but are considerably longer and purer. Considering the lower coverage combined with the relatively short rules of the (h_{Lap}, h_{Lap}), the conventional Laplace heuristic cannot match the quality of its inverted variant.

```
[149| 0] ischemia :- b.i. < 60.0001.
[140| 0] ischemia :- b.i. < 70.0001, fibrin. >= 3.8.
[137| 0] ischemia :- b.i. < 75.0001, fibrin. >= 3.9.
```

(a) using (h_m, h_m)

```
[165| 4] ischemia :- b.i. < 70.0001.
[181|18] ischemia :- b.i. < 80.0001.
[158| 8] ischemia :- b.i. < 85.0001, fibrin. >= 3.7.
```

(b) using (h_{WRA}, h_{WRA})

```
[147| 0] ischemia :- rrrrdyast.. >= 70, fibrin. >= 2.8, b.i. < 60.0001.
[139| 0] ischemia :- age >= 58, rrrrdyast.. >= 80, b.i. < 60.0001.
[107| 0] ischemia :- rrrrdyast.. >= 80, fibrin. >= 3.5, b.i. < 65.0001, chol. >= 5.2.
```

(c) using $(h_{Lap}, \mathbf{q}_{Lap})$

Fig. 4. Three best rules learned for the class ischemia in the *brain ischemia* dataset including true and false positives.

Figure 4 shows another exemplary comparison, this time on the *brain ischemia* dataset. Again we obtain high coverage rules with both (h_m, h_m) as well as (h_{WRA}, h_{WRA}). The difference is that on this dataset, the algorithm variant $(h_{Lap}, \mathbf{q}_{Lap})$ appears to generate considerably more descriptive rules without significant coverage loss: the two best rules cover almost as many examples, yet they include attributes not referenced by the rules learned by the standard algorithms. All algorithms seem to associate the *Barthel index* (*b.i.*), a 0-100 scale to measure performance in activities of daily living, as well as the *fibrinogen* value (*fibrin.*), where healthy values are between 2.0 and 3.7, with a brain stroke. However, the rules found by $(h_{Lap}, \mathbf{q}_{Lap})$ appear to offer more insight. For instance, they refer to the diastolic blood pressure (denoted as *rrrrdyast..*, normally below 89 mmHg), the total cholesterol value (referred to as chol., normally between 3.6 and 5.0 mmol/L) as well as the age of the patient. None of these are referenced by the rules learned with standard configurations. More notably, the value ranges covered by the conditions found in the rules learned by our modified variant seem to be semantically interesting: according to the best rules found, a brain stroke is related to old age, a relatively high diastolic blood pressure and a cholesterol value out of the healthy value range. While we are not experts in the medical domain, this information does seem to carry some relevance, and it was not included by the unmodified algorithms.

```
[9|4] umemp_low :- landuse_total >= 16942.3,
                   RnD_personel_percent_of_act_pop >= 0.78.
```

(a) using (h_m, h_m) or (h_{WRA}, h_{WRA}) (both learn the same rule)

```
[5|0] umemp_low :- hospital_beds_per100000hab >= 760.1,
                   unemployment_rate_total < 7.8001,
                   landuse_total >= 17589.3.
[5|0] umemp_low :- unemployment_rate_total < 8.4001,
                   disposable_income < 17003.0001,
                   avg_annual_population_growth < 0.8901,
                   hospital_beds_per100000hab >= 807.7.
[4|0] umemp_low :- landuse_total >= 16942.3,
                   RnD_personel_percent_of_act_pop >= 0.78,
                   electricity_production_capacity_MWh < 1248.7001.
```

(b) using (h_{Lap}, q_{Lap})

Fig. 5. Best rules learned for class low unemployment in the *regions-eurostat* dataset including true and false positives.

In Fig. 5 we look at rules learned for the *low unemployment* class in the *regions-eurostat* dataset. Most notably, both weighted relative accuracy as well as the m-Estimate will learn a rule with relatively high coverage (including false positives) consisting of two conditions. Using (h_{Lap}, q_{Lap}) we eliminate an equal amount of false and true positives, reducing the total coverage somewhat, but adding *potentially* semantically interesting attribute queries. For instance, while all algorithms appear to associate a low unemployment rate with a large total land use as well as a certain percent of population working in *RnD* (possibly describing a metropolitan area), some additional info is included with the more specialized rules produced by the latter algorithm, such as a good medical infrastructure (above 8 hospital beds per 100 inhabitants) and information on power generation, disposable income and population growth. This seems plausible although it would require domain knowledge in order to evaluate it properly.

4.6   Inverted Heuristics and Sparse Data

A particularly interesting dataset is *deptmts-dbpedia-qrel*, where we can see that inverted heuristics learn rules that have 20 or more conditions whereas conventional heuristics learn rules with 10 or fewer conditions (Table 2). On the other hand, these rules cover considerably more examples than the other cases (Table 3). Because of this, all positive examples are often only covered with a single rule (Table 4).

The reason for this lies in the fact that this dataset has a very large number of attributes compared to the instance count $(A >> E)$, which results in some interesting properties of the output rules when using an inverted heuristic for refinement purposes. We have already shown that inverted refinement heuristics will cause the rule construction process to prefer *completeness* over *consistency*, i.e. such algorithm configurations will avoid excluding true positives early on, instead attempting to take smaller steps in coverage space. Because the aforementioned datasets are encoded in the WEKA [9] *sparse ARFF format* where

```
[33|0] unemp_low :- LiguE2Players = 0, YagoGeoEntity <= 98, FrenchPoliticians = 0,
                    owl#Thing = 0, OlympicGoldMedalistsForFrance = 0,
                    StoneBridges = 0, FrenchPrincesses = 0,
                    FrenchNaziCollaborators = 0, ExecutedFrenchPeople = 0,
                    _Feature = 0, Writer =0, StateGovernor = 0,
                    SmallForwards = 0, PeopleOfTheParisCommune = 0,
                    SSOfficers = 0, RugbyUnionProps = 0, PeopleFromTarbes = 0,
                    MediterraneanPortCitiesAndTownsInFrance = 0,
                    CommunesOfTerritoireDeBelfort = 0.
```

Fig. 6. A long rule learned for class **low unemployment** in the *deptmts-dbpedia-qrel* dataset using $(h_{\text{Lap}}, q_{\text{Lap}})$

missing values have to be explicitly encoded as such, a substantial number of attribute values will be zero for every data point. As these are then common to all instances of this class, inverted heuristics will tend to add them to the rule, as can be seen from the rule shown in Fig. 6. Each of these conditions will barely exclude any instances and can thus be steered towards maintaining *completeness* while only slowly improving *consistency*. This will result in substantially longer rules than their standard counterparts have learned. Learning such a rule will often cover the entire concept (positive class) with just one rule by excluding all false positives in tiny steps.

However, in this case, the semantic quality of these longer rules is indeed questionable because they essentially focus on listing all properties that are *not* known to be the case for a département with low unemployment rate. Many of them will also be the case for other regions, but this particular conjunction is only true for this group, as the high number of covered positive examples and the fact that no negatives are covered shows.

5   Conclusions

The main message that we would like to communicate with this position paper is that shorter rules are not necessarily preferable to longer rules. While previous work [20] has already demonstrated this in a quantitative evaluation for a predictive setting, this paper focusses on the descriptive quality of the found rules. Rules may have the same or very similar statistical properties (like amount of covered examples) but may vary considerably with respect to the number and nature of the conditions that are included in the rule formation, and, in consequence, with respect to their understandability.

On a more technical side, we were able to show that the use of inverted heuristics will produce longer descriptive rules in a variety of medical and Semantic Web datasets. Rules learned by *weighted relative accuracy*, a measure commonly used in subgroup discovery, show a somewhat higher coverage, but also include a substantial fraction of false positives. On the other hand, rules learned with inverted heuristics exhibit a substantially lower false positive count with only slightly lower total coverage and considerably longer rules. It is interesting to note that the *m-Estimate* would include only a very small number of false positives while covering more true positives than the rules created by the inverted algorithm variants, which always exclude all false positives.

We also observed a potential disadvantage of inverted heuristics on sparse datasets. Given their preference for maintaining completeness, these heuristics will often learn a single rule per class on sparse data, covering all true positives and no false positives, only by adding a large number of conditions that state properties that are not true for examples of this class, each excluding only a few false positives.

When taking a look at the semantics of the learned rules, the most interesting trait of inverted heuristics in subgroup discovery is that the most significant subgroups often query interesting attributes in addition to those included by the other algorithm configurations. This behaviour was most apparent concerning the *brain ischemia* medical dataset, where attributes and associated values that make sense in a semantic context (brain strokes being related to old age, high cholesterol and high blood pressure) were only found in the best rules learned by one of the inverted algorithm variants. To make a concluding statement concerning the usefulness of our approach in descriptive rule discovery, it would be necessary to conduct a proper evaluation from experts in the application domains, with the goal of estimating the semantic value of the attribute queries contained in the found rules.

Acknowledgements. We would like to thank Dragan Gamberger, Nada Lavrač, and Heiko Paulheim for letting us play with their data.

References

1. Bensusan, H.: God doesn't always shave with Occam's Razor - learning when and how to prune. In: Nédellec, C., Rouveirol, C. (eds.) Proceedings of the 10th European Conference on Machine Learning (ECML 1998), pp. 119–124 (1998)
2. Blumer, A., Ehrenfeucht, A., Haussler, D., Warmuth, M.K.: Occam's Razor. Inf. Process. Lett. **24**, 377–380 (1987)
3. Domingos, P.: The role of Occam's Razor in knowledge discovery. Data Min. Knowl. Discovery **3**(4), 409–425 (1999)
4. Fürnkranz, J.: Separate-and-conquer rule learning. Artif. Intell. Rev. **13**(1), 3–54 (1999)
5. Fürnkranz, J., Flach, P.A.: ROC 'n' rule learning - towards a better understanding of covering algorithms. Mach. Learn. **58**(1), 39–77 (2005)
6. Fürnkranz, J., Gamberger, D., Lavrač, N.: Foundations of Rule Learning. Springer, Heidelberg (2012)
7. Gamberger, D., Lavrač, N.: Active subgroup mining: a case study in coronary heart disease risk group detection. Artif. Intell. Med. **28**(1), 27–57 (2003)
8. Ganter, B., Wille, R.: Formal Concept Analysis - Mathematical Foundations. Springer, Heidelberg (1999)
9. Hall, M., Frank, E., Holmes, G., Pfahringer, B., Reutemann, P., Witten, I.H.: The weka data mining software: an update. SIGKDD Explor. **11**(1), 10–18 (2009)
10. Janssen, F., Fürnkranz, J.: On the quest for optimal rule learning heuristics. Mach. Learn. **78**(3), 343–379 (2010)

11. Kralj, P., Lavrač, N., Gamberger, D., Krstačić, A.: Contrast set mining through subgroup discovery applied to brain ischaemina data. In: Zhou, Z.-H., Li, H., Yang, Q. (eds.) PAKDD 2016. LNCS (LNAI), vol. 4426, pp. 579–586. Springer, Heidelberg (2007). doi:10.1007/978-3-540-71701-0_61

12. Kralj Novak, P., Lavrač, N., Webb, G.I.: Supervised descriptive rule discovery: a unifying survey of contrast set, emerging pattern and subgroup mining. J. Mach. Learn. Res. **10**, 377–403 (2009)

13. Lavrač, N., Kavšek, B., Flach, P., Todorovski, L.: Subgroup discovery with CN2-SD. J. Mach. Learn. Res. **5**, 153–188 (2004)

14. Michalski, R.S.: On the quasi-minimal solution of the general covering problem. In: Proceedings of the 5th International Symposium on Information Processing (FCIP 1969), pp. 125–128, Bled, Yugoslavia (1969)

15. Michalski, R.S.: A theory and methodology of inductive learning. Artif. Intell. **20**(2), 111–162 (1983)

16. Mitchell, T.M.: The Need for Biases in Learning Generalizations. Technical report, Computer Science Department, Rutgers University, New Brunswick, MA (1980)

17. Murphy, P.M., Pazzani, M.J.: Exploring the decision forest: an empirical investigation of Occam's Razor in decision tree induction. J. Artif. Intell. Res. **1**, 257–275 (1994)

18. Paulheim, H., Fürnkranz, J.: Unsupervised generation of data mining features from linked open data. In: Proceedings of the International Conference on Web Intelligence and Semantics (WIMS 2012) (2012)

19. Ristoski, P., Paulheim, H.: Analyzing statistics with background knowledge from linked open data. In: Proceedings of the 1st International Workshop on Semantic Statistics (SemStats-2013). CEUR workshop proceedings, Sydney, Australia (2013)

20. Stecher, J., Janssen, F., Fürnkranz, J.: Separating rule refinement and rule selection heuristics in inductive rule learning. In: Calders, T., Esposito, F., Hüllermeier, E., Meo, R. (eds.) ECML PKDD 2014. LNCS (LNAI), vol. 8726, pp. 114–129. Springer, Heidelberg (2014). doi:10.1007/978-3-662-44845-8_8

21. Stumme, G., Taouil, R., Bastide, Y., Pasquier, N., Lakhal, L.: Computing iceberg concept lattices with Titanic. Data Knowl. Eng. **42**(2), 189–222 (2002)

22. Webb, G.I.: Further experimental evidence against the utility of Occam's Razor. J. Artif. Intell. Res. **4**, 397–417 (1996)

23. Wille, R.: Restructuring lattice theory: an approach based on hierarchies of concepts. In: Rival, I. (ed.) Ordered Sets, pp. 445–470. Reidel, Dordrecht-Boston (1982)

24. Zaki, M.J., Hsiao, C.J.: CHARM: an efficient algorithm for closed itemset mining. In: Grossman, R.L., Han, J., Kumar, V., Mannila, H., Motwani, R. (eds.) Proceedings of the 2nd SIAM International Conference on Data Mining (SDM-02), pp. 457–473. Arlington, VA (2002)

Exploiting Spatial Correlation of Spectral Signature for Training Data Selection in Hyperspectral Image Classification

Annalisa Appice[1,2(✉)] and Pietro Guccione[3]

[1] Dipartimento di Informatica, Università Degli Studi di Bari Aldo Moro,
via Orabona, 4, 70126 Bari, Italy
annalisa.appice@uniba.it
[2] Consorzio Interuniversitario Nazionale per l'Informatica - CINI, Bari, Italy
[3] Dipartimento di Ingegneria Elettrica ed Informazione, Politecnico di Bari,
via Orabona, 4, 70125 Bari, Italy
guccione@poliba.it

Abstract. Supervised classification is commonly used to produce a thematic map from hyperspectral data. A classifier is learned from training pixels and used to assign a known class (theme) to each pixel (imagery data example). However, supervised classification requires a sufficient number of representative training samples to be accurate. These samples are usually selected by expert visual inspection or field survey. Consequently, collecting representative samples is a very challenging task due to the high cost of true sample selecting and labeling. This paper introduces an unsupervised learning schema, where the most suitable pixels to train the classifier are selected via image segmentation. This reduces the expert effort required for choosing training samples. In our proposal, clustering is performed by accounting for the property of spatial correlation of pixel-level spectral information, so that thematic objects can be retrieved via unsupervised learning and representative training data can be sampled throughout clusters. Experimental results highlight that the pixel classification accuracy outperforms the results of a random selection scheme.

1 Introduction

Remote sensing is attracting growing interest in applications such as urban planning, agriculture, forestry and monitoring [1]. Recent advances in hyperspectral imaging technology allow nowadays the simultaneous measurement of hundreds of spectral bands for each image pixel. This high spectral resolution increases the possibility of more accurate classification of materials of interest in the spectral domain. In this scenario, hyperspectral image classification can be performed, in order to produce thematic maps from hyperspectral data.

A thematic map represents the Earth's surface objects. Its construction implies that themes or categories, selected for the map, are distinguished in the remotely sensed image. Classification assigns a known class (theme) to each pixel

© Springer International Publishing Switzerland 2016
T. Calders et al. (Eds.): DS 2016, LNAI 9956, pp. 295–309, 2016.
DOI: 10.1007/978-3-319-46307-0_19

(imagery data example). Every pixel is expressed with a vector space model that represents the spectral signature as a vector of numeric features (namely spectral features) and is also associated with a specific position in a uniform grid, which describes the spatial arrangement of the scene. It is assigned a certain (possibly unknown) spectral response, i.e. class label.

The automatic classification of hyperspectral data is not a trivial task. It is made complex by the high cost of true sample labeling coupled with the high number of spectral features. The human-supervised effort needed to collect only few labeled imagery pixels, properly distributed among the classes, makes the definition of the proper training set for learning an imagery classifier still an open challenge [19]. On the other hand, the low number of collected ground truth labels, compared to the high number of spectral features (also known as Hughes's phenomenon [11]), is not always sufficient for a reliable estimate of the classifier parameters.

In the last decades, Support Vector Machines (SVMs) have been widely investigated in hyperspectral classification, in order to deal with Hughes's phenomenon. In fact, they address large feature spaces and produce solutions from sparsely labeled data [15]. In alternative, dimensionality reduction techniques have been adopted, in order to mitigate Hughes's phenomenon as the dimensionality of the spectral data is high [23]. In any case, in these studies, training pixels are still picked randomly over the remotely sensed data. Therefore, they can lie in non interesting areas and consequently some classes can be missed. Even the most recent studies, that investigate active learning in hypersectral classification (e.g. [17,18]), start learning from an initial pool of randomly selected, initial training samples. However, from the classifier point of view, the most representative samples should be smartly identified and labeled by the expert, in order to construct training set where all classes are appropriately represented. At the same time, the expert action is expected to be minimized in the labeling of the training samples.

To the best of our knowledge, the problem of automatically and smartly selecting optimal training samples for learning an accurate classifier has attracted a few attention in remotely sensed imaging literature. Rajadell et al. [20] have recently authored a seminal study in this field. This study presents an unsupervised learning scheme, namely clustering, to sample the representative pixels, that are manually labeled from the expert and used to learn the classifier. A clustering model is constructed on the spectral signature of a remotely sensed image as a means to identify the imagery pixels that potentially belong to the same (unknown) theme. By accounting for the discovered clustering model, training pixels can be selected to be properly distributed among clusters. Under the hypothesis that the clustering model is accurate (i.e. every cluster actually groups pixels belonging to the same (unknown) theme), selecting training pixels throughout clusters may contribute to properly distribute training sample among the various thematic objects hidden in the remotely sensed data.

In this paper, we advance on this idea by leveraging the power of the expected spatial correlation of the spectral signature, in order to improve the clustering accuracy and, consequently, the ability of selecting properly distributed, suitable training pixels. In particular, we propose to accommodate the spatial autocorrelation of the spectral signature into the clustering process as, in this way, new accuracy can be gained when detecting thematic objects over the remotely sensed data.

The paper is organized as follows. The next section illustrates the motivations and contributions of this study. Section 3 introduces basic concepts of this study, while Sect. 4 illustrates the proposed unsupervised learning scheme. Section 5 describes the data sets, the experimental setup and reports the results. Finally, in Sect. 6 some conclusions are drawn and future work is outlined.

2   Motivations and Contributions

The spatial correlation of the spectral signature refers to the relation (or dependence) between spectral signatures of pixels due to their spatial proximity. Intuitively, spatial correlation means that the features for a specific pair of points are more (less) similar than would be expected for a random pair of points [12]. In the case of hyperspectral images of geographical areas, spatial correlation exists in the positive form, as there is a slowly progressive spatial variation in the spectral signature [13]. This means that by picturing the spatial variation of the observed features in a map, we may observe regions where the distribution of values is smoothly continuous, with some boundaries possibly marked by sharp discontinuities. These discontinuities are due to the bounds of the thematic objects. The analysis of the property of spatial correlation of spectral data poses specific issues.

One issue is that most of the models that represent and learn data with spatial autocorrelation are based on the assumption of spatial stationarity. This means that possible significant variabilities in correlation dependencies throughout the sensed space are overlooked. The variability could be caused by a different underlying latent structure of the space, which varies among its portions in terms of scale of values or density of measures. As pointed out by [2], when correlation varies significantly throughout space, it may be more accurate to model the dependencies locally rather than globally.

Another issue is that the spatial correlation analysis is frequently decoupled from the multivariate analysis. In this case, a learning process accounts for the spatial correlation of univariate data, while dealing with distinct variables separately. Ignoring complex interactions among multiple variables may overlook interesting insights into the correlation of potentially related variables at any site [4]. Based upon this idea, a multivariate view of the concept of spatial correlation is presented in [3]. For each variable, the spatial dependence of the measured univariate data is computed as the variance of the local indicator of the spatial correlation. Hence, the multivariate analysis is performed by exploring the mean of the variance of the local spatial correlation computed on the separate variables.

In this paper, we develop an approach to modeling non stationary spatial correlation of hyperspectral data by using partition-based clustering techniques.

Clusters of pixels that share a similar spectral signature at nearby locations are identified. During clustering, the spatial correlation is coupled with a multivariate analysis by accounting for the spatial dependence of data and their multivariate dissimilarity, simultaneously [8]. This is done by minimizing a cluster cost function of the local indicators of the spatial correlation computed for hyperspectral data of pixels partitioned in the same cluster. Finally, training pixels are sampled to be properly distributed among these clusters.

The specific contributions in this paper are:

1. the investigation of the property of spatial autocorrelation in the hyperspectral signature of remotely sensed data;
2. the development of an unsupervised example selection schema that accommodates a local indicator of the spatial autocorrelation property of spectral signature into the clustering process as, in this way, more representative training examples can be chosen throughout the detected clusters to be manually labeled for supervised classification.

3   Basic Concepts

Introductory concepts of this study include hyperspectral data, local indicators of spatial autocorrelation and partition-based clustering.

3.1   Hyperspectral Data

Let $\mathcal{D}$ be a *hyperspectral imagery dataset*, that is, a set of pixels (examples). Every pixel represents a region of around a few square meters of the Earth's surface (i.e. it is a function of the sensor's spatial resolution). It is associated with the spatial coordinates XY, as well as with the m-dimensional vector of spectral features $\mathbf{S} = S_1, S_2, \ldots, S_m$. Every spectral feature S_i is a numeric feature that expresses how much radiation is reflected, on average, across the pixel region, at the i-th band of the considered spectral profile. Pixels of $\mathcal{D}$ are labeled according to an unknown target function, whose range is a finite set of classes $C = \{C_1, C_2, \ldots, C_k\}$. Every class C_i represents a distinct theme (i.e. type of Earth's surface). In general, pixels are equally distributed in space over a regular grid, so that a hyperspectral dataset can be represented by a matrix. Thus, the spatial coordinate X is associated with the row index, while the spatial coordinate Y is associated with the column index of the matrix. Based on this premise, let $p(x, y)$ be a pixel located at the (x, y) row-column position of the imagery matrix. A *spatial neighborhood* is a set of pixels q (task-relevant pixels) surrounding p (target pixel) in the matrix. In the imagery analysis literature, spatial neighborhoods frequently have a square shape [19], although alternative shapes like a circle or a cross can be also considered. Formally, let R be a positive, integer-valued radius, the *square-shaped* spatial neighborhood $\mathcal{N}(p, R)$ of pixel p is the set of imagery pixels $q(x + I, y + J)$, so that $-R \leq I, J \leq +R$.

3.2   Local Indicators of Spatial Autocorrelation

Local indicators look for "local patterns" of spatial dependence within the study region (see [5] for a survey). They return one value for each sampled location of a variable; this value expresses the degree to which that location is part of a spatial cluster. Widely speaking, a local indicator of spatial autocorrelation allows us to discover deviations from global patterns of spatial association, as well as hot spots like local clusters or local outliers. Several local indicators of spatial autocorrelation are formulated in the literature. In particular, the standardized Getis and Ord local GI* [10] is a local indicator of spatial autocorrelation, which has gained wide acceptance in the literature coupled with the cluster analysis [3,10]. It is formulated as follows:

$$GI^*(i) = \frac{1}{\sqrt{\frac{S^2}{n-1}\left(n\sum_{j=1,j\neq i}^{n}\lambda(ij)^2 - \Lambda(i)^2\right)}}\left(\sum_{j=1,j\neq i}^{n}\lambda_{ij}z(j) - \bar{z}\,\Lambda(i)\right), \quad (1)$$

where $\lambda(ij)$ is a spatial (Gaussian or bi-square) weight between the locations of i and j, $\Lambda(i) = \sum_{j=1,j\neq i}^{n}\lambda(ij)$ and $S^2 = \dfrac{\sum_{j=1}^{n}(z(j)-\bar{z})^2}{n}$. A positive value for $GI^*(i)$ indicates clusters of high values around i, while a negative value for $GI^*(i)$ indicates clusters of low values around i.

3.3   Partition-Based Clustering

A partition-based clustering algorithm can be considered, in order to partition n imagery pixels around k representative medoids (i.e. representative pixels of the image). These medoids are selected so that total dissimilarity of all pixels to their nearest medoid is minimal. As a partition-based algorithm, we consider PAM (Partitioning Around Medioids, see [21,24] for details). Our choice is motivated by the fact that it accepts dissimilarity data (i.e. a dissimilarity matrix collecting the dissimilarities computed between each pair of pixels), which are the sole required data input to determine the clusters. It uses a greedy search which may not find the optimum solution, but it is faster than exhaustive search. It is robust to the presence of noise and outliers. It includes an algorithm to select initial medoids. Therefore, unlike partition-based clustering algorithms (such as k-means and k-medoids), PAM doe not need initial random guesses for the cluster centers, at the cost of a mild additional complexity [16].

More specifically, the considered algorithm inputs: an integer k, that is, the number of clusters to discover and a $|n| \times |n|$ dissimilarity matrix D, where $\mathrm{diss}(p_i, p_j) = \mathrm{diss}(p_j, p_i)$ measures the "difference" or dissimilarity between pixels $p_i \in \mathcal{D}$ and $p_j \in \mathcal{D}$. It outputs a clustering pattern $\mathcal{P}(\mathcal{C})$, that is, a set of

k clusters $C_1, C_2, \ldots, C_k$, such that: $\emptyset \notin \mathcal{P}(\mathcal{C})$, $\bigcup_{C_i \in \mathcal{P}(\mathcal{C})} C_i = \mathcal{D}$ and $\forall\ C_i, C_j \in \mathcal{P}(\mathcal{C})$, with $i \neq j$, then $C_i \cap C_j = \emptyset$.

In particular, PAM can be used to partition the geo-located pixels $\mathcal{D}$ around a subset of medoid pixels $\mathcal{M} = \{m_1, \ldots, m_k\} \subset \{p_1, p_2, \ldots, p_n\} = \mathcal{D}$, which minimize the objective function:

$$\sum_{i=1}^{n} \min_{\ell=1,2,\ldots,k} \mathrm{diss}(p_i, m_\ell). \tag{2}$$

The search for the medoid set $\mathcal{M}$ takes the form of the steepest ascent hill climber, beginning with a deterministic building phase in which a "reasonable" initial set of k medoids is constructed. In each iteration of the subsequent swap phase, PAM attempts to improve the set of selected medoids and, therefore, to increase the quality of the clustering. This is done by selecting pairs $(p_i, m_j) \in (\mathcal{D} - \mathcal{M}) \times \mathcal{M}$ that produce the best decrease in the objective function when their roles are switched. Each $p_i \in \mathcal{D}$ is assigned to the cluster corresponding to the nearest medoid $m_l \in \mathcal{M}$.

4   Learning Process

We describe a two-phased learning process (see Fig. 1) that comprises a spatial-aware unsupervised phase (clustering) and a supervised phase (classification). The entire process inputs the unclassified hyperspectral image $\mathcal{D}$, the number of expected classes k and the percentage $p\%$ of imagery pixels of $\mathcal{D}$ to be manually labeled by the expert for the inductive process. The dataset $\mathcal{D}$ is the set on n pixels. For each pixel, both the spatial coordinates XY and the spectral signature $\mathbf{S}$ are known (see details in Sect. 3.1); the class C is unknown. The process outputs a classified map of the entire image, where each imagery pixel is assigned to a specific class of the class set C.

Initially, the unsupervised phase is performed, in order to identify a few *representative* training samples, which are manually labeled by the expert. It is four-stepped:

1. For every imagery pixel $p_i \in \mathcal{D}$, for every spectral attribute $S_h \in \mathbf{S}$, the local standardized Getin and Ord indicator of spatial autocorrelation, namely $GI^*(p_i, S_h)$, is computed according to Formula 1 (see the definition in Sect. 3.2). To explore the spatial autocorrelation structure of the spectral signatures, we consider square neighborhoods with radius R (see the definition in Sect. 3.1) and define the spatial weights λ_{ij} (see Formula 1) as follows:

$$\lambda_{ij} = \begin{cases} \frac{1}{d(p_i, p_j)} & if\ p_j \in \mathcal{N}(p_i, R) \\ 0 & otherwise \end{cases}, \tag{3}$$

where $d(p_i, p_j)$ is the Euclidean distance computed between spatial coordinates of imagery pixels p_i and p_j (see details in Sect. 3.1). We note that this

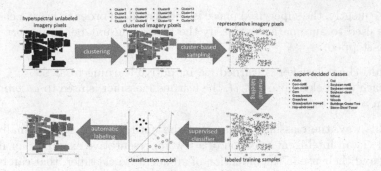

Fig. 1. Two-phased learning schema: unsupervised phase (clustering) to identify representative training samples, which are manually labeled by the expert, and supervised phase (classification) to learn a classifier, which is used to automatically label the imagery pixels.

weighting schema, that is commonly used in spatial data mining, is coherent with the Tobler's first Law of Geography [25] as it contributes to consider the near neighbors more autocorrelated than the distant ones.

2. For every pair of pixels (p_i, p_j), the dissimilarity $\mathrm{diss}(p_i, P_j, GI^*)$ is computed as the Eucludean distance between the local indicator of the spatial autocorrelation of the spectral signatures associated to both p_i and p_j, respectively. Formally

$$\mathrm{diss}(p_i, p_j, GI^*) = \sum_{S_h \in \mathbf{S}} (GI^*(p_i, S_h) - GI^*(p_j, S_h))^2 \qquad (4)$$

We note that, in this way, we are able to construct the $n \times n$ spatial-aware dissimilarity matrix $\mathsf{D}_{\mathsf{GI}^*}$, where $\mathsf{D}_{\mathsf{GI}^*}(i, j) = \mathrm{diss}(p_i, p_j, GI^*)$.

3. The PAM algorithm (see details in Sect. 3.3) processes the spatial-aware dissimilarity matrix $\mathsf{D}_{\mathsf{GI}^*}$, in order to group n imagery pixels of $\mathcal{D}$ into k clusters $\mathcal{P}(\mathcal{C}) = \{\mathcal{C}_1, \mathcal{C}_2, \ldots, \mathcal{C}_k\}$. As the clustering phase is performed on the local indicators of the spatial autocorrelation of the spectral signature of the imagery pixels in $\mathcal{D}$, it reasonable that the detected clusters would group pixels exhibiting similar signatures at close locations.[1]

4. For each detected cluster $\mathcal{C}_i \in \mathcal{P}(\mathcal{C})$, $p\%$ of the imagery pixels clustered in $\mathcal{C}_i$ are randomly sampled. These pixels are *manually* labeled by the expert, that assigns each sampled pixel $p_i \in sample(\mathcal{C}_i, p\%)$ to one class of $C_i \in C$. Therefore, this operation allows us to construct the labeled training set $\mathcal{L} \subseteq \mathcal{D} = \bigcup_{\mathcal{C}_i \in \mathcal{P}(\mathcal{C})} randomSample(\mathcal{C}_i, p\%)$, that is spanned on $\mathbf{S} \times C$. The unlabeled set is $\mathcal{U} = \mathcal{D} - \mathcal{L}$, that is spanned on $\mathbf{S}$.

---

[1] This expected property of the presented clustering procedure is, empirically, investigated in Sect. 5.2 of this study.

Subsequently, the supervised phase is performed, in order to learn a classifier, which is used to automatically classify the remaining unlabeled imagery pixels. It is two-stepped:

1. A multi-class classifier is learned via inductive learning from $\mathcal{L}$.
2. For each unlabeled pixel $p_i \in \mathcal{U}$, the learned classifier is used to *automatically* assign a class to p_i.

In this way, the classification map of the entire image can be, finally, constructed via an intelligent selection of a few representative examples for manual labeling and their use in the induction of an accurate classifier, that can be used to derive an automatic classification of the remaining pixels.

5   Empirical Study

The presented learning process is evaluated by considering two benchmark hyperspectral images, namely Indian Pines and Pavia University (http://www.grss-ieee.org/community/technical-committees/data-fusion/data-sets/). These data sets are selected for the following reasons: (1) They have a high spatial resolution. (2) They contain rich spectral information (100–200 bands) and a high number of classes (9–16 classes). (3) They correspond to different scenarios. (4) Ground truths are available for these data. Additionally, they are considered benchmark data in hyperspectral image classification (e.g. [19]).

The empirical study is performed, in order to seek answers to the following questions:

1. Is clustering performed by accounting for the dissimilarity defined on the local spatial autocorrelation of the spectral signature more accurate than clustering performed by accounting for the dissimilarity defined on the spectral signature (see Sect. 5.2)?
2. How does the accuracy of the classification change by deciding to sample training pixels randomly throughout the imagery clusters rather than randomly throughout the full image (see Sect. 5.3)?

Before we proceed to present the empirical results (see Subsects. 5.2 and 5.3), we provide a description of the datasets used (see Subsect. 5.1).

5.1   Hyperspectral Data

AVIRIS Indian Pines was obtained by the Airborne Visible Infrared Imaging Spectrometer (AVIRIS) sensor over the Indian Pines region in Northwestern Indiana in 1992. The image contains 220 spectral bands, but 20 spectral bands have been removed due to the noise and water absorption phenomena. The spatial resolution is of 20 m and the spatial size is of 145×145 pixels, which are classified into 16 mutually exclusive classes (see Fig. 2(a)). This data set represents a very challenging land-cover classification scenario, in which the primary crops of the area (mainly corn and soybeans) were very early in their growth cycle, with only about

5 % canopy cover [19]. Discriminating among the major crops under these circumstances can be a very difficult task. This scenario is also made more complex by the imbalanced number of available labeled pixels per class.

ROSIS Pavia University was obtained by the Reflective Optics System Imaging Spectrometer (ROSIS) sensor during a flight campaign over the Engineering School at the University of Pavia, in 2003. Water absorption bands were removed, and the original 115 bands were reduced to 103 bands. It has a spatial resolution of 1.3 m. The image has a spatial size of 610×340 pixels, which are classified into 9 classes (see Fig. 2(d)).

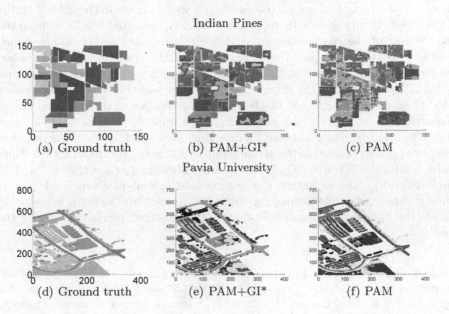

Fig. 2. The AVIRIS data of Indian Pines and the ROSIS data of Pavia University: ground truths ((a) and (d)), clusters detected by PAM+GI* ((b) and (e)) and clusters detected by PAM ((c) and (f)). Clusters are detected with number of clusters k equal to the actual number of distinct classes in the ground truths of the image (i.e. $k = 16$ for Indian Pinese and $k = 9$ for Pavia University).

5.2   Clustering Performance Analysis

We compare the quality of clusters discovered by the spatial clustering (denoted as PAM + GI*) described in Sect. 4 and dealing with the property of spatial autocorrelation with the quality of clusters discovered by the baseline, traditional clustering (denoted as PAM) performed by neglecting the spatial autocorrelation. In this study, for each pixel p, for each spectral attribute S, the local standardized Getis and Ord indicator $GI^*(p, S)$ is computed on the square-shaped spatial neighborhood $\mathcal{N}(p, R)$ with $R = 5$. On the other hand, the competitor PAM processes dissimilarity data that represent the Euclidean distance straightly computed between the imagery spectral signatures.

The quality of clusters is measured in terms of clustering purity [14]. This metric is computed by assigning each cluster to the class that is most frequent in the cluster, and then the accuracy of this assignment is measured by counting the number of correctly assigned pixels and dividing by number of imagery pixels. Purity is measured in the range [0,1]. The higher the purity, the more accurate the clustering model.

The clustering model is discovered by ranging the number of detected clusters between k, $k-1$ and $k+1$ with k the number of actual classes in the ground truths of the image. Purity results are reported in Table 1. Collected results confirm the effectiveness of the spatial-aware methodology, presented in this study, which accounts for the local spatial autocorrelation of the spectral signatures when clustering the imagery pixels. In particular, the purity of the clusters detected by PAM+GI* is, in general, greater than the purity of the clusters detected by PAM. This result is confirmed by the spatial distribution of the clustered pixels. In particular, we analyze the clustering maps built by both PAM+GI* (see Figs. 2(b) and (e)) and PAM (see Figures Figs. 2(c) and (f)) when the detected number of cluster is equal to the actual number of classes ($k = 16$ for Indian Pines and $k = 9$ for Pavia University) in the dataset. These maps show that PAM+GI* can, effectively, take advantage of the presented spatial-aware methodology. It gains accuracy when discriminating objects of interest on the map, by reducing visibly the salt-and-pepper distribution of the clustered pixels throughout the image.

Table 1. Clustering analysis (AVIRIS data of Indian Pines and ROSIS data of Pavia University): purity of clusters discovered by PAM+GI* and PAM (baseline) by varying the number of detected clusters between $k - 1$, k (actual number of distinct classes in the ground truths of the image) and $k + 1$ with $k = 16$ for Indian Pines and $k = 9$ for Pavia University. The highest purity is in bold.

Algorithm	Indian Pines			Pavia University		
	k-1	k	k+1	k-1	k	k+1
PAM+GI*	**0.596**	**0.583**	**0.590**	0.707	**0.753**	**0.755**
PAM	0.536	0.568	0.554	**0.747**	0.744	0.730

5.3 Classification Performance Analysis

We compare the performance of the classifier induced with the cluster-based construction of the training set to the performance of the classifier induced with the baseline random-based construction of the training set. For the cluster-based construction of the training set, we consider the imagery clusters discovered by

PAM+GI* with k (number of clusters) equal to the number of real classes in the considered dataset. We randomly select $p\%$ representative training pixels throughout the discovered clusters (see details in Sect. 4). Differently, for the random-based construction of the training set, we randomly select $p\%$ representative training pixels throughout the entire image (this corresponds to consider all imagery pixels grouped in a single cluster).

We perform this analysis by generating training labeled sets with $p\%$ varying among 2.5 %, 5 %, 10 % and 20 %. Ten training trials have been generated with both the cluster-based approach and the random-based approach for each value $p\%$. For each trial, the classification step is performed by resorting to the inductive Support Vector Machine (SVM) [7]. This choice is motivated by several studies reported in the literature (e.g. [6,9,19]), which show that inductive SVMs are applied to hyperspectral image classification with great success, outperforming several other inductive classifiers. As the hyperspectral classification problem is a multi-class problem, we learn multi-class SVMs with the "one-against-all" strategy. SVMs are learned with the Gaussian kernel rule, while parameters are optimally selected according to a grid-search method and a three-fold cross validation of the labeled set.

Table 2. Classification analysis (AVIRIS data of Indian Pines and ROSIS data of Pavia University): average accuracy (AA), overall accuracy (OA) and Cohen's kappa coefficient κ (mean ± standard deviation) of multi-class SVMs learned by considering the 2.5 %, 5 %, 10 % and 20 % of the ground truth as the labeled set of the inductive phase. Representative training samples are generated for 10 trials by resorting to both the random-based approach (RANDOM) and the cluster-based approach (PAM+GI*). Accuracy metrics are computed on the unlabeled part of each trial. The highest accuracy is in bold.

	Labeled set%	Indian Pines		Pavia Univerisity	
		RANDOM	PAM+GI*	RANDOM	PAM+GI*
AA	2.5 %	.5103 ± 0185	**.5901** ± .0187	.8473 ± .0213	**.8590** ± .0144
	5 %	.6119 ± .0178	**.6831** ± .0120	.8778 ± .0030	**.8834** ± .0065
	10 %	.7124 ± .1400	**.7942** ± .0218	.8866 ± .0040	**.8979** ± .0011
	20 %	.8163 ± .0100	**.8527** ± .0084	.8942 ± .0019	**.9024** ± .0045
OA	2.5 %	.6361 ± .0097	**.6956** ± .0079	.8843 ± .0077	**.9010** ± .032
	5 %	.7190 ± .0189	**.7621** ± .0066	.9072 ± .0046	**.9154** ± .0010
	10 %	.8062 ± .0051	**.8218** ± .0043	.9159 ± .0028	**.9241** ± .0010
	20 %	.8573 ± .0032	**.8671** ± .0019	.9254 ± .0023	**.9306** ± .0017
κ	2.5 %	.6108 ± .0261	**.6377** ± .0144	.8452 ± .0110	**.8676** ± .0042
	5 %	.6775 ± .0213	**.7271** ± .0078	.8759 ± .0063	**.8871** ± .0014
	10 %	.7844 ± .0088	**.7878** ± .0099	.8877 ± .0037	**.8988** ± .0014
	20 %	.8370 ± .0036	**.8482** ± .0022	.8984 ± .0009	**.9076** ± .0023

The classification performance is evaluated in terms of accuracy of the subsequent classification process. The accuracy of the classification phase is measured in terms of overall accuracy (OA), average accuracy (AA) and Cohen's kappa coefficient (κ) [22] computed on the unlabeled part. The metrics are averaged on the various trials and standard deviation is computed. These metrics are selected as they are, usually, considered by the hyperspectral image classification community. For each accuracy metric (OA, AA, as well as κ), the higher the metric, the more accurate the classifier. In particular, these metrics are defined in terms of elements x_{ij} of the error matrix associated to the classified imagery pixels. Each element x_{ij} denotes the number of imagery pixels with ground truth c_j, which are labeled with class c_i. C is the number of distinct classes in the image. Let us consider $x_i = \sum_i x_{ii}$, $x_{i+} = \sum_j x_{ij}$, $x_{+j} = \sum_i x_{ij}$ and $N = \sum_i \sum_j x_{ij}$.

Then, $OA = \frac{x_i}{N}$ $AA = \frac{1}{C} \sum_i \frac{x_{ii}}{X_{+i}}$ $\kappa = \dfrac{N x_i - \sum_i x_{i+} x_{+i}}{N^2 - \sum_i x_{i+} x_{+i}}$.

Accuracy results are reported in Table 2 for all values of $p\%$, while the classification maps constructed in one trial with $p\% = 10\%$ are shown in Figs. 3(b) and (c) for Indian Pines and Figs. 3(e) and (f) for Pavia University. Collected results highlight that the classification accuracy achieved by using the spatial-aware cluster-based selection approach, in order to smartly sample the

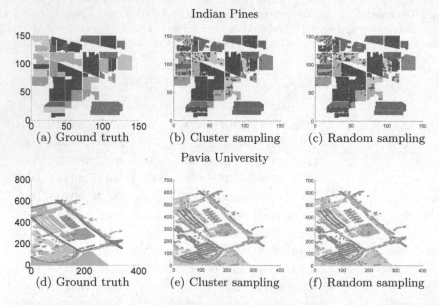

Fig. 3. The AVIRIS data of Indian Pines and the ROSIS data of Pavia University: ground truths ((a) and (d)), classification maps computed with the cluster-based strategy ((b) and (e)) and random-base strategy ((c) and (f)).Classification maps are constructed with training sets collecting $p\% = 10\%$ of imagery pixels).

representative training samples outperforms the accuracy of classifiers learned from fully random sampled labeled sets. This accuracy performance can be observed independently on the labeled set size, although the ability of smartly selecting representative samples for labeling contributes to achieve the higher classification performance in the most critical cases, i.e. when fewer samples are labeled by the expert for the inductive process. This is confirmed by the analysis of the ratio of the AA of the classifier computed with the cluster-based approach on the AA of the classifier computed with the random-based approach (see Fig. 4(a) for Indian Pines and Fig. 4(b) for Pavia University). We note that the lower the size of the labeled set (i.e. 2.5 % sampled examples for training), the higher the ratio (i.e. the accuracy improvement).

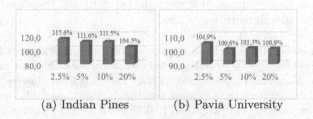

(a) Indian Pines   (b) Pavia University

Fig. 4. The AVIRIS data of Indian Pines and the ROSIS data of Pavia University: the ratio (Y axis) of the AA of the classifier computed with the cluster-based approach on the AA of the classifier computed with the random-based approach by varying the labeled set size (X axis).

6   Conclusion

This paper addresses the challenging problem of smartly sampling the optimal representative training pixels of an hyperspectral image. Sampled pixels are manually labeled by the expert and processed via an inductive learning process, in order to automatically derive accurate classifications of unlabeled pixels in the image. We describe a learning approach that, novelly, accounts for the property of spatial autocorrelation of the spectral data, in order to sample the representative training samples. In particular, we propose to resort to unsupervised learning, namely clustering, in order to detect the thematic objects of a map. Representative training pixels are then sampled throughout the discovered clusters/objects, in order to avoid missing classes in the training set. To improve the quality of the discovered thematic objects, clusters are computed by considering the local indicators of the spatial autocorrelation (i.e. stardardized Getis and Ord) of the spectral data. This allows us the detect clusters of pixels, which exhibit similar spectral signature at near locations, that is, a plausible characterization of thematic objects. The empirical study proves that the consideration of the spatial autocorrelation improves the accuracy of detected clusters and, in general, the presented sampling methodology outperforms the traditional random sampling methodology, which is, commonly, adopted in the hyperspectral image classification literature. Some directions for further work are still to be explored. Choosing the correct number of classes(/clusters) to be considered during the unsupervised learning phase is an open problem. In the present study, we have assumed that the number of classes is known apriori. Estimating this number in an unsupervised setting requires further investigations. In addition, the smart selection of examples for labeling is a crucial step of active learning, where this selection is repeated during the iterative process. We can explore new active learning algorithms that apply a combination of clustering and local indicators of the spatial correlation of the spectral signature, in order to determine representative examples for the active labeling. On the other hand, it would be interesting to investigate the use some big data technologies (e.g. MapReduce) in the design of the proposed algorithm, in order to be able to evaluate the performances when large volume of hypersepctral data are processed via specific parallel processing architectures.

Acknowledgments. Authors thank Giuseppe Lorusso for his support in developing the algorithm presented. This work is carried out in partial fulfillment of the research objectives of the European project "MAESTRA - Learning from Massive, Incompletely annotated, and Structured Data (Grant number ICT-2013-612944)" funded by the European Commission, as well as the ATENEO 2012 project "Mining Complex Patterns" and the ATENEO 2014 project "Mining of network data" funded by University of Bari "Aldo Moro".

References

1. Ablin, R., Sulochana, C.: A survey of hyperspectral image classification in remote sensing. Int. J. Adv. Res. Comput. Commun. Eng. **2**(8), 2986–3000 (2013)
2. Angin, P., Neville, J.: A shrinkage approach for modeling non-stationary relational autocorrelation. In: Proceedings of 8th IEEE International Conference on Data Mining, pp. 707–712. IEEE Computer Society (2008)
3. Appice, A., Malerba, D.: Leveraging the power of local spatial autocorrelation in geophysical interpolative clustering. Data Min. Knowl. Discov. **28**(5–6), 1266–1313 (2014)
4. Bailey, T., Krzanowski, W.: An overview of approaches to the analysis and modelling of multivariate geostatistical data. Math. Geosci. **44**(4), 381–393 (2012). http://dx.doi.org/10.1007/s11004-011-9360-7
5. Boots, B.: Local measures of spatial association. Ecoscience **9**(2), 168–176 (2002)
6. Chen, C., Li, W., Su, H., Liu, K.: Spectral-spatial classification of hyperspectral image based on kernel extreme learning machine. Remote Sens. **6**(6), 5795–5814 (2014)
7. Cortes, C., Vapnik, V.: Support-vector networks. Mach. Learn. **20**(3), 273–297 (1995)
8. Dray, S., Jombart, T.: Revisiting guerry's data: introducing spatial constraints in multivariate analysis. Ann. Appl. Stat. **5**(4), 2278–2299 (2011)
9. Fauvel, M., Tarabalka, Y., Benediktsson, J., Chanussot, J., Tilton, J.: Advances in spectral-spatial classification of hyperspectral images. Proc. IEEE **101**(3), 652–675 (2013)

10. Getis, A., Ord, J.K.: The analysis of spatial association by use of distance statistics. Geogr. Anal. **24**(3), 189–206 (1992)
11. Hughes, G.: On the mean accuracy of statistical pattern recognizers. IEEE Trans. Inf. Theor. **14**(1), 55–63 (1968)
12. Legendre, P.: Spatial autocorrelation: trouble or new paradigm? Ecology **74**(6), 1659–1673 (1993)
13. Li, M., Zang, S., Zhang, B., Li, S., Wu, C.: A review of remote sensing image classification techniques: the role of spatio-contextual information. Eur. J. Remote Sens. **47**, 389–411 (2014)
14. Manning, C., Raghavan, P., Schutze, H.: Introduction to Information Retrieval. Cambridge University Press, Cambridge (2008)
15. Melgani, F., Bruzzone, L.: Classification of hyperspectral remote sensing images with support vector machines. IEEE Trans. Geosci. Remote Sens. **42**(8), 1778–1790 (2004)
16. Mondal, B., Choudhury, J.: A comparative study on k means and pam algorithm using physical characters of different varieties of mango in india. Int. J. Comput. Appl. **5**(78), 21–24 (2013)
17. Pasolli, E., Melgani, F., Tuia, D., Pacifici, F., Emery, W.J.: SVM active learning approach for image classification using spatial information. IEEE Trans. Geosci. Remote Sens. **52**(4), 2217–2233 (2014)
18. Pasolli, E., Yang, H.L., Crawford, M.M.: Active-metric learning for classification of remotely sensed hyperspectral images. IEEE Trans. Geosci. Remote Sens. **54**(4), 1925–1939 (2016)
19. Plaza, A., Benediktsson, J.A., Boardman, J.W., Brazile, J., Bruzzone, L., Camps-Valls, G., Chanussot, J., Fauvel, M., Gamba, P., Gualtieri, A., Marconcini, M., Tilton, J.C., Trianni, G.: Recent advances in techniques for hyperspectral image processing. Remote Sens. Environ. **113**(Supplement 1(0)), S110–S122 (2009)
20. Rajadell, O., Garcia-Sevilla, P., Dinh, V.C., Duin, R.P.W.: Semi-supervised hyperspectral pixel classification using interactive labeling. In: 2011 3rd Workshop on Hyperspectral Image and Signal Processing: Evolution in Remote Sensing (WHISPERS), pp. 1–4 (2011)
21. Reynolds, A., Richards, G., de la Iglesia, B., Rayward-Smith, V.: Clustering rules: a comparison of partitioning and hierarchical clustering algorithms. J. Math. Modell. Algorithms **5**(4), 475–504 (2006)
22. Richards, J.A.: Remote Sensing Digital Image Analysis: An Introduction, 2nd edn. Springer, New York (1993)
23. Stearns, S.D., Wilson, B.E., Peterson, J.R.: Dimensionality reduction by optimal band selection for pixel classification of hyperspectral imagery. In: Proceedings of the SPIE, Applications of Digital Image Processing XVI, vol. 2028, pp.118–127 (1993)
24. Struyf, A., Hubert, M., Rousseeuw, P.: Clustering in an object-oriented environment. J. Stat. Softw. **1**(4), 1–30 (1997)
25. Tobler, W.: A computer movie simulating urban growth in the Detroit region. Econ. Geogr. **46**(2), 234–240 (1970)

A Comparison of Different Data Transformation Approaches in the Feature Ranking Context

Matej Petković[1,2(✉)], Panče Panov[1], and Sašo Džeroski[1,2,3]

[1] Jožef Stefan Institute, Jamova Cesta 39, Ljubljana, Slovenia
{matej.petkovic,pance.panov,saso.dzeroski}@ijs.si
[2] Jožef Stefan International Postgraduate School,
Jamova Cesta 39, Ljubljana, Slovenia
[3] CIPKeBiP, Jamova Cesta 39, Ljubljana, Slovenia

Abstract. Due to the omnipresence of high-dimensional datasets, feature selection and ranking are very important steps in data preprocessing. In this work, we propose three transformations for real-valued features. The transformations are based on estimating the probability densities of the features. Originally, we propose modified distance measures for the ReliefF algorithm, which is one the most prominent feature ranking algorithms. To enable their comparison with the other feature ranking algorithms, we present data transformations that are mathematically equivalent to the modified distance measures. Finally, we evaluate our proposed transformations used in combination with several feature ranking methods on a set of benchmark datasets.

1 Introduction

In predictive modeling, we frequently encounter high-dimensional problems. One approach for dealing with the high-dimensional input space of a dataset S, is to perform feature selection. First, we find the subset $\mathcal{F}$ of those features that influence the target the most. Then, a predictive model is built from the filtered dataset, using only the features from the set $\mathcal{F}$. Another case when feature selection comes in handy is when a machine learning expert works in collaboration with a domain expert: Predictive models such as decision trees, are easier to understand and interpret when a small number of features is used.

Due to the omnipresence of high-dimensional datasets, there is a plethora of methods for feature selection, including filter, wrapper and embedded methods [3]. Some of these methods directly select a feature subset, while others, called filter methods, sort the features with respect to some relevance measure, i.e., compute a feature ranking, and then select a set of top-ranked features [21]. The filter methods are usually the simplest and the fastest, since they directly evaluate the relevance of each feature. In addition, there is a large number of methods based on correlation and information theory. These include Info Gain and ReliefF [15]. Wrapper methods, explore the space of all possible feature subsets, guided by a classifier's performance on the current subset of features. Finally, in embedded methods, the feature selection is part of the learning process [21].

© Springer International Publishing Switzerland 2016
T. Calders et al. (Eds.): DS 2016, LNAI 9956, pp. 310–324, 2016.
DOI: 10.1007/978-3-319-46307-0_20

Each feature selection method can have some limitations. For example, ReliefF can have a problem with estimating the relevance of numeric features [15]. Recursive feature elimination via support vector machines [11] can face a problem when the classes are not well separated, when the models we learn through the procedure are weak, and the ranking is sub-optimal.

In the context of predictive modelling, it is often advisable to scale or smooth the data [13] when the features are numeric and take values from different ranges. Such transformations can boost the performance of the predictive model. In addition, some probability-distribution-based modifications of a dataset can be equivalently described as a modification of the distance measure used in the algorithm itself, as is the case with ReliefF.

We will transform a dataset with two different goals. In the first case, we will transform the dataset, then apply feature ranking, and finally, build predictive model using the modified dataset. In the second case, we will transform the dataset just to obtain a better feature ranking and then proceed to build a predictive model with the selected features, and using them as they appear in the original dataset.

In this paper, we compare two known transformations, used by Cao and Obradovic [7], with a kernel-estimation-based transformation. For the considered transformations, we investigate whether the first possibility leads to a better predictive model and whether the second possibility leads to better rankings. Additionally, we investigate which combination of a feature ranking method and a data transformation method works the best.

This paper is organized as follows. In Sect. 2, we give a description of the related work, including the ReliefF algorithm. In Sect. 3, we describe our contributions, i.e., the proposed feature transformations used within ReliefF. In Sect. 4, the experimental design is fully described. In Sect. 5, the experimental results are given. Finally, in Sect. 6, we present the conclusions and further work.

2   Background and Related Work

In this section, we give a description of the background that serves as the basis for our work. First, we describe the notation used throughout the paper and formalize the feature ranking task. Then, we proceed to the description of the ReliefF algorithm. Finally, we describe the transformations of the feature space.

2.1   Preliminaries

Notation. In this work, we assume that the elements (data examples or instances) $r = (x, y) \in S$ consist of a descriptive part x and a target value y. The descriptive part is a vector of n feature values x_i, which are either numeric or nominal. In the first case, the domain of feature x_i is a subset of $\mathbb{R}$, whereas in the latter, x_i can take values from an arbitrary finite set. We will use the notation $r_i = x_i$ for the input values and r_y for the target value. Finally, we assume a nominal target, i.e., a classification task.

Feature ranking. As briefly mentioned in Sect. 1, feature ranking can be seen as a special case of feature selection. It applies when the dimension n of the input space is too high for a predictive model to be built, when we want to discard redundant features, or when understandability of the model is of high importance.

One way to obtain the subset $\mathcal{F}$ of all features $\mathcal{X}$, is to apply a feature ranking algorithm. The input for a feature ranking algorithm is a dataset S, whereas the output is the list

$$\boldsymbol{R} = [x_{i_1}, x_{i_2}, \ldots, x_{i_n}] \tag{1}$$

where $\boldsymbol{R}$ is ordered by decreasing relevance of the features x_{i_j}. The main part of such an algorithm is a relevance measure r, which maps features x_i to a relevance score $r(x_i)$. After computing the relevance scores for all the features, we can select only features x_{i_j} for $1 \leq j \leq k$, where k is a predefined number.

2.2 Feature Ranking with ReliefF

The ReliefF algorithm is considered as one of the best feature ranking algorithms. Its predecessor Relief [14] can handle only binary target variables, while ReliefF [15] can handle an arbitrary nominal target. In the following, we first give the motivation which Relief builds upon, and then, a detailed description.

Suppose we are given two instances, $\boldsymbol{r}$ and $\boldsymbol{s}$, which are near to each other. If they belong to the same class, the distance between the values $\boldsymbol{r}_i$ and $\boldsymbol{s}_i$ should also be (relatively) small. Otherwise, feature x_i cannot have a large influence on the target. On the other hand, if $\boldsymbol{r}$ and $\boldsymbol{s}$ belong to different classes and the distance between $\boldsymbol{r}_i$ and $\boldsymbol{s}_i$ is noticeable, we should consider the feature x_i relevant, since a change in its values is associated to a change in the target.

Another important aspect of Relief(F) is that the differences between the nearest neighbours are compared. To capture the feature interactions without explicitly considering all possible pairs of features, one can slightly perturb an instance and inspect how the feature values change locally. Since the nearest neighbour of an instance is its slightest known perturbation in a dataset, in ReliefF we consider the differences between neighbouring pairs.

The distance between the values $\boldsymbol{r}_i$ and $\boldsymbol{s}_i$ is defined as

$$d_i(\boldsymbol{r}, \boldsymbol{s}) = \begin{cases} \mathbf{1}[\boldsymbol{r}_i = \boldsymbol{s}_i] & : x_i \text{ discrete} \\ \frac{|\boldsymbol{r}_i - \boldsymbol{s}_i|}{M_i - m_i} & : x_i \text{ numeric} \end{cases} \tag{2}$$

where $\mathbf{1}$ is the indicator function, and M_i and m_i are the maximal and minimal value of x_i in the dataset. The nearest neighbours are computed with respect to the distance $d(\boldsymbol{r}, \boldsymbol{s}) = \sum_i d_i(\boldsymbol{r}, \boldsymbol{s})$.

In Algorithm 1, we present the pseudocode of the standard version of ReliefF. In the pseudocode, given an instance $\boldsymbol{r}$, other instances are referred to as hits if they are in the same class as $\boldsymbol{r}$, and as misses otherwise. The input of the algorithm consists of a dataset S, a number of iterations J, and a number of nearest neighbours k. The latter is used to increase the robustness. The output

Algorithm 1. ReliefF(S, J, k) [15]

1: $w \leftarrow$ vector of n zeros
2: **for** $j = 1, 2, \ldots, J$ **do**
3:      $r \leftarrow$ randomly chosen record from S
4:      $h_1, \ldots, h_k \leftarrow k$ nearest hits
5:      **for all** $c \in Y \backslash \{r_y\}$ **do**
6:          $m_{c;1}, \ldots, m_{c;k} \leftarrow k$ nearest misses of class c
7:      **end for**
8:      **for** $i = 1, 2, \ldots, m$ **do**
9:          $w_i \leftarrow w_i + \sum_{c \neq r_y} \frac{P(c)}{1-P(r_y)} \sum_{\ell=1}^{k} d_i \left(m_{c;\ell}, r \right) / Jk - \sum_{\ell=1}^{k} d_i \left(h_\ell, r \right) / Jk$
10:      **end for**
11: **end for**
12: **return** w

of the algorithm is a vector of feature weights w. The larger the weight, the higher the importance of the feature.

2.3   Transformations of the Feature Space

Normalization and z-score. The two most widely used methods for transformation of numeric variables are normalization and z-score. In the case of normalization, we linearly transform the feature values x_i, so that $x_i \in [0, 1]$: $x_i \mapsto \frac{x_i - m_i}{M_i - m_i}$, where $M_i = \max_{r \in S} r_i$ and $m_i = \min_{r \in S} r_i$. In the case of z-score, the feature x_i is looked upon as a random variable and is standardized, so that the new mean $E[x_i]$ and the variance $\text{Var}[x_i]$ equal (approximately) zero and one respectively: $z(x_i) = (x_i - \mu_i)/\sigma_i$, where $\mu_i = (\sum_{r \in S} r_i)/|S|$ and $\sigma_i^2 = (\sum_{r \in S} (r_i - \mu_i)^2)/(|S| - 1)$ are unbiased estimates for the mean and the variance of x_i. If the number of examples in the dataset is small, the estimated values of mean and variance can considerably differ from the real values, i.e., the z-scores can be easily affected by outliers. As mentioned by Cao and Obradovic [7], the same holds for normalization. The second, possibly undesired property of the new scores, is that their values can be unbounded.

Transformation with CDF. If we want to have bounded values of numeric features, transforming the feature x_i via its cumulative distribution function (CDF) F is more desirable, since F maps all real numbers onto the interval $[0, 1]$: $F(t) = P(x_i < t) = \int_{-\infty}^{t} f(t) \mathrm{d}t$. Here, f is the probability density function (PDF). It is also well known that the random variable $F(x_i)$ is distributed uniformly on the interval $[0, 1]$. Cao and Obradovic [7] argue that this could lead to a better predictive model, since the feature values that are originally close to each other, become more distant after the transformation.

Empirical CDF. The problem in the case of feature transformation with CDF is that the distribution, hence the CDF of x_i, is unknown. To overcome this, the empirical CDF (ECDF) can be used. This is a step function, defined as

$$\hat{F}(t) = \sum_{r \in S} \mathbf{1}[r_i \le t], \tag{3}$$

where $\mathbf{1}$ is the indicator function. Even though ECDF is not a continuous function, it possesses some favorable properties, since the Glivenko–Cantelli theorem assures that ECDF converges to CDF uniformly, as the number of observations grows [6,10]. However, in the case of high dimensional microarray datasets (used in our experiments later on), the number of data examples can be rather small and the data can be noisy [3]. Consequently, in this case the ECDF might not be the best approximation of the CDF.

Sigmoid function. Bowling et al. [5] report that the sigmoid function $s(t) = 1/(1 + \exp(-t))$ can be used as an accurate approximation of the CDF of the standard normal distribution [5]. Since normality cannot be usually assumed, Cao and Obradovic [7] allow for a generalized sigmoid function, defined as

$$s_G(t) = \frac{1}{\sqrt[\nu]{1 + Q \exp(-B(t - M))}}. \tag{4}$$

The low number of the parameters (Q, B, M and ν) prevents overfitting. This is the reason why fitting the curve (4) is more robust than using the ECDF (3). The (locally) optimal generalized sigmoid curve s_G^i for feature x_i is computed with respect to the loss function

$$\ell(s_G) = \sum_{r \in S} \left| \hat{F}_{x_i}(r_i) - s_G(r_i) \right|^2, \tag{5}$$

where $\hat{F}_{x_i}$ is the ECDF of the feature x_i. To optimize the parameters, we can use gradient descent algorithm.

3   Using Different Data Transformation Approaches with the ReliefF Algorithm

In this section, we propose three novel modifications of the ReliefF algorithm, where we modify the distance measure (2). The first two are based on transformations using the ECDF and the generalised sigmoid function, whereas the third is based on kernel density estimation (KDE). Next, we show how these modifications of a dataset are equivalent to the three modifications of the ReliefF algorithm. The section concludes with a description of the implementation.

3.1   Proposed ReliefF Modifications

ECDF and sigmoid. Motivated by Cao and Obradovic [7], who compared the ECDF (3) and generalized sigmoid function (4) transformations of data in the context of predictive modeling, we propose the modifications ReliefFecdf and ReliefFsigmoid. For `ReliefFecdf`, we define

$$d_i(\boldsymbol{r}, \boldsymbol{s}) = \left| \hat{F}_{x_i}(\boldsymbol{r}_i) - \hat{F}_{x_i}(\boldsymbol{s}_i) \right|, \tag{6}$$

whereas for `ReliefFsigmoid`, we have

$$d_i(\boldsymbol{r}, \boldsymbol{s}) = \left| s_G^i(\boldsymbol{r}_i) - s_G^i(\boldsymbol{s}_i) \right|. \tag{7}$$

Note that the distances (6) and (7) already map to the interval $[0, 1]$. These two distance measures can be seen as the measures that take into account the distribution of x_i, since CDF F_{x_i} and its approximation s_G^i are part of the definition of d_i. The next transformation is related to these two, but instead on CDF, it is based on probability density function (PDF).

KDE. The goal of KDE is to find a good approximation of the PDF f_{x_i} of the feature x_i. When we obtain a KDE approximation $\hat{f}_{x_i}$, we define the distance

$$d_i(\boldsymbol{r}, \boldsymbol{s}) = \left| \int_{\boldsymbol{r}_i}^{\boldsymbol{s}_i} \hat{f}_{x_i}(t) \mathrm{d}t \right|. \tag{8}$$

It is important to note two facts. First, in the ideal case, when one of the functions F_{x_i} and f_{x_i} (hence both) is known, the definitions (6) and (8) are the same, since PDF is the derivative of CDF. The second, maybe more important fact is, that the modification (8) is not biased against or towards any distribution, since the following theorem can be proven.

Theorem. *Let X and Y be two i.i.d. random variables with PDF f. The distribution of $Z = \left| \int_X^Y f(t) \mathrm{d}t \right|$ does not depend on f.*

Additionally, we can see that in the ideal case, when we will replace $\hat{f}_{x_i}$ by f_{x_i}, the definitions (2) and (8) coincide for the uniformly distributed x_i.

To be able to approximate f and compute the integral (8) efficiently, we restrict ourselves to a parametric family of *Gaussian mixtures* $\hat{f}(t) = \sum_{k=1}^{K} \alpha_k N_{\mu_k, \sigma_k}(t)$, which are a good compromise between the favourable computational properties of the normal distribution and the number of degrees of freedom of a distribution. Here, the *kernel functions* are defined as normal densities N_{μ_k, σ_k}, and the *prior probabilities* $\alpha_k > 0$ add up to 1. Since the estimation of the parameters of a Gaussian mixture is a hard problem to approach analytically, we could use the popular expectation-maximization (EM) algorithm of Dempster et al. [8]. This is an iterative algorithm that provably converges to the locally optimal solution (see [22]).

However, when we used the EM-algorithm in our previous work [18], where we modified ReliefF in order to alleviate its underestimation of numeric features as compared to nominal ones, we faced some issues: the number of kernels K cannot be easily determined in an automated fashion, the rate of convergence can be low, and there is a strong dependence of the obtained values on the initialization of the parameters. Hence, we rather use the following form of KDE estimate

$$\hat{f}_i(t) = \frac{1}{|S|} \sum_{r \in S} \frac{1}{h_i} N\left(\frac{t - r_i}{h_i} \right), \tag{9}$$

where $N = N_{1,0}$ is the standard normal kernel. The optimal bandwidth h_i^* that is at the end plugged in (9) is found by the algorithm of Botev et al. [4].

Once we have $\hat{f}_i$, we should find an efficient and accurate approximation of the integral (8). It suffices to find an approximation of the standard normal CDF. Abramowitz and Stegun [2] give a list of useful approximations, among which we choose the one denoted with 7.1.26. It is of the form $\hat{F}(x) = 1 - \exp(-x^2) \sum_{k=1}^{5} a_k/(1 + px)^k$ and has an error smaller than $1.8 \cdot 10^{-7}$.

3.2 Equivalent Data Transformations

Since we want to compare ReliefF and its modifications of ReliefF to two other feature ranking algorithms (SVM-RFE and IG that will be described later on), we do not use the modified versions of ReliefF directly. We rather transform the dataset and then apply the algorithms. This makes the three algorithms comparable, while the results of ReliefF remain intact.

ECDF and sigmoid. The modifications that correspond to the ECDF (6) and sigmoidal (7) version of the distance are rather straight-forward: We simply use the formulas (3) and (4).

KDE. The modification in the case of KDE is a bit more complex. For an arbitrary CDF F and a corresponding PDF f, it holds that $F(b) - F(a) = \int_a^b f(t)\mathrm{d}t$. Hence, if we apply the transformation $r_i \mapsto \int_{-\infty}^{r_i} \hat{f}_{x_i} \mathrm{d}t$, then the absolute difference between two transformed values of feature x_i equals (8).

However, the time complexity for transforming the feature x_i (if h_i^* is given), is $\mathcal{O}(|S|^2)$, since we could not find any better approaches than evaluating the $|S|$-term sum (9) in $|S|$ points (e.g., Fast Fourier Transform [19] needs only $\mathcal{O}(n \log n)$ time to evaluate a polynomial of degree n in n points) directly. Nevertheless, the time complexity of ReliefKDE remains the same (provided that computing h_i^* does not take a considerable amount of time), since the time complexity of the original ReliefF is already $\mathcal{O}(|S|^2 n)$, if the usual value $J = |S|$ is used.

3.3 Implementation

For transforming the datasets by using ECDF and KDE, we used the standard Python library and the `numpy` module. The ECDF transformation is implemented in a straight-forward way. For the KDE transformation, we used Lewis's implementation [16] of the algorithm of Botev et al. [4] for finding the optimal bandwidth.

For the sigmoid transformation, we used Octave's function `fminunc`, where we optimized the function (5), starting from the initial values, as suggested by Cao and Obradovic [7]. The Jacobian matrix was also a part of the input. However, it turns out that this initialization can lead to a solution $(\hat{Q}, \hat{M}, \hat{B}, \hat{\nu})$ that is far from optimal. The main problem was the value $\hat{\nu}$, which was sometimes quite large or even `Inf`. Since $\lim_{\nu \to \infty} \sqrt[\nu]{|x|} = 1$, for all x, this caused a degenerated sigmoid function. In that case, we found the optimal $\hat{\nu}^*$ for a fixed $(\hat{Q}, \hat{M}, \hat{B})$ (using `fminunc`), and then repeated the first step, starting from $(\hat{Q}, \hat{M}, \hat{B}, \hat{\nu}^*)$.

4   Experimental Design

In this section, we describe the experimental design. First, we discuss the experimental questions we want to answer. Next, we describe the algorithms and the datasets used in the experiments. Finally, we describe the evaluation measures and experimental procedure.

Experimental questions. Our experimental design is constructed in such a way to answer several lines of inquiry. (1) We want to see if the transformation of the input dataset helps to obtain better performance in a feature ranking context as compared to the use of the original data. (2) We want to examine whether it is better to compute the feature ranking on the transformed or the original dataset. (3) We want to compare the different combinations of dataset transformations and ranking algorithms, in terms of ranking quality.

Algorithms. In this paper, we consider and compare three different types of feature ranking algorithms. These include ReliefF, Info Gain and Recursive Feature Elimination via Support Vector Machines (SVM-RFE) [11]. All three ranking algorithms were used together with three dataset transformation methods (ECDF, KDE, sigmoid), resulting in nine different ranking/transformation pairs. Since we already described the ReliefF algorithm in Sect. 2.2, here we just briefly describe the other two algorithms.

Info Gain (IG) is a fast and well known feature ranking algorithm, where the relevance of a nominal feature is defined with the formula $r(x_i) = H(y) - H(y \mid x_i)$, where H is the Shannon entropy. If the feature is numeric, it is first discretized. Since all transformations φ of the datasets are monotonic functions $(a \geq b \Rightarrow \varphi(a) \geq \varphi(b))$, we do not expect that there will be many considerable differences among the transformations for the case of IG due to the discretization.

Recursive Feature Elimination via Support Vector Machines (SVM-RFE) builds SVM models by using linear kernels. The procedure starts with using the complete set of features $\mathcal{C} = \mathcal{X}$ to train a model on a given dataset. Thus, we obtain the normal $\boldsymbol{n}$ and the bias b of the hyperplane that defines the decision rule $y(\boldsymbol{x}) = \text{sign}(\boldsymbol{x}\boldsymbol{n} - b)$. Since the features x_i, for which $|\boldsymbol{n}_i|$ is small, do not have much influence on $y(\boldsymbol{x})$, we can consider them to be irrelevant. Hence, we eliminate the feature $x_i = \text{argmin}_j |\boldsymbol{n}_j|$ from our dataset and recursively call the procedure on the reduced set of features $\mathcal{C} \setminus \{x_i\}$. After $n = |X|$ steps, we obtain a ranking: the later the feature is eliminated, the greater its importance. Optionally, we can discard more than one feature at time. In that case, the features that are discarded at the same step, have the same importance. We use Weka's implementation [12] of the three considered feature ranking algorithms. For ReliefF and IG, the default parameter values are used. For SVM-RFE, `percentToEliminatePerIteration` was set to 10 and `percentTreshold` was set to 50 (if the number of features is greater than 50, we eliminate 10 % of the current features; we eliminate one feature otherwise).

Datasets. We considered 28 datasets from the UCI repository [17], where the number of features ranges between 4 and 279, and the number of instances ranges between 131 and 5000, see Table 1. These datasets come from various domains, but are rather small in terms of the number of features, hence we included ten high-dimensional cancer gene expression data sets [1], where only numeric features are present (see Table 2).

Table 1. Description of *small* datasets (name, number of features and instances). The sets, on which the SVM-RFE algorithm could not be run, are marked in bold.

dataset	features	instances	dataset	features	instances
arrhythmia	279	452	**heart-c**	13	303
australian	14	690	**heart-h**	13	294
balance	4	625	hepatitis	19	155
biodeg-p1-discrete	31	328	hypo	25	3163
biodeg-p2-discrete	61	328	image	19	2310
contraceptive	9	1473	ionosphere	34	351
diabetes	8	768	iris	4	150
diatoma_vulgare-t14200	9	1060	**reuma-1**	22	462
dis	28	3772	**reuma-2**	16	462
diversity_all	86	292	sea_cucumber-leuco	10	128
echo	6	131	sonar	60	208
german	20	1000	water_all	80	292
glass	9	214	waveform	21	5000
heart	13	270	wine	13	178

Evaluation measures. For the evaluation of the rankings, we follow the methodology developed by Slavkov [20]. An algorithm for feature relevance estimation produces a feature ranking R (1). We define $\mathcal{X}_j$, $1 \le j \le n$, as sets of j topmost features. For each j, we build a predictive model for a data set S, considering only the features from $\mathcal{X}_j$. Let α_j denote the accuracy of the j-th model. The points $(|\mathcal{X}_j|, \alpha_j)$ form a *feature addition curve*. We can also consider the features from the complement of $\mathcal{X}_j$ (denoted $\mathcal{X}_j^C$). The numbers $\bar{\alpha}_j$ denote the accuracy of these models and the points $(|\mathcal{X}_j^C|, \bar{\alpha}_j)$ form a *feature removal curve*. Both types of curves are illustrated in Fig. 1, where we show the results for the ReliefF and Info Gain rankings on the original version of the Arrhythmia dataset.

The motivation behind the two-fold approach is the following. For higher α_j's and lower $\bar{\alpha}_j$'s, more relevant features are positioned at the beginning of the ranking R and less relevant features are positioned at its end. To evaluate the quality of the ranking, we define

$$\alpha = \left(\sum_j w_j \alpha_j\right)/w \quad \bar{\alpha} = \left(\sum_j w_{n-j+1}\bar{\alpha}_j\right)/w \quad \text{and} \quad \delta = (\alpha - \bar{\alpha})/2 \qquad (10)$$

where $w = \sum_j w_j$. In our experiments, we choose $w_j = 1/j$. A good ranking should have high α- and low $\bar{\alpha}$-score, hence δ rewards rankings with quickly increasing α's and punishes rankings with slowly decreasing $\bar{\alpha}$'s.

Table 2. Description of *big* datasets (name, number of features and instances).

Name	#features	#instances
amlPrognosis	12625	54
bladderCancer	5724	40
breastCancer	12625	24
childhoodAll	8280	110
cmlTreatment	12625	28
dlbcl	7070	77
leukemia	5174	72
mll	12533	72
prostate	12533	102
srbct	2308	83

Experimental procedure. For each original dataset, we create its 3 transformed variants (ECDF, KDE and sigmoid). To increase the stability of the results, 10-times 10-fold cross-validation is used.

First of all, we randomly partition the dataset into 10 folds. Then, these folds are used for each of the four versions on the dataset (the original and the three transformed). For each test fold, a feature ranking is computed on the complementary group of the nine training folds. Then, the predictive models from the previous section are built, as described above (for feature subsets of different size) using linear support vector machines (SMO in Weka). We build those models on the current version of the dataset and on its original version (in line with the experimental questions). Each of the two groups of models is than evaluated on the corresponding test fold.

After performing the 10-times 10-fold cross-validation, the confusion matrices of the models from the different iteration steps are averaged. Using the

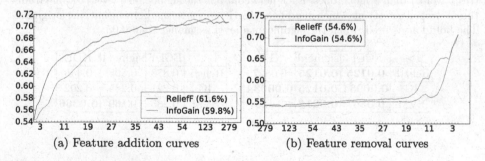

(a) Feature addition curves   (b) Feature removal curves

Fig. 1. Feature addition/removal curves for the set Arrhythmia (original version), obtained from the ReliefF and Info Gain rankings. The α (a) and $\bar{\alpha}$ (b) scores of the rankings are given in the brackets.

confusion matrices, we compute the scores (10) for each ranking. Hence, we have three indicators of the quality of the feature ranking for each dataset. For each indicator, we perform Friedman's test, to see, whether all the considered ranking algorithms perform equally well. If this hypothesis is rejected, we also perform Nemenyi's post-hoc test. When we compare only two algorithms (Question (2) in Sect. 4), we perform Wilcoxon test. All the tests are described by Demšar [9].

5   Results and Discussion

In this section, we present the results of the performed experiments, which answer our experimental questions. First, we analyze if the transformation of the input dataset helps to obtain better performance in a ranking context, as compared to the use of the original data. Next, we examine if it is better to compute the feature ranking on the transformed or on the original dataset. Finally, we perform a comparison[1] of the different combinations of dataset transformations and ranking algorithms, in terms of ranking quality.

Question (1). To answer the question whether is better to use the original or the transformed version of a dataset, we simply compare the rankings, i.e., corresponding models, that were computed and tested on the transformed datasets, to the rankings/models obtained from the original datasets. Hence, for each of the nine algorithm/transformation pairs, where the algorithms include ReliefF, SVM and IG, and the transformations include ECDF, sigmoid and KDE, we perform the Wilcoxon test between (algorithm, transformation) and (algorithm, original). The results for *big* datasets are given in Table 3.

Table 3. A comparison of the performances of the algorithms that were run and tested on a particular transformed version of the dataset, to the performances of the algorithms on the original datasets. The results are for the *big* datasets and based on the α-scores (left table) and $\bar{\alpha}$-scores (right table). The symbol $>$ (respectively $<$) in the upper left corner of a table denotes, whether the modified algorithms performed better (resp. worse) than original ones. Each field contains the significance level (p-value from the Wilcoxon's test) at which modified algorithm performs better that the original. The bold font is used, when the significance level is smaller than 0.05.

$>$	ECDF	sigmoid	KDE
ReliefF	**0.0125**	**0.0125**	0.0527
IG	**0.00934**	**0.0125**	**0.00934**
SVM	**0.0367**	**0.0218**	0.126

$<$	ECDF	sigmoid	KDE
ReliefF	0.878	0.508	0.445
IG	0.221	0.285	0.202
SVM	0.285	0.169	**0.0366**

---

[1] Due to its implementation in Weka, the SVM-RFE algorithm could not be applied to datasets with non-binary nominal features, hence the results of SVM-RFE are based on 19 (and not 28) *small* datasets. From now on, we refer to SVM-RFE as SVM.

We can see that, in all cases, the modified version of the algorithm performs better than the original one, in terms of α-scores. The difference is statistically significant at the level 0.05 in 7 out of 9 cases.

On the other hand, the scores $\bar{\alpha}$ are worse for the modified versions of the data, although the differences are usually far from being statistically significant (except for the (SVM, KDE) combination). Since the comparisons of the versions are consistent (either all modified version perform better or all perform worse), we can conclude that a predictive model itself can in fact benefit from the transformations; The accuracies α_j and $\bar{\alpha}_j$ on the modified datasets are higher than those on the original datasets, which leads to better feature addition and worse feature removal curves. Additionally, we can report here that the δ-scores show that the transformed datasets yield better performance, although the differences are significant only for ReliefF and SVM, coupled with ECDF and sigmoid.

For *smaller* datasets, the transformed versions also perform better, although without significant differences (either at level 0.10 or 0.05) for every of the scores α, $\bar{\alpha}$ and δ. For *all* datasets combined, the transformed methods are always better at the significance level 0.10, many of them also at 0.05, for scores α. The scores $\bar{\alpha}$ give significant differences only for sigmoid and KDE version of SVM with p-values smaller than 0.05. The score δ shows no significant differences.

Question (2). In the previous section, we have seen that modifying the dataset leads to better predictive models. However, we do not yet know, whether this is due to better rankings. It might be that learning models from the modified datasets is easier than learning models on the original dataset.

To determine, whether it is better to compute the feature ranking on the modified dataset, when the original version of the data is used for building a model, we fix the ranking algorithm, obtain the rankings on the original and transformed versions of a dataset, and evaluate the rankings only on the original dataset. Friedman's and (possibly) Nemenyi's statistical tests are used to inspect whether there are any differences in performance and where do they occur.

For ReliefF, there are no statistically significant differences among the transformations. However, in none of the nine combinations of dataset group (*small, big* and *all*) and type of score (α, $\bar{\alpha}$ and δ), the original dataset is ranked first: (Relief, KDE) wins in 7 cases and (Relief, ECDF) in the remaining two.

For SVM rankings, significant differences between the methods occur. However, a somehow surprising discovery is that the original method is the best in 6 of the nine (dataset group, score) combinations. We conclude that different features are (more) important in the original and in the modified versions of the datasets. In Fig. 2, we present the statistically significant differences on a critical distance diagram, where the results of the Nemenyi's post-hoc test are presented. The groups of algorithms, among which no significant differences occur, are connected by a red line.

In the case of IG, the differences are far from significant. As mentioned before, this is expected, since all the transformations are monotone, hence the discretization method that is used by IG should produce almost the same nominal attributes, regardless of the particular transformation applied. The extreme

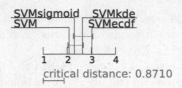

critical distance: 0.8710

Fig. 2. The average rank diagram that is based on the results of Nemenyi's test for δ-scores of different versions of SVM, on *all* datasets (at level 0.05). (Color figure online)

result is obtained in the case of $\bar{\alpha}$ for *big* datasets: all of the four versions (original + 3 modified) have the same average rank 2.5.

Question (3). Here, we discuss which combination (ranker, transformation) is the best. We already know that results that are obtained by evaluating the same ranking on the transformed and the original version of a dataset are somewhat contradictory, i.e., one of the two compared rankings has typically a better α-score, while the other has a better $\bar{\alpha}$-score. Hence, we will not compare this two options of evaluation, and only 12 combinations (3 algorithms and $3+1$ modifications) will be compared.

We start with the results where the rankings from the modified dataset are evaluated on its original version. Again, the comparisons were made by using Friedman's and Nemenyi's tests. The differences are not always statistically significant, but in 7 out of 9 considered scenarios (*small*, *big*, *all* datasets, and α, $\bar{\alpha}$ and δ scores) the combination that performs best is (ReliefF, KDE). In the case when α-scores are compared on the *big* datasets, (SVM, KDE) performs best, whereas in the case when δ-scores are compared on the *big* datasets, (ReliefF, ECDF) is the best.

Typically, the four versions of the algorithms are grouped together, which is expected, since the previous results showed, that the results of the versions of one algorithm may not differ much. Here, we explicitly show only the results for the *big* datasets. Since the differences in terms of α-scores are not statistically significant, we only give the comparisons that are based on the $\bar{\alpha}$-scores and δ-scores (Fig. 3). The results for *small* and *all* datasets are almost the same regarding the order of algorithms (top four algorithms are from Relief-group, whereas Info Gain and SVM-RFE are mixed) and the significance of the results (no significant differences only in the case of the α-scores).

In the case when the rankings were evaluated on the transformed datasets, there is one noteworthy difference with respect to the previous case: the opposition between the α and $\bar{\alpha}$ results causes (among other differences in the ordering) that (SVM, KDE) and (SVM, ECDF) perform worst in terms of $\bar{\alpha}$-scores and the best in terms of α-scores. But, in general, the combination (ReliefF, KDE) produces the best results in terms of ranking quality.

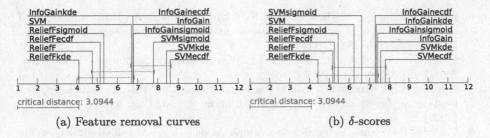

(a) Feature removal curves   (b) δ-scores

Fig. 3. The average rank diagram that is based on the results of Nemenyi's test for *big* datasets for $\bar{\alpha}$-score (a) and δ-score (b), at the level of significance 0.05.

6   Conclusions and Future Work

In this paper, we presented a comparison of different data transformation approaches in the context of feature ranking. We have shown that it is beneficial to transform a dataset using the presented approaches and then use only the transformed version, in terms of predictive performance. On the other hand, it seems that in most cases that the transformation does not lead to a better ranking for the original dataset. Even worse, ranking obtained by using SVM-RFE for the original version of the data performed best. However, it is reassuring that the combination of feature ranking with ReliefF and feature transformation by KDE performs best in the vast majority of cases.

We suggest that KDE transformation is used, since it exhibits good predictive performance. Also, we did not face nearly as many convergence problems as in the sigmoid case. Due to its simplicity, using ECDF is also a legitimate choice. However, on contrast to Cao and Obradovic [7], we would discourage the use of the sigmoid transformation, for the aforementioned computational issues.

The starting motivation for this work was the improvement of ReliefF. We have made a step forward with the new approach to estimating the PDF of a feature moving from the EM-algorithm to that of Botev et al. [4]. However, the differences are still not significant. Hence, we believe that there is still room for improvement of the estimation of the numeric features, which will be addressed in our further work.

Acknowledgements. We would like to acknowledge the support of the EC through the projects: MAESTRA (FP7-ICT-612944) and HBP (FP7-ICT-604102), and the Slovenian Research Agency through a young researcher grant and the program Knowledge Technologies (P2-0103).

References

1. Visualization-based cancer microarray data classification analysis. http://www.biolab.si/supp/bi-cancer/projections. Accessed 04 Oct 2015
2. Abramowitz, M., Stegun, I.: Handbook of Mathematical Functions (1972)

3. Bolón-Canedo, V., Sánchez-Maroño, N., Alonso-Betanzos, A., Benítez, J.M., Herrera, F.: A review of microarray datasets and applied feature selection methods. Inf. Sci. **282**, 115–135 (2014)
4. Botev, Z., Grotowsky, J., Kroese, D.P.: Kernel density estimation via diffusion. Ann. Stat. **38**(5), 2916–2957 (2010)
5. Bowling, S.R., Khasawneh, M.T., Kaewkuekool, S., Cho, B.R.: A logistic approximation to the cumulative normal distribution. J. Ind. Eng. Manag. **2**, 114–127 (2009)
6. Cantelli, F.P.: Sulla determinazione empirica delle leggi di probabilita. Giornale dell'Istituto Italiano degli Attuari **4**, 421–424 (1933)
7. Cao, X.H., Obradovic, Z.: A robust data scaling algorithm for gene expression classification. In: Proceedings of the 15th IEEE International Conference on Bioinformatics and Bioengineering (2015)
8. Dempster, A.P., Laird, N.M., Rubin, D.B.: Maximum likelihood from incomplete data via the EM algorithm. J. Roy. Stat. Soc. Ser. B (Methodol.) **39**, 1–38 (1977)
9. Demšar, J.: Statistical comparisons of classifiers over multiple data sets. J. Mach. Learn. Res. **7**, 1–30 (2006)
10. Glivenko, V.I.: Sulla determinazione empirica delle leggi di probabilita. Giornale dell'Istituto Italiano degli Attuari **2**, 92–99 (1933)
11. Guyon, I., Weston, J., Barnhill, S., Vapnik, V.: Gene selection for cancer classification using support vector machines. Mach. Learn. **46**(1), 389–422 (2002)
12. Hall, M., Frank, E., Holmes, G., Pfahringer, B., Reutemann, P., Witten, I.H.: The WEKA data mining software: an update. ACM SIGKDD Explor. Newsl. **11**(1), 10–18 (2009)
13. Han, J., Kamber, M., Pei, J.: Data Mining: Concepts and Techniques, 3rd edn. MorganKaufmann Publishers Inc., San Francisco (2011)
14. Kira, K., Rendell, L.A.: The feature selection problem: traditional methods and a new algorithm. In: Proceedings of the Tenth National Conference on Artificial Intelligence, AAAI 1992, pp. 129–134 (1992)
15. Kononenko, I., Robnik-Šikonja, M.: Theoretical and empirical analysis of ReliefF and RReliefF. Mach. Learn. J. **53**, 23–69 (2003)
16. Lewis, A.: Getdist. https://github.com/cmbant/getdist. Accessed 27 May 2016
17. Lichman, M.: UCI machine learning repository (2013)
18. Petković, M., Panov, P., Džeroski, S.: Improved ranking of numeric features with ReliefF. Presented at the Workshops on Machine Learning in Computational Biology (MLCB) & Machine Learning in Systems Biology (MLSB) (2015)
19. Rao, K.R., Kim, D.N., Hwang, J.J.: Fast Fourier Transform - Algorithms and Applications, 1st edn. Springer Publishing Company, Incorporated, Heidelberg (2010)
20. Slavkov, I.: An Evaluation Method for Feature Rankings. Ph.D. thesis, Mednarodna podiplomska šola Jožefa Stefana, Ljubljana (2012)
21. Stańczyk, U., Jain, L.C. (eds.): Feature Selection for Data and Pattern Recognition. Studies in Computational Intelligence, vol. 584. Springer, Heidelberg (2015). doi:10.1007/978-3-662-45620-0
22. Wu, C.: On the convergence properties of the EM algorithm. Ann. Stat. **11**, 95–103 (1983)

On Selection Bias with Imbalanced Classes

Gert Jacobusse[1] and Cor Veenman[1,2(✉)]

[1] Digital and Biometric Traces, Netherlands Forensic Institute, Laan van Ypenburg 6,
2497 GB The Hague, The Netherlands
`g.jacobusse@nfi.minvenj.nl`
[2] Leideu Institute of Advanced Computer Science, Leiden University, Niels Bohrweg
1, Leiden, The Netherlands
`corveenman@gmail.com`

Abstract. In various applications, such as law enforcement and medical
screening, one class outnumbers the other, which is called class imbalance. The inspection to recognize targets from the minority class is usually driven by experience and expert knowledge. In that way, targets can
be found way above the base rate to make the inspection process feasible. In order to make the search for targets more efficient, the inspected
samples can serve as training set for a learning method. In this study,
we show how the introduced selection bias can be remedied in several
ways using unlabeled data. With a synthetic dataset and a real-world
law enforcement dataset, we show that adding unlabeled data to the
non-targets strongly improves ranking performance. Importantly, completely leaving out the labeled non-targets and using only the unlabeled
data as non-targets gives the best results.

Keywords: Classification · Class imbalance · Selection bias · Unlabeled
data · Supervised learning · Semi-supervised learning · PU learning

1 Introduction

In real-life screening problems, there is usually a relatively small group of targets that needs to be detected among a large group of non-targets [17]. These
applications range from medical screening [23], screening of job applicants [29],
biochemical screening [20] to law enforcement [30] and fraud detection [11,28].
In such class imbalance problems, the screening procedure must be informed
and directed towards the target group to make the procedure feasible, that is, to
restrict the number of unsuccessful inspections. We consider class imbalance of
the order of 1 in 100 between targets and non-targets. In such scenarios random
sampling would lead to a success rate of one percent, which can hardly ever be
justified. Therefore, in an expert knowledge driven way, a screening procedure is
designed with parameters that are known to correlate with the target phenomenon, such as age for medical cancer screening or a restaurant category in law
enforcement for illegal workers. The data, that is collected, inspected and labeled
this way, is strongly biased towards the target group. Moreover, the subset of

© Springer International Publishing Switzerland 2016
T. Calders et al. (Eds.): DS 2016, LNAI 9956, pp. 325–340, 2016.
DOI: 10.1007/978-3-319-46307-0_21

non-targets is similar to these targets. Yet, the class imbalance that was the reason for the selective search is no longer pronounced in the inspected dataset or even absent. This is what we consider selection bias with imbalanced classes, which is the subject of this study.

In order to make a clear distinction between class imbalance in the underlying distributions and class imbalance in the available selected dataset, we will define two separate notions:

- class prior imbalance: the imbalance in the priors of the underlying distributions of the classes;
- class sample imbalance: the imbalance in the available samples of the classes involved.

In Fig. 1, this is illustrated in a one-dimensional example. This dataset has clear class prior imbalance: the volume of targets is 5 % of the volume of non-targets. Beyond the selection threshold $t = 3$, there are more or less as many targets (minority class) as non-targets (majority class). So, samples that are selected through such a selection bias would not have class sample imbalance. Clearly, the bulk of the non-targets are not sampled in the selected dataset.

In Fig. 2, we show a two dimensional example of two Gaussians with class prior imbalance of 1 (targets) in 10 (non-targets). A random sample from these distributions is shown in Fig. 2(a) with 10 targets and 100 non-targets. When a well-informed selection process retrieves only samples that have $y > 1.5$, the targets and non-targets are more or less balanced, see Fig. 2(b). Or, there is more or less no class sample imbalance.

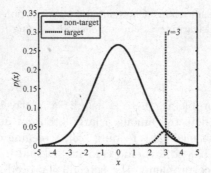

Fig. 1. Illustration of the distributions of classes with clear class prior imbalance. When the samples with $x > 3$ are selected through a biased inspection process, the resulting training set is more or less balanced.

In order to be more efficient, the collected dataset can be exploited to obtain a data driven predictive model to find likely targets among the samples that have not yet been inspected, i.e. the unlabeled samples. In a law enforcement scenario, the possible targets from the unlabeled data can be ordered from the

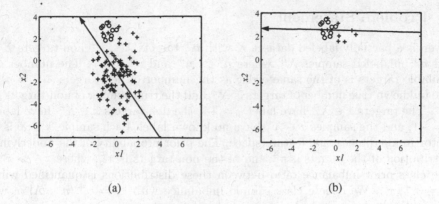

(a)   (b)

Fig. 2. Example of the result of selection bias towards the features of the targets ('o'). Non-targets are displayed with '+'s. In (b) the samples from (a) are shown for which $y > 1.5$. The derived classification/ranking model is displayed in both figures with an arrow that points towards the targets.

most likely to the least likely target, i.e. a ranking problem [30]. In that way, the capacity of the agencies can be scheduled towards maximal success in finding targets[1]. In the example of Fig. 2(a), a model based on all data would recognize the few targets to be situated in the upper left corner of the 2-D space, which is indicated with the direction of the arrow. The selected subset of the data in Fig. 2(b), gives rise to quite a different image of the distinction between targets and non-targets. The targets seem to be situated at the left part of the space, which is clearly suboptimal considering the distribution of the complete dataset.

In this work, we propose a sampling based method to deal with the selection bias in case of class prior imbalance, where the unlabeled data play a central role. We tested the method on a synthetic dataset for which we simulate class prior imbalance and selection bias. In addition, we tested the method on a real-life law enforcement dataset. In contrast to the other datasets, this dataset has no obvious ground truth available to validate the method. We propose to use a subset of the dataset as validation set, because of its specific properties.

The layout of the paper is as follows. In the next section, we first introduce the formal problem definition and terminology. Then, we overview the different lines of related work. Next, we propose our methods of exploiting unlabeled data to remedy selection bias with imbalanced classes. We describe the experiments we did to test the method on synthetic and real-life data and conclude with discussions and outlook from the work.

---

[1] A strict application of this scenario would make it hard to discover new phenomena or trends, since the goal is to find targets similar to previous cases. This exploratory aspect is, however, not the subject of this study.

2   Problem Statement

Given is a partially labeled dataset X with n^+ targets X^+, n^- non-targets X^- and n^0 unlabeled samples X^0, where $n^0 \gg n^+$ and $n^0 \gg n^-$. The number of available targets is of the same order as the number of non-targets: $n^+ \approx n^-$. The unknown true number of targets is N^+ and the true number of non-targets is N^-. The targets $x \in X^+$ have label $y = +1$, the non-targets $x \in X^-$ have label $y = -1$, and the samples $x \in X^0$ have no known label. Each sample $x \in X$ is a vector in a p dimensional feature space. The prior probability of the underlying distribution of the targets is π^+ and of the non-targets is π^-, where $\pi^- \gg \pi^+$. The class prior imbalance ratio between these distributions is quantified with $c_p = \pi^+/\pi^-$. We denote class sample imbalance with $c_s = n^+/n^-$. Also, we use $c_p(X)$ to denote the class prior ratio of targets versus non-targets in the unlabeled set X.

The labeled samples have been selected ($s = 1$) through a biased process that is driven only by the feature vector of the sample from the probability distributions $P(x|y = +1)$ and $P(x|y = -1)$. In other words, the selection of a sample is independent of its label given its feature vector: $P(s|x, y) = P(s|x)$ [32].

The goal is to correctly rank the unlabeled samples from the most likely to the least likely targets. The required ordering function $F(x)$ gives the highest score to the most likely target samples. The quality of the ordering is measured with the Area Under the ROC Curve (AUC) [1]:

$$\text{AUC}(X, F(\cdot)) = \frac{1}{n^+ \cdot n^-} \sum_{x_1 \in X^+} \sum_{x_2 \in X^-} \delta(x_1, x_2), \tag{1}$$

where

$$\delta(x_1, x_2) = \begin{cases} 0 & \text{if } F(x_1) \geq F(x_2) \\ 1 & \text{if } F(x_1) < F(x_2) \end{cases} \tag{2}$$

3   Related Work

The research that we explore in this work touches many fields of research. Below we describe the essential elements of these fields and stress the similarities and differences. We divide the related work up into the way the training data is drawn from the underlying data distributions. The training sample can be representative, undersampled and biased. Clearly, this separation is not as strict as suggested. For instance, some selection bias can and will always be present, while also some unlabeled data is often available. Further, in all cases, while we use the general term *learning*, we consider *classifier* learning only.

3.1   Representative Sample

First, a representative dataset with labeled targets and non-targets is typically a starting point for pattern recognition and supervised learning. Representativeness means that the probability of selecting a sample is independent of its feature vector and its label, i.e. $P(s|x,y) = P(s) = c$. The fundamental assumption that the given dataset is a representative sample from the underlying distributions is clearly violated in this study.

Class Imbalance. Many machine learning methods optimize global criteria like error or accuracy. In case the classes are harder to differentiate, such models are principally unsuitable for class imbalance problems. The criterion of these methods is then optimized by a degenerate model that always decides for the majority class. In this way the samples will be ordered randomly. Several approaches have been published to deal with class (sample) imbalance, such as undersampling the majority class or oversampling the minority class [5,13,14,31]. In our problem, class imbalance is present in the underlying distributions, but not visible in the labeled training data. These solutions are therefore not applicable as such.

3.2 Small Sample

When labeling of the data is expensive or at least labor intensive, it can be beneficial to use unlabeled data in addition to the labeled data for model learning. Typically, the unlabeled samples strongly outnumber the labeled samples.

Semi-supervised Learning. In the semi-supervised setting [3,34], it is often assumed that the underlying distributions are undersampled more or less randomly, i.e., $P(s|x,y) = P(s) = c$. The structure of the data $P(x)$ can be derived from all data, i.e. including the unlabeled data. Additional assumptions are used to benefit from the unlabeled samples such as a stronger smoothness assumptions than with ordinary supervised learning, a cluster assumptions and a manifold assumption [3]. In our case with strong class prior imbalance and class overlap, the smoothness assumption and cluster assumption will expectedly not hold, see for instance Fig. 1.

PU (Positive and Unlabeled) Learning. The Positive and Unlabeled (PU) learning problem is a special case of the semi-supervised learning problems [15]. In PU learning there are no labeled examples from the non-target class in the training set ($X^- = \emptyset$). The unlabeled set X^0 contains both targets and non-targets and the targets X^+ are often assumed to be sampled representatively [4], or at least randomly [9,19,33] from all targets. That is, $P(s|x,y=1) = P(s|y=1) = c$ (and $P(s|x,y=-1) = P(s|y=-1) = 0$). Several researchers cast this problem as a cost-sensitive learning problem [8,9,25]. Like in a PU learning problem, we do have labeled targets and a substantial amount of unlabeled data. But the assumption that the targets are randomly sampled from their underlying distribution is definitely violated in our problem scenario. Moreover, in our dataset there are labeled non-targets available that we can exploit.

3.3 Biased Sample

Clearly, our research is grounded in the selection bias literature. In [32], an overview of types of selection bias is given. Like this study, most research in the field deals with bias on features only. That is, the selection of a sample is independent of its label given its feature vector: $P(s|x,y) = P(s|x)$ [32]. Several approaches deal with such bias by reweighting the sample loss [6, 22, 24].

Our problem situation comes closest to the work of [16]. They also assume the availability of unlabeled samples. In that work, the unlabeled samples are used to better estimate the sample weights. Further, the selection bias is covariate shift [27], which assumes that $P(y|x)$ for training and test conditions is the same. We do not assume such properties. Furthermore, our approach allows for the application of many different learners. Our proposed data sampling method can be seen as a training set preprocessing step. We will work that out in the next section.

4 Method

The method we propose is sampling oriented. The available dataset consists of three groups of samples. The labeled targets X^+, the labeled non-targets X^- and the unlabeled mixture of targets and non-targets X^0.

For the sampling based method, we exploit that in the unlabeled samples X^0 the class sample imbalance is present, or:

Proposition 1

$$c_p(X^0) \ll 1 \tag{3}$$

By definition:

$$c_p(X^0) = \frac{N^+ - n^+}{N^- - n^-} \tag{4}$$

and

$$N^+ + N^- = n^0 + n^+ + n^- = |X|. \tag{5}$$

Further, the labeled and unlabeled data together X is a random sample, so:

$$\frac{N^+}{N^-} \approx c_p, \text{ and } N^- \gg N^+. \tag{6}$$

Given that also $n^0 \gg n^+, n^-$, it follows:

$$n^0 \approx N^-, \text{ and } N^- \gg n^-, \tag{7}$$

so

$$\frac{N^+ - n^+}{N^- - n^-} \approx \frac{N^+ - n^+}{N^-} = c_p - \frac{n^+}{N^-}. \tag{8}$$

Then, $c_p(X^0) < c_p \ll 1$.

4.1   Selection Bias

The labeled targets and non-targets have been selected using a biased selection strategy using features that individually or jointly correlate with the class label. Without loss of generality, we assume selection strategies that select a sample $(s = 1)$ for inspection if and only if it is at the right ('target') side of the selection line:

$$\begin{cases} s = 1, & S(x) > t \\ s = 0, & S(x) \leq t \end{cases} \tag{9}$$

where $S(x)$ is the selection bias function and t is the threshold.

We define a *perfect selection bias function* S_p as a function with the following property: if averaged over all possible threshold values t, a perfect selection bias function has the highest probability of picking a target over an non-target sample given the features. This comes down to that function S that has maximal AUC.

Alternatively, and usually, the selection strategy is suboptimally based on experience of inspectors or screeners. We denote such a biased strategy with selection function S_b.

4.2   Training Set Construction

Both selection strategies lead to non-representative training sets that are available for learning a model. In the problem statement the goal is formulated as to obtain an optimal ordering of the unlabeled data. The model $F(x)$ with optimal ordering of the data is the model with maximal AUC. This is the same as the perfect selection function S_p would achieve. The function S_p is, unfortunately, generally unknown.

Now the goal is to learn $F(x)$ using the available data. We propose three ways in which X^+, X^- and X^0 can be exploited for this purpose to obtain an effective training set. In our proposal, the targets in the training set X_{tr}^+ always equals the whole set of labeled targets X^+. The non-target set in the training set X_{tr}^- consists of a primary set X_{tr1}^- and a secondary sampled set X_{tr2}^-. The secondary set is subsampled from a large set in order to prevent the curse of strong class imbalance during learning [18]. For instance tree learners are known to suffer from it [18]. The complete training set is then:

$$X_{tr} = X_{tr}^+ \cup X_{tr1}^- \cup X_{tr2}^- \tag{10}$$

For the construction of the primary (X_{tr1}^-) and secondary non-target training set (X_{tr2}^-), we propose the following three strategies.

Positive, Negative, and Sampled from Unlabeled (PN[U]). The application of a classifier with as training set the labeled X^+ and X^- seems straightforward. These are the only samples with ground truth. Further, it is not uncommon to have an order of $100 - 1000$ targets and non-targets, which are reasonable numbers for supervised model learning. However, the question is to what extent the selection bias, either S_b or even S_p hampers the model learning. That is, expectedly, from the non-targets and even from the targets only samples from the tail of the distribution are available. Many classifiers may suffer from such a degenerate data sample. As shown earlier, with good target oriented bias, the selected samples will be in the 'margin' of the classes, while the high density areas of the non-target class are not sampled.

Therefore, in addition to the labeled non-targets, we propose to subsample from the unlabeled samples X^0 to obtain a larger set of mostly non-targets, see Proposition 1. In this way, the high density area of the non-target class is also represented in the training set. Indeed at the cost of a number of wrongly labeled targets. However, we consider problems where the class prior imbalance is of order 1 in 100, or $c_p = 0.01$. Following Proposition 1, the number of label errors introduced this way is limited. Ultimately, the resulting training set will become strongly imbalanced, which may harm certain classifiers [17,31]. Restricting the number of samples from X^0 may therefore be beneficial. The ratio of samples drawn into X_{tr2}^- is the sampling ratio r_s.

$$X_{tr1}^- = X^-, \ X_{tr2}^- \subseteq X^0, \text{ and } r_s = \frac{|X_{tr2}^-|}{|X^0|} \tag{11}$$

Positive, and Sampled from Negative and Unlabeled (P[NU]). The set X^- is a very specific set of non-target samples. Through the feature vector driven selection bias, these non-targets are not representative at all for the non-target distribution $P(x|y = -1)$. Different from normal labeled data sets, in this case we know which samples are probably in the margin between the classes. Stated differently, the vast majority of the unlabeled samples are likely further away from the target class than the labeled non-target samples. This is a phenomenon that we could exploit. In this proposal, we sample from the mix of labeled non-targets and unlabeled samples. Such a non-target training set X_{tr}^- is probably a better representation of the non-targets even with a small sample from X^0.

$$X_{tr1}^- = \emptyset, \ X_{tr2}^- \subseteq X^- \cup X^0, \text{ and } r_s = \frac{|X_{tr2}^-|}{|X^- \cup X^0|} \tag{12}$$

Positive, and Sampled from Unlabeled (P[U]). Finally, the most rigorous proposal is to completely discard the labeled set X^-. Since the samples from X^- are presumably in the margin, they can be confusing for the model. Moreover, in domains such as law enforcement the labeling is typically noisy. If during an inspection the violation of the law cannot be proven, the sample will be labeled as non-target. In this respect, resources and legal possibilities are limited. Therefore, we propose a training set construction strategy where the non-target training samples X_{tr}^- are sampled solely from the unlabeled labeled samples X^0.

$$X_{tr1}^- = \emptyset, \ X_{tr2}^- \subseteq X^0, \text{ and } r_s = \frac{|X_{tr2}^-|}{|X^0|} \tag{13}$$

5   Experiments

In this section, we describe the experiments we did to evaluate the training set construction strategies PN[U], P[NU], and P[U] to obtain the best ranking model. A major challenge is how to test and compare the learned models using the various training sets. In order to have controlled conditions, we started with a synthetic dataset generated from known distributions and ground truth labels. Then, we applied the proposed method to a true law enforcement dataset, which was the motivation for this research.

We applied an extensive set of linear and non-linear classifiers for all the experiments described, such as the Fisher classifier [7], logistic regression [7], L1 and L2 regularized linear SVM [10], L1 and L2 regularized logistic regression [10], gradient boosting [12], and random forests [2]. In order to focus on the general message of this study, we only report the results of the Fisher classifier. Some other classifiers had higher performance for certain problem parameter settings, but needed parameter tuning of several hyper parameters. The trends and general results were the same for all methods employed.

5.1   Synthetic Data

The experiments with synthetic data were run using two Gaussian distributions where the distance between the class means is 1 in one dimension and 0 in the 99 other dimensions, and the variance per class $\sigma^2 = 1$ in all $p = 100$ dimensions. In this controlled experiment, we additionally varied the class prior imbalance to measure the influence of this parameter on the proposed three training set construction methods.

Biased Imbalanced Dataset Construction. We created the biased, imbalanced, and partially labeled dataset as follows. First, we generated a large Gaussian test dataset (100,000 samples) with class prior imbalance ratio $c_p = 0.01$ to obtain strong class imbalance according to the assumptions of this study. The minority class was the target class. Then, we generated a reference dataset with the same properties as the test set with in total 10,000 samples. For convenience, we estimated $S_p(x)$ with a classifier learned with this reference dataset resulting in $\hat{S}_p(x)$. Alternatively, we could have derived it straight from the data generation prescription. We applied this simulated selection bias function $\hat{S}_p(x)$ to the reference dataset and ranked it according to the output of function $\hat{S}_p(x)$. We set the threshold t, such that for the first 500 samples it holds $\hat{S}_p(x) > t$. These samples were used to construct X^+ and X^-, with their original labels. All remaining samples were put in the set of unlabeled samples X^0. Accordingly, the labeled targets and non-target samples are at the tails of their respective distributions in the same direction in feature space, similar to the examples in Fig. 1.

Results. We applied the three training set construction strategies PN[U], P[NU], and P[U] for various subsample ratios r_s for the secondary training set X_{tr2}^-. In Fig. 3, we show the AUC performance of the Fisher classifier for this experiment. The figure shows that with very few unlabeled samples, the method PN[U] has higher performance than P[NU] and P[U]. This could be expected, because PN[U] has at least a substantial set of true negatives from X^-, and the training set construction strategies P[NU] and P[U] have hardly any negative training samples. When the number of samples in X_{tr2}^- increases to 1% (i.e. $|X_{tr}^+| \approx |X_{tr2}^-|$), both P[NU] and P[U] outperform PN[U]. Finally, when all data has been used, PN[U] and P[NU] converge to the same AUC performance, since then both have as training set $X_{tr}^- = X^- \cup X^0$. Importantly, discarding the samples from X^- altogether according to strategy P[U] gives a small ($\approx 1.0\%$) improvement with respect AUC performance over PN[U] and P[NU]. The PN[U] curve in the figure also suggests improvement of discarding the samples from X^-. That is, towards the right side of the figure, the PN[U] curve is still strongly increasing, while adding unlabeled samples. The P[NU] curve that is based on a mixture of negative and unlabeled samples already flattens off. It follows that making the imbalance between unlabeled samples and non-target samples larger by discarding X^- indeed helps.

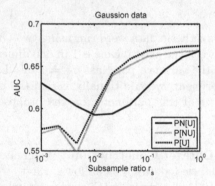

Fig. 3. AUC performance as function of the amount of unlabeled data for the synthetic 100-dimensional Gaussian distributions. The results are shown for the different training conditions: PN[U], P[NU], and P[U]. The class prior imbalance ratio $c_p = 0.01$.

Variable Class Imbalance. Next, we varied the class prior imbalance c_p to see if discarding the non-target class is always beneficial. We let c_p range from $0.0001 \ldots 1.0$. As in the first experiment, we took the top 500 samples that were ranked by the biased model. Again, these samples had their original labels, while the lower ranked samples were considered unlabeled. Because we always took the top 500 samples, not only the imbalance ratio of the reference and test set changed, but also the class sample imbalance ratio c_s of the labeled sets X^+ and X^-. With every setting we ran the experiment 30 times. In Table 1, we show the gain in average AUC between PN[U], P[NU], and P[U] when all unlabeled data are used.

Table 1. AUC performance with various class prior imbalance ratios c_p. The table shows the difference between scenarios PN[U], P[NU], and P[U], when all unlabeled data is used. Note that PN[U] and P[NU] are the same when all unlabeled data is used.

c_p	n^+	n^-	$PN[U] = P[NU]$	$P[U]$	Δ
0.001	9	491	0.575	0.576	0.001
0.003	19	481	0.628	0.634	0.006
0.01	39	461	0.669	0.676	0.007
0.03	76	424	0.710	0.714	0.004
0.1	212	288	0.741	0.743	0.002
0.3	390	110	0.751	0.751	0.000
1.0	484	16	0.750	0.750	0.000

As can be seen in Table 1, the gain in AUC is negligible at the extremes, i.e. when $c_p = 0.001$ and when $c_p = 1.0$. When the class imbalance $c_p = 0.01$ then the AUC gain is maximal, approximately 1 %. These results can be explained by the different effects that the decrease in size of X^- and at the same time increase in size of X^+ may have on the results in the experiments. These are hard to separate. However, in these experiments it never hurts to discard the labeled non-targets.

5.2 Law Enforcement Data

The motivation for this research was an investigation project for the Inspectorate SZW of the Ministry of Social Affairs and Employment, The Hague. The objective of the project was to find companies that are likely to violate the Foreign Nationals Employment Act. The dataset consists of 65,487 companies from which various information is stored in a collected database, such as:

- the internal database of the Inspectorate SZW: name of company, postal code, city, phone number, main branch of company;
- tax information over three years: such as turnover tax, tax return, total salaries, possible tax fines;
- chamber of commerce information: sector of company activities, personnel, legal form.

Not all database values were available for all companies. Since we do not know the cause of missingness, being Missing At Random (MAR), Missing Completely At Random (MCAR), or Missing Not At Random (MNAR) [21], we encoded all relevant variables with an additional variable indicating whether the value of the variable was available or not [26]. This is especially important in case the data are MNAR, because then the information in the missingness itself can be exploited.

In addition to the given data, other features were engineered, such as: a set of features that dummy code most frequent cities and sectors of activity, the number of similar companies in the neighborhood, ratios such as salary/turnover, variations over the years in minimum, mean, and maximum of various tax data. From the given data fields, in total $p = 272$ features were derived.

Validation. The inspectorate has roughly two different reasons for doing an inspection at a company. The first is that based on their knowledge and expertise a company with certain properties is more likely to violate the Foreign Nationals Employment Act, which we call Inspectorate Bias Selection (IBS). The other reason is that a third party has made a report indicating that such a violation is suspected. We call this type of selection Third Party Selection (TPS).

The IBS inspections can be considered according to the selection bias of the inspectorate. For these samples holds: $P(s|x,y) = P(s|x)$. This is the type of selection bias we assumed throughout this research. Among the companies, 5,458 have been inspected in the last 5 years as a result of the knowledge and expertise of the inspectorate. From these, 610 were fined for one or more violations of the Foreign Nationals Employment Act. These data can serve as training data, where $n^+ = 610$, $n^- = 4,848$, and $n^0 = 60,029$.

To test the performance of the strategies PN[U], P[NU], and P[U], a suitable test set with ground truth is required. For this real-world dataset, the ground truth for the unlabeled data is not readily available. Alternatively, we propose to use the TPS samples. The inspections based on third party reports will not be according to the selection bias from the inspectorate themselves. The cases have probably been reported for other reasons than the in the data available information. The reports are from governmental organizations like the police, the Netherlands Food and Consumer Product Safety Authority, the immigration office, or ordinary citizens. Therefore, we consider these diverse reports to representatively cover the population of targets. That is, the presence of a report is independent of the features given the labels: $P(s|x, y = +1) = P(s|y = +1)$ [32]. This implies that we can use these targets as test set *targets*. Further, it is given that most companies are non-targets. Among the unlabeled samples also most samples are non-target. We therefore use the unlabeled samples as non-targets in the test set, while we know we introduce limited label noise this way.

Now, the non-targets X^- and the unlabeled data X^0 play a role in both the training phase and test phase. Moreover, the *test targets count as unlabeled in the training set* and the *training targets as unlabeled in the test set*. To handle training and test sets properly, a suitable cross-validation scheme is required. We used 10-fold cross-validation in the following way. We collected indexes from the whole dataset in the standard 9/10th training and 1/10th test scheme as usual with cross-validation. Then, the training indexes select targets from the IBS targets and non-targets from the unlabeled set, which includes targets from TPS targets. The test set contains target indexes from the TPS targets and also non-targets from unlabeled set, which in this case include IBS targets. We repeated the 10-fold experiments for strategy PN[U], P[NU], and P[U] three times.

Results. Fig. 4 shows the averaged AUC performance curves of the tests performed. The figure displays very similar behavior with respect to AUC performance, compared to the synthetic experiment in Fig. 3. Adding unlabeled samples led to substantial increase in AUC for all three training set construction strategies. With very little unlabeled samples $r_s \approx 0.0002$, strategy P[NU] and P[U] already outperform PN[U] in AUC. This comes down to approximately only 10 samples in X_{tr}^- for strategy P[NU] and P[U]. The gain of discarding X^- (P[U] compared to P[NU]) is approximately 2 % in AUC. Again, at the right side of the figure, the PN[U] curve is still strongly increasing, while adding unlabeled samples. The P[NU] curve that is based on a mixture of non-target and unlabeled samples already flattens off. Indeed, discarding X^- helps.

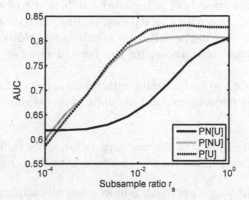

Fig. 4. Performance as function of the amount of unlabeled data for the Inspectorate SZW dataset with training conditions PN[U], P[NU], and P[U].

6   Conclusions

In this paper we studied selection bias with imbalanced classes. This specific problem arises in screening situations such as law enforcement and medical screening. Because of the strong class imbalance, biased selection strategies are employed to make the screening procedure feasible. We proposed three training set construction strategies that exploit unlabeled data to learn ranking models to make the screening more efficient. The proposed strategies exploit labeled targets, labeled non-targets and unlabeled samples as follows:

– Positive, Negative, and Sampled from Unlabeled (PN[U]);
– Positive, and Sampled from Negative and Unlabeled (P[NU]);
– Positive, and Sampled from Unlabeled (P[U]).

With synthetic and real-life law enforcement data, we tested the proposed strategies on their ranking abilities to find targets among the unlabeled samples.

We explored the problem with many different classifiers ranging from the Fisher classifier, logistic regression and SVM to random forests and gradient boosting. For the sake of computational speed and clarity of the message, we selected the Fisher classifier for the experiments.

The synthetic experiments were performed with biased imbalanced data from Gaussian distributions. With class imbalance between targets and non-targets of at least 1–10, the AUC performance of the models improved strongly by adding unlabeled data to the non-target class. Importantly, completely discarding the labeled non-target samples led to the best AUC performance.

For the real-world law enforcement dataset, we proposed to use a set of targets as validation set that were selected through third party reports of suspected violations of the Foreign Nationals Employment Act. These targets have the property that the selection bias is (mostly) independent of the feature vector given the class label. Also with this dataset, adding unlabeled data to the non-targets improved the ranking performance significantly. Moreover, also with this real-life dataset, completely discarding the labeled non-targets was the most effective.

For both synthetic and real-life data with (1) class prior imbalance, (2) biased selected targets and non-targets, and (3) clear class overlap, the following can be concluded:

– Adding unlabeled samples to the non-targets improves ranking performance,
– Leaving out the labeled non-targets results in even better performance.

The proposed validation approach is more generally applicable. Also in other screening situations, samples can be recognized as target through other processes than the standard selection bias. For instance, the population can be screened for diseases based on age and gender, while patients pro-actively go to the doctor in case of medical symptoms. For this research, such targets were useful to test our method of dealing with selection bias with imbalanced classes. Clearly, in case such target samples are available, these should be used as training set targets to have a better (less biased) representation of the target class.

Acknowledgements. We would like to thank Gerard Meester and his colleagues from the Inspectorate SZW of the Ministry of Social Affairs and Employment, The Hague, The Netherlands, for their support and making their data available for our research. We also thank Dr. Marco Loog for his critical review of and extensive comments on a previous version of the paper.

References

1. Bradley, A.: The use of the area under the ROC curve in the evaluation of machine learning algorithms. Pattern Recogn. **30**(7), 1145–1159 (1997)
2. Breiman, L.: Random forests. Mach. Learn. **45**(1), 5–32 (2001)
3. Chapelle, O., Schölkopf, B., Zien, A.: Semi-supervised Learning (2006)

4. Chaudhari, S., Shevade, S.: Learning from positive and unlabelled examples using maximum margin clustering. In: Huang, T., Zeng, Z., Li, C., Leung, C.S. (eds.) ICONIP 2012. LNCS, vol. 7665, pp. 465–473. Springer, Heidelberg (2012). doi:10.1007/978-3-642-34487-9_56

5. Chen, C., Liaw, A., Breiman, L.: Using Random Forest to Learn Imbalanced Data. Technical report, Department of Statistics, University of Berkeley (2004)

6. Cortes, C., Mohri, M., Riley, M., Rostamizadeh, A.: Sample selection bias correction theory. In: Freund, Y., Györfi, L., Turán, G., Zeugmann, T. (eds.) ALT 2008. LNCS (LNAI), vol. 5254, pp. 38–53. Springer, Heidelberg (2008). doi:10.1007/978-3-540-87987-9_8

7. Duda, R., Hart, P., Stork, D.: Pattern Classification. John Wiley and Sons Inc., New York (2001)

8. Elkan, C.: The foundations of cost-sensitive learning. In: Proceedings of the 17th International Joint Conference on Artificial Intelligence, IJCAI 2001, vol. 2, pp. 973–978 (2001)

9. Elkan, C., Noto, K.: Learning classifiers from only positive and unlabeled data. In: Proceedings of the 14th ACM SIGKDD International Conference on Knowledge Discovery and Data Mining, KDD 2008, pp. 213–220. ACM, New York (2008)

10. Fan, R.E., Chang, K.W., Hsieh, C.J., Wang, X.R., Lin, C.J.: LIBLINEAR: a library for large linear classification. J. Mach. Learn. Res. **9**, 1871–1874 (2008)

11. Fawcett, T., Provost, F.: Adaptive fraud detection. Data Min. Knowl. Disc. **1**, 291–316 (1997)

12. Friedman, J.H.: Greedy function approximation: a gradient boosting machine (2000)

13. Guo, X., Yin, Y., Dong, C., Yang, G., Zhou, G.: On the class imbalance problem. In: 2008 Fourth International Conference on Natural Computation, ICNC 2008, vol. 4, pp. 192–201. IEEE (2008)

14. He, H., Garcia, E.: Learning from imbalanced data. IEEE Trans. Knowl. Data Eng. **21**(9), 1263–1284 (2009)

15. Hu, H., Sha, C., Wang, X., Zhou, A.: A unified framework for semi-supervised PU learning. World Wide Web **17**(4), 493–510 (2014)

16. Huang, J., Smola, A., Gretton, A., Borgwardt, K., Scholkopf, B.: Correcting sample selection bias by unlabeled data. In: Advances in Neural Information Processing Systems, vol. 19, p. 601 (2007)

17. Japkowicz, N., Stephen, S.: The class imbalance problem: a systematic study. Intell. Data Anal. **6**(5), 429–449 (2002)

18. Kubat, M., Matwin, S.: Addressing the curse of imbalanced training sets: one-sided selection. In: Proceedings of the Fourteenth International Conference on Machine Learning (ICML), pp. 179–186. Morgan Kaufmann (1997)

19. Li, H., Chen, Z., Liu, B., Wei, X., Shao, J.: Spotting fake reviews via collective positive-unlabeled learning. In: IEEE International Conference on Data Mining (ICDM 2014) (2014)

20. Li, Q., Wang, Y., Bryant, S.: A novel method for mining highly imbalanced high-throughput screening data in PubChem. Bioinformatics **25**(24), 3310–3316 (2009)

21. Little, R.J.A., Rubin, D.B.: Statistical Analysis with Missing Data. Wiley, New York (2002)

22. Liu, A., Ziebart, B.: Robust classification under sample selection bias. In: Advances in Neural Information Processing Systems 27: Annual Conference on Neural Information Processing Systems 2014, Montreal, Quebec, Canada, 8–13 December 2014, pp. 37–45 (2014)

23. Malof, J., Mazurowski, M., Tourassib, G.: The effect of class imbalance on case selection for case-based classifiers: an empirical study in the context of medical decision support. Neural Netw. **25**(1), January 2012

24. Mansour, Y., Mohri, M., Rostamizadeh, A.: Domain adaptation: learning bounds and algorithms. CoRR

25. du Plessis, M., Niu, G., Sugiyama, M.: Analysis of learning from positive andun-labeled data. In: Advances in Neural Information Processing Systems 27: Annual Conference on Neural Information Processing Systems, Montreal, Quebec, Canada, 8–13 December 2014, pp. 703–711 (2014)

26. Ramoni, M., Sebastiani, P.: Robust learning with missing data. Mach. Learn. **45**(2), 147–170 (2001)

27. Shimodaira, H.: Improving predictive inference under covariate shift by weighting the log-likelihood function. J. Stat. Plann. Infer. **90**(2), 227–244 (2000)

28. Van Vlasselaer, V., Akoglu, L., Eliassi-Rad, T., Snoeck, M., Baesens, B.: Guilt-by-constellation: fraud detection by suspicious clique memberships. In: 2015 48th Hawaii International Conference on System Sciences (HICSS), pp. 918–927. IEEE, January 2015

29. Varshney, K., Chenthamarakshan, V., Fancher, S., Wang, J., Fang, D., Mojsilović, A.: Predicting employee expertise for talent management in the enterprise. In: Proceedings of the 20th ACM SIGKDD International Conference on Knowledge Discovery and Data Mining, KDD 2014, pp. 1729–1738. ACM, New York (2014)

30. Veenman, C.: Data base investigation as a ranking problem. In: Proceedings of the European Intelligence and Security Informatics Conference (EISIC), Odense, Denmark, 21–24 August 2012

31. Visa, S., Ralescu, A.: Issues in mining imbalanced data sets - a review paper. In: Proceedings of the Sixteen Midwest Artificial Intelligence and Cognitive Science Conference, pp. 67–73 (2005)

32. Zadrozny, B.: Learning and evaluating classifiers under sample selection bias. In: Proceedings of the Twenty-First International Conference on Machine Learning, ICML 2004, p. 114. ACM, New York (2004)

33. Zhou, J., Pan, S., Mao, Q., Tsang, I.: Multi-view positive and unlabeled learning. In: Proceedings of the 4th Asian Conference on Machine Learning, ACML 2012, Singapore, Singapore, 4–6 November 2012, pp. 555–570 (2012)

34. Zhu, X.: Semi-supervised learning literature survey. Technical report (2006)

A Framework for Classification in Data Streams Using Multi-strategy Learning

Ali Pesaranghader[1]([⊠]), Herna L. Viktor[1], and Eric Paquet[1,2]([⊠])

[1] School of Electrical Engineering and Computer Science,
University of Ottawa, Ottawa, Canada
{apesaran,hviktor,eric.paquet}@uottawa.ca
[2] Information and Communications Technologies,
National Research Council of Canada, Ottawa, Canada
eric-paquet@nrc-cnrc.gc.ca

Abstract. Adaptive online learning algorithms have been successfully applied to fast-evolving data streams. Such streams are susceptible to concept drift, which implies that the most suitable type of classifier often changes over time. In this setting, a system that is able to seamlessly select the type of learner that presents the current "best" model holds much value. For example, in a scenario such as user profiling for security applications, model adaptation is of the utmost importance. We have implemented a *multi-strategy* framework, the so-called TORNADO environment, which is able to run multiple and diverse classifiers simultaneously for decision making. In our framework, the current learner with the highest performance, at a specific point in time, is selected and the corresponding model is then provided to the user. In our implementation, we employ an Error-Memory-Runtime (EMR) measure which combines the *error-rate*, the *memory usage* and the *runtime* of classifiers as a performance indicator. We conducted experiments on synthetic and real-world datasets with the Hoeffding Tree, Naive Bayes, Perceptron, K-Nearest Neighbours and Decision Stumps algorithms. Our results indicate that our environment is able to adapt to changes and to continuously select the best current type of classifier, as the data evolve.

Keywords: Data stream mining · Classification · Concept drift · Parallel learning · Multi-strategy learning · Adaptive learning

1 Introduction

Online and adaptive learning from evolving data streams have attracted the attention of many researchers during the last decade. Online learning algorithms continuously update their classification models by processing instances one-by-one to avoid any performance degradation [1,15]. Adaptive learning algorithms not only update classification models online but also monitor any change in the data distribution (the so-called *concept drift*), by using drift detectors. Subsequently, the

© Springer International Publishing Switzerland 2016
T. Calders et al. (Eds.): DS 2016, LNAI 9956, pp. 341–355, 2016.
DOI: 10.1007/978-3-319-46307-0_22

models are updated once a drift is detected [2–6]. Online and adaptive learning algorithms have been successfully applied in machine learning applications, robotics, recommender systems, and intrusion detection systems [7–9].

Recently, there has been a surge of interest in so-called *pocket data mining*, where the aim is to run data mining models on small devices [13]. In this setting, the models produced by data stream mining algorithms should not only have low error-rates. Rather, the learning techniques should also be lightweight and efficient in terms of runtime and other resource allocations. Specifically, it has been shown that running even a lightweight decision tree classifier will often crash a mobile device [13,14]. It then becomes important to consider issues such as storage space and memory utilization, when building models within a streaming environment.

However, in most streaming studies [2–9] the error-rate (or accuracy) is used as the defining measure of performance for evaluating adaptive learners. The interplay with memory usage and runtime considerations has only been investigated in a few studies. For example, Bifet et al. [10] considered memory, time and accuracy to compare the performances of ensembles of classifiers. Here, each measure was individually used for evaluation. In [11], Bifet et al. introduced the RAM-Hour measure, where every RAM-Hour equals to 1 GB of RAM occupied for one hour, to compare the performance of three versions of perceptron-based Hoeffding Trees. In addition, Zliobaite et al. [12] proposed the return on investment (ROI) measure for probing whether adaptation of a learning algorithm is beneficial. They concluded that adaptation should only take place if the expected gain in performance, measured by accuracy, exceeds the cost of other resources (e.g. memory and time) required for adaptation. In a recent study by Olorunnimbe et al. [13], the ROI measure is employed to adapt the number of classifiers used in an ensemble setting.

In this paper, we introduce the TORNADO framework that addresses the above-mentioned issues. In our environment, we implemented a number of very diverse learners that are run in *parallel* against a single data stream. A dynamic, *multi-strategy learning* approach is followed, where diverse classifiers co-exist. All of the learners access the stream at the same time and proceed to construct their individual models in *parallel*. This framework thus stands in contrast to existing environments, such as MOA [15] and Weka [16], where the different types of classifiers are run independently and/or where an ensemble is employed. We also introduce an Error-Memory-Runtime (EMR) measure that is used to continuously assess and rank the performance of the different types of classifiers. This ranking is subsequently used to dynamically select the model constructed by the current "best" learner and present it to the users.

The remainder of this paper is organized as follows: The next section introduces the EMR measure that we use to select the "best" model at a specific time. We detail our TORNADO framework in Sect. 3. Section 4 describes our experimental evaluation, and Sect. 5 concludes the paper.

2   Balancing Performance Measures

As stated above, error-rate is most often used to evaluate the performance of classifiers against a data stream that is susceptible to concept drift. Intuitively, as illustrated in Fig. 1, a change in the distribution of the data, as caused by concept drift, may lead to the error-rates of different types of classifiers to increase or decrease. In this figure, we simulate a number of drift points and show that, as a drift occurs, the classifier with the lowest error-rate changes. Based on this observation, it follows that a multi-strategy learning system where different types of classifiers co-exist and where the model, from the current "best" learner is provided to the users, may hold much value.

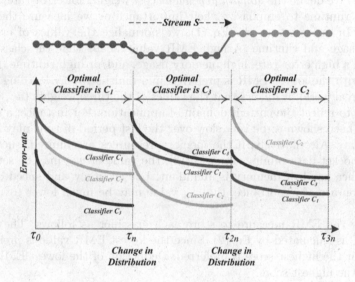

Fig. 1. Illustration of concept drift and error-rate interplay

However, following an *"error-rate-only"* approach is not beneficial in all settings. For instance, in an emergency response setting, the response time, i.e. the time it takes to present a model to the users, may be the most important criterion. That is, users may be willing to sacrifice accuracy for speed and partial information. Further, reconsider the area of pocket (or mobile) data mining, which has much application in areas such as defense and environmental impact assessment [14]. Here, the memory resources may be limited, due to connection issues, and thus reducing the memory footprint is also of importance.

In such a pocket data mining setting, consider two classifiers C_1 and C_2 which are used for classification over stream S. Suppose that the current model constructed by classifier C_1 has an 8.0 % error-rate, 20 MB memory usage, and 100 s runtime. On the other hand, the model constructed by classifier C_2 has the same error-rate, 1.5 MB memory usage, and 10,000 s runtime. It follows that these two classifiers have the same level of error-rate. However, the memory usage

of the first classifier may terminate the program, when running on a mobile device
[14]. Further, the second classifier may not be suitable in an emergency response
setting, where the goal is to optimize *just-in-time* decision making.

Based on these observations, we introduce the Error-Memory-Runtime
(EMR) measure. The EMR measure is defined as:

$$EMR_C = \frac{w_e \cdot E_C + w_m \cdot M_C + w_r \cdot R_C}{w_e + w_m + w_r} \tag{1}$$

where w_e is the error-rate cost weight, E_C is the error-rate of classifier C,
w_m refers to the memory usage cost weight, M_C denotes the memory usage of
classifier C, w_r depicts the runtime cost weight and R_C refers to the runtime of
classifier C. We define the *domain dependent* cost weights for error-rate, memory
usage and runtime to emphasize their importance as we measure the cost of
classifiers. In order to use the Eq. (1), we normalize the values of error-rate,
memory usage and runtime. A high EMR value means that the classifier has
resulted in a high error-rate, high memory usage, and/or high runtime. Hence, a
classifier with the lowest EMR is preferred in a multi-strategy learning setting.

In the real world, it follows that the values of the three weights (w_e, w_m, w_r)
will be set to reflect the current domain of application. For instance, a classifier
which has been shown to be very slow over the last period of time may be made
to fade away. Alternatively, if the memory resources are limited, such as the
case in a pocket data mining scenario [13], the value of w_m may be set higher.
On the other hand, if memory is abundant, but accuracy and speed of model
construction are of importance, the w_m value may be much lower (or even set
to zero).

We use the EMR measure to score each classifier as follows. The score of
classifier C is calculated by Eq. (2). Since the lowest EMR value is preferred, it
follows that the highest score is preferred. Deduction of the lowest EMR from 1
results in the highest score.

$$Score_C = 1 - EMR_C \tag{2}$$

The average EMR which represents the average cost of a classifier over time
is defined as follow:

$$\overline{EMR_C} = \frac{1}{T} \cdot \sum_{\tau=1}^{T} EMR_C{}^{\tau} \tag{3}$$

In this equation, T is the total number of EMRs calculated for the classifier
C so far and $EMR_C{}^{\tau}$ is the τth EMR of that classifier. In addition, the average
score of the classifier C is defined as:

$$\overline{Score_C} = 1 - \overline{EMR_C} \tag{4}$$

3   The Tornado Framework

In this section, we introduce the TORNADO framework, coded in Python, as outlined in Fig. 2. Recall that, in our framework, a number of diverse classifiers are

executed in *parallel*, against the same stream. In our system, a number of diverse base classifiers co-exist. That is, we include classifiers with different learning styles such as rule-based learners, decision trees, and nearest neighbour approaches in our Framework. This choice is based on the observation that employing diverse learning styles often leads to highly accurate and stable models.

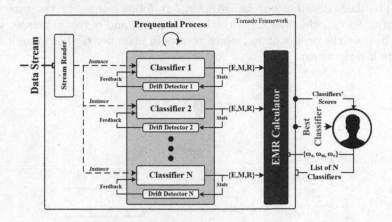

Fig. 2. The TORNADO framework

We focus on the STREAM READER, CLASSIFIERS, DRIFT DETECTORS and EMR CALCULATOR components of the framework in this section. The inputs are a Stream, the error-rate weight w_e, the memory usage weight w_m, and the runtime weight w_r. Our framework operates based on the *prequential* approach where instances are first tested and then used for training, allowing for a maximum usage of the data [1,18].

The framework's workflow is as follows. Initially, the STREAM READER reads instances from the stream one-by-one, and sends each to the learners for model construction. Each learner builds incremental models. That is, it first tests the instance and then trains from the latter for updating its model. At the same time, CLASSIFIERS send their statistics (e.g. error-rate and total number of instances observed since the last drift), to their respective DRIFT DETECTOR in order to detect a potential occurrence of a drift. Also, each one of the classifiers sends the error-rate, memory usage and runtime (including testing and training time) to the EMR CALCULATOR in order to evaluate the scores which are subsequently ranked. As a result, the current model with the highest rank is presented to the user. This selection may change, as we learn incrementally from the stream and concept drift occurs. This process continues until either a pre-defined condition is met or when all the instances in the stream are observed.

The following observation is noteworthy. Our multi-strategy framework contains different types of base learners, i.e. it is different from an ensemble of classifiers. In this framework, the individual learners proceed independently to construct their models. The reason for this design decision is that we aim to utilize diverse learning strategies that potentially address concept drifts differently. However, future work may include incorporating lightweight ensembles, i.e. ensembles with lower memory utilization, into our Framework.

Figure 3 gives an example showing how TORNADO runs the classifiers simultaneously and then recommends the best one by considering their scores at each time interval. During τ_0 to τ_n, classifier C_1 is considered as being the best classifier. The data distribution change at τ_n. It follows that, in the interval in between τ_n to τ_{2n}, classifier C_3 is the best classifier and is recommended to the user. Then, a drift occurs at τ_{2n} which results in classifier C_2 becoming the one with the highest score.

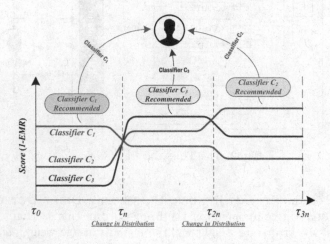

Fig. 3. Recommending the classifier with the highest score

The pseudocode of our approach is shown in Algorithm 1. The inputs are a Stream, a list of Classifiers, the error-rate weight w_e, the memory usage weight w_m, and the runtime weight w_r. The outputs are the *scores of classifiers* and the *classifier recommended*. The first instance is used to train all classifiers. Future instances are tested and used for the training by the classifiers. If there is a drift, the classifier and the corresponding drift detector are reset. When an instance is processed, the CALCULATESCORES function is invoked in order to calculate the scores associated with the classifiers. Within this function, the NORMALIZELIST function normalizes the error-rate, the memory usage and the runtime. Subsequently, the CALCULATEEMR function calculates the EMRs associated with the various classifiers. Finally, all of the classifiers scores are shown to the users and the classifier with the current lowest EMR is recommended. Recall that this process continues until the process is terminated by the user, or the end of the stream is met.

Algorithm 1. High-level Pseudocode of the TORNADO Framework

```
 1: inputs:
      stream                                          ▷ Stream of Instances
      classifiers                                     ▷ List of Classifiers
      w_e, w_m, w_r              ▷ Error-rate, Memory usage, Runtime Weights
 2: outputs:
      scores                                          ▷ Scores of Classifiers
      classifier_recommended                          ▷ Classifier Recommended

 3: for each instance in stream do
 4:     X = instance.X                                ▷ Values of Attributes
 5:     y = instance.y                                ▷ Real Class
 6:     for each classifier in classifiers do
 7:         y' = classifier.test(X)                   ▷ Predicted Class
 8:         classifier.update_error_rate(y, y')
 9:         if classifier.detector.drift is True then
10:             classifier.reset()       ▷ Resetting the Classifier and its Detector
11:         else
12:             classifier.train(X)
13:         end if
14:     end for
15:     scores = CALCULATESCORES(classifiers, w_e, w_m, w_r)
16:     classifier_recommended = GETBESTCLASSIFIER(scores)
17:     PRINTOUT(scores)                      ▷ Printing out the Scores of Classifiers
18:     RECOMMEND(classifier_recommended)        ▷ Printing out the Best Classifier
19: end for

20: function CALCULATESCORES(classifiers, w_e, w_m, w_r)
21:     errors, memories, runtimes, scores = []
22:     for each classifier in classifiers do
23:         errors.append(classifier.error)
24:         memories.append(classifier.memory)
25:         runtimes.append(classifier.runtime)
26:     end for
27:     NORMALIZELIST(errors, memory, runtimes)
28:     for i in range(0, classifiers.length) do
29:         emr = CALCULATEEMR(errors[i], memories[i], runtimes[i], w_e, w_m, w_r)
30:         scores.append(1 - emr)
31:     end for
32:     return scores
33: end function

34: function CALCULATEEMR(e, m, r, w_e, w_m, w_r)
35:     return (w_e · e + w_m · m + w_r · r)/(w_e + w_m + w_r)
36: end function

37: function GETBESTCLASSIFIER(scores)
38:     return classifiers[scores.index(scores.max())]
39: end function
```

---

4   Experimental Evaluation

In this section, we describe our experimental evaluations against seven synthetic and five real-world datasets from the data stream mining domain [3,10,11,17]. The Hoeffding Tree (HT), Naive Bayes (NB), Perceptron (PR), K-NN, and Decision Stump (DS) learning algorithms are available in our framework. In all experiments, the parameters of the Hoeffding Tree are set to $\delta = 10^{-7}$ (the allowed chance of error), $\tau = 0.05$ (the tie) and $n_{min} = 200$ (the minimum number of observation at each leaf), as determined in [19]. The Perceptron's learning rate is set to $\alpha = 0.01$, as determined in [11]. From the preliminary experiments, we found that 5-NN is the most suitable classifier of the K-NN family in terms of error-rate, memory usage and runtime results.

In our current experimental setup, the error-rate weight w_e, the memory usage weight w_m and the runtime weight w_r are all set to one in order to avoid any bias. We used the Drift Detect Method (DDM) [3] for drift detection, as it is less subject to false alarm and requires less memory [1,3,6]. We report the error-rate, the memory usage (in kilobytes) and the runtime (in seconds) of the classifiers at every thousand instances (this value was set by inspection). All experiments were executed on an Intel Core i7-3770 CPU Quad Core 3.4 GHz with 16 GB of RAM running on Windows 10.

4.1   Experiments on Synthetic Datasets

We generated seven synthetic datasets, as described in [3,17], for our experiments. Each dataset contains two class labels of *positive* and *negative*, and 100,000 (one hundred thousand) instances. Concept drifts occur at every 20,000 instances in the SINE1, SINE2 and MIXED, at every 25,000 instances in the CIRCLES, and at every 33,333 instances in the STAGGER datasets.

- SINE1 · *with abrupt concept drift*: It consists of two attributes x and y uniformly distributed within the interval $[0, 1]$. Instances under the curve $sin(x)$, i.e. the classification function, are classified as positive and other instances as negative. At a drift point, the classification paradigm is reversed, e.g. instances under the curve are classified as negative while the remaining instances are classified as positive.
- SINE2 · *with abrupt concept drift*: Like SINE1, it has two attributes x and y which are uniformly distributed in between 0 and 1. The classification function is $0.5 + 0.3 * sin(3\pi x)$. Instances under the curve are classified as positive while the other instances are classified as negative. At a drift point, the classification paradigm is inverted.
- CIRCLES · *with gradual concept drift*: It involves two attributes x and y which are uniformly distributed. The classification is performed with four disk functions. The boundary of each disk is defined by a circle $<(x_c, y_c), r_c>$ that is further described by the function $(x - x_c)^2 + (y - y_c)^2 - r_c^2 = 0$, where (x_c, y_c) is the center and r_c is the radius. The parameters for the four disks are $<(0.2, 0.5), 0.15>$, $<(0.4, 0.5), 0.2>$, $<(0.6, 0.5), 0.25>$ and $<(0.8, 0.5), 0.3>$ respectively. A change in the classification function results in a drift.

- STAGGER · *with abrupt concept drift*: It contains only three nominal attributes of *size {small, medium, large}*, *color {red, green}* and *shape {circular, non-circular}*. Before the first drift point, instances are labelled positive if *color = red∧size = small*. Then after this point and before the second drift, instances are classified positive if *color = green∨shape = circular*, and finally after this second drift point, instances are classified positive only if *size = medium ∨ size = large*.
- MIXED · *with abrupt concept drift*: It has two Boolean attributes v and w, and two numeric attributes uniformly distributed in the interval [0, 1]. The instances are classified as positive, if at least two of the three following conditions are satisfied, namely two of v, w and/or $y < 0.5 + 0.3 * sin(3\pi x)$ hold. The classification paradigm is reversed, after each drift point.
- $\text{SINE1}_1^{100,000}$ • $\text{SINE2}_1^{100,000}$ (S1•S2): We generated this stream by creating the union of SINE1 and SINE2.
- $\text{STAGGER}_1^{100,000}$ • $\text{MIXED}_1^{100,000}$ • $\text{SINE1}_1^{100,000}$ (S•M•S1): We generated this stream through concatenating all the instances in STAGGER, MIXED, and SINE1. That is, we appended all attributes and all classes.

Next, we discuss our experimental results when applying our multi-strategy learning framework to these data streams. The reader should recall that the main aim of these experiments was to illustrate how the "best" type of classifier changes over time, in the presence of concept drift.

We illustrated the evolution of classifiers' scores, error-rates, memory usage and runtime over time for the SINE1•SINE2 and STAGGER•MIXED•SINE1 data streams in Figs. 4 and 5, respectively. The "current best" classifier is shown at the top of the Figures.

Figure 4 shows that Perceptron initially performs better than the other classifiers in terms of the combined error-rate, memory usage, and runtime. However, Naive Bayes, Decision Stump and 5-NN are better choices towards the end of the stream. Specifically, Naive Bayes takes over as the best classifier until the end of the stream is reached. Despite the fact that the Hoeffding Tree and 5-NN classifiers have a lower error-rate than Decision Stump, their global performances are lower because of their memory usages and runtimes, respectively.

In Fig. 5, we compare the performances of classifiers as instances arrive over time for S•M•S1. At first, Perceptron globally dominates the other classifiers for the first 20 % of the instances. Then, it is outperformed by Decision Stump until 35 % of the instances have been observed and to be, in turn, outdone by Naive Bayes for the remainder of the stream, with Perceptron a close second.

We present the average scores for the five types of classifiers in Table 1. These values are included to depict the general behaviours of the classifiers against our synthetic data streams. Our analysis shows that there is no clear winner, in terms of overall performance. However, we notice that the Perceptron and Naive Bayes classifiers are the best at balancing the three indicators. Further, memory usages of Hoeffding trees are prohibitive, while the average runtimes of K-NN are the highest. Thus, the Hoeffding tree algorithm would be better suited in an

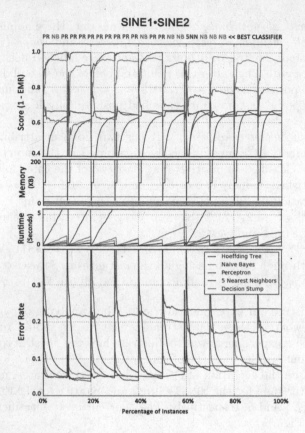

Fig. 4. S1•S2 Data stream: Scores, error-rates, memory usages and runtimes

environment where memory is not a critical resource, while K-NN seems to be unsuitable for building just-in-time models.

In this subsection we conducted experiments on the synthetic datasets and we confirmed, as expected, that the "best" classifier changes over the time, as drifts occur. In the next subsection we continue our experiments over real-world datasets.

4.2 Experiments on Real-World Datasets

In this section, we first describe the real-world datasets used in our experiments. These datasets were obtained from the UCI Machine Learning Repository [20]:

- NURSERY [21]: It consists of eight nominal attributes, five class labels and 12,960 instances. This dataset was used to recommend nursery to families.
- SHUTTLE dataset has nine numeric attributes, seven class labels and 58,000 instances. It was designed to predict suspicious states during a NASA shuttle mission. We use the training set which consists of 43,500 instances.

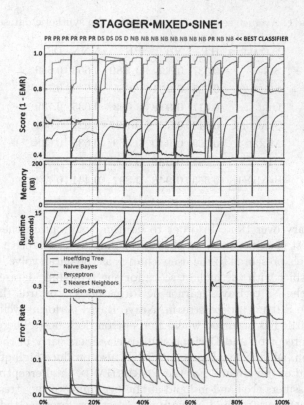

Fig. 5. S•M•S1 Data stream: scores, error-rates, memory usages and runtimes

- ELECTRICITY (ELEC) [22]: This dataset was collected from the Australian New South Wales Electricity company and has 45,312 instances with one nominal attribute, seven numeric attributes, and two class labels.
- POKERHAND [23]: The dataset consists of 1,000,000 instances and eleven attributes. Each record of the POKERHAND dataset is an example of a hand consisting of five playing cards drawn from a standard card deck of 52. Each card is described using two attributes (the suit and rank), for ten predictive attributes.
- ADULT [24]: It has six numeric and eight nominal attributes, two classes, and 48,842 instances. It allows predicting if a person has an annual income higher than $50,000.

We again followed the *prequential* learning method, as explained in Sect. 3, and proceed to incrementally construct models against these real-world datasets. Figures 6 and 7 show the performances of the various classifiers against the data streams. Note that the "*spikes*" in performance correspond to drift points. These figures clearly illustrate the fact that the performances of the various classifiers

Table 1. Average scores. of classifiers for the synthetic datasets

DATASETS	HT	NB	PR	5-NN	DS
SINE1	0.537	0.938	**0.980**	0.655	0.639
SINE2	0.595	**0.951**	0.665	0.649	0.786
CIRCLES	0.517	**0.918**	0.663	0.635	0.739
STAGGER	0.613	0.879	**0.997**	0.621	0.601
MIXED	0.569	**0.940**	0.869	0.635	0.648
S1•S2	0.567	**0.932**	0.809	0.656	0.692
S•M•S1	0.566	**0.948**	0.892	0.441	0.714

considerably vary over time. In order to continuously obtain the best performances, the best classifier must be chosen, leading to an adaptive environment where the models presented to the users change as the data evolve. For example, the initial classifier with the highest score for the NURSERY dataset is Decision Stump, while the Naive Bayes learner performs better towards the end of the stream. For the SHUTTLE data stream, Naive Bayes performs well, except just after concept drift has occurred where Perceptron does best. There is interplay between the Perceptron and Decision Stump classifiers early in the ELECTRICITY data stream, while Naive Bayes and 5-NN also produce high quality models towards the end of the stream. For the POKERHAND, the Perceptron and Decision Stump classifiers are hand-in-hand at the beginning of the stream, while the Perceptron, Hoeffding tree and Naive Bayes classifiers produce the best models later on. The ADULT data stream is the only one that has a clear winner, namely the Naive Bayes learner. Thus, the use of multi-strategy learning does not seem advantageous on this particular dataset.

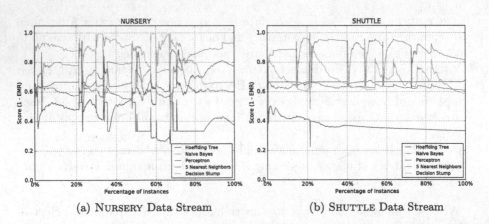

(a) NURSERY Data Stream   (b) SHUTTLE Data Stream

Fig. 6. The behaviors (overall scores) of classifiers for NURSERY and SHUTTLE data streams

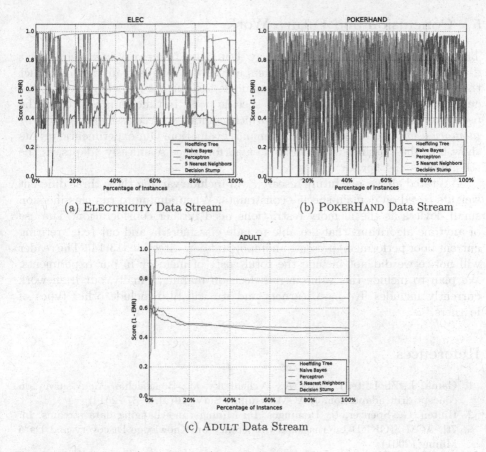

(a) ELECTRICITY Data Stream

(b) POKERHAND Data Stream

(c) ADULT Data Stream

Fig. 7. The behaviors (overall scores) of classifiers for ELECTRICITY, POKERHAND and ADULT data streams

Recall that the aim of our TORNADO framework is to create a multi-strategy environment in which classifiers with diverse learning strategies co-exist, and where the most suitable model is presented to the user at any given time. Our system currently includes five different learning algorithms. In the future, we also plan to extend this framework to include other types of classifiers. We are especially interested in including light-weight versions of algorithms. It further follows that the importance of the various weights will heavily depend on the domain of application. We will investigate the optimization of the various weights associated with the measures. Our future work will also focus on dynamically adapting the membership based on the performance of the individual classifiers. We envisage a system where new types of classifiers are added while poorer performing classifiers (e.g. ones that performed well on older data, or never performed well) retire. The valuable retirees are maintained, offline, in order to handle re-occurring concept drifts. Significantly, this will enable the system to revisit history and to adapt to new drifts that follow past patterns.

5   Conclusion and Future Work

This paper presented the multi-strategy TORNADO framework in which various diverse learners co-exist. These learners access the same stream and construct their models in *parallel*. In our environment, the current model created by the current "best" type of classifier, at a specific point in time, is presented to the user. We utilized an EMR measure which combines the classifiers' *error-rate*, *memory usage* and *runtime* for performance evaluation. Our experimental results show how the classifier with the highest performance seamlessly adapts, as the stream ebbs and flows.

As stated above, our future research will include varying these three different weights based on domain-specific constraints. When aiming to run classifiers on small devices as the memory restrictions need further consideration. The use of anytime algorithms that are able to fade classifiers in and out (e.g., retiring current poor performers) presents an important new direction [13,14]. The reader will notice we did not include the total costs of memory in our experiments. We plan to include this value in our decision making. Finally, our framework currently includes five base learners and we will also include other types of learners.

References

1. Gama, J., Zliobaite, I., Bifet, A., Pecheniziky, M., Bouchachia, A.: A survey on concept drift adaptation. J. ACM Comput. Surv. **46**(4), 1–37 (2014)
2. Hulten, G., Spencer, L., Domingos, P.: Mining time-changing data streams. In: 7th ACM SIGKDD International Conference on Knowledge Discovery and Data Mining (2001)
3. Gama, J., Medas, P., Castillo, G., Rodrigues, P.: Learning with drift detection. In: Bazzan, A.L.C., Labidi, S. (eds.) SBIA 2004. LNCS (LNAI), vol. 3171, pp. 286–295. Springer, Heidelberg (2004). doi:10.1007/978-3-540-28645-5_29
4. Gama, J., Fernandes, R., Rocha, R.: Decision trees for mining data streams. J. Intell. Data Anal. **10**(1), 23–45 (2006)
5. Bifet, A., Gavalda, R.: Learning from time-changing data with adaptive windowing. In: SIAM International Conference on Data Mining, pp. 443–448 (2007)
6. Huang, D.T.J., Koh, Y.S., Dobbie, G., Bifet, A.: Drift detection using stream volatility. In: Appice, A., Rodrigues, P.P., Santos Costa, V., Soares, C., Gama, J., Jorge, A. (eds.) ECML PKDD 2015. LNCS (LNAI), vol. 9284, pp. 417–432. Springer, Heidelberg (2015). doi:10.1007/978-3-319-23528-8_26
7. Koren, Y.: Collaborative filtering with temporal dynamics. In: 15th ACM SIGKDD International Conference on Knowledge Discovery and Data Mining, pp. 447–456 (2009)
8. Lee, W., Stolfo, S.J., Mok, K.W.: Adaptive intrusion detection: A data mining approach. J. Artif. Intell. Rev. **14**(6), 533–567 (2000)
9. Stavens, D., Hoffmann, G., Thrun, S.: Online speed adaptation using supervised learning for high-speed, off-road autonomous driving. In: 20th International Joint Conference on Artificial Intelligence, pp. 2218–2224 (2007)

10. Bifet, A., Holmes, G., Pfahringer, B., Kirkby, R., Gavalda, R.: New ensemble methods for evolving data streams. In: 15th ACM SIGKDD International Conference on Knowledge Discovery and Data Mining, pp. 139–148 (2009)
11. Bifet, A., Holmes, G., Pfahringer, B., Frank, E.: Fast perceptron decision tree learning from evolving data streams. In: Zaki, M.J., Yu, J.X., Ravindran, B., Pudi, V. (eds.) PAKDD 2010. LNCS (LNAI), vol. 6119, pp. 299–310. Springer, Heidelberg (2010). doi:10.1007/978-3-642-13672-6_30
12. Zliobaite, I., Budka, M., Stahl, F.: Towards cost-sensitive adaptation: when is it worth updating your predictive model? Neurocomputing 150, 240–249 (2015)
13. Olorunnimbe, M.K., Viktor, H.L., Paquet, E.: Intelligent adaptive ensembles for data stream mining: A high return on investment approach. In: Ceci, M., Loglisci, C., Manco, G., Masciari, E., Ras, Z.W. (eds.) NFMCP 2015. LNCS (LNAI), vol. 9607, pp. 61–75. Springer, Heidelberg (2016). doi:10.1007/978-3-319-39315-5_5
14. Gaber, M., Stahl, F., Gomes, J.B.: Pocket Data Mining: Big Data on Small Devices. Studies in Big Data. Springer, Heidelberg (2014)
15. Bifet, A., Holmes, G., Kirkby, R., Pfahringer, B.: MOA: Massive online analysis. J. Mach. Learn. Res. 11, 1601–1604 (2010)
16. Hall, M., Frank, E., Holmes, G., Pfahringer, B., Reutemann, P., Witten, I.H.: The weka data mining software: An update. ACM SIGKDD Explor. Newsl. 11(1), 10–18 (2009)
17. Kubat, M., Widmer, G.: Adapting to drift in continous domain. In: 8th European Conference on Machine Learning, pp. 307–310. Springer, Heidelberg (1995)
18. Gama, J., Sebastiao, R., Rodrigues, P.P.: On evaluating stream learning algorithms. J. Mach. Learn. 90(3), 317–346 (2013)
19. Domingos, P., Hulten, G.: Mining high-speed data streams. In: 6th ACM SIGKDD International Conference on Knowledge Discovery and Data Mining, pp. 71–80 (2000)
20. Lichman., M.: UCI Machine Learning Repository. University of California Irvine, School of Information and Computer Science (2013)
21. Zupan, B., Bohanec, M., Bratko, I., Demsar, J.: Machine learning by function decomposition. In: International Conference on Machine Learning (ICML), pp. 421–429 (1997)
22. Harries, M.: Splice-2 Comparative Evaluation: Electricity Pricing. Technical Report, University of New South Wales, Australia (1999)
23. Cattral, R., Oppacher, F., Deugo, D.: Evolutionary data mining with automatic rule generalization. In: Recent Advances in Computers, Computing and Communications pp. 296–300 (2002)
24. Kohavi, P.: Scaling up the accuracy of naive-bayes classifiers: A decision-tree hybrid. In: 2nd International Conference on Knowledge Discovery and Data Mining (1996)

Networks

Anomaly Detection in Networks
with Temporal Information

Fabrizio Angiulli$^{(\boxtimes)}$, Fabio Fassetti, and Estela Narvaez

DIMES, University of Calabria, Rende, Italy
{f.angiulli,f.fassetti,e.narvaez}@dimes.unical.it

Abstract. We present a technique for node anomaly detection in networks where arcs are annotated with time of creation. The technique aims at singling out anomalies by taking simultaneously into account information concerning both the structure of the network and the order in which connections have been established. The latter information is obtained by timestamps associated with arcs. A set of temporal structures is induced by checking certain conditions on the order of arc appearance denoting different kinds of user behaviors. The distribution of these structures is computed for each node and used to detect anomalies. The anomaly score measures the deviation from the expected number of structures associated with each node on the basis of the correlation between nodes degree and numerousness of exhibited structures. The resulting algorithm has low computational cost and is applicable to large networks. We present experimental results on some real-life networks showing the reliability of the approach.

1 Introduction

The large use of social networks supplies a huge amount of data which provides much information about individuals and individual behaviors reflecting human relationship in the real world. Such behaviors can be model as relational structures among the actors of the social network.

Among the interesting hidden knowledge that can be mined by analyzing node behaviors, a relevant role is played by the anomaly discovery, where the aim is to find those individuals that can be considered as outliers, since they assume exceptional behaviors. The problem of finding malicious nodes in networks is of interest in many areas such as fake account detection, spammer node detection, ddos attacks in computer networks, and many others. Much work has been made to detect anomalous nodes mostly based on detecting anomalous structures around the individual [1].

However, in many scenarios, the exceptional behavior of an individual has not to be searched only in the structural composition of its neighborhood but

This research has been partially supported by the PRIN project 20122F87B2 titled "Compositional Approaches for the Characterization and Mining of Omics Data" co-finance by the Italian Ministry of Education, University and Research.

T. Calders et al. (Eds.): DS 2016, LNAI 9956, pp. 359–375, 2016.
DOI: 10.1007/978-3-319-46307-0_23

the exceptional behavior is characterized by the temporal sequence of connection establishments. Thus, taking into account the time dimension sheds interesting lights on individuals' behaviors. As such, our approach is orthogonal to the works aimed at mining structural properties of large static networks.

Consider, for example, the individuals registered to the Facebook social network and arcs between them defined as follows: if a marks b as a friend there is an arc from a to b and, vice versa, there is an arc from b to a if b marks a as a friend. Thus, the arc from a to b represents that either a sends a Facebook request to b or a accepts a Facebook request coming from b.

Consider, now, an individual a with five hundreds friends then with five hundreds other individuals there is a connection from a and a connection towards a. Clearly, it is not anomalous since such a number of friends is not so exceptional. But, if a is always the first to send Facebook friend requests and all the five hundreds just accept this request (and, then, for any b friends of a the arc from a to b always precedes the arc from b to a), a becomes a clear outlier.

The problem tackled with in the rest of the paper can be defined as follows:

Anomaly detection in timed networks. *Given a timed network, that is a network where each arc is equipped with a timestamp denoting the time of creation of the corresponding link, find the nodes in the network that are considerably dissimilar with respect to the rest of the network nodes when both the structure of their neighborhood and the order in which the structure has been established are taken into account.*

Different approaches have been proposed in the literature that search for anomalies in dynamic networks, among them [2–6].

We point out that the approach here proposed is substantially different from techniques dealing with dynamic networks. Indeed, our aim is not to determine the points in time in which a certain portion of the networks (typically a community or a subgraph) exhibited a significant change, as usually done by dynamic-graph anomaly detection techniques. Rather, our primary aim is to analyze each single node by taking simultaneously into account its temporal footprint.

In this sense our approach can be regarded as a static-graph anomaly detection technique in which temporal information has a privileged role in characterizing the behavior of network nodes.

The rest of the work is organized as follows. Section 2 reports preliminary notions and describe how the individual behavior is modeled; subsequent Sect. 3 illustrates the specific behavior models we retrieve to detect outliers; Sect. 4 is devoted to discuss the outlier score and its properties; Sect. 5 presents the several phases of the mining algorithm; Sect. 6 depicts the experiments we conduct on real datasets; finally, Sect. 7 concludes the work.

2   Behaviors on Timed Networks

In this section, we report the preliminary definitions and the notations employed throughout the paper. We aim at modeling the behavior of a node in the network through the way the node has interacted with its neighborhood during the time.

Thus, first of all we introduce the model of network equipped with time information tackled by the proposed technique.

Definition 1 (Timed Network). *A timed network (or, simply, network) is a triple* $\mathcal{N} = (V, E, \tau)$, *where* $V = \{v_1, \ldots, v_n\}$ *is a set of* nodes, $E = \{e_1, \ldots, e_m\}$ *is a set of* arcs, *with each* $e_i = \langle s_i, d_i \rangle$ *an ordered pair of nodes in* V, *and* τ *a function associating each arc* $\langle s, d \rangle$ *in* E *with a timestamp representing the instant of time in which the connection from* s *to* d *is established.*

Moreover, given a node v, we refer to the set of nodes v' such that there is an arc from v to v' as the set of *outgoing neighbors* of v (or, simply, neighbors) and we denote it as $\overrightarrow{N}(v)$. Vice versa, the set of nodes v' such that there is an arc from v' to v is referred to as the set of *ingoing neighbors* of v and we denote it as $\overleftarrow{N}(v)$. Finally, the total number of outgoing and ingoing neighbors is denoted as $\deg(v)$, then $\deg(v) = |\overrightarrow{N}(v)| + |\overleftarrow{N}(v)|$.

Next, we provide formal definition of *contact* and *awareness* between nodes that are exploited for modeling interactions.

Given a network $\mathcal{N} = (V, E, \tau)$ and two nodes v and v' in V, we say that v *contacts* v' at time t if $\langle v, v' \rangle \in E$ and $\tau(\langle v, v' \rangle) = t$. Also, in this case, we say that v' is a contact of v starting from the instant of time t.

For example, in Fig. 1a, v contacts v' at time $t = 10$ and, hence, starting from that time, v' is a contact of v.

An *interaction* between two nodes v and v' is fired (or established) at time t if either v contacts v' at time t or v' contacts v at time t. In such a case, the established contact is said to be the *contact associated with the interaction*.

Given an interaction i between two nodes v and v', the inverse interaction of i is the establishment of the contact inverse with respect to the contact associated with i.

Thus, in Fig. 1a and b, there is an interaction between v and v' fired at time $t = 10$ having as associated contact the arc from v to v' and, then, the inverse interaction between v' and v is fired at time $t = 20$, having as associated contact the arc from v' to v.

Next we provide the definition of *awareness* which intends to model the intuition that an individual knows another individual, which is not one of its contacts, due to the presence of a common friend.

Definition 2 (Awareness). *Given a network* $\mathcal{N} = (V, E, \tau)$ *and two nodes* v *and* v' *in* V, *we say that the node* v *is* mediately aware *(or, simply, aware) of* v' *at time* t_a *if there exists a node* v'' *in the network* $\mathcal{N}$, *such that* v *contacts* v'' *at time* t_1, v'' *contacts* v' *at time* t_2, $\max\{t_1, t_2\} \le t_a$ *and (i) either* $\langle v, v' \rangle$ *is not in* E *or (ii)* $t_a \le \tau(\langle v, v' \rangle)$. *Moreover, we call* intermediary *the node* v'' *responsible of the awareness.*

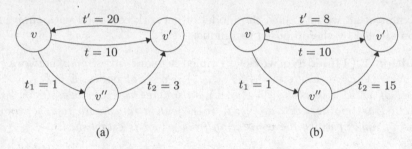

Fig. 1. Example

Note that, according to our definition, v is no more aware of v' once v' becomes a contact of v, since with the awareness we want to model the mediated knowledge between individuals.

In Fig. 1a, v is aware of v' at each instant of time in the range $[3, 9]$. Starting from the instant of time $t = 10$, v' becomes a contact of v and, then, v is no more aware of v'. Conversely, in Fig. 1b, v is never aware of v' and, starting from the instant of time $t = 10$, v' becomes a contact of v.

The notions of *contacts* and *awareness* are next exploited to model the behavior of a node within its neighborhood and, in particular, we distinguish between two families of behaviors: *action behavior* and *reaction behavior*.

Definition 3 (Action Behavior). *Let $\mathcal{N}$ be a network, an action is an interaction between two nodes v and v' of the network fired before that the inverse interaction is fired.*

Let v be a node of a network $\mathcal{N}$ and let v' be one of its neighbor. According to the above definition, the action behaviors involving v can be both the establishing of a connection from v to v' *preceding* a possibly connection from v' to v and the establishing of a connection from v' to v *preceding* a possibly connection from v to v'.

Consider Fig. 1a. There are two actions involving v: (*i*) the contact from v to v' which is accomplished before that v' contacts v and (*ii*) the contact from v to v''. Consider, now, Fig. 1b. There are again two actions involving v: (*i*) the contact from v' to v which is accomplished before that v contacts v' and (*ii*) the contact from v to v''.

Definition 4 (Reaction Behavior). *Let $\mathcal{N}$ be a network, a reaction is an interaction between two nodes v and v' of the network fired after that the inverse interaction is fired.*

The reaction behaviors involving v are both the establishing of a connection from v to v' *succeeding* a connection from v' to v and the establishing of a connection from v' to v *succeeding* a connection from v to v'.

Consider Fig. 1a. There is one reaction involving v: the contact from v' to v which is accomplished after that v' contacts v. Consider, now, Fig. 1b. There is again one reaction involving v: the contact from v' to v which is accomplished after that v contacts v'.

After having defined the concepts of actions and reactions, we can distinguish among different kinds of actions and reactions on the basis of the properties holding at the instant of time in which they are performed.

For example, Figs. 1a and b depict two different kinds of actions:

i. the node (v) is aware of the other node (v') when performs the action of contacting it (Fig. 1a);
ii. the node (v') is not aware of the other node (v) when performs the action of contacting it (Fig. 1b).

In the following Sect. 3 we will describe in details which kinds of actions and reactions are addressed in this work.

The technique we propose aims at detecting outliers on the basis of their behavior taking simultaneously into account and suitably combining actions and reactions. Specifically, each action–reaction couple models a different *scenario* and we will denote it with the expression $A \leftrightarrow R$, where A is the action and R the reaction.

For example, consider the Twitter social network and consider the scenario in which an individual v starts to follow another individual v' and v' does not follow back v. There, we can individuate an action performed by v and received by v' and a reaction performed by v' (the decision of not following back v) and received by v.

For each scenario there are several involved performers: there is the performer who makes the action, the performer who receives the action, the performer who makes the reaction, the performer who receives the reaction and, in some cases, the performer involved as intermediary. Thus, on each scenario a node can play different roles. We call *structure* the coupling of role and scenario. Each structure defines a precise role played on a precise scenario and is referred to a single node called *actor* of the structure.

The previous example induces, then, four structures:

s_1: node making action (who decides on its own initiative to follow another node);
s_2: node receiving action (who is followed by another node on its own initiative);
s_3: node making reaction (who decides to do not follow back a node by which it was followed);
s_4: node receiving reaction (who is not followed back by a followed node).

For structures s_1 and s_4 the actor is v while for the structures s_2 and s_3 the actor is v'.

Hence, once actions and reactions have been defined, those draw several scenarios and relative roles played. Scenarios and associated roles induce a set of S structures which encodes the node behavior. In particular, evaluating how frequently a node v plays each possible role on each scenario (then, how frequent v is the actor of each structure) leads to the building of the vector $\phi(v) = (\phi_{s_1}(v), \dots, \phi_{s_k}(v))$ which represents the distribution of the roles played by the node on the different scenarios and $\phi_{s_i}(v)$ represents how frequently v is the actor of the structure s_i. This distribution semantically encodes the *behavior* of the node in the network and can be effectively exploited to find anomalous individuals, as detailed in the Sect. 4.

3 Modeled Behaviors

This section is devoted to present the behaviors considered. In particular, we present the kinds of actions and reactions we capture to gather information for modeling the overall node behavior. However, the approach is easily extensible to cover other kinds of actions/reactions.

Given a node v and one of its neighbor v', next we present the actions taken into account to model the behavior of v and start by summarizing the notation employed:

t the instant of time $\tau(\langle v, v' \rangle)$ associated with $\langle v, v' \rangle$;

t' the instant of time $\tau(\langle v', v \rangle)$ if $\langle v', v \rangle \in E$; if $\langle v', v \rangle$ is not in E then t' is set to -1, meaning that it is not defined;

t_M the greatest instant of time smaller than t such that v is aware of v' at time t_M due to the intermediary v'', and we refer as v_M the node v''; if v was not aware of v' when the connection from v and v' has been established then t_M is set to -1, meaning that it is not defined.

For the sake of readability, we employ $\mathtt{def}(t)$ for indicating that $t \neq -1$ and $\mathtt{undef}(t)$ for indicating that $t = -1$.

For each considered action/reaction, we discuss the semantic behavior associated with it together with the conditions to be checked in order to verify if the behavior under analysis is actually assumed.

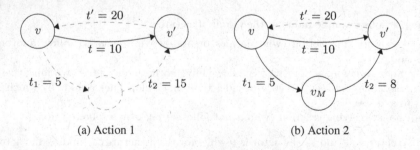

(a) Action 1   (b) Action 2

Fig. 2. Action behaviors

(Action 1) **a node contacts another node on its own initiative**.

This action represents that v contacts v' before that v' contacts v and without v being aware of v' (see Fig. 2a).

Condition: $(\mathtt{undef}(t') \vee t \leq t') \wedge \mathtt{undef}(t_M)$

(Action 2) **a node contacts another node due to an intermediary**.

This structure means that v contacts v' before that v' contacts v but after that v becomes aware of v' (see Fig. 2b).

Condition: $(\mathtt{undef}(t') \vee t \leq t') \wedge \mathtt{def}(t_M) \wedge t_M < t$

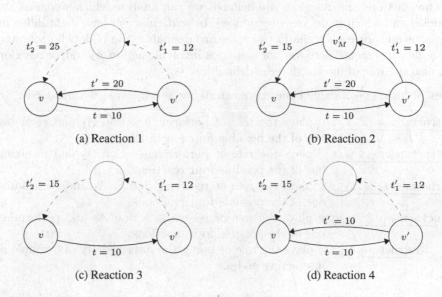

(a) Reaction 1   (b) Reaction 2

(c) Reaction 3   (d) Reaction 4

Fig. 3. Reaction behaviors

As for the reactions, we capture four kinds of reactions for v. Note that, since these are reactions, we assume that either a connection from v to v' or a connection from v' to v has already been fired.

(Reaction 1) **a node directly replies to the node who contacts him**:

This reaction models that v' contacts v since v contacts v' and not because v' becomes aware of v (see Fig. 3a).

Condition: $\mathtt{def}(t) \wedge \mathtt{def}(t') \wedge t < t' \wedge (\mathtt{undef}(t'_M) \vee t'_M < t)$

(Reaction 2) **a node replies to the node who contacts him due to an intermediary**:

This reaction represents that v' contacts v after that v contacts v' but only after that v' becomes aware of v (see Fig. 3b).

Condition: $\mathtt{def}(t) \wedge \mathtt{def}(t') \wedge \mathtt{def}(t'_M) \wedge t < t'_M < t'$

(Reaction 3) **a node does not reply to the node who contacts him**: This is not an actual reaction since it represents that v' does not react to the action performed by v (see Fig. 3c).

Condition: $\mathtt{def}(t) \bigwedge \mathtt{undef}(t')$

(Reaction 4) **a node contacts the node who contacts him independently**: This is not an actual reaction since it represents that v' contacts v on its own initiative (see Fig. 3d).

Condition: $\mathtt{def}(t) \bigwedge \mathtt{def}(t') \bigwedge t' = t \bigwedge \mathtt{undef}(t'_M)$

Once actions and reactions are defined, we can analyze which scenarios are modeled and which structures are induced. In particular, we have eight different scenarios and for each scenario two roles are definable, the node who acts and the node who reacts. Moreover, for scenarios involving action A_2 and/or reaction R_2, also the role of intermediary is definable.

Thus, focusing on a single node v, we can define seventeen structures for it:

structures $s_1 \ldots s_4$: v plays the role of performing action A_1 and receiving one of the possible four reactions;

structures $s_5 \ldots s_8$: v plays the role of performing action A_2 and receiving one of the possible four reactions;

structures $s_9 \ldots s_{12}$: v plays the role of receiving action A_1 and performing one of the possible four reactions;

structures $s_{13} \ldots s_{16}$: v plays the role of receiving action A_2 and performing one of the possible four reactions;

structure s_{17} : v plays the role of being the intermediary of a couple of interacting nodes.

For example, consider the scenario depicted in Fig. 1a again: *v contacts v' after viewing that v' is a contact of v'' which is one of its contact. v' reacts to this contacting back v.* The scenario is then $A_2 \leftrightarrow R_1$ and the roles are: (1) who makes the action A_2 and receives the action R_1, (2) who receives the action A_2 and makes the reaction R_1, and (3) who plays the role of intermediary. Thus, after analyzing this scenario, for the node v we have to update the structure associated with (1), that is s_5; for the node v' the structure associated with (2), that is s_{13}; and for the node v'' the structure associated with (3), that is s_{17}.

4   Anomaly Score

The distribution $\phi_s(v)$ encodes the behavior of the node v in terms of how much frequently it is involved in the different structures. We can then exploit the distributions ϕ_s in order to determine how typical is the behavior of each node with respect to the whole population. With this aim, an anomaly score is assigned to each node.

Given the network $\mathcal{N} = (V, E, \tau)$, for each structure $s \in \mathcal{S}$ the regression line of the set of points $P_s(\mathcal{N}) = \{(\deg(v_i), \phi_s(v_i)) \mid v_i \in V\}$ is computed.

Let α_s and β_s denote be the parameters of the estimated line. The anomaly score of the node v_i with respect the structure s is defined as:

$$\mathrm{sc}_s(v_i) = \left| \frac{\phi_s(v_i) - [\alpha_s \cdot \deg(v_i) + \beta_s]}{\log_2(1 + \deg(v_i))} \right| \tag{1}$$

The numerator of Eq. (1) represents the deviation of the observed number of structures $\phi_s(v_i)$ from the expected one y_i, according to the value predicted by the regression curve $y_i = \alpha_s \cdot \deg(v_i) + \beta_s$. As for the denominator, it serves the purpose of taking into account the cardinality of the neighborhood of v_i, while the absolute value is needed to capture both the upper tail and the lower tail of resulting distribution.

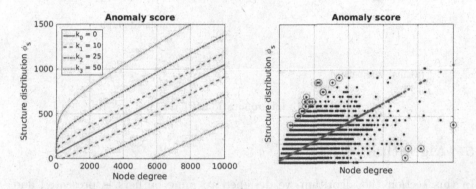

Fig. 4. Example of bandwidths associated with the score of a structure (left) and example highlighting the top twenty anomalies on a real dataset (right).

Figure 4 on the left reports a regression line (the solid curve for $k_0 = 0$, having parameters $\alpha = 0.1$ and $\beta = 50$) and the bandwidths associated with different values of anomaly score (specifically, for $k_1 = 10$, $k_2 = 25$, and $k_3 = 50$; e.g., the score associated with points falling within the dashed bandwidth will be not greater than k_1). Figure 4 on the right reports the structure distribution associated with a real dataset and the top twenty anomalies according to Eq. (1).

Scores associated with each single structure are then normalized in order to make them homogeneous

$$\widehat{\mathrm{sc}}_s(v_i) = \frac{\mathrm{sc}_s(v_i)}{\mathrm{std}(\{\mathrm{sc}_s(\mathbf{v})\})} \tag{2}$$

and, hence, expressed in terms of number of standard deviations.

The anomaly score of a node v_i is

$$\mathrm{sc}(x_i) = \sum_s \mathrm{sc}_s(x_i), \tag{3}$$

that is obtained by combining the scores computed with respect to the single structures.

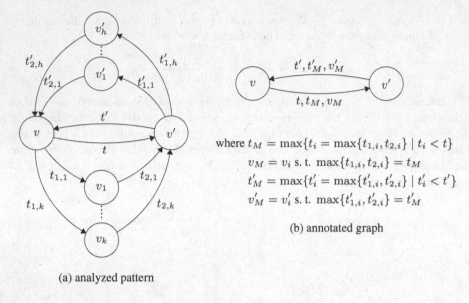

where $t_M = \max\{t_i = \max\{t_{1,i}, t_{2,i}\} \mid t_i < t\}$
$v_M = v_i$ s. t. $\max\{t_{1,i}, t_{2,i}\} = t_M$
$t'_M = \max\{t'_i = \max\{t'_{1,i}, t'_{2,i}\} \mid t'_i < t'\}$
$v'_M = v'_i$ s. t. $\max\{t'_{1,i}, t'_{2,i}\} = t'_M$

(b) annotated graph

(a) analyzed pattern

Fig. 5. Graph annotation (Phase 1)

5   Algorithm

In this section the algorithm we designed to mine outliers is presented and its properties are discussed. The algorithm consists in three main phases each accounted next.

Phase 1. This phase has the intent of enriching the information associated with the arcs (see Fig. 5). In order to retrieve behaviors illustrated in Sect. 3 we need to find, for each arc $\langle v, v' \rangle$ with associated timestamp t, if v is aware of v' at time t, namely we have to search for a node $\hat{v}$ such that both edge $e_1 = \langle v, \hat{v} \rangle$ and edge $e_2 = \langle v, \hat{v} \rangle$ exist and the timestamps t_1 and t_2 associated with these edges are both strictly smaller than t. Among these nodes, we are interested in the node v_M which is the most recent responsible of the fact that v is aware of v'. Finally, the edge $e = \langle v, v' \rangle$ is annotated with the node v_M^e and the time t_M^e which is the instant of time starting by which v is aware of v' due to v_M^e; in formula t_M^e is the maximum between the time associated with the arc $\langle v, v_M^e \rangle$ and the time associated with the arc $\langle v_M^e, v' \rangle$. Concluding, given a network $\mathcal{N} = (V, E, \tau)$, the phase returns the annotated network $\mathcal{N}^+ = (V, E, \tau^+)$ where $\tau^+(e)$ returns the triple $(\tau(e), t_M^e, v_M^e)$.

Computational Complexity of Phase 1. As for the cost of this phase, let $\mathcal{N} = (V, E, \tau)$ be the analyzed network, let $n = |V|$ and let $m = |E|$. We iterate over the set of edges and for each arc $e = \langle v, v' \rangle$ in E we iterate over the set $\overrightarrow{N}(v)$ of neighbors of v in order to search v_M and, then, for each neighbor $\hat{v}$ of v we have to check if there exists an arc from it to v'. This latter operation can be performed through a binary search in the list of the outgoing arcs of $\hat{v}$. Thus, the overall cost is

$$\sum_{\langle v, v' \rangle} \sum_{\hat{v} \in \overrightarrow{N}(v)} \log \overrightarrow{N}(\hat{v}) = O(m \cdot \overline{n} \cdot \log \overline{n}) \tag{4}$$

where $\overline{n}$ denotes the mean number of neighbors of nodes in the networks.

Phase 1. Network information enrichment

    Input: A network $\mathcal{N} = (V, E, \tau)$

    Output: The annotated network $\mathcal{N}^+$

1 **foreach** edge $e = \langle v, v' \rangle$ in E **do**

2   let $t = \tau(e)$ be the timestamp associated with the edge from v to v';

3   set t_M to -1;

4   set v_M to $\emptyset$;

5   **foreach** edge $\hat{e} = \langle v, \hat{v} \rangle$ **do**

6   let $t_1 = \tau(\hat{e})$ be the timestamp associated with the edge from v to $\hat{v}$;

7   **if** $\langle \hat{v}, v' \rangle$ belongs to E **then**

8   let $t_2 = \tau(\langle \hat{v}, v' \rangle)$ be the timestamp associated with the edge from $\hat{v}$ to v';

9   let $\hat{t}$ be $\max\{t_1, t_2\}$;

10   **if** $t_M < \hat{t} < t$ **then**

11   set t_M to $\hat{t}$;

12   set v_M to $\hat{v}$;

13   associate t_M and v_M with the e;

      // then substitute $\tau(e) = t$ with $\tau^+(e) = (t, t_M, v_M)$– see Fig. 5

Phase 2. This phase has the intent of mining the behavior of each individual in the network, starting from the annotated network coming from the previous phase. Then, given an annotated network $\mathcal{N}^+ = (V, E, \tau^+)$ we iterate over the set of nodes V and for each node v in V we iterate over the set of neighbors and for each neighbor v' in $\overrightarrow{N}(v)$ the behaviors depicted in Sect. 3 are evaluated. In particular, through the information provided by $\mathcal{N}^+$ the conditions associated with the behaviors are checked and, according to the result of the check, the behavior counters are updated. The result of this phase is then the distribution of the behaviors for each node.

Computational Complexity of Phase 2. As for the cost of this phase, let $\mathcal{N}^+ = (V, E, \tau^+)$ be the analyzed network, let $n = |V|$ and let $m = |E|$. Iterating over the set of nodes and for each node v iterating over the set of neighbors $\overrightarrow{N}(v)$ corresponds to iterating over the set of edges. For each edge, in constant time we can obtain the required information by $\mathcal{N}^+$ and we can evaluate all the conditions. Since the number of conditions is fixed, also this latter step can be accomplished in constant time. Thus, the overall cost of this phase is $O(m)$.

Phase 2. Structures computation

Input: An annotated network $\mathcal{N}^+ = (V, E, \tau^+)$
Output: The distribution of structures ϕ_v for each node v

1 **foreach** node v in V **do**
2   set $\phi_s(v)$ to 0 for each structure s;
3   **foreach** neighbor v' of v **do**
4   extract tuple $T = (\tau^+(\langle v, v'\rangle), \tau^+(\langle v', v\rangle))$;
       // T contains then $(t, t_M, v_M, t', t'_M, v'_M)$– see Fig. 5b
5   **foreach** action a **do**
6   **foreach** reaction r compatible with a **do**
7   let s be the structure associated with the pair $a \leftrightarrow r$;
8   **if** $C_a(T)$ **and** $C_r(T)$ **then**
9   update $\phi_s(v)$;
10   **if** $a \leftrightarrow r$ involves node v_M **then**
11   update $\phi_{\hat{s}}(v_M)$;

Phase 3. This phase has the intent of detecting outlier individuals. Starting from the distribution of behaviors computed by the previous phase, we have to compute the outlier score as defined in Sect. 4. The first step consists in computing, for each considered structure, the regression line. The second step consists in computing for each structure s and for each node v the score $sc_s(v)$ achieved by node v on the structure s by means of Eq. (1). Next the standard deviation of the scores assumed by nodes on structure s is computed and, then, this value is exploited to normalize the scores (lines 6–7).

After that all the structures have been analyzed, the outlier score of each node v is computed by properly aggregating the score achieved by v on each single structure by means of Eq. (3).

Computational Complexity of Phase 3. As for the cost of this phase, let $\mathcal{N}^+ = (V, E, \tau^+)$ be the analyzed network, let $n = |V|$ and let $m = |E|$. Computing the regression line has a cost linear with respect to the number n of nodes. Next, for each node we had to compute the score. Since Eq. (1) is computable in constant time, also this step has a cost linearly dependent from n. Normalizing the scores costs $O(n)$ as well and, finally, also computing the overall outlier score costs $O(n)$ since Eq. (3) iterates over a fixed number of structures.

6 Experimental Results

In this section experimental results concerning the introduced technique are presented. All the datasets employed are from the *Online Social Networks Research*. All the dataset underwent a preprocessing during which multiple and self links have been removed. The *Digg friends* dataset[1] contains data about stories

---

[1] http://www.isi.edu/integration/people/lerman/downloads.html.

Phase 3. Outlier mining

Input: The distribution of structures ϕ_v for each node v
Output: The overall outlier score for each node v

1 **foreach** structure s **do**
2 | compute the regression line of the observations $(\deg(\mathbf{v}), \phi_\mathbf{s}(\mathbf{v}))$;
3 | **foreach** node v in $\mathcal{N}$ **do**
4 | | compute the score $sc_s(v)$ of v for structure s through Eq. (1);
5 | compute the standard deviation of the score $std(\{sc_s(\mathbf{v})\})$;
6 | **foreach** node v in $\mathcal{N}$ **do**
7 | | compute the normalized score of v for structure s through Eq. (2);

8 **foreach** node v in $\mathcal{N}$ **do**
9 | compute the overall score of v through Eq. (3);

promoted to Digg's front page[2] over a period of a month in 2009. The dataset contains Digg users who have voted for a story. We considered the voters' friendship links, where a link `user_id` → `friend_id` means that `user_id` is watching the activities of (is a fan of) `friend_id`. User identifiers are available already anonymized. The network analyzed consists of $279,631$ nodes and $2,251,166$ arcs. The *Facebook wall* dataset[3] contains a list of all of the wall posts from the Facebook New Orleans network. Each line contains two anonymized user identifiers, meaning the second user posted on the first user's wall. The third column is the times of the wall post. The network analyzed consists of $45,813$ nodes and $264,004$ arcs. The *Wikipedia growth*[4] dataset contains links between Wikipedia pages and the time when these links were first created. The dataset represents the complete history of the network over a period of 826 days, between January 1st, 2005 and April 6th, 2007. The datasets is anonymized to protect the privacy of page authors. The network analyzed consists of $1,870,709$ nodes and $39,953,145$ arcs.

The following table reports the total number of the structures mined, for each of the structures s_i in the set $\mathcal{S}$. Notice that the total counts associated with the pairs of structures (s_i, s_{i+8}) are identical, since these structures are induced by symmetric roles in the same scenario.

Clearly, this does not mean that they are redundant, since the respective counts per node differ in general, being different the number of times in which the single node plays each role in the same scenario. In the *Facebook wall* dataset the structures s_4, s_8, s_{12}, and s_{16}, capturing the simultaneity of the response, are not present since timestamps are almost all different (only 846 timestamps appear more than once in the dataset) and it is never the case that two users

---

[2] *Digg* is a news aggregator (http://digg.com) aiming to select stories for the Internet audience such as science, trending political issues, and viral Internet issues. It allows people to vote web content up or down.
[3] http://socialnetworks.mpi-sws.org/data-wosn2009.html.
[4] http://socialnetworks.mpi-sws.org/data-wosn2008.html.

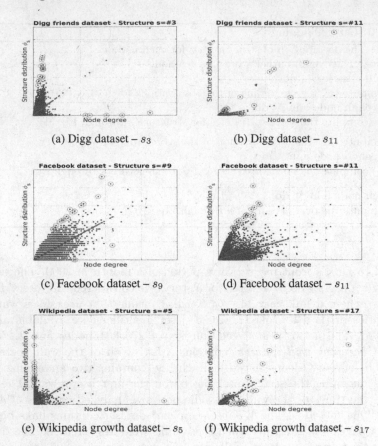

(a) Digg dataset – s_3　　　　(b) Digg dataset – s_{11}

(c) Facebook dataset – s_9　　　　(d) Facebook dataset – s_{11}

(e) Wikipedia growth dataset – s_5　　(f) Wikipedia growth dataset – s_{17}

Fig. 6. Structure count distribution. (Color figure online)

Structure	Digg friends	Facebook wall	Wikipedia growth
s_1, s_9	61, 122	57, 476	2, 030, 550
s_2, s_{10}	8, 307	4, 167	641, 677
s_3, s_{11}	683, 282	74, 268	22, 416, 947
s_4, s_{12}	6, 294	0	7, 596
s_5, s_{13}	65, 844	16, 319	451, 128
s_6, s_{14}	44, 301	2, 629	293, 043
s_7, s_{15}	681, 317	28, 552	10, 694, 970
s_8, s_{16}	152	0	56
s_{17}	1, 009, 571	80, 042	14, 052, 936
Total	7, 574, 125	446, 864	87, 124, 870

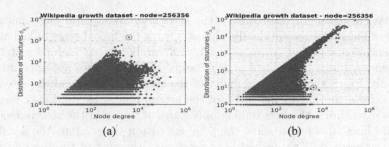

Fig. 7. Notable structure distributions for a top outlier node

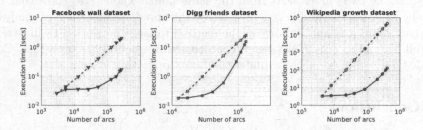

Fig. 8. Scalability analysis.

simultaneously make a post on the respective wall. In general, these structures are among the less numerous in these datasets due to the fine granularity of the temporal scale.

Figure 6 shows the scatter plots of the node degree versus the structure count for some of the structures mined. Each plot reports also the regression line of the points (represented by the red points) and highlights the top 20 anomalies (the red circled points) according to the anomaly score of Eq. (1).

Next, we discuss on the knowledge mined for one of the top outlier in a dataset in order to highlight how the proposed technique is able to provide not only the outlier score but also an in intelligible interpretation of the score, shedding light on semantic underlying the decision made by the technique of signaling a node as anomalous.

Specifically, we focus on the *Wikipedia growth* dataset and on the node *out* having id $256, 356$. We consider the two structures that mostly contribute to the large score achieved by *out*. Figure 7 reports how the number of times in which *out* is actor of structure s_1 and s_{11} places the node with respect to the number of times in which the other nodes perform as actors of those structures.

It is clear by the plots that *out* is located in both cases at the margin of the distribution. In particular, *out* performs as actor of structure s_1 much more times than other nodes, while performs as actor of structure s_{11} much less times than nodes having a similar neighborhood cardinality.

From a semantic point of view, these plots naturally lead to a description of the outlierness that could explain the exceptionality of the node. Since *out* very often plays as actor for structure s_1, the associated Wikipedia page has a high number of links and, exceptionally, almost always the linked page links back the page associated with *out*. Moreover, since *out* rarely plays as actor for structure s_{11}, whenever *out* is linked by a page p, rarely a link to p does not appear in *out*.

Finally, Fig. 8 shows the scalability of the method. We varied the size of the datasets, from the 1 % to the 100 % of the original data, by randomly sampling nodes and retaining the arcs linking only pairs of nodes in the selected sample. The solid lines in the plot show the total execution time versus the number of arcs of the network. The dashed lines represents the cost of the method in the average case, as reported in Eq. (4), with a constant prefactor computed in a way such that the two curves start from the same point. The curves show that the actual cost of the method is generally below that predicted by the cost analysis, thus confirming the applicability of the method to large networks. To illustrate, the full *Wikipedia growth* dataset was processed in about 120 s on a Intel Core i7 2.40 GHz equipped machine under the Linux operating system.

7   Conclusions

We considered the anomaly detection in timed networks problem whose goal is to single out anomalies by taking into account simultaneously information concerning both the structure of the network and the order in which connections have been established. Our primary aim is to analyzing each single node by taking simultaneously into account its temporal footprint. We defined a set of spatio-temporal structures is induced by checking certain conditions on the order of arc appearance denoting different kinds of user behaviors and exploited their distribution to detect anomalies. We presented a scalable algorithm and experimental results showing the peculiarity of the knowledge mined by our technique and its applicability to the analysis of large networks.

References

1. Akoglu, L., Tong, H., Koutra, D.: Graph based anomaly detection anddescription: a survey. Data Min. Knowl. Disc. **29**(3), 626–688 (2015). doi:10.1007/s10618-014-0365-y
2. Chen, Z., Hendrix, W., Samatova, N.F.: Community-based anomaly detection in evolutionary networks. J. Intell. Inf. Syst. **39**(1), 59–85 (2012)
3. Gupta, M., Gao, J., Sun, Y., Han, J.: Community trend outlier detection using soft temporal pattern mining. In: Flach, P.A., Bie, T., Cristianini, N. (eds.) ECML PKDD 2012. LNCS (LNAI), vol. 7524, pp. 692–708. Springer, Heidelberg (2012). doi:10.1007/978-3-642-33486-3_44
4. Ji, T., Yang, D., Gao, J.: Incremental local evolutionary outlier detection for dynamic social networks. In: Blockeel, H., Kersting, K., Nijssen, S., Železný, F. (eds.) ECML PKDD 2013, Part II. LNCS (LNAI), vol. 8189, pp. 1–15. Springer, Heidelberg (2013). doi:10.1007/978-3-642-40991-2_1

5. Mongiovi, M., Bogdanov, P., Ranca, R., Singh, A.K., Papalexakis, E.E., Faloutsos, C.: Netspot: spotting significant anomalous regions on dynamic networks. In: Proceedings of the 13th SIAM international conference on data mining (SDM), Texas-Austin, TX. SIAM (2013)

6. Wang, T., Fang, C.V., Lin, D., Wu, S.F.: Localizing temporal anomalies in large evolving graphs. In: Proceedings of the 2015 SIAM International Conference on Data Mining, pp. 927–935. SIAM (2015)

Accelerating Computation of Distance Based Centrality Measures for Spatial Networks

Kouzou Ohara[1]([✉]), Kazumi Saito[2], Masahiro Kimura[3], and Hiroshi Motoda[4,5]

[1] Department of Integrated Information Technology,
Aoyama Gakuin University, Sagamihara, Japan
`ohara@it.aoyama.ac.jp`
[2] School of Administration and Informatics, University of Shizuoka, Shizuoka, Japan
`k-saito@u-shizuoka-ken.ac.jp`
[3] Department of Electronics and Informatics, Ryukoku University, Otsu, Japan
`kimura@rins.ryukoku.ac.jp`
[4] Institute of Scientific and Industrial Research, Osaka University, Ibaraki, Japan
`motoda@ar.sanken.osaka-u.ac.jp`
[5] School of Computing and Information Systems,
University of Tasmania, Hobart, Australia

Abstract. In this paper, by focusing on spatial networks embedded in the real space, we first extend the conventional step-based closeness and betweenness centralities by incorporating inter-nodes link distances obtained from the positions of nodes. Then, we propose a method for accelerating computation of these centrality measures by pruning some nodes and links based on the cut links of a given spatial network. In our experiments using spatial networks constructed from urban streets of cities of several types, our proposed method achieved about twice the computational efficiency compared with the baseline method. Actual amount of reduction in computation time depends on network structures. We further experimentally show by examining the highly ranked nodes that the closeness and betweenness centralities have completely different characteristics to each other.

Keywords: Closeness centrality · Betweenness centrality · Spatial network · Distance-based centrality · Cut link

1 Introduction

Studies of the structure and functions of large complex networks have attracted a great deal of attention in many different fields such as sociology, biology, physics and computer science [15]. Our final goal of this research is to develop methodologies and technologies that are accurate, efficient and scalable to analyze complex networks which are now ubiquitous in almost every field of science and to discover useful knowledge from there. For this we think it is critical to be able to assess and utilize inherent and intrinsic characteristics of such networks.

© Springer International Publishing Switzerland 2016
T. Calders et al. (Eds.): DS 2016, LNAI 9956, pp. 376–391, 2016.
DOI: 10.1007/978-3-319-46307-0_24

One common approach to analyze large complex networks is investigating their characteristics through a measure called centrality. Various kinds of centralities are used according to what we want to know. For example, if our goal is to know the topological characteristics of a network, degree, closeness, and betweenness centralities [11] can be used. If it is to know the importance of nodes that constitute a network, HITS [5] and PageRank [3] centralities are often used. Influence degree centrality [12] is another one to measure the importance of nodes. Among these conventional centralities, we focus on the closeness and betweenness centralities in this work because they are closely related to real world problems such as location planning of commercial or evacuation facilities in a wide area. Here, note that closeness and betweenness centralities usually approximate the distance between two distinct nodes by the number of links traversed to get to one node from another. This approximation may not be realistic when analyzing networks such as real traffic networks, one of the real world problems. Thus, as a particular class, we focus on spatial networks embedded in the real space, like urban streets, whose nodes occupy a precise position in two or three-dimensional Euclidean space, and whose links are real physical connections [10]. Analyzing and characterizing the structure of such large spatial networks will play an important role for understanding and improving the usages of these networks, as well as discovering new insights for developing and planning city promotion, trip tours and so on. To facilitate such research work, in this paper, we propose techniques useful to accelerate their computations based on network pruning. This is motivated by the fact that the computation time to calculate the value of such conventional *step-based* centralities for every single node in a network becomes larger as the size of the network gets larger because of the necessity to traverse each link in the network multiple times.

In this paper, we first extend the conventional step-based closeness and betweenness centralities by incorporating inter-nodes link distances obtained from the positions of nodes. Hereafter, we refer to these extended ones as distance-based centrality measures. Then, we propose a method for accelerating computation of these centrality measures by pruning some nodes and links that are related to the cut links of a given spatial network, where each cut link divides the network into two connected components by its removal. Here, we should note that our approach is applicable to a wider range of spatial networks, and potentially to complex networks with inter-nodes link distances, although in this paper, we focus on spatial networks constructed from urban streets by mapping the intersections of streets into nodes and the streets between the nodes into links. In the experiments we evaluate the performance of our proposed acceleration techniques for the spacial networks constructed from cities of several types, and discuss the characteristics of the distance-based centralities.

This paper is organized as follows. After first explaining the related work in Sect. 2, we describe the details of our proposed methods in Sect. 3. In Sect. 4, we evaluate the characteristics of networks obtained by our proposed methods both qualitatively and quantitatively, and give our conclusion in Sect. 5.

2   Related Work

As mentioned earlier, analyzing and characterizing the structure of large spatial networks like urban streets will play an important role for understanding and improving the usages of these networks embedded in the real space. Thus, the structure and functions of spatial networks have also been studied by many researchers [4,10,13,17,18,20]. From structural viewpoints, centrality measures have been widely used to analyze spatial networks [10,18], especially by extending the conventional notions of centrality measures on simple networks into those of weighted networks based on road usage frequency of urban streets [13,17]. From functional viewpoints, traffic usage patterns in urban streets have been investigated [4,20]. In this paper, unlike these previous studies, we focus on extending the conventional centralities by incorporating inter-nodes link distances obtained from the positions of nodes, and propose a method for accelerating the calculation of these centrality measures.

To find important nodes in a network, several centrality measures have been presented in the field of social and information network analysis. Representative centrality measures include degree, closeness and betweenness centralities [11], HITS (hub and authority) centrality [5], and PageRank centrality [3]. These measures are also closely related to quantifying how influential each node is in the context of information diffusion, and influence degree centrality can be defined by evaluating the influence of each node. Unlike centrality measures derived only from network topology, influence degree centrality exploits a dynamical process on the network as well. An efficient method of simultaneously estimating the influence degrees of all the nodes was presented under the SIR model setting [12]. We note that influence degree centrality can also be employed for identifying super-mediators in the social network [19]. In this paper, as our first step to develop these centrality notions for spatial networks, we focus on closeness and betweenness centralities, which have been widely used in the field of social network analysis. Our proposed method in this paper shares the same basic idea of the previous method presented under the SIR model setting [12] in that redundant nodes and links are pruned for accelerating the computation of the centrality measures.

For some centrality measures such as betweenness centrality, their computation becomes harder as the network size increases, since it needs to take the global network structure into account. Thus, several researchers presented methods of approximating such centralities [1,6–8,16]. For instance, a bottom-k sketch [7,8] is obtained by associating with each node in a network an independent random rank value drawn from a probability distribution. The bottom-k sketch of each node in the network can be quite efficiently calculated by orderly assigning the rank values from the smallest one to those nodes reachable by reversely following links over the network. Based on this framework, a greedy Sketch-based Influence Maximization (SKIM) algorithm has been proposed, and it has been shown that the SKIM algorithm scales to graphs with billions of edges, with one to two orders of magnitude speedup over the best greedy methods [9]. Here in this paper, unlike the above approximation approaches, we rather focus on exactly computing the centrality measures whereby there is no need to worry about the approximation performance. Note that these exact solutions can be used as the ground-truth for evaluating the approximation performance.

3   Proposed Method

Let $G = (\mathcal{V}, \mathcal{E})$ be a spatial network consisting of a single connected component without self-loops, where $\mathcal{V} = \{u, v, w, \cdots\}$ and $\mathcal{E} = \{e, \cdots\} \subset \mathcal{V} \times \mathcal{V}$ are sets of nodes and undirected links, respectively. For each link $e = (u, v)$, we express the distance between nodes u and v by $d(u, v)$, where we can obtain these distances from the positions of nodes in the spatial network. For each pair of nodes $u, w \in \mathcal{V}$ without the direct connection, we define the distance $d(u, w)$ as the geodesic distance over the network, as usual. Then, for each node $u \in \mathcal{V}$, we can define the following distance based closeness centrality measure:

$$
DC(u) = \left(\sum_{w \in \mathcal{V}} d(u, w) \right)^{-1}. \tag{1}
$$

Note that the distance based closeness centrality $DC(u)$ is a natural extension to the conventional step based closeness centrality $SC(u)$ because $DC(u)$ reduces to $SC(u)$ by setting $d(u, v) = 1$ for each link $(u, v) \in \mathcal{E}$. Similarly, for each node $v \in \mathcal{V}$, we can define the following distance based betweenness centrality measure:

$$
DB(v) = \sum_{u \in \mathcal{V} \setminus \{v\}} \sum_{w \in \mathcal{V} \setminus \{u, v\}} \frac{\sigma(u, w; v)}{\sigma(u, w)} \tag{2}
$$

where $\sigma(u, w)$ is the total number of the paths with the smallest distance between node u and node w in G and $\sigma(u, w; v)$ is the number of those paths between node u and node w in G that passes through node v. Again, note that the distance based betweenness centrality $DB(v)$ is a natural extension to the conventional step based betweenness centrality $SB(v)$ because $DB(v)$ also reduces to $SB(v)$ by setting $d(u, v) = 1$ for each link $(u, v) \in \mathcal{E}$. By applying the best-first search algorithm starting from each node $u \in \mathcal{V}$ with respect to distance $d(u, w)$, we can calculate these centrality measures, $DC(u)$ and $DB(v)$, for all the nodes in G. As mentioned earlier, when calculating closeness and betweenness centrality measures, their computation becomes harder as the network size increases. Below we propose a method of improving the computational efficiency to calculate these centrality measures.

3.1   Pruning Techniques for Closeness Centrality

We say that a link $e \in \mathcal{E}$ is a cut link if the network G is divided into two connected components by eliminating the link e. For a given cut link $e = (u, v)$, let $\mathcal{V}(u \setminus v)$ and $\mathcal{V}(v \setminus u)$ be the sets of nodes in the two connected components, each of which includes the nodes u and v, respectively, i.e., $u \in \mathcal{V}(u \setminus v)$, $v \in \mathcal{V}(v \setminus u)$, $\mathcal{V}(u \setminus v) \cap \mathcal{V}(v \setminus u) = \emptyset$ and $\mathcal{V}(u \setminus v) \cup \mathcal{V}(v \setminus u) = \mathcal{V}$. Let $\eta(u)$ and $\delta(u)$ be the number of the nodes and the accumulated value of distances obtained by our pruning process relating to the node u, respectively. Then, after initializing $\eta(u) \leftarrow 1$ and $\delta(u) \leftarrow 0$ for each node $u \in \mathcal{V}$, and setting $\mathcal{W} \leftarrow \mathcal{V}$ and $\mathcal{F} \leftarrow \mathcal{E}$, we consider calculating $DC(x)^{-1}$ for each node $x \in \mathcal{W}$ by using the following formula:

$$DC(x)^{-1} = \sum_{w \in \mathcal{W}} (\eta(w)d(x,w) + \delta(w)) \tag{3}$$

Now, let $e = (u, v)$ be a cut link; then, for a node $x \in \mathcal{V}(u \setminus v)$, we can decompose $DC(x)^{-1}$ as follows:

$$DC(x)^{-1} = \sum_{w \in \mathcal{V}(u \setminus v)} d(x,w) + \sum_{w \in \mathcal{V}(v \setminus u)} (d(x,u) + d(u,v) + d(v,w))$$

$$= \sum_{w \in \mathcal{V}(u \setminus v)} d(x,w) + |\mathcal{V}(v \setminus u)|d(x,u) + |\mathcal{V}(v \setminus u)|d(u,v) + \sum_{w \in \mathcal{V}(v \setminus u)} d(v,w). \tag{4}$$

By using the above decomposition shown in Eq. (4), we can derive a recursive TP (Top-down Pruning) technique for closeness centrality. More specifically, we consider updating $\eta(u)$ and $\delta(u)$ by using the following formulae:

$$\eta(u) \leftarrow 1 + |\mathcal{V}(v \setminus u)|,$$

$$\delta(u) \leftarrow |\mathcal{V}(v \setminus u)|d(u,v) + \sum_{w \in \mathcal{V}(v \setminus u)} d(v,w), \tag{5}$$

Then, after pruning the node set $\mathcal{V}(v \setminus u)$ and the cut link $e = (u, v)$ from the network $G = (\mathcal{V}, \mathcal{E})$ as $\mathcal{W} \leftarrow \mathcal{V} \setminus \mathcal{V}(v \setminus u)$ and $\mathcal{F} \leftarrow \mathcal{E} \setminus \{e\}$, we can equivalently calculate $DC(x)^{-1}$ by using Eq. (3) for each node $x \in \mathcal{V}(u \setminus v)$ as obtained by Eq. (1). Clearly, we can apply the same arguments to the case of $x \in \mathcal{V}(v \setminus u)$. Therefore, by generalizing the update formulae of Eq. (5) as follows:

$$\eta(u) \leftarrow \eta(u) + \sum_{w \in \mathcal{W}(v \setminus u)} \eta(w),$$

$$\delta(u) \leftarrow \delta(u) + d(u,v) \sum_{w \in \mathcal{W}(v \setminus u)} \eta(w) + \sum_{w \in \mathcal{W}(v \setminus u)} \eta(w)d(v,w), \tag{6}$$

for all the node $x \in \mathcal{V}$, we can exactly calculate the closeness centrality measure $DC(x)$ by using the above recursive TP update shown in Eq. (6), where $\mathcal{W}(u \setminus v)$ and $\mathcal{W}(v \setminus u)$ denote the sets of nodes in the two connected components, respectively, which are obtained by the removal of the cut link $e = (u, v)$, just like those defined for $\mathcal{V}$.

As a special case, by focusing on the fact that each link of degree-one node is a cut link, we derive a recursive BP (Bottom-up Pruning) technique algorithm for closeness centrality. More specifically, for a degree-one node v and its cut link

$e = (u, v)$, we consider updating $\eta(u)$ and $\delta(u)$ as follows:

$$\eta(u) \leftarrow \eta(u) + \eta(v),$$
$$\delta(u) \leftarrow \delta(u) + \eta(v)d(u, v) + \delta(v) \tag{7}$$

Then, after pruning the node v and its cut link $e = (u, v)$ from the network $G = (\mathcal{W}, \mathcal{F})$ as $\mathcal{W} \leftarrow \mathcal{W} \setminus \{v\}$ and $\mathcal{F} \leftarrow \mathcal{F} \setminus \{e\}$, and updating $\eta(u)$ and $\delta(u)$ by Eq. (7), we can also equivalently calculate $DC(x)^{-1}$ by using Eq. (3) for each node $x \in \mathcal{W}$ as obtained by Eq. (1). Here note that after obtaining the value of $DC(u)^{-1}$, we can calculate $DC(v)^{-1}$ as follows:

$$DC(v)^{-1} \leftarrow DC(u)^{-1} + (|\mathcal{V}| - 2\eta(v))d(u, v)$$

Therefore, for all the node $x \in \mathcal{V}$, we can exactly calculate the closeness centrality measure $DC(x)$ by using the above recursive TP and BP techniques.

3.2 Pruning Techniques for Betweenness Centrality

We derive the recursive pruning techniques for calculating the betweenness centrality measure $DB(x)$. Let $\phi(u)$ be the number of node pairs obtained by our pruning process relating to the node u. Again, after initializing $\eta(u) \leftarrow 1$ and $\phi(u) \leftarrow 0$ for each node $u \in \mathcal{V}$, and setting $\mathcal{W} \leftarrow \mathcal{V}$ and $\mathcal{F} \leftarrow \mathcal{E}$, we consider calculating $DB(x)$ for each node $x \in \mathcal{W}$ by using the following formula:

$$DB(x) = \phi(x) + \sum_{w \in \mathcal{W} \setminus \{x\}} \eta(w)\xi(x; w). \tag{8}$$

Here, $\xi(x; w)$ stands for the number of descendant nodes of x obtained by the best first search starting from the node w, i.e.,

$$\xi(x; w) = \eta(x) - 1 + \sum_{z \in C(x; w)} (\xi(z; w) + 1)\frac{\sigma(w, x)}{\sigma(w, z)},$$

where $C(x; w)$ means the set of direct child nodes of x for the search from node w. Now, let $e = (u, v) \in \mathcal{E}$ be a cut link; then, for a node $w \in \mathcal{V}(u \setminus v)$, we can calculate $\xi(u; w)$ as follows:

$$\xi(u; w) = |\mathcal{V}(v \setminus u)| + \sum_{z \in C(u; w) \setminus \{v\}} (\xi(z; w) + 1)\frac{\sigma(w, u)}{\sigma(w, z)}, \tag{9}$$

and $\xi(u; w) = |\mathcal{V}(u \setminus v)| - 1$ for $w \in \mathcal{V}(v \setminus u)$. By using the above calculation shown in Eq. (9) and noting that the node u mediates any node pair (w, y) starting from $w \in \mathcal{V}(v \setminus u)$ and ending at $y \in \mathcal{V}(u \setminus v) \setminus \{u\}$, but no pair starting from $w \in \mathcal{V}(v \setminus u)$ and ending at $y \in \mathcal{V}(v \setminus u)$, we can derive a recursive TP technique for betweenness centrality. More specifically, we consider updating $\eta(u)$ and $\phi(u)$ by using the following formulae:

$$\eta(u) \leftarrow 1 + |\mathcal{V}(v \setminus u)|,$$
$$\phi(u) \leftarrow \mathcal{V}(v \setminus u)|(|\mathcal{V}(u \setminus v)| - 1)|, \tag{10}$$

Then, after pruning the node set $\mathcal{V}(v \setminus u)$ and the cut link $e = (u, v)$ as $\mathcal{W} \leftarrow \mathcal{V} \setminus \mathcal{V}(v \setminus u)$ and $\mathcal{F} \leftarrow \mathcal{E} \setminus \{e\}$, we can equivalently calculate $DB(x)$ by using Eq. (8) for each node $x \in \mathcal{V}(u \setminus v)$ as obtained by Eq. (2). Again, we can apply the same arguments to the case of $x \in \mathcal{V}(v \setminus u)$. Therefore, by generalizing the update formulae of Eq. (10) as follows:

$$\eta(u) \leftarrow \eta(u) + \sum_{w \in \mathcal{W}(v \setminus u)} \eta(w),$$

$$\phi(u) \leftarrow \phi(u) + \sum_{w \in \mathcal{W}(v \setminus u)} \eta(w) \left(\sum_{w \in \mathcal{W}(u \setminus v)} \eta(w) - 1 \right), \tag{11}$$

for all the node $x \in \mathcal{V}$, we can also exactly calculate the betweenness centrality measure $DB(x)$ by using the above recursive TP update shown in Eq. (11).

Similarly to the case of the closeness centrality, by focusing on the fact that each link of degree-one node is a cut link, we derive a recursive BP technique algorithm for betweenness centrality. More specifically, for a degree-one node v and its cut link $e = (u, v)$, we consider updating $\eta(u)$, $\phi(u)$ and $\phi(v)$ as follows:

$$\eta(u) \leftarrow \eta(u) + \eta(v),$$
$$\phi(u) \leftarrow \phi(u) + \eta(v)(|\mathcal{V}| - \eta(v) - 1),$$
$$\phi(v) \leftarrow \phi(v) + (\eta(v) - 1)(|\mathcal{V}| - \eta(v)). \tag{12}$$

Then, after pruning the node v and its cut link $e = (u, v)$ from the network $G = (\mathcal{W}, \mathcal{F})$ as $\mathcal{W} \leftarrow \mathcal{W} \setminus \{v\}$ and $\mathcal{F} \leftarrow \mathcal{F} \setminus \{e\}$, and updating $\eta(u)$, $\phi(u)$ and $\phi(v)$ by Eq. (12), we can equivalently calculate $DB(x)$ by using Eq. (8) for each node $x \in \mathcal{W}$ as obtained by Eq. (2). Here note that for any pruned nodes v, we calculate its measure by $DB(v) \leftarrow \phi(v)$. Therefore, for all the node $x \in \mathcal{V}$, we can exactly calculate the closeness centrality measure $DB(x)$ by using the above recursive TP and BP techniques.

3.3   Summary of Proposed Method

Hereafter, we consider simultaneously calculating both the closeness and betweenness centrality measures for all the node $v \in \mathcal{V}$. In our proposed method, the BP technique is applied before the TP techniques, because we can easily know the degree-one nodes in our network G. Although we can individually incorporate these techniques into the baseline method that do not employ our proposed pruning techniques, we only consider the proposed method without the TP technique, which is referred to as the BP method. Since it is difficult to analytically examine the effectiveness of these techniques, we empirically evaluate the computational efficiency of these three methods in comparison to the baseline method without the proposed pruning techniques, which is referred to as the BL method.

4   Experiments

In this section, after explaining the dataset used in our experiments, we evaluate the performance of our proposed acceleration algorithm, and discuss the characteristics of the distance based centralities.

4.1   Dataset

We used OSM (OpenStreetMap) data of eight cities in our experiments, i.e., Barcelona (Spain, Europe), Bologna (Italy, Europe), Brasilia (Brazil, South America), Cairo (Egypt, Africa), Washington D.C. (United States, North America), New Delhi (India, Asia), Richmond (United States, North America), and San Francisco (United States, North America). These are a subset of cities studied in [10]. In August, 2015, we obtained the OSM data of these eight cities from Metro Extracts[1]. Here note that in our experiments, the area of each city is more than 100 times larger than those of the previous study [10].

From the OSM data of each city, we extracted all highways and all nodes appearing in them, and constructed each spatial network by mapping the ends, intersections and curve-fitting-points of streets into nodes and the streets between the nodes into links. Then, based on GRS80 [14], we calculated each inter-node link distance from the positions of the nodes, each of which is described by a pair of latitude and longitude. Table 1 shows the basic statistics of the networks for the eight selected cities, where deg means the average degree of nodes calculated by $deg = 2|\mathcal{E}|/|\mathcal{V}|$, and avg, $s.d.$ and max stands for the average, standard deviation and maximum of inter-node link distances, respectively. From this table, we can see that although the area and the numbers of nodes and links, $|\mathcal{V}|$ and $|\mathcal{E}|$, are substantially different, the average degrees are quite similar as common characteristics of these spatial networks. On the other hand, it seems that each city has its own characteristics about the statistics of link distances.

In order to more closely study the characteristics of link distances, we examined the frequency of link distances obtained by

$$f(k) = |\{e = (u,v) \in \mathcal{E} \; : \; \epsilon_k \leq d(u,v) < \epsilon_{k+1}\}|, \tag{13}$$

where k is a non-negative integer, and ϵ_k is set to k meter. Figure 1 shows our analysis results of the eight cities shown in Table 1, where the horizontal and vertical axes stand for the distance and frequency, respectively. From these results plotted as log-log graphs, we can observe that each frequency curve is reasonably approximated by a power-law-like distribution. This suggests that since some link distances have relatively quite larger values, the distance-based centrality measures may have significantly different characteristics in comparison to those of the step-based centrality measures. In this paper, we experimentally evaluate these differences by focusing on the closeness and betweenness centralities.

---

[1] https://mapzen.com/data/metro-extracts.

Table 1. Basic statistics as network.

No.	Name	Area	$|\mathcal{V}|$	$|\mathcal{E}|$	deg	avg	s.d.	max
1	Barcelona	45×30 km	344,095	378,074	2.2	27.0	34.7	1,396.4
2	Bologna	60×45 km	262,839	281,294	2.1	34.2	59.3	4,341.1
3	Brasilia	120×104 km	197,829	240,546	2.4	78.6	135.6	8,369.9
4	Cairo	87×86 km	195,228	225,049	2.3	69.8	123.2	8,078.4
5	Washington D.C	23×18 km	114,758	128,746	2.2	25.3	33.7	1,058.5
6	New Delhi	109×75 km	284,964	336,183	2.4	66.2	99.7	10,968.3
7	Richmond	54×35 km	371,174	394,161	2.1	28.5	45.5	15,359.4
8	San Francisco	90×50 km	492,266	541,162	2.1	33.9	48.3	7,241.0

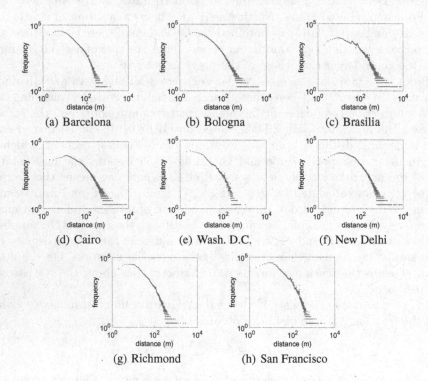

Fig. 1. Frequency of link distances.

4.2 Computational Efficiency

As described earlier, we evaluated the efficiency of the proposed method which simultaneously calculates $DC(v)$ and $DB(v)$ for each node $v \in \mathcal{V}$, by comparing the computation time of the baseline (BL), only bottom-up pruning (BP), and the proposed (PR) methods. We implemented the BL method based on Brandes's algorithm [2] known as the standard and efficient technique for computing the betweenness centrality of each node in a network. Figure 2 shows the computation

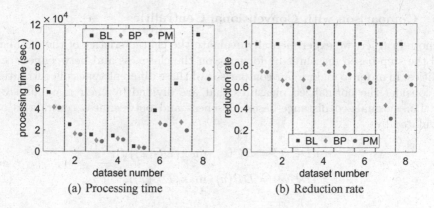

Fig. 2. Computation time comparison.

time of each method, where the dataset numbers shown in the horizontal axis are identical to those of Table 1. Figure 2(a) compares the actual processing time of these methods, where our programs implemented in C were executed on a computer system equipped with two Xeon X5690 3.47 GHz CPUs and 192 GB main memory with a single thread within the memory capacity. Figure 2(b) compares the reduction rates of computation time for these methods from the BL method.

From Figs. 2(a) and (b), we can see that for all the networks, the BP method steadily improves the computational efficiency of the BL method, and the PR method slightly improve that of the BP method. These results demonstrate the effectiveness of the proposed techniques. More specifically, as expected, from Fig. 2(a) and Table 1, we can see that the processing time of the BL method is almost proportional to the size of network. In contrast, from Fig. 2(b), we can see that the reduction rates of the BP and PR methods depend on the network, i.e., around from 0.4 to 0.8 for the BP method and around from 0.3 to 0.7 for the PR method. These results indicate that the effects of our pruning techniques depend on the networks.

On the other hand, we note that the improvement rates of the PR method over the BP method are modest, i.e., the reduction rates by the TP technique are not so remarkably effective. This must be partly because the TP technique requires additional computation costs for detecting cut links. Overall, we can conjecture that the proposed method combining both the BP and TP techniques is more reliable than the other two methods in terms of computation time because it produced the best performance for all of the eight networks. In short, reduction of computation time depends on network structures, but overall we can say that use of both techniques can increase the computational efficiency by nearly twice of the BL method.

4.3   Comparison with Conventional Centralities

As noted earlier, we experimentally evaluate the characteristics of the distance-
and the step-based measures by focusing on the closeness and betweenness centralities. In order to make a fair comparison of these different types of centralities,
we consider the normalized measures that are divided by their maximum values, i.e., normalized distance-based closeness and betweenness centralities are
calculated by

$$nDC(v) = DC(v)/\max_{x \in \mathcal{V}}\{DC(x)\},$$
$$nDB(v) = DB(v)/\max_{x \in \mathcal{V}}\{DB(x)\}, \tag{14}$$

respectively. Here, let $v(k_{DC})$ be the k-th node which has the k-th largest value
in the distance-based closeness centrality values $\{DC(v) : v \in \mathcal{V}\}$. Similarly,
we can define the k-th node for the other centrality measures as $v(k_{SC})$, $v(k_{DB})$
and $v(k_{SB})$, respectively. Recall that we denote the step-based closeness and
betweenness centrality measures as $SC(v)$ and $SB(v)$, respectively. Then, we
consider characterizing the top-k nodes of these measures by plotting the following quartets:

$$\{(k, nDC(v(k_{DC}))), (k, nDC(v(k_{SC}))), (k, nDB(v(k_{DB}))), (k, nDB(v(k_{SB})))\}. \tag{15}$$

Namely, we evaluate the top-k nodes of the distance- and the step-based closeness centralities by the normalized measures of the distance-based closeness centrality and those of the distance- and the step-based betweenness centralities by
the normalized measures of the distance-based betweenness centrality.

Figure 3 shows our experimental results, where the horizontal and vertical
axes stand for the rank k up to top-100 and the normalized distance-based closeness or betweenness centrality measure, respectively. We can see that the closeness and betweenness centralities have completely different characteristics. In
fact, the top-100 values of the distance- and the step-based closeness centralities
(DC = $nDC(v(k_{DC}))$ and SC = $nDC(v(k_{SC}))$) are almost the same, while the
situation is completely different for the two betweenness centralities. The top-100
values of the distance-based betweenness centrality substantially decrease (DB =
$nDB(v(k_{DB}))$), while the top-100 values of the step-based betweenness centrality have no regularity and violently change (SB = $nDB(v(k_{SB}))$). Namely, these
results indicate that the ranking by the distance- and the step-based closeness centralities is almost the same, while the ranking by the distance-based betweenness
centrality is substantially different from the ranking by the step-based betweenness centrality. These experimental results suggest that for arbitrary pairs of
nodes, most of the paths with the minimum distances are different between the
distance- and the step-based centralities, while the distances of these paths are
relatively close to each other. On the other hand, we can see that the curves for
the top-100 nodes are somewhat different from each other across different network
datasets, although their global behaviors are generally quite similar regardless of
any pairs of the centrality measures and the networks. Thus, it is expected that

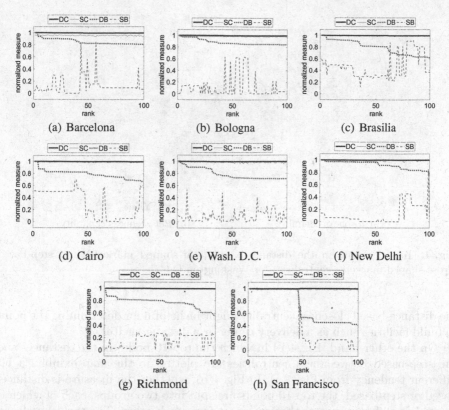

Fig. 3. Comparison of top-100 nodes.

the normalized measure curves shown in Fig. 3 may be able to uncover some characteristics of these networks, e.g., the nodes within the top-50 in San Francisco are quite similar for both the closeness and betweenness centralities.

Next, we investigate how differently the top ranked nodes in each centrality measure are distributed on a spatial map. To this end, we plotted the top-10 nodes in each centrality measure on the street map for the Washington D.C. dataset as shown in Fig. 4, where the top-10 nodes in the distance-based centrality are denoted by a star-shaped marker, and those in the step-based centrality are denoted by a cross-shaped marker. From Fig. 4(a), it is found that the top-10 locations in the closeness centrality are concentrated in a small area for both the distance-based and the step-based centralities. It seems natural that close neighboring nodes have similar values for the closeness centrality because the distances to another node could be similar regardless of the criterion whether it be distance-based or step-based. Here, remember that the rankings by the distance- and the step-based closeness centralities tend to be almost the same in Fig. 3. But, the actual areas in which the top-10 nodes are located are different in the distance- and the step-based centralities. Especially, the area is more limited and closer to the central area of the city for the distance-based centrality. This characteristic of

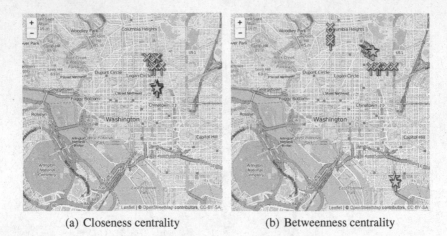

(a) Closeness centrality   (b) Betweenness centrality

Fig. 4. Top-10 points in the distance-based (star-shaped markers) and step-based (cross-shaped markers) centralities for Washington D.C.

the distance-based closeness centrality might be helpful for determining the point to build facilities such as a delivery center or a disaster shelter.

On the other hand, Fig. 4(b) in which the top-10 nodes in the distance- and the step-based betweenness centralities are plotted in the map exhibits a bit different tendency from Fig. 4(a). In Fig. 4(b), whether the measure is distance-based or step-based, the top-10 points are split into two groups, each of which is located on different streets that are apart from each other. The top-10 nodes in the distance-based betweenness centrality have a higher degree of concentration than those in the step-based one, similarly to the case of closeness centrality in Fig. 4(a). Further, the streets along which the top-10 points in the distance-based betweenness centrality are plotted are away from the streets along which the top-10 points in the step-based betweenness centrality are plotted. This tendency coincides with what we observed in Fig. 3 that the ranking by the distance-based betweenness centrality is substantially different from the ranking by the step-based betweenness centrality. Interestingly, the street that is close to the bottom-right corner and has three star-shaped markers is connecting to a freeway via a bridge, which is different from the other streets with markers. Indeed, one of the three points corresponds to the top node in the distance-based betweenness centrality. These points might be more critical in this urban traffic network than the others because it seems difficult to find alternative routes having a similar distance-based betweenness centrality value from the nearby street. If the street is closed and the traffic is blocked, we would have to go a very long way around. The distance-based betweenness centrality might be helpful to find such critical points in a traffic network.

We further plotted the top-10 nodes in each centrality measure on the street map for the Cairo dataset in the same fashion as in Fig. 4. The resulting maps are shown in Fig. 5. It is found that the findings we obtained above from Fig. 4

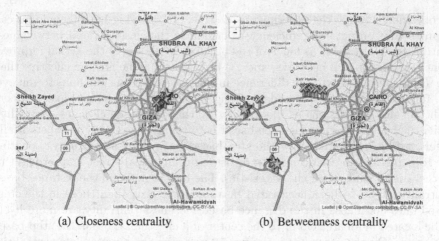

(a) Closeness centrality   (b) Betweenness centrality

Fig. 5. Top-10 points in the distance-based (star-shaped markers) and step-based (cross-shaped markers) centralities for Cairo

also hold in Fig. 5 although the two cities have different distributions of link distances as shown in Table 1. Note that the zoom level in Fig. 5 is different from the one in Fig. 4. It shows a wider area than Fig. 4 does. Thus, the groups of the top-10 nodes found in Fig. 5(b) are more distant from each other than those in Fig. 4(b). This is due to the difference between Washington D.C. and Cairo in the distribution of link distances.

5 Conclusion

In this paper, we first extended the conventional step-based closeness and betweenness centralities to analyze spatial networks. Unlike these conventional centralities that adopt the number of links to be traversed to reach one node from another as the distance between them, the extended distance-based closeness and betweenness centralities take into account the inter-nodes link distances obtained from the positions of nodes. They are natural extensions of the conventional centralities and general enough to include their definitions as a special case. Second, we have proposed two novel techniques to improve the computational efficiency to compute the distance-based centralities. Both are based on graph cut and recursively applied to a network in order to reduce scanning size needed for computing the centralities. The TP (Top-down pruning) technique recursively decomposes a network into two disjoint sub-networks that are connected by a cut link, while the BP (Bottom-up pruning) technique recursively removes a degree-one node by eliminating a cut link adjacent to it before the TP technique is applied. Note that it is straightforward to extend these techniques so that they can deal with directed networks.

We conducted extensive experiments using real-world road networks of eight different cities that have different distributions of link distances. Major findings we obtained through the experiments are that (1) applying both techniques can improve the computation efficiency to calculate the closeness and betweenness centralities about twice as much of the baseline method, (2) the actual improvement rates depend on networks, and (3) the BP technique is more effective than the TP technique due to the difference in the cost of finding cut links to remove. We further found that the node ranking by the distance-based closeness centrality is almost the same as the ranking by the step-based closeness centrality, while the ranking by the distance-based betweenness centrality is substantially different from the ranking by the step-based betweenness centrality. Locating the top-10 nodes in each centrality measure on an actual street map brought us further insights into their characteristics. The top-10 nodes in the distance-based centralities tend to be concentrated in a narrower area than those in the step-based ones do. Those in the distance-based closeness centrality are more likely to be located in an area closer to the center of a city than those in the step-based one are so. On the other hand, the top-10 nodes in the betweenness centralities tend to be split into two groups whether or not the measure is distance-based or step-based, and these two groups are geographically apart from each other. The distance-based closeness centrality seems more helpful to find locations to build facilities easy to access in a given area, while the distance-based betweenness centrality seems more useful to detect critical points in a traffic network. Finding these points in a network in an efficient way is one of our future directions. In addition, since centrality values of individual nodes can be computed independently, it is possible to parallelize those computations. This synergistically work with the proposed methods, leading to further improvement in efficiency and scalability. We are planning to empirically confirm this effect, too.

Acknowledgments. This material is based upon work supported by the Air Force Office of Scientific Research, Asian Office of Aerospace Research and Development (AOARD) under award number FA2386-16-1-4032, and JSPS Grant-in-Aid for Scientific Research (C) (No. 26330261).

References

1. Boldi, P., Vigna, S.: In-core computation of geometric centralities with hyperball: a hunderd billion nodes and beyond. In: Proceedings of the 2013 IEEE 13th International Conference on Data Mining Workshops (ICDMW 2013), pp. 621–628 (2013)
2. Brandes, U.: A faster algorithm for betweenness centrality. J. Math. Sociol. **25**, 163–177 (2001)
3. Brin, S., Page, L.: The anatomy of a large-scale hypertextual web search engine. Comput. Netw. ISDN Syst. **30**, 107–117 (1998)
4. Burckhart, K., Martin, O.J.: An interpretation of the recent evolution of the city of Barcelona through the traffic maps. J. Geogr. Inf. Syst. **4**(4), 298–311 (2012)
5. Chakrabarti, S., Dom, B., Kumar, R., Raghavan, P., Rajagopalan, S., Tomkins, A., Gibson, D., Kleinberg, J.: Mining the web's link structure. IEEE Comput. **32**, 60–67 (1999)

6. Chierichetti, F., Epasto, A., Kumar, R., Lattanzi, S., Mirrokni, V.: Efficient algorithms for public-private social networks. In: Proceedings of the 21st ACM SIGKDD International Conference on Knowledge Discovery and Data Mining (KDD 2015), pp. 139–148 (2015)
7. Cohen, E.: Size-estimation framework with applications to transitive closure and reachability. J. Comput. Syst. Sci. **55**, 441–453 (1997)
8. Cohen, E.: All-distances sketches, revisited: HIP estimators for massive graphs analysis. In: Proceedings of the 33rd ACM SIGMOD-SIGACT-SIGART Symposium on Principles of Database Systems, pp. 88–99 (2015)
9. Cohen, E., Delling, D., Pajor, T., Werneck, R.F.: Sketch-based influence maximization and computation: scaling up with guarantees. In: Proceedings of the 23rd ACM International Conference on Information and Knowledge Management, pp. 629–638 (2014)
10. Crucitti, P., Latora, V., Porta, S.: Centrality measures in spatial networks of urban streets. Phys. Rev. E **73**(3), 036125 (2006)
11. Freeman, L.: Centrality in social networks: conceptual clarification. Soc. Netw. **1**, 215–239 (1979)
12. Kimura, M., Saito, K., Ohara, K., Motoda, H.: Speeding-up node influence computation for huge social networks. Int. J. Data Sci. Anal. **1**, 1–14 (2016)
13. Montis, D.A., Barthelemy, M., Chessa, A., Vespignani, A.: The structure of interurban traffic: a weighted network analysis. Environ. Plann. B Plann. Des. **34**(5), 905–924 (2007)
14. Moritz, H.: Geodetic reference system 1980. J. Geodesy **74**(1), 128–133 (2000)
15. Newman, M.E.J.: The structure and function of complex networks. SIAM Rev. **45**, 167–256 (2003)
16. Ohara, K., Saito, K., Kimura, M., Motoda, H.: Resampling-based framework for estimating node centrality of large social network. In: Džeroski, S., Panov, P., Kocev, D., Todorovski, L. (eds.) DS 2014. LNCS (LNAI), vol. 8777, pp. 228–239. Springer, Heidelberg (2014). doi:10.1007/978-3-319-11812-3_20
17. Opsahl, T., Agneessens, F., Skvoretz, J.: Node centrality in weighted networks: generalizing degree and shortest paths. Soc. Netw. **32**(3), 245–251 (2010)
18. Park, K., Yilmaz, A.: A social network analysis approach to analyze road networks. In: Proceedings of the ASPRS Annual Conference (2010)
19. Saito, K., Kimura, M., Ohara, K., Motoda, H.: Super mediator - a new centrality measure of node importance for information diffusion over social network. Inf. Sci. **329**, 985–1000 (2016)
20. Wang, P., Hunter, T., Bayen, A.M., Schechtner, K., Gonzalez, M.C.: Understanding Road Usage Patterns in Urban Areas. Scientific Reports 2 (2012)

A Semi-supervised Approach to Measuring User Privacy in Online Social Networks

Ruggero G. Pensa[(⊠)] and Gianpiero Di Blasi

Department of Computer Science, University of Torino, Turin, Italy
ruggero.pensa@unito.it

Abstract. During our digital social life, we share terabytes of information that can potentially reveal private facts and personality traits to unexpected strangers. Despite the research efforts aiming at providing efficient solutions for the anonymization of huge databases (including networked data), in online social networks the most powerful privacy protection is in the hands of the users. However, most users are not aware of the risks derived by the indiscriminate disclosure of their personal data. With the aim of fostering their awareness on private data leakage risk, some measures have been proposed that quantify the privacy risk of each user. However, these measures do not capture the objective risk of users since they assume that all user's direct social connections are close (thus trustworthy) friends. Since this assumption is too strong, in this paper we propose an alternative approach: each user decides which friends are allowed to see each profile item/post and our privacy score is defined accordingly. We show that it can be easily computed with minimal user intervention by leveraging an active learning approach. Finally, we validate our measure on a set of real Facebook users.

Keywords: Privacy metrics · Active learning · Online social networks

1 Introduction

Online social networks are among the main traffic sources in the Internet. At the end of 2014, they attracted more than 31 % of the worldwide internet traffic towards the Web. Facebook, the most famous social networking platform, drives alone 25 % of the whole traffic. As a comparison, Google search engine represents just over 37 % of the global traffic[1]. More than two billions people are estimated to be registered in at least one of the most popular social media platforms (Facebook hits the goal of one billion users in 2012). Overall, the number of active "social" accounts are more than two billions. The famous "six degrees of separation" theory has been far exceed in Facebook, where an average degree of 3.57 has been recently observed[2]. Consequently, social network users are constantly exposed to privacy leakage risks. Although most users do not disclose

---

[1] Source: http://www.alexa.com/.

[2] https://research.facebook.com/blog/three-and-a-half-degrees-of-separation/.

© Springer International Publishing Switzerland 2016
T. Calders et al. (Eds.): DS 2016, LNAI 9956, pp. 392–407, 2016.
DOI: 10.1007/978-3-319-46307-0_25

very sensitive facts (private life events, diseases, political ideas, sexual preferences, and so on), they are simply not aware of the risks due to the disclosure of less sensitive information, such as GPS tags, photos taken during a vacation period, page likes, or comments on news. As an example, the research project myPersonality [14] carried out at the University of Cambridge has shown that, by leveraging Facebook user's activity (such as "Likes" to posts or fan pages) it is possible to "guess" some very private traits of the user's personality. According to another study, it is even possible to infer some user characteristics from the attributes of users who are part of the same communities [18]. As a consequence, privacy has become a primary concern among social network analysts and Web/data scientists. Also, in recent years, many companies are realizing the necessity to consider privacy at every stage of their business. In practice, they have been turning to the principle of *Privacy by Design* [5] by integrating privacy requirements into their business model.

Despite the huge research efforts aiming at providing efficient solutions to the anonymization of huge databases (including networked data) [3,25], in online social networks the most powerful privacy protection is in the hands of the users: they, and only they, decide what to publish and to whom. Even though social networking sites (such as Facebook), notify their users about the risks of disclosing private information, most people are not aware of the dangers due to the indiscriminate disclosure of their personal data when they surf the net. Some social media provide advanced tools for controlling the privacy settings of the user's profile [24]. However, yet a large part of Facebook content is shared with the default privacy settings and exposed to more users than expected [17]. According to Facebook CTO Bret Taylor, even though most people have modified their privacy settings[3], in 2012, still "13 million users [in the United States] said they had never set, or didn't know about, Facebook's privacy tools[4]".

Some studies try to foster risk perception and awareness by "measuring" users' profile privacy according to their privacy settings [16,23]. These metrics usually require a *separation-based* policy configuration: in other terms, the users decide "how distant" a published item may spread in the network. Typical separation-based privacy policies for profile item/post visibility include: visible to no one, visible to friends, visible to friends of friends, public. However, this policy fails when the number of user friends becomes large. According to a well-known anthropological theory, in fact, the maximum number of people with whom one can maintain stable social (and cybersocial) relationships (known as Dunbar's number) is around 150 [10,20], but the average number of user friends in Facebook is more than double[5]. This means that many social links are weak (offline and online interactions with them are sporadic), and a user who sets the

---

[3] http://www.zdnet.com/article/facebook-cto-most-people-have-modified-their-privacy-settings/.

[4] http://www.consumerreports.org/cro/magazine/2012/06/facebook-your-privacy/index.htm.

[5] http://www.pewresearch.org/fact-tank/2014/02/03/6-new-facts-about-facebook/.

privacy level of an item to "visible to friends" probably is not willing to make that item visible to *all* her friends.

To address this limitation, in this paper we propose a *circle-based* formulation of the privacy score proposed by Liu and Terzi [16]. We assume that a user may set the visibility of each action and profile item separately for each other user in her friend list. For instance, a user u may decide to allow the access to all photo albums to friends f_1 and f_2, but not to friend f_3. In our score, the sensitivity and visibility of profile item i published by user u are computed according to the set of u's friends that are allowed to access the information provided by i. Since the expression of explicit allow/deny policy for each friend and each item may require huge labeling efforts, we also propose an active learning labeling approach to limit the number of manual operations. We show experimentally that (i) our circle-based definition of privacy score better capture the real privacy leakage risk and (ii) the active learning approach provides accurate results in terms of both predicted privacy settings and final privacy score.

The remainder of the paper is organized as follows: we briefly review the related literature in Sect. 2; the overview and the theoretical details of our score are presented in Sect. 3; the active learning approach is presented in Sect. 4; Sect. 5 provides the report of our experimental validation; finally, we draw some conclusions in Sect. 6.

2   Related Work

Most research efforts in social network privacy are devoted to the identification and formalization of privacy breaches and to the anonymization of networked data [25]. All these works focus on how to share social networks owned by companies or organizations masking the identities or the sensitive connections of the individuals involved. However, increasing attention is being paid to the privacy risk of users caused by their information-sharing activities (e.g., posts, likes, shares). In fact, since disclosing information on the web is a voluntary activity, a common opinion is that users should care about their privacy during their interaction with other social network users. Thus, another branch of research has focused on investigating strategies and tools to enhance the users' privacy awareness and help them act more safely during their day-to-day social network activity. In [6] the authors present an online game, called Friend Inspector, that allows Facebook users to check their knowledge of the visibility of their shared personal items and provides recommendations on how to improve privacy settings. Instead, Fang and LeFevre [11] propose a social networking privacy wizard based on active learning. The wizard iteratively asks the user to allow or deny the visibility of profile items to selected friends and assign privileges to the rest of the user's friends using a classifier. [4] presents a tool to detect unintended information loss in online social networks by quantifying the privacy risk attributed to friend relationships in Facebook. The authors show that a majority of users' personal attributes can be inferred from social circles. In [22] the authors present a privacy protection tool that measures the inference probability of sensitive attributes from friendship links. In addition, they suggest self-sanitization actions to regulate the amount of leakage. [12], instead, introduces a machine learning technique to monitor users' privacy settings and recommend reasonable privacy options. Other approaches to privacy control in social networks investigate the problem of the risk perception. In [1,2], for instance, the authors propose to provide users with a measure of how much it might be risky to have interactions with them, in terms of disclosure of private information. They use an active learning approach to estimate user risk from few required user interactions.

The privacy measure we propose in this paper is closely related to the work of Liu and Terzi [16]. They propose a framework to compute a privacy score measuring the users' potential risk caused by their participation in the network. This score takes into account the sensitivity and the visibility of the disclosed information and leverages the item response theory as theoretical basis for the mathematical formulation of the score. Another privacy measure has been proposed in [23] where the authors introduce a privacy index to measure the user privacy exposure in a social network. This index, however, strongly relies on pre-defined sensitivity values for users' items. Furthermore, in both proposals, the privacy measures are computed by leveraging separation-based privacy policies. Differently from the above mentioned papers, our proposal considers circle-based policy settings that better suits the real user visibility preferences.

3   A Circle-Based Definition of Privacy Score

In this section we introduce our circle-based privacy score aiming at supporting the users participating in a social network in assessing their own privacy leakage risk. Most social networking platforms (such as Facebook or Google+), provide an adequate flexibility in configuring privacy of profile items and user's actions. Moreover, they offer some advanced facilities, such as the possibility of grouping friends into special lists or social circles. But privacy is not just a matter of users' preferences; it also relies on the context in which an individual is immersed: the position within the network (very central users are more exposed than marginal users), her or his own attitude on disclosing very private facts, and so on. Hence, we propose a privacy score that takes all these aspects into account and fits the real user expectations about the visibility of profile items.

Before entering the technical details of our approach, we briefly introduce some basic mathematical notation required to formalize the problem.

3.1   Preliminaries and Notation

Here we introduce the mathematical notation we will adopt in the rest of our paper. We consider a set of n users $\mathcal{U} = \{u_1, \ldots, u_n\}$ corresponding to the individuals participating in a social network. Each user is characterized by a set of m properties or profile items $\mathcal{P} = \{p_1, \ldots, p_m\}$, corresponding, for instance, to personal information such as gender, age, political views, religion, workplace, birthplace and so on. Hence, each user u_i is described by a vector $\boldsymbol{p}^i =< p_{i1}, \dots, p_{im} >$.

Users are part of a social network. Without loss of generality, we assume that the link between two users is always reciprocal (if there is a link from u_j to u_j then there is also a link from u_j to u_i). Hence, the social network here is represented as an undirected graph $G(V, E)$, where V is a set of n vertices $\{v_1, \dots, v_n\}$ such that each vertex $v_i \in V$ is the counterpart of user $u_i \in \mathcal{U}$ and E is a set of edges $E = \{(v_i, v_k)\}$. Given a pair of users $(u_i, u_k) \in \mathcal{U}$, $(v_i, v_k) \in E$ iif users u_i and u_k are connected (e.g., by a friendship link).

For any given vertex $v_i \in V$ we define the neighborhood $\mathcal{N}(v_i)$ as the set of vertices v_k directly connected to the vertex v_i, i.e., $\mathcal{N}(v_i) = \{v_k \in V \mid (v_i, v_k) \in E\}$. Conversationally speaking, $\mathcal{N}(v_i)$ is the set of friends (also known as *friend-list*) of user u_i, hence we use $\mathcal{N}(v_i)$ or $\mathcal{N}(u_i)$ interchangeably. Given a user u_i and her friend-list $\mathcal{N}(u_i)$, we also define the *ego network* centered on user u_i as the graph $G_i(V_i, E_i)$, where $V_i = \mathcal{N}(v_i) \cup \{v_i\}$ and $E_i = \{(v_k, v_l) \in E \mid v_k, v_l \in V_i\}$.

Finally, for any user u_i we introduce a *privacy policy matrix* $\boldsymbol{M}_i \in \{0,1\}^{n_i \times m}$ (with $n_i = |\mathcal{N}(u_i)|$) defined as follows: for any element m_{kj}^i of $\boldsymbol{M}_i$, $m_{kj}^i = 1$ iif profile item $p_j \in \mathcal{P}$ is visible to user $u_k \in \mathcal{N}(u_i)$ (0 otherwise, i.e., iif user u_k is not allowed to access profile item p_j).

It is worth noting that our framework can be easily extended to the case of directed social networks (such as Twitter): in this case, the privacy policies are defined only on inbound links.

3.2 Privacy Score

Our measure is inspired by the privacy score defined by Liu and Terzi [16]. It measures the user's potential risk caused by his or her participation in the network. A $n \times m$ response matrix $\boldsymbol{R}$ is associated to the set of n users $\mathcal{U}$ and the set of m profile properties $\mathcal{P}$. In [16], each element r_{ij} of $\boldsymbol{R}$ contains a privacy level that determines the willingness of user u_i to disclose information associated with property p_j. In the binomial case $r_{ij} \in \{0,1\}$: $r_{ij} = 1$ (resp. $r_{ij} = 0$) means that user u_i has made the information associated with profile item p_j publicly available (resp. private). In the multinomial case, entries in $\boldsymbol{R}$ take any non-negative integer values in $\{0, 1, \dots, \ell\}$, where $r_{ij} = h$ (with $h \in \{0, 1, \dots, \ell\}$) means that user u_i discloses information related to item p_j to users that are at most h links away in the social network G (e.g., if $r_{ij} = 0$ user u_i wants to keep p_j private, if $r_{ij} = 1$ user u_i is willing to make p_j available to all friends, if $r_{ij} = 2$ user u_i is willing to make p_j available to the friends of her or his friends, and so on). For this reason, we call this policy *separation-based*. However, in this work, we adopt a different meaning for the entries r_{ij} of $\boldsymbol{R}$: in our framework r_{ij} is directly proportional to the number of friends to whom u_i is willing to disclose the information of profile property p_j. Hence, we can compute $\boldsymbol{R}$ according to the *circle-based* privacy policies defined by matrices $\boldsymbol{M}_i$'s using this formula:

$$r_{ij} = \left\lfloor \ell \cdot \frac{1}{|\mathcal{N}(u_i)|} \sum_{k=1}^{|\mathcal{N}(u_i)|} m_{kj}^i \right\rfloor \tag{1}$$

where $\mathcal{N}(u_i)$ is the set of friends of user u_i, m^i_{kj} denotes the visibility of user u_i's profile item p_j for friend u_k, and $\lfloor \cdot \rfloor$ is the floor function. As a consequence, $r_{ij} = \ell$ iif $\forall u_k \in \mathcal{N}(u_i)$, $m^i_{kj} = 1$. Our definition is conceptually different from the original one, since the latter does not take into account the possibility of disclosing personal items to just a part of friends.

In the following, we use $\boldsymbol{R}^S$ when we refer to the response matrix computed with the original separation-based policy approach defined in [16]. We use $\boldsymbol{R}^C$ when we refer to our circle-based definition of response matrix.

Using the response matrix it is possible to compute the two main components of the privacy function: the sensitivity β_{jh} of a profile item p_j for a given privacy level h, and the visibility V_{ijh} of a profile item p_j due to u_i for a given level h. The sensitivity of a profile item p_j depends on the item itself (attribute "sexual preferences" is usually considered more sensitive than "age"). The visibility, instead, captures to what extent information about profile item p_j of user u_i spreads in the network. For the computation details of β_{jh} and V_{ijh} we invite the reader to refer to [16], where a mathematical model based on item response theory (a well known theory in psychometrics) is used to compute sensitivity and visibility. Intuitively, sensitivity β_j is such that the more users adopt at least privacy level h for privacy item p_j, the less sensitive p_j is w.r.t. level h. Instead, visibility V_{ijh} is higher when the sensitivity of profile items is low and when users have the tendency to disclose lots of their profile items. Moreover, it depends on the position of user u_i within the network and can be computed by exploiting any information propagation models [13].

The privacy score $\phi_p(u_i, p_j)$ for any user u_i and profile property p_j is computed as follows:

$$\phi_p(u_i, p_j) = \sum_{h=0}^{\ell} \beta_{jh} \cdot V_{ijh}. \tag{2}$$

and the overall privacy score $\phi_p(u_i)$ for any user u_i is given by

$$\phi_p(u_i) = \sum_{j=1}^{m} \phi_p(u_i, p_j). \tag{3}$$

From Eqs. 2 and 3 it is clear that users that have the tendency to disclose sensitive profile properties to a wide public are more prone to privacy leakage. Intuitively, $\phi_p(u_i) = 0$ means that, in each element of the summation, either $\beta_{jh} = 0$ (the profile item p_j is not sensitive at all), or $V_{ijh} = 0$ (the profile item p_j is kept private). On the contrary, the privacy score is maximum when a user discloses to all her or his friends ($V_{ijh} = 1$) all sensitive information ($\beta_{jh} = 1$).

In this paper, we use ϕ_p^S when we refer to the score computed using the original separation-based response matrix $\boldsymbol{R}^S$; we use ϕ_p^C when we refer to the privacy score leveraging our circle-based definition of response matrix $\boldsymbol{R}^C$.

4   Semi-supervised Privacy Policy Definition

Our definition of privacy score requires the availability of visibility preferences for all user friends. However, setting them correctly is often an annoying and frustrating task and many users may prefer to adopt simple but extreme strategies such as "visible-to-all" (exposing themselves to the highest risk), or "hidden-to-all" (wasting the positive social and economic potential of social networking websites). In this section we present a semi-supervised approach to minimize the user's intervention while computing the circle-based privacy policy matrices M_i. The classification model should be as accurate as possible in predicting those privacy preferences not explicitly set by the users. Moreover, the model should be easily updatable when the user sets more privacy preferences or adds new users. Our choice is to use a Naive Bayes classifier [19], which is simple and converge quickly even with few training data. Moreover, it can be easily embedded in an active learning framework using, for instance, uncertainty sampling [9] thus minimizing the intervention of the user in the model training phase.

Table 1. Example of Input Dataset for the Classification Task

Friend ID	Age	Gender	Hometown	Community	No. of friends	C_{work}	C_{photos}	$C_{politics}$
102030	"21–30"	Male	Rome	C10	"501–700"	allow	allow	deny
203040	"31–40"	Female	Madrid	C5	"201–300"	allow	deny	deny
304050	"15–19"	Female	Paris	C7	"101–200"	allow	deny	deny
405060	"41–50"	Female	Berlin	C5	"701–1000"	allow	deny	deny
506070	"51–60"	Male	Rome	C10	"501–700"	allow	allow	deny
607080	"21–30"	Female	Rome	C10	"301–500"	?	?	?
708090	"41–50"	Male	Madrid	C5	"301–500"	?	?	?

For any given user $u_i \in \mathcal{U}$ and any given profile item $p_j \in \mathcal{P}$ we define a classification problem in which we have a set of $|\mathcal{N}(u_i)|$ instances $D = \{d_1, \ldots, d_{|\mathcal{N}(u_i)|}\}$ corresponding to all friends of u_i. Each instance d_k is characterized by a set of p attributes $\{A_1, \ldots, A_p\}$ with discrete values and m class variables $\{C_1, \ldots, C_m\}$ that take values in the domain $\{allow, deny\}$: $C_j = allow$ (resp. $C_j = deny$) means that friend u_k is allowed (resp. is not allowed) to access the information of profile item p_j of user u_i. The values of attributes $\{A_1, \ldots, A_p\}$ are partly derived from the profile vector $\boldsymbol{p}^k = <p_{k1}, \ldots, p_{km}>$ of users u_k, partly from the ego network $G_i(V_i, E_i)$ of user u_i (see Sect. 3.1). For instance, they may contain information such as the workplace and home-town of u_k, or the communities in G_i u_k belong to. Table 1 is an example of possible small dataset for a generic user consisting of five training instances and two test instances with three profile-based attributes, two network-based attributes and three class variables.

The Naive Bayes classification task can be regarded as estimating the class posterior probabilities given a test example d_k, i.e., $Pr(C_j = allow|d_k)$ and $Pr(C_j = deny|d_k)$. The class with the highest probability is assigned to the example d_k. Given a test example d_k, the observed attribute values are given by the vector $\boldsymbol{d}^k = \{a_1^k, \ldots, \ldots, a_p^k\}$, where a_s^k is a possible value of A_s, $s = 1, \ldots, p$. The prediction is the class c ($c \in \{allow, deny\}$) such that $Pr(C_j = c | A_1 = a_1^k, \ldots, A_p = a_p^k)$ is maximal. By Bayes' theorem, the above quantity can be expressed as

$$Pr(C_j = c | A_1 = a_1^k, \ldots, A_p = a_p^k)$$

$$= \frac{Pr(A_1 = a_1^k, \ldots, A_p = a_p^k | C_j = c) Pr(C_j = c)}{Pr(A_1 = a_1^k, \ldots, A_p = a_p^k)}$$

$$= \frac{Pr(A_1 = a_1^k, \ldots, A_p = a_p^k | C_j = c) Pr(C_j = c)}{\sum_{c_x} Pr(A_1 = a_1^k, \ldots, A_p = a_p^k | C_j = c_x) Pr(C_j = c_x)} \tag{4}$$

where, $Pr(C_j = c)$ is the class prior probability of c, which can be estimated from the training data. If we assume that conditional independence holds, i.e., all attributes are conditionally independent given the class $C_j = c$, then

$$Pr(A_1 = a_1^k, \ldots, A_p = a_p^k | C_j = c) = \prod_{s=1}^{p} Pr(A_s = a_s^k | C_j = c) \tag{5}$$

and, finally

$$Pr(C_j = c | A_1 = a_1^k, \ldots, A_p = a_p^k) =$$

$$= \frac{Pr(C_j = c) \prod_{s=1}^{p} Pr(A_s = a_s^k | C_j = c)}{\sum_{c_x} Pr(C_j = c_x) \prod_{s=1}^{p} Pr(A_s = a_s^k | C_j = c_x)} \tag{6}$$

Thus, given a test instance d_k, its most probable class is given by:

$$c = \arg \max_{c_x} \left\{ Pr(C_j = c_x) \prod_{s=1}^{p} Pr(A_s = a_s^k | C_j = c_x) \right\} \tag{7}$$

where the prior probabilities $Pr(C_j = c_x)$ and the conditional probabilities $Pr(A_s = a_s^k | C_j = c_x)$ are estimated from the training data.

To predict all C_j's accurately without requesting too much labeling work to u_i, we adopt an *active learning* approach named *uncertainty sampling* [15] based on the *maximum entropy* principle [9]. In an active learning settings the learning algorithm is able to interactively ask the user for the desired/correct labels of unlabeled data instances. A way to reduce the amount of labeling queries to the users is to sample only those data instances whose predicted class is most uncertain. Different measures of uncertainty have been proposed in the literature, e.g., least confidence [8], smallest margin [21] and maximum entropy [9], but for binary classification tasks they are equivalent. Hence, we decide to adopt the maximum entropy principle. According to this principle, the most uncertain data instance d_u is given by:

$$d_u = \arg \max_{d_k} \left\{ -\sum_{c_x} Pr(C_j = c_x | d_k) \log Pr(C_j = c_x | d_k) \right\} \tag{8}$$

Since probabilities $Pr(C_j = c_x|d_k)$ are exactly those computed by the Naive Bayes classifier to take its decision, this principle can be easily adapted to our classification task.

Once all friends' labels are predicted, each entry of the policy matrix M_i can be updated as follows:

$$\forall u_k \in \mathcal{N}(u_i), \ m_{kj}^i = \begin{cases} 1, & \text{if } C_j = allow \text{ for } u_k \\ 0, & \text{if } C_j = deny \text{ for } u_k. \end{cases} \tag{9}$$

The entries of M_i are then used to compute the response matrix R^C as described in Sect. 3. Note that the original separation-based definition of privacy score can not take advantage of this active learning strategy.

5   Experimental Results

In this section we report and discuss the results of an online experiment that we conducted on real Facebook users. The main objectives of our experiment are: (i) to study the relationship between the separation-based privacy policies and our circle-based policy definition; (ii) to analyze the relationship between the separation-based privacy score ϕ_p^S defined in [16] and our circle-based score ϕ_p^C; (iii) to assess the performances of our active learning approach in terms of classification accuracy and privacy score robustness.

The section is organized as follows: first, we describe the data and how we gathered them; then we provide the details of our experimental settings; finally we report the results and discuss them.

5.1   Dataset

Our online experiments were conducted in two phases. In the first phase we promoted the web page of the experiment[6] where people could voluntarily grant us access to some data related to their own Facebook profile and friends' network. We were not able to access any other information rather than what we asked the permission for, i.e.: email (needed to contact the users for the second phase of our experiment), public profile, friend list, gender, age, work, education, hometown, current location and pagelikes. The participants were perfectly aware about the data we asked for and the purpose of our experiment. In this first phase, data were gathered through a Facebook application developed in Java JDK 8, using Version 1.0 of Facebook Graph API. From March to April 2015, we collected the data of 185 volunteers, principally from Europe, Asia and Americas. The social network consisting of all participants plus their friends is an undirected graph with 75,193 nodes and 1,377,672 edges.

---

[6] http://kdd.di.unito.it/privacyawareness/.

Q1	Which people would you like to tell that you have just changed job?
Q2	If your relationship status changed, which friends would you like to tell?
Q3	After a nice holiday, which friends would you share your photos with?
Q4	With whom would you like to share a comment on current affairs/politics?
Q5	With whom would you like to share your mood or something personal that happened to you?

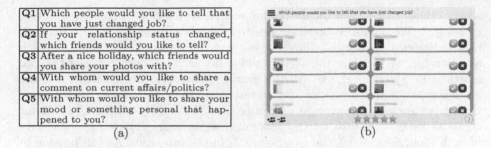

(a)   (b)

Fig. 1. The five questions (a) and the graphical interface (b) of our online survey

During the second phase, all the remaining participants were contacted for the interactive part of our experiment. First, the participants had to indicate to which level (0=no one, 1=close friends, 2=friends except acquaintances, 3=all friends, 4=friends of friends, 5=everyone on Facebook) they were willing to allow the access to five personal profile topics. The topics were proposed in form of direct questions (see Fig. 1(a)) with different levels of sensitivity. We used the answers to fill the response matrix R^S. Then, to each participant, we proposed a list of 60 randomly chosen friends and 6 randomly chosen friends of friends (when available). The participants had to indicate to which people they were willing to allow the access to the same five topics. For this phase, we developed a Java JDK 8 mobile-friendly web application leveraging Version 2.0 of Facebook Graph API. Figure 1(b) provides a screenshot of our online survey. We used the answers on friends to fill the response matrix R^C. From May 2015 to February 2016, 74 out of 185 participants answered all questions of two surveys. Hence, in our experiments, we consider the network data provided by all 185 participants and the survey data related to the 74 participants who completed the questionnaire. All the data have been anonymized to preserve volunteers' privacy. The entries in the two resulting 74×5 matrices R^S and R^C take values in $\{0, \ldots, 5\}$.

5.2 Separation-Based vs. Circle-Based Policies

As a preliminary analysis, we measure how the perception of topic sensitivity changes when the two policies (separation-based and circle-based) are presented to the participants. To this purpose we compare the two response matrix R^S and R^C in several ways. First, we measure the Pearson's correlation coefficient between the two matrices. Given two series of n values $X = x_1 \ldots, x_n$ and $Y = y_i, \ldots, y_n$, the Pearson's coefficient is computed as:

$$\rho(X, Y) = \frac{\sum_{i=1}^{n} (x_i - \overline{x})(y_i - \overline{y})}{\sqrt{\sum_{i=1}^{n} (x_i - \overline{x})^2} \sqrt{\sum_{i=1}^{n} (y_i - \overline{y})^2}} \tag{10}$$

where $\overline{x} = \sum_{i=1}^{n} x_i/n$ and $\overline{y} = \sum_{i=1}^{n} y_i/n$. It basically captures the correlation between the two series of values and ranges between -1 (for inversely correlated sets of values) and $+1$ (for the maximum positive correlation). In our experiment, $n \doteq 74 \cdot 5$. We obtain a moderate positive correlation ($\rho(\boldsymbol{R}^S, \boldsymbol{R}^C) = 0.4632$), that indicates a substantial difference between the two policies. Then, for each question Q_j, we measure the average difference between each entry of the two matrices as $\sum_i (r_{ij}^d - r_{ij}^b)/n$. All the average differences are positive, i.e., the given separation-based policies are less restrictive than circle-based ones. In particular, we measure an average difference of 0.54 for Q_1, 0.43 for Q_2, 0.32 for Q_3, 0.35 for Q_4 and 0.15 for Q_5. Moreover, we measure the overall sensitivity of each topic as $\beta_j = \sum_h^\ell \beta_{jh}$ (see Sect. 3.2) in the two cases. As can be seen in Fig. 2(a), all sensitivity values increase when the circle-based policy is adopted. The improved sensitivity perception is confirmed when we look at the users' policies more deeply. In particular, for each question Q_j, we count:

- the number **A** of participants that, in the separation-based test, have made Q_j at least visible to friends of their friends ($r_{ij}^S \geq 4$), but have denied the access to Q_j to some of the friends of their friends in the circle-based test;
- the number **B** of users that have granted the access to some of the friends of their friends in the circle-based test while $r_{ij}^S < 4$ in the separation-based test;
- the number **C** of participants that, in the separation-based test, have made Q_j visible at least to all friends ($r_{ij}^S \geq 4$), but have denied the access to Q_j to some of their friends in the circle-based test $r_{ij}^C < 5$;
- the number **D** of participants that, in the circle-based test, have made Q_j visible to all friends ($r_{ij}^C = 5$), but have denied the access to Q_j to some of their friends in the separation-based test $r_{ij}^S < 3$.

The results in Table 2 indicate that the major differences are on questions Q_3 and Q_4, that are the less sensitive according to Fig. 2(a). However, then passing from a separation-based policy to a circle-based one, many users have reviewed their choices in a more restrictive way for question Q_1 and Q_2 as well.

Table 2. Policy differences in visibility

Measure	Q1	Q2	Q3	Q4	Q5
A	2	2	4	9	1
B	0	0	4	9	1
C	20	5	19	21	4
D	0	0	4	9	1

Finally, we also compute the privacy scores $\phi_p^S(u_i, p_j)$ and $\phi_p^C(u_i, p_j)$ for each question Q_j and each participant u_i. The average score values are given in Fig. 2(b). Interestingly, although the circle-based policy increases the perception of topic sensitivity, the related privacy scores are sensibly smaller than those computed within the separation-based hypothesis, i.e., the participants have a safer behavior w.r.t. the visibility of the topics. For the sake of completeness,

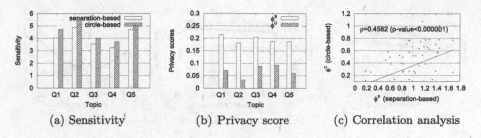

(a) Sensitivity   (b) Privacy score   (c) Correlation analysis

Fig. 2. Comparative results (separation-based approach vs. circle-based approach)

we perform a correlation analysis between the values of $\phi_p^S(u_i)$ and $\phi_p^C(u_i)$ in Fig. 2(c). The value of the Pearson's ρ coefficient (0.4582) shows moderate positive correlation between the two series of scores.

5.3 Assessment of the Active Learning Approach

To measure the performances of the active learning approach, we generate 74×5 datasets (one for each pair of users and questions) that we use to train and test the Naive Bayes classifier. These datasets contain, for each friend u_k of a user u_i, the following attributes: *gender* and *age* of u_k, *countryman* (true, if u_k and u_i were born in the same place, *fellow_citizen* (true, if u_k and u_i live in the same place), *coworker* (true, if u_k and u_i work or have worked in the same place), *schoolmate* (true, if u_k and u_i are or have studied in the same school/college/university), and the *Jaccard similarity of page likes* of u_i and u_k. All attribute values are derived from the information extracted by the Facebook profiles, when available. Additionally, we also consider the *list of communities* u_k is part of. To this purpose, we execute a community detection algorithm on the so called "ego-minus-ego" networks (the subgraph induced by the vertex set $\mathcal{N}(u_i) \setminus \{u_i\}$) of all 74 users. We use *DEMON* [7], a local-first approach based on a label propagation algorithm that is able to discover overlapping communities. The algorithm requires two parameters as input: the minimum accepted size for a community (*minCommunitySize*) and a parameter ϵ that determines the minimum overlap two communities should have in order to be merged. In our experiments, we set *minCommunitySize* $\doteq 3$ (to discard very small communities) and $\epsilon = 0.5$ (to admit an average overlap degree). Finally, each friend has a class variable that takes values in the set $\{allow, deny\}$.

We conduct the experiment as follows. To simulate the active learning framework, for each user and question, (i) we start with just five (randomly chosen) labeled friends with which we train the Naive Bayes classifier described in Sect. 4; (ii) we test the classifier on the remaining 55 friends and (iii) choose the friend whose prediction is the most uncertain, following the maximum entropy criterion (see Eq. 8 in Sect. 4); (iv) we assign to this friend the same label declared by the participant and (v) we re-train the classifier on $5 + 1$ instances (friends); (vi) finally, we test the new classifier on the remaining 54 instances. We repeat iteratively the last four steps until there are no test instances left.

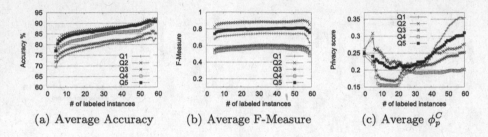

(a) Average Accuracy   (b) Average F-Measure   (c) Average ϕ_p^C

Fig. 3. Prediction Accuracy vs. Privacy function: average results

At the end of each prediction step, we measure the following performance parameters: (i) the *Accuracy* of the predictions; (ii) the *F-Measure* of the predictions, computed as $F\text{-}Measure = 2 \cdot (precision \cdot recall)/(precision + recall)$ where precision and recall are computed by considering the *deny* class as the positive one; (iii) the privacy score (Eq. 3) computed by considering both given and predicted $\{allow, deny\}$ labels for all 74 users and applying Eq. 9 to calculate matrices M_i and Eq. 1 to compute the response matrix R^C.

The values of the three parameters are averaged on all 74 users and 30 runs. In each run, the first five labeled friends are chosen randomly. The initial value of the privacy function (when no labels are given) is computed by assigning random labels to all 60 friends.

The results are provided in Fig. 3. The values of the three parameters are reported for each question separately. As a general observation, the accuracy of the prediction increases significantly with the number of labeled friends (see Fig. 3(a)). The growth of the F-Measure is less sharp, instead (Fig. 3(b)). We recall that both measures are computed on the test instances only. The small drop of Accuracy and F-Measure in the last steps can be explained by the fact that misclassification errors of few test instances (less than 5 samples) are more likely to happen. Interestingly, predictions are more accurate for the two most sensitive questions (Q2 and Q5). As for the privacy scores (Fig. 3(c)), they start to decrease when few friends (5 to 15) are labeled, then they start to grow almost monotonically and the differences among them are more emphasized. This behavior can be partially explained by noting that, as the amount of labeled friends increases, the sensitivity perceived by the users gets closer to the realistic sensitivity of the five topics.

5.4   Reliability of the Predictions

We also study the robustness of the approach by extending the prediction to all participants' friends. Since we do not have the correct labels for friends who do not belong to the list proposed to the participants, we can only measure the privacy scores computed on the basis of the predicted set of labels. We compare these measures with those computed by just considering the labeled friends.

To do that, we first compare the sensitivity values in the two cases (see Fig. 4(a)). All questions are subject to an increase of their sensitivity, but when

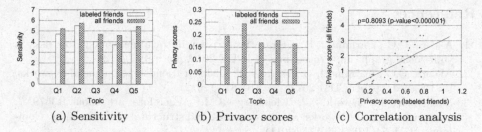

(a) Sensitivity   (b) Privacy scores   (c) Correlation analysis

Fig. 4. Privacy scores computed with labeled friends only vs. privacy scores computed on all friends

looking at the average privacy scores (Fig. 4(b)) we note that all scores are higher than those computed when considering only labeled friends. This means that the visibility of the topics is high. Hence, we perform a correlation analysis in order to check whether the behavior of scores is coherent in the two cases and measure the Pearson's ρ coefficient on the two series of privacy score values. We obtain a Pearson's coefficient of $\rho = 0.8093$ with a p-value $p < 0.000001$ (see Fig. 4(c)) denoting high positive correlation. This result confirm that: (i) the experiments on the limited set of 60 friends per user are significant enough and that, (ii) the approach is reliable even for users with a realistic number of friends and few given labels. Note that the overall number of friends of the participants spans between 120 and 1558 (with an average of 435).

6   Conclusions

With the final goal of fostering users' privacy awareness in the Web, we have proposed a privacy score based on an active learning approach to provide the users of online social networks with a measure of their privacy leakage. We have validated experimentally our metrics on an original dataset obtained through an online survey on real Facebook users. The experiments have shown the effectiveness and the reliability of our approach. In particular, we have shown that state-of-the-art metrics are based on a distorted perception of sensitivity of published items. Based on these results, we believe that our framework can be easily plugged into any domain-specific or general-purpose social networking platforms. Furthermore, it may inspire the design of privacy-preserving social networking components for *Privacy by Design* compliant software [5].

Acknowledgments. The work presented in this paper has been co-funded by Fondazione CRT (grant number 2015-1638). The authors wish to thank all the volunteers who participated in the survey.

References

1. Akcora, C.G., Carminati, B., Ferrari, E.: Privacy in social networks: How risky is your social graph? In: Proceedings of ICDE 2012, pp. 9–19 (2012)
2. Akcora, C.G., Carminati, B., Ferrari, E.: Risks of friendships on social networks. In: Proceedings of ICDM 2012, pp. 810–815 (2012)
3. Backstrom, L., Dwork, C., Kleinberg, J.M.: Wherefore art thou R3579X?: anonymized social networks, hidden patterns, and structural steganography. Commun. ACM **54**(12), 133–141 (2011)
4. Becker, J., Chen, H.: Measuring privacy risk in online social networks. In: Proceedings of Web 2.0 Security and Privacy (W2SP) (2009)
5. Cavoukian, A.: Privacy by design [leading edge]. IEEE Technol. Soc. Mag. **31**(4), 18–19 (2012)
6. Cetto, A., Netter, M., Pernul, G., Richthammer, C., Riesner, M., Roth, C., Sänger, J.: Friend inspector: A serious game to enhance privacy awareness in social networks. In: Proceedings of IDGEI 2014 (2014)
7. Coscia, M., Rossetti, G., Giannotti, F., Pedreschi, D.: Uncovering hierarchical and overlapping communities with a local-first approach. TKDD **9**(1), 6:1–6:27 (2014)
8. Culotta, A., McCallum, A.: Reducing labeling effort for structured prediction tasks. In: Proceedings of AAAI 2005, pp. 746–751 (2005)
9. Dagan, I., Engelson, S.P.: Committee-based sampling for training probabilistic classifiers. In: Proceedings of ICML 1995, pp. 150–157 (1995)
10. Dunbar, R.I.M.: Do online social media cut through the constraints that limit the size of offline social networks? Roy. Soc. Open Sci. **3**(1), 50292 (2016)
11. Fang, L., LeFevre, K.: Privacy wizards for social networking sites. In: Proceedings of WWW 2010 (2010)
12. Ghazinour, K., Matwin, S., Sokolova, M.: Monitoring and recommending privacy settings in social networks. In: Proceedings of 2013 EDBT/ICDT Workshop, pp. 164–168 (2013)
13. Kempe, D., Kleinberg, J.M., Tardos, É.: Maximizing the spread of influence through a social network. In: Proceedings of SIGKDD 2003, pp. 137–146 (2003)
14. Kosinski, M., Stillwell, D., Graepel, T.: Private traits and attributes are predictable from digital records of human behavior. PNAS **110**(15), 5802–5805 (2013)
15. Lewis, D.D., Gale, W.A.: A sequential algorithm for training text classifiers. In: Proceedings of SIGIR 1994, pp. 3–12 (1994)
16. Liu, K., Terzi, E.: A framework for computing the privacy scores of users in online social networks. TKDD **5**(1), 6 (2010)
17. Liu, Y., Gummadi, P.K., Krishnamurthy, B., Mislove, A.: Analyzing facebook privacy settings: user expectations vs. reality. In: Proceedings of SIGCOMM IMC 2011, pp. 61–70 (2011)
18. Mislove, A., Viswanath, B., Gummadi, P.K., Druschel, P.: You are who you know: inferring user profiles in online social networks. In: Proceedings of WSDM 2010, pp. 251–260 (2010)
19. Mitchell, T.M.: Machine learning. McGraw-Hill, New York (1997)
20. Roberts, S.G.B., Dunbar, R.I.M., Pollet, T.V., Kuppens, T.: Exploring variation in active network size: Constraints and ego characteristics. Soc. Netw. **31**(2), 138–146 (2009)
21. Scheffer, T., Decomain, C., Wrobel, S.: Active hidden markov models for information extraction. In: Hand, D.J., Kok, J.N., Berthold, M.R. (eds.) IDA 1999. LNCS, vol. 1642, pp. 309–318. Springer, Heidelberg (2001). doi:10.1007/3-540-44816-0_31

22. Talukder, N., Ouzzani, M., Elmagarmid, A.K., Elmeleegy, H., Yakout, M.: Privometer: Privacy protection in social networks. In: Proceedings of M3SN 2010, pp. 266–269 (2010)
23. Wang, Y., Nepali, R.K., Nikolai, J.: Social network privacy measurement and simulation. In: Proceedings of ICNC 2014, pp. 802–806 (2014)
24. Wu, L., Majedi, M., Ghazinour, K., Barker, K.: Analysis of social networking privacy policies. In: Proceedings of 2010 EDBT/ICDT Workshops (2010)
25. Zheleva, E., Getoor, L.: Privacy in social networks: A survey. In: Aggarwal, C.C. (ed.) Social Network Data Analytics, pp. 277–306. Springer, Heidelberg (2011)

On Using Temporal Networks to Analyze User Preferences Dynamics

Fabíola S.F. Pereira[1]($\boxtimes$), Sandra de Amo[1], and João Gama[2]

[1] Federal University of Uberlândia, Uberlândia, Brazil
{fabiola.pereira,deamo}@ufu.br
[2] LIAAD INESC TEC, University of Porto, Porto, Portugal
jgama@fep.up.pt

Abstract. User preferences are fairly dynamic, since users tend to exploit a wide range of information and modify their tastes accordingly over time. Existing models and formulations are too constrained to capture the complexity of this underlying phenomenon. In this paper, we investigate the interplay between user preferences and social networks over time. We propose to analyze user preferences dynamics with his/her social network modeled as a temporal network. First, we define a temporal preference model for reasoning with preferences. Then, we use evolving centralities from temporal networks to link with preferences dynamics. Our results indicate that modeling Twitter as a temporal network is more appropriated for analyzing user preferences dynamics than using just snapshots of static network.

Keywords: Temporal networks · User preferences · Evolving centralities

1 Introduction

What drives people's preferences dynamics? Modeling users' preferences and needs is one of the most important personalization tasks in the information retrieval domain. User preferences are fairly dynamic, since users tend to exploit a wide range of items and modify their tastes accordingly over time. Moreover, all the time users are facing others' opinions and being socially influenced. This scenario dispatches several research efforts to investigate the interplay between *user preferences* and *social networks* [1,2].

In this paper, we are interested in user preferences dynamics, i.e., the observation of how a user evolves his/her preferences over time. In our context, user preference is a specific type of opinion, that establishes an order relation between two objects. For example, when a user says: "I prefer sports than religion", we clearly identify his preference to sports subjects over religion.

Although social networks are fairly dynamic, traditional approaches in online social networks analysis consider nodes and edges as being persistent over time [3]. We go a step beyond and discuss social networks as *temporal networks*,

© Springer International Publishing Switzerland 2016
T. Calders et al. (Eds.): DS 2016, LNAI 9956, pp. 408–423, 2016.
DOI: 10.1007/978-3-319-46307-0_26

where the times when edges are active are an explicit element of the representation [4]. A classical example of temporal networks is on disease contagion. Usually, the spreading of diseases occurs through contact between two people. Considering the order of these contacts is important and more meaningful than just analyzing an aggregate static network that ignores when these contacts occurred. In the context of social networks, our topic of interest, temporal networks are being used to represent users interactions [5,6].

We hypothesize that centralities properties of temporal networks can reveal interesting patterns of users' preferences dynamics, especially when handling change detection. The running example below illustrates the preference dynamics problem we are investigating in this paper and its link with temporal networks.

Motivating example. Let us consider the most popular microblogging social network, Twitter. In Twitter, a user A usually follows another B if A is interested in what B posts. A common behavior is, for example, if A is fan of soccer, probably he is following sports channels and personalities like Messi and Neymar. Let us imagine, now, the context of an e-commerce recommendation system able to infer user preferences from Twitter in order to improve recommendations quality. In Fig. 1(a) we have a sequence of preferences inferred for user A about different themes (sport, TV, religion and music) that he likes to follow, post, share or read about in Twitter. An edge (a_1, a_2) indicates that a_1 is preferred to a_2 (edges inferred by transitivity are not represented in the figure). Analyzing A's preferences, we notice that in the first week of January, his preferences keep stable, just appearing a preference of *TV* over *music* themes and disappearing a preference of *sport* over *music*. However, on day 9, A changed his mind and is preferring music over TV. This can be an indicative of *changes* on his preferences and interests.

Now, in Fig. 1(b) we have snapshots of A's Twitter network. One edge from node A to node B indicates that A is followed by B (the flow of information). It is well know that Twitter network is fairly dynamic, and people start and end relationships (follows) all the time [7]. According to our hypothesis, many aspects of A's network structure may have influenced on his preferences change. For instance:

- the structural *position* of A in the network can be changed, as well as the global network topology. This means that he could be held a close position to music influential users;
- from second week of January, according to A's *position*, he is not receiving any information about religion;
- A's *position* is always close to sport celebrities.

It is an essential point for our hypothetical recommendation system to detect and predict A's preferences evolution over time and their changes. This could improve the sales of CDs, for example, when recommending to A. We show that the temporal-topological Twitter structure is strongly correlated with user preferences dynamics.

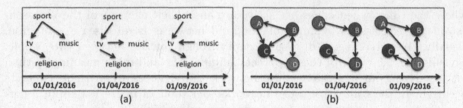

Fig. 1. (a) Sequence of A's preferences. (b) Snapshots from A's Twitter network

Contributions. The main contributions of this paper can be summarized as follows: (1) proposal of a temporal preference model for representing and reasoning with preferences over time; (2) a preference change detection algorithm for detecting changes events in preferences; (3) a centrality change detection algorithm; (4) proposal of a correlation between preferences and centrality measures in temporal networks; (5) a set of experiments validating our proposals, specially against static networks counterpart.

Organization of the paper. First, we discuss some related work. Then, we formalize user preferences dynamics by proposing a temporal preference model able to detect changes on preferences. After that, we present social temporal networks, discussing temporal aspects that were adopted in our analysis. In the Methodology Section, we mainly describe how we elicited preferences from Twitter dataset to validate our proposal. Then, we present the experimental results and conclude the paper.

2   Related Work

We summarize related literature according to two different aspects that we are addressing in this paper: (1) user preferences on social networks and (2) temporal networks.

2.1   User Preferences on Social Networks and Dynamics

Social networks are playing an important role as a type of source knowledge for mining (or inferring) user preferences in the Preference Learning area. The work [1], for example, combines user attributes and social network ties to discover user preferences. In [2], social influence has been used on classifiers to predict users' preferences. In this paper, we go a step beyond, from the point that preference relations between objects are already known and social networks are not used on the learning process, but on the new Preference Dynamics field. The work [8] investigates the music listening histories of Last.fm users focusing on the changes in their preferences based on their choices for different artists at different points in time. The modeling has been done by the survival analysis statistical method. In [9] the focus is on preferences shift (or change) over time. The authors propose a new measure of user preference dynamics (UPD) that captures the shifting rate of preferences. The work [10] is orthogonal to ours because it combines the growth of social links with the generation of user interaction events on those links. The focus is on social interactions, while ours is on user preferences. The topic of preference change has also been studied on sociology field [11,12].

2.2 Temporal Networks

The literature of temporal networks, focused on social media, is essentially concentrated on understanding patterns of information diffusion [13,14]. Especially, the ideas presented in [5] motivated us to develop this research. The author reviews methods to analyze temporal networks that seek to identify key mediators and how temporal and topological structure of interaction affects spreading processes. In the same way, [15] analyzes temporal centrality metrics. However, our focus is not on information diffusion, but on user preferences dynamics.

Researches on graph metrics for temporal networks essentially address this issue. [6,16] discuss that there are various concepts of *shortest* path for temporal graphs and propose efficient algorithms to compute them. In [17] graph metrics are revisited for temporal networks in order to take into account the effects of time ordering on causality.

Remarking on detecting changes on networks, according to [18], the concept of *change* in a graph sequence falls into one of three main categories according to the structure level of interest: global structure, local structure and the community-level structure. Our focus is on local changes. Global and communities' changes do not hold in our experiments.

3   User Preferences Dynamics

There is no consensus on the definition of preferences dynamics [19]. We adopt the following definition:

Definition 1 (User Preferences Dynamics (UPD)). *UPD refer to the observation of how a user evolves his/her preferences over time.*

First, we propose a temporal preference model able to represent and reasoning with preferences on time. Then, we describe how to detect events of changes on temporal preferences.

3.1 Temporal Preference Model

A *preference* is an *order relation* between two objects. For example, when a user says: *"I prefer sports to politics"*, if we order *sports* and *politics* in a ranking, we can clearly identify that *sports* will be in the top position.

Definition 2 (Temporal Preference Relation $\succ_t$). *A temporal preference relation (or temporal preference, for short) on a finite set of objects $A = \{a_1, a_2, ..., a_n\}$ is a strict partial order over A inferred on time t, i.e., a binary relation $R \subseteq A \times A$ satisfying the irreflexivity and transitivity properties on t. Typically, a strict partial order is represented by the symbol $\succ$. Considering $\succ_t$ as a temporal preference relation, we denote by $a_1 \succ_t a_2$ the fact that a_1 is preferred to a_2 on t.*

Definition 3 (User Temporal Profile Γ_t^u). *User temporal profile Γ_t^u is the transitive closure (TC) of all temporal preferences of user u on t.*

Example 1. Let $A = \{sport, tv, religion, music\}$ be the set of objects in our running domain representing themes of interest of user A. Figure 1(a) illustrates the temporal preferences of A on days 1, 4 and 9 through better-than graphs. Remark that an edge (a_1, a_2) indicates that a_1 is preferred to a_2 and edges inferred by transitivity are not represented. We have: $\Gamma_1^A = \{sport \succ_1 tv, tv \succ_1 religion, sport \succ_1 religion, sport \succ_1 music\}$, $\Gamma_4^A = \{sport \succ_4 tv, tv \succ_4 religion, sport \succ_4 religion, tv \succ_4 music, sport \succ_4 music\}$ and $\Gamma_9^A = \{sport \succ_9 tv, tv \succ_9 religion, sport \succ_9 religion, music \succ_9 tv, music \succ_9 religion\}$.

3.2 Detecting Changes on Temporal Preferences

A key property of temporal preferences is the irreflexivity. We say that a temporal profile Γ_t^u is *inconsistent* when there is a preference $a_1 \succ_t a_1 \in \Gamma_t^u$. It would means that *"I prefer X better than X!"*, which does not hold for a strict partial order.

Our proposal for detecting preference change is based on the *consistency* of user temporal profiles. The idea is to compute the union of user profiles collected over time, infer temporal preferences by transitivity considering all timestamps and verify if there is any inconsistency on the resulted set of preferences. If yes, we detect an event of preference change. These concepts are formalized in what follows.

Definition 4 (Temporal Profile Union Ω_t^u). *Two temporal preferences of the type $a_1 \succ_{t-1} a_2$ and $a_2 \succ_t a_3$, can unite to infer a third temporal preference $a_1 \succ_{t'} a_3$, once considering transitivity of both, temporal preference relation and timestamp order. A temporal profile union Ω_t^u is the transitive closure (TC) of all irreflexive relations given by $\Gamma_{t-1}^u \cup \Gamma_t^u$.*

Definition 5 (Preference Change δ_t^u). *If there is a temporal preference inconsistency in Ω_t^u a preference change has been detected on time t for user u. In other words, a preference change δ_t^u is defined as:*

$$\delta_t^u = \begin{cases} 1, & \text{if there is a temporal preference inconsistency in } \Omega_t^u \\ 0, & \text{otherwise} \end{cases} \tag{1}$$

Remarking on Example 1, let us consider $W = \{1, 4, 9\}$ the set of intervals time stamps. The temporal profile union $\Omega_9^A = \{..., tv \succ_4 music, music \succ_9$

$tv, tv \succ_{9'} tv, ...\}$ contains the inconsistency $tv \succ_{9'} tv$. So, a preference change
has been detected on time 9 ($\delta_9^A = 1$). Intuitively, we have that on day 1, for
example, A prefers to read/post/share on his social network news about *sport*,
but between tv and *religion* he is in the mood for tv. On the following days,
A's preferences practically do not change, just appearing a preference of tv over
music. However, on day 9, A's presented a preference change, as *music* became
preferred over tv.

The size of the intervals time stamps in W determines if we are tracking
short-term or long-term preference events. As example of real events, we can
cite new product releases and special personal occasions such as birthdays [20].

In order to formalize the detection of changes in our temporal preference
model, we propose the *PrefChangeDetection* algorithm. The intuition of this
algorithm is to analyze *better-than graphs* of a user during the observation intervals in W from social networks. If the resulting graph has at least one cycle
(meaning an inconsistency) we have detected a preference change. The Algorithm 1 formalizes this idea.

Algorithm 1. PrefChangeDetection

Input: User u, set of intervals W, a vector Γ^u of size $|W|$ containing u's temporal
profiles for each $t \in W$ extracted from the social network $G = (V, E)$
Output: A vector δ^u of size $|W|$ containing u's preference changes for each $t \in W$
1: $BG_{prev}^u \leftarrow \emptyset$
2: **for all** $t \in W$ **do**
3:     build better-than graph BG_t^u from $\Gamma^u[t]$
4:     $BG^u \leftarrow BG_{prev}^u \cup BG_t^u$
5:     **if** BG^u is not acyclic **then**
6:         $\delta^u[t] = 1$
7:     **else**
8:         $\delta^u[t] = 0$
9:     $BG_{prev}^u \leftarrow BG_t^u$
10: **return** δ^u

Remarking on the complexity analysis, the time to build a better-than graph
(line 3) varies according to the preference mining algorithm used. In Sect. 5 we
present the algorithm we use in this paper for mining preferences from social net-
works. In worst case, its complexity is $O(|V|)$, where $|V|$ is the number of nodes.
The time to detect if a directed graph is acyclic (line 5) is $O(|A| + |\Omega_t^u|)$ where
A is the set of objects in the domain (the nodes) and Ω_t^u is the temporal profile
union containing the preference orders (the edges). Hence, *PrefChangeDetection*,
in the worst case, has complexity of $O(|W||V|(|A| + |\Omega_t^u|))$.

4   Social Networks Evolution

We introduce the background of temporal networks, leveraging terms like temporal networks, temporal graphs, static networks and evolving graphs. Next, we propose an algorithm for detecting changes on centralities metrics.

4.1   Temporal Networks vs. Static Networks

In this paper, we explore two different representation of social networks: as a static graph structure and as a temporal graph [4,6]. The static graph structure is a traditional approach where the temporal aspects are aggregated and the evolution is analyzed just as a set of graphs snapshots over time [6]. On the other hand, in temporal graphs (or temporal networks) the information of when interactions between nodes happen is taken into account. Let us formalize these concepts.

Definition 6 (Static Networks). *A network $G_s = (V, E)$ is static (or aggregate) if there is not any time reference in the edges.*

Definition 7 (Temporal Networks). *Temporal networks or temporal graphs $G_t = (V, E)$ are graphs with temporal edges, i.e., each edge contains the information of when it has been created and when it has been deleted* [6].

Example 2. Consider the temporal and the aggregate graphs in Fig. 2. They represent a social network, where the nodes are users and the edges are interactions (for example, tweets) between two users. Suppose that node A has a high impact information to spread in the network. If we analyze the network from the aggregate graph perspective, the information will reach node F. This is not true for the temporal network, as A just interacts on time t_3 with B and after that, it is not possible to reach F from B.

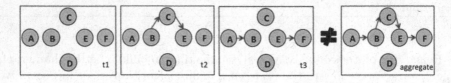

Fig. 2. Temporal network vs. static (aggregate) network

The centrality metrics analysis is inherent to what network representation we are using. Remarking on Example 2, there is a path between nodes A and F in aggregate graph, but not in temporal graphs. This implies in different values of centralities for these nodes. The problem of evolving centralities in temporal networks is addressed in [6]. In Sect. 6 we show that the *betweenness* and *closeness* centralities have different behaviors according to the network representation and, consequently, they correlate with user preferences in different ways.

4.2 Detecting Changes on Centrality Metrics

In order to detect changes on temporal metrics, and consequently on graph structure, we defined a baseline approach founded on change-point in rankings [21]. The idea is to maintain a ranking R of all the nodes in the graph according to their metrics values for each interval time stamp t_i inside the observation set of intervals W. Based on the variations of metrics values and ranking positions from t to $t + 1$, we detect changes.

Definition 8 (Ranking Position). *Let $G = (V, E)$ be a graph and $W = \{t, t+ 1, ...\}$ a set of intervals. We define C_t^u as the centrality value of u on time t, for $u \in V$ and $t \in W$. Let us consider a ranking R_t where the centrality values of all nodes in V are ranked in descending order on time t. We define pos_t^u as the position of node u in R_t, i.e., $C_t^u > C_t^v$ iff $pos_t^u > pos_t^v$, for $u, v \in V$.*

Definition 9 (Temporal Metric Change λ_t^u). *We define Λ_t^u as the acceleration of node u in centrality ranking position from time $t - 1$ to time t:*

$$\Lambda_t^u = \frac{|pos_t^u - pos_{t-1}^u|}{max(pos_t^u, pos_{t-1}^u)} \tag{2}$$

A temporal metric change λ_t^u is detected when Λ_t^u is greater than a threshold θ:

$$\lambda_t^u = \begin{cases} 1, & \Lambda_t^u > \theta \\ 0, & otherwise \end{cases} \tag{3}$$

The algorithm *CentralityChangeDetection* (Algorithm 2) implements change detection in centrality metrics based on above definitions. The complexity time is given by the computation of centralities (line 3). In the worst case, the betweenness centrality involves calculating the shortest paths between all pairs of vertices on a graph, which takes $O(|V|^3)$. Hence, the total complexity is $O(|W||V|^3)$.

5 Methodology

In our methodology we chose as temporal social network the Twitter follower/followee network and extract preferences based on the structure of this network.

5.1 Dataset

The Twitter dataset from [6] was used to validate our proposal. In order to correlate user preferences and the structure of social networks, from evolution viewpoint, we need a dataset (1) containing the information of *when* relationships start and end in the network and (2) some semantic information about the nodes from which it is possible to extract preferences.

The temporal information of Twitter has the following meaning: each node is a user and an edge $(u, v, t_{init}, t_{end})$ indicates that v starts following u at t_{init} and

---

Algorithm 2. CentralityChangeDetection

Input: User u, set of intervals W, social network $G = (V, E)$, threshold θ
Output: A vector λ^u of size $|W|$ containing temporal metrics changes for each $t \in W$
1: $R_{t-1} \leftarrow \emptyset$
2: **for all** $t \in W$ **do**
3:     calculate centralities from G and build ranking R_t
4:     **if** $R_{t-1} = \emptyset$ **then**
5:         $\Lambda_t^u = 0$
6:     **else**
7:         $\Lambda_t^u \leftarrow \frac{|pos_t^u - pos_{t-1}^u|}{max(pos_t^u, pos_{t-1}^u)}$
8:     **if** $\Lambda_t^u > \theta$ **then**
9:         $\lambda^u[t] = 1$
10:     **else**
11:         $\lambda^u[t] = 0$
12:     $R_{t-1} \leftarrow R_t$
13: **return** λ^u

---

unfollows u at t_{end+1} (v follows u during $[t_{init}, t_{end}]$). As we are dealing with a real dynamic social network, a user can follows and unfollows another user all the time. This is the most interesting aspect that we are investigating: how following relationships on Twitter can allow us to understand user preferences dynamics? The dataset contains 144975 users and 1222118 temporal edges, observed from 08/28/2015 to 12/15/2015. In [6] there are more insights about time-changing characteristics of data.

5.2   Preference Mining

The dataset we used was crawled from Twitter based on a "seed celebrity" policy: choose s celebrities users as seeds and for each seed, select his/her followers. These seeds play an important role for the extraction of users preferences from data. We labeled the 27 seeds based on 9 themes that they represent as celebrities. The themes are *politics, sport, religion, news, music, humor, TV, fashion and health*. For example, Neymar is a sport celebrity, Gisele Bündchen from fashion and the pope is a religion representative.

The 9 themes adopted are the domain of preferences. The intuition in this preference mining process is: if user u follows a lot of religions personalities and does not follow anyone from fashion field, then u has more interest in religion than in fashion. Thus, we mined preferences of the type: $religion \succ_u^t fashion$.

We consider the preference strength w as the number of seeds of the same theme that a user u follows. For example, if u follows 3 *religion* seeds and 1 *news* seed, we have $religion \succ_u^t news$. If u does not follow any *health* seed, we have that the remaining themes are preferred over health. These situations are illustrated in Fig. 3. This method is solely language-agnostic and based on the semantics inferred by the structure of the network [22].

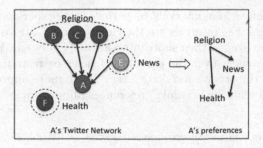

Fig. 3. Mining preferences from Twitter

There are many drawbacks on mining preferences in this way. We can infer that a user prefer a topic even if the posts are badly written. Or, that soccer players always post about soccer, which is not necessary true. As this paper is a pioneer quantitative analysis of preferences dynamics in Twitter, we chose this preference mining approach as baseline. In [23] a technique to extract preferences from tweets has been proposed, but we did not use it because our dataset do not contain users tweets. We just have the network topology.

5.3 Discussion

There are many directions to explore from the concepts presented in this paper: (i) networks are related to preferences to what extent? (ii) Any social network can be used to analyze UPD? (iii) What are the best centrality metrics? (iv) What is the best network modeling to analyze UPD: static or temporal? In this paper, we perform experiments to validate the direction (iv).

We analyze how effective are temporal networks to track user preferences dynamics against static networks in Twitter dataset. The analysis is founded on the correlation strength of preferences changes and centralities metrics changes over time. According to our methodology, the preferences are elicited from the same graph structure that we perform centralities measures analysis. Hence, preference change and centrality change are naturally correlated. Our focus here is to show that there is a significant difference between correlations strength obtained with temporal networks representation in relation to static networks counterpart. So, the methodology does not imply a bias in the experiments.

6 Experimental Results

We show that temporal networks are better representations than static networks for the analysis of user preferences dynamics.

6.1 Experimental Environment

Intervals. The solutions we are proposing for the problem of preferences and centralities events detection are highly sensitive to the granularity of observation window W. We define four intervals to perform the experiments, described in Table 1. Remark that the intervals are the elements of W. If we are interested in tracking short-term events, then short intervals like Daily and Weekly fit better. For instance, preferences over the domains of news or restaurants have a high changing rate. On the other hand, long intervals are more appropriate when the events are not frequent, for example preferences about movies and politics.

Table 1. Intervals inside observation window $W = [08/28/2015, 12/15/2015]$

Period	# of intervals	Values
Daily (DA)	110	$DA_1 = [08/28, 08/28], ..., DA_{110} = [12/15, 12/15]$
Weekly (WE)	15	$WE_1 = [09/01, 09/07], WE_2 = [09/08, 09/14], ..., WE_{15} = [12/08, 12/14]$
Fortnightly (FO)	7	$FO_1 = [09/01, 09/15], FO_2 = [09/16, 09/30], ..., FO_7 = [11/30, 12/14]$
Monthly (MO)	3	$MO_1 = [09/01, 09/30], MO_2 = [10/01, 10/31], MO_3 = [11/01, 11/30]$

Users. The dataset contains 144975 users. For computing the events along time intervals, the values λ_t^{AVG} and δ_t^{AVG} correspond to the average across all users for each time interval t.

Centrality Metrics. *Betweenness* and *closeness* metrics are used in our experiments. Betweenness considers how important nodes are in connecting other nodes. In closeness centrality the intuition is that the more central nodes are, the more quickly they can reach other nodes [24]. Thus, these metrics are related with nodes that play *influence* and *spreading* roles in the network, respectively. These are potential features for understanding users preferences dynamics. For centrality events detection we vary the threshold $\theta = \{0.2, 0.4, 0.6\}$.

Social Network Representation. We compare the behavior of *temporal graphs* and *static graphs* structures in relation to preferences and centralities changes over time.

6.2 Analyzing Change Events and Correlations

Q1: What are the changes behaviors of users (preferences) and nodes (centralities)? Are these variables really dynamics?

We verify the change rate for both preferences and centralities over different time intervals. The results are illustrated in Fig. 4. The analysis of daily intervals is not interesting here as there is no difference between temporal and static networks for daily intervals (the temporal network has granularity of one day). The parameter θ has been fixed as 0.4 in this analysis, corresponding to the intermediate value of our range. In next analysis we show that it does not affect the correlations behaviors.

On average, the rate of users that change their preferences are 30.39 %, 32.17 % and 40.11 % for week, fortnight and month intervals, respectively. For temporal betweenness centrality, the averages of changes are 43.56 %, 46.44 % and 46.39 %; and 36.42 %, 35.07 %, 45.01 % for static betweennes. Temporal closeness changes averages are 38.85 %, 38.78 % and 41.23 %; and 24.16 %, 26.28 % and 33.78 % for static closeness. Generally, as the interval size increases, the change rate increases as well. This occurs due to the trade off between domain of preferences and social network. We are investigating the domain of users' preferences to post/share/read in Twitter. According to our analysis, this is very dynamic and even short-term data (week intervals) have relevancy.

The most important observation is that the curves of preferences and temporal centralities have similar behaviors for all scenarios, different from static centralities. This observation indicates that if we consider temporal modeling of Twitter network to track evolving betweenness and closeness nodes centralities, we have a better notion of users respective preferences dynamics than if considering static modeling.

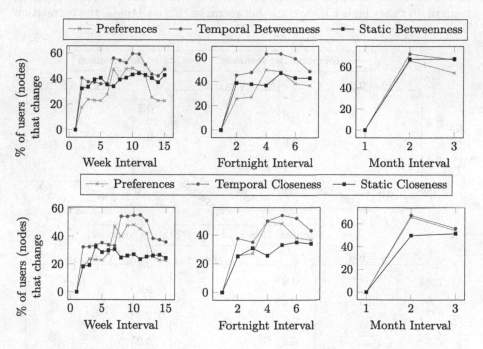

Fig. 4. Change rates for betweenness (up) and closeness (down) centralities

Q2: Are temporal networks more appropriate to analyze user preferences dynamics than static networks?

We use Pearson Correlation Coefficient (PCC) metric to evaluate which network model better represents preferences dynamics. As previously appointed, we mine *preferences* from the same graph structure that we perform *centralities* measures analysis. It is expected that both variables are correlated. As a matter of fact, in this analysis, we explore the *correlation strength* difference between temporal and static networks representations in our context.

The results are illustrated in Fig. 5. $PCC(\delta_t^{AVG}, \lambda_t^{AVG})$ has been calculated considering betweenness and closeness centralities. Each scenario has three periods – Day, Week and Fortnight. The month intervals are not illustrated due the small size of the series (only three values). For each centrality we vary the parameter θ. This parameter indicates that the closer to 1, more significant are the centralities changes that are being considered.

Both centralities metrics correlate significantly (as compared to the corresponding critical values – in all scenarios critical values are lower than 0.1). Two random variables (with no correlation) would have a 95 % probability of PCC greater than a critical value or lower. As expected, we observe a high correlation between the change events in user preferences and in centrality metrics.

We highlight the difference between temporal and static values. Considering the analyzed scenarios, on average, temporal betweenness has a correlation strength 40 times higher than static betweenness. For closeness, the correlation

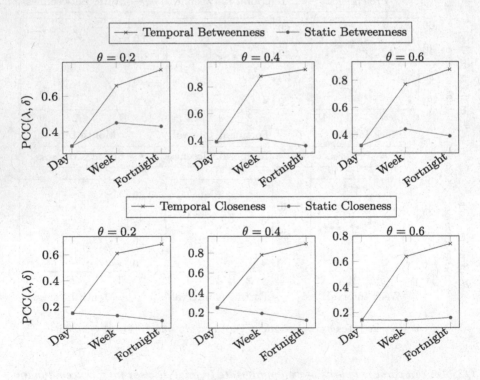

Fig. 5. PCC between centralities and preferences

strength is 59 times higher. This corroborates our investigation that changes in temporal metrics indicate changes on users preferences. The correlation difference between temporal and static metrics is an evidence that dynamics of preferences fit better in temporal networks representation.

Q3: Why temporal networks are good social network representation in the analysis of user preferences dynamics?

The results obtained so far can be explained by the phenomena of information propagation and inherent consequences of homophily and influence. The main difference between temporal and static networks, as discussed in Sect. 4, is that temporal networks take into account the contact sequence between nodes to compute paths [6] and this has an impact on different centralities measures. The related work [14] discuss about the relation of preferences and information propagation in social networks. The aspects described on motivating example (Sect. 1) could illustrate that preferences are directed by information flow in the social network. Finally, temporal networks represent information flow more realistically.

7   Conclusion

In this paper we have investigated the interplay between *user preferences dynamics* and *evolving social networks*. We have introduced a new temporal preference model able to describe dynamics of user preferences through user profiles and change detection. We have presented a social network analysis correlating centralities metrics evolution with user preferences dynamics over Twitter data. Our findings have shown that there is a high correlation between changes on temporal centrality metrics – *betweenness* and *closeness* – and changes on user preferences, against static centrality metrics counterpart.

Acknowledgments. This work was supported by the research project "TEC4Growth - Pervasive Intelligence, Enhancers and Proofs of Concept with Industrial Impact / NORTE-01-0145-FEDER-000020", financed by the North Portugal Regional Operational Programme (NORTE 2020), under the PORTUGAL 2020 Partnership Agreement, and through the European Regional Development Fund (ERDF) and by European Commission through the project MAESTRA (Grant number ICT-2013-612944). Fabiola Pereira is financed by the ERDF – European Regional Development Fund through the Operational Programme for Competitiveness and Internationalization - COMPETE 2020 Programme within project POCI-01-0145-FEDER-006961, and by National Funds through the FCT – Fundação para a Ciência e a Tecnologia (Portuguese Foundation for Science and Technology) as part of project UID/EEA/50014/2013. This work was also supported by the Brazilian Research Agencies CAPES and CNPq.

References

1. Li, J., Ritter, A., Jurafsky, D.: Inferring user preferences by probabilistic logical reasoning over social networks. arXiv preprint (2014). arXiv:1411.2679
2. Abbasi, M.A., Tang, J., Liu, H.: Scalable learning of users' preferences using networked data. In: Proceedings of the 25th ACM Conference on Hypertext and Social Media, pp. 4–12. ACM (2014)
3. da Costa, L.F., Rodrigues, F.A., Travieso, G., Villas Boas, P.R.: Characterization of complex networks: a survey of measurements. Adv. Phys. **56**(1), 167–242 (2007)
4. Holme, P., Saramäki, J.: Temporal networks. Phys. Rep. **519**(3), 97–125 (2012)
5. Holme, P.: Analyzing temporal networks in social media. Proc. IEEE **102**(12), 1922–1933 (2014)
6. Pereira, F.S.F., Amo, S., Gama, J.: Evolving centralities in temporal graphs: a twitter network analysis. In: First Workshop on High Velocity Mobile Data Management Co-Located with 17th IEEE International Conference on Mobile Data Management (MDM), pp. 43–48 (2016)
7. Arias, M., Arratia, A., Xuriguera, R.: Forecasting with twitter data. ACM Trans. Intell. Syst. Technol. (TIST) **5**(1), 8 (2013)
8. Kapoor, K., Srivastava, N., Srivastava, J., Schrater, P.: Measuring spontaneous devaluations in user preferences. In: 19th ACM SIGKDD International Conference on Knowledge Discovery and Data Mining, pp. 1061–1069. ACM (2013)
9. Rafailidis, D., Nanopoulos, A.: Modeling the dynamics of user preferences in coupled tensor factorization. In: Proceedings of the 8th ACM Conference on Recommender systems, pp. 321–324. ACM (2014)
10. Yang, Z., Xue, J., Wilson, C., Zhao, B.Y., Dai, Y.: Process-driven analysis of dynamics in online social interactions. In: Proceedings of the 2015 ACM on Conference on Online Social Networks, pp. 139–149. ACM (2015)
11. Liu, F.: Preference change and information processing. Technical report, ILLC, University of Amsterdam (2006)
12. Liu, F.: Preference change a quantitative approach. Stud. Logic **2**(3), 12–27 (2009)
13. Tang, J., Musolesi, M., Mascolo, C., Latora, V.: Temporal distance metrics for social network analysis. In: Proceedings of the 2nd ACM Workshop on Online Social Networks, pp. 31–36. ACM (2009)
14. Guille, A., Hacid, H.: A predictive model for the temporal dynamics of information diffusion in online social networks. In: Proceedings of the 21st International Conference Companion on World Wide Web, pp. 1145–1152. ACM (2012)
15. Tang, J., Musolesi, M., Mascolo, C., Latora, V., Nicosia, V.: Analysing information flows and key mediators through temporal centrality metrics. In: 3rd Workshop on Social Network Systems, p. 3. ACM (2010)
16. Wu, H., Cheng, J., Huang, S., Ke, Y., Lu, Y., Xu, Y.: Path problems in temporal graphs. Proc. VLDB Endowment **7**(9), 721–732 (2014)
17. Nicosia, V., Tang, J., Mascolo, C., Musolesi, M., Russo, G., Latora, V.: Graph metrics for temporal networks. In: Holme, P., Saramäki, J. (eds.) Temporal Networks. Understanding Complex Systems, pp. 15–40. Springer, Heidelberg (2013)
18. Koujaku, S., Kudo, M., Takigawa, I., Imai, H.: Community change detection in dynamic networks in noisy environment. In: 24th International Conference on World Wide Web Companion, pp. 793–798 (2015)
19. Liu, F.: Reasoning about Preference Dynamics. Synthese Library. Springer, Netherlands (2011)

20. Xiang, L., Yuan, Q., Zhao, S., Chen, L., Zhang, X., Yang, Q., Sun, J.: Temporal recommendation on graphs via long-and short-term preference fusion. In: 16th ACM SIGKDD International Conference on Knowledge Discovery and Data Mining, pp. 723–732 (2010)
21. Wei, W., Carley, K.M.: Measuring temporal patterns in dynamic social networks. ACM Trans. Knowl. Disc. Data (TKDD) **10**(1), 9 (2015)
22. Klochko, M.A., Ordeshook, P.C.: Endogenous Time Preferences in Social Networks. Edward Elgar Publishing, Cheltenham (2005)
23. Pereira, F.S.F.: Mining comparative sentences from social media text. In: Second Workshop on Interactions between Data Mining and Natural Language Processing Co-Located with European Conference on Machine Learning and Principles and Practice of Knowledge Discovery in Databases, pp. 41–48 (2015)
24. Zafarani, R., Abbasi, M.A., Liu, H.: Social Media Mining: An Introduction. Cambridge University Press, New York (2014)

Kernels and Deep Learning

Soft Kernel Target Alignment for Two-Stage Multiple Kernel Learning

Huibin Shen[✉], Sandor Szedmak, Céline Brouard, and Juho Rousu

Department of Computer Science, Aalto University and Helsinki Institute
for Information Technology, 02150 Espoo, Finland
{huibin.shen,sandor.szedmak,celine.brouard,juho.rousu}@aalto.fi

Abstract. The two-stage multiple kernel learning (MKL) algorithms gained the popularity due to their simplicity and modularity. In this paper, we focus on two recently proposed two-stage MKL algorithms: ALIGNF and TSMKL. We first show through a simple vectorization of the input and target kernels that ALIGNF corresponds to a non-negative least squares and TSMKL to a non-negative SVM in the transformed space. Then we propose ALIGNF+, a soft version of ALIGNF, based on the observation that the dual problem of ALIGNF is essentially a one-class SVM problem. It turns out that the ALIGNF+ just requires an upper bound on the kernel weights of original ALIGNF. This upper bound makes ALIGNF+ interpolate between ALIGNF and the uniform combination of kernels. Our experiments demonstrate favorable performance and improved robustness of ALIGNF+ comparing to ALIGNF. Experiments data and code written in python are freely available at github (https://github.com/aalto-ics-kepaco/softALIGNF).

Keywords: Multiple kernel learning · Kernel target alignment · Soft margin SVM · One-class SVM

1 Introduction

The choice of the kernels is critical for good performance in kernel based learning algorithms. In recent years, the traditional practise of choosing the 'best' kernel by cross-validation has given way to Multiple Kernel Learning (MKL) where, instead of using the single best one, one seeks for the best combination of the kernels.

A significant body of work on MKL exists for classification tasks [7]. These algorithms can be broadly divided into two kinds: one-stage models integrate learning the kernel combinations with learning the classifier [12,14], and two-stage models learn the kernels combination without referring to the subsequent classifier learning phase [1,2,4,11]. Two-stage methods are appealing in practical applications due to their relative simplicity and modularity, and competitive accuracy compared to one-stage methods. A prime example is the centered kernel target alignment based algorithm ALIGNF [2] which is based on maximizing the alignment between a combination of centered kernels and the ideal target kernel.

© Springer International Publishing Switzerland 2016
T. Calders et al. (Eds.): DS 2016, LNAI 9956, pp. 427–441, 2016.
DOI: 10.1007/978-3-319-46307-0_27

The kernel target alignment problem is essentially a linear regression in a transformed input output space, as also demonstrated by [22]. Recent advancement on two stage MKL relies on replacing the regression problem with a classification problem [3,13]. With a simple transformation, the regression based kernel target alignment and the classification based two-stage method can be unified in one setting.

In this paper, we propose ALIGNF+, a soft formulation of ALIGNF, that combines the ideas of one-class SVM and soft margin SVM [18], and is shown to interpolate between two competitive MKL algorithms. The experiments demonstrate favorable empirical performance of ALIGNF+ and ability to resist label noise when coupled with SVM in the second stage. For multilabel datasets, we also compared ALIGNF and ALIGNF+ when coupled with a multilabel method in the second stage, namely operator-valued kernel regression (OVKR) [16], where we also found that ALIGNF+ is more robust than ALIGNF.

The paper is organized as follows: Sect. 2 revisits the ALIGNF algorithm and shows it can be cast into a regression problem through a simple transformation, also relied upon by the classification framework proposed by [13]. Section 3 first derives a dual problem of ALIGNF, and shows that is essentially a one-class SVM problem and then introduces the soft ALIGNF algorithm based on this observation. Experiments are described in Sect. 4 followed by conclusions in Sect. 5.

2   Unified Setting for Two-Stage MKL Methods

In this section, we first review the ALIGNF algorithm, then we show it is equal to solving a non-negative least squares problem in a transformed space where the works of [3,13] naturally fit in. The basic setting is a binary classification problem with the training instances of (x, y) drawn from a distribution P over $\mathcal{X} \times \mathcal{Y}$ where $x \in \mathbb{R}^d$ and $y \in \{+1, -1\}$. p positive semi-definite (PSD) kernel functions are defined on $\mathcal{X}$: $K_1, \ldots, K_p$. For a set of n training instances, the input and label can be represented as $\mathbf{X} \in \mathbb{R}^{n \times d}$ and $y \in \mathbb{R}^n$ and the corresponding kernel matrices are $\mathbf{K}_1, \ldots, \mathbf{K}_p \in \mathbb{R}^{n \times n}$. We use $|| \cdot ||$ to denote the ℓ_2 norm throughout the paper.

2.1   Centered Kernel Target Alignment

The target kernel matrix computed from labels is typically $\mathbf{K}_y = yy^T$ [4]. The *kernel target alignment* between two kernel matrices $\mathbf{K}$ and $\mathbf{K}'$ [4] is defined as follows:

$$\hat{A}(\mathbf{K}, \mathbf{K}') = \frac{\langle \mathbf{K}, \mathbf{K}' \rangle_F}{\sqrt{\langle \mathbf{K}, \mathbf{K} \rangle_F \langle \mathbf{K}', \mathbf{K}' \rangle_F}}, \tag{1}$$

where $\langle \cdot, \cdot \rangle_F$ is the Frobenius inner product.

Let $\mathbf{I}$ be the identity matrix, and $\mathbf{1}$ be a unit vector. Given $\mathbf{U} = [\mathbf{I} - \frac{\mathbf{1}\mathbf{1}^T}{n}]$, the centered version of a kernel matrix can be computed with the simple formula:

$$\mathbf{K}_c = \mathbf{U}\mathbf{K}\mathbf{U}.$$

Centered kernel target alignment is found to be better correlated with the performance of the kernels [2], which is simply defined by applying (1) with the centered kernel:

$$\hat{\rho}(\mathbf{K}, \mathbf{K}') = \frac{\langle \mathbf{K}_c, \mathbf{K}'_c \rangle_F}{\sqrt{\langle \mathbf{K}_c, \mathbf{K}'_c \rangle_F \langle \mathbf{K}'_c, \mathbf{K}'_c \rangle_F}}. \tag{2}$$

The two-stage MKL method ALIGNF [2] seeks a non-negative linear combination of kernels, i.e. $\mathbf{K}_\mu = \sum_{k=1}^p \mu_k \mathbf{K}_k$ with the constraint $\mathcal{M} = \{\|\mu\| = 1, \mu \geq 0\}$ such that $\hat{\rho}(\mathbf{K}_\mu, \mathbf{K}_y)$ is maximized:

$$\max_{\mu \in \mathcal{M}} \hat{\rho}(\mathbf{K}_\mu, \mathbf{K}_y). \tag{3}$$

Let $\mathbf{M}$ and a be defined by:

$$(\mathbf{M})_{k\ell} = \langle \mathbf{K}_{k_c}, \mathbf{K}_{\ell_c} \rangle_F, \quad k, \ell = 1, \ldots, p,$$
$$(a)_k = \langle \mathbf{K}_{k_c}, \mathbf{K}_y \rangle_F, \quad k = 1, \ldots, p.$$

The following proposition, proved by [2], shows that the optimization (3) can be solved via quadratic programming (QP).

Proposition 1. *Let v^* be the solution of the following QP:*

$$\min_{v \geq 0} v^T \mathbf{M} v - 2v^T a. \tag{4}$$

Then, the solution μ^* of the Optimization (3) is given by $\mu^* = v^*/\|v^*\|$.

We now assume that all the kernel matrices are centered without mentioning, i.e. the notation $\mathbf{K}$ is interchangeably used to denote the corresponding centered version $\mathbf{K}_c$.

2.2 Reformulation Through Vectorized Kernel Matrices

A simple vectorization on the input kernel matrices and target kernel matrix unifies two previously proposed two-stage MKL methods, ALIGNF [2] and TSMKL [13]. The following proposition shows that the optimization in (4) is equivalent to solve a non-negative least squares problem.

Proposition 2. *Let $vec(\cdot)$ stack all the columns of a matrix into a vector and $\hat{x}_k = vec(\mathbf{K}_k), k = 1, \ldots, p$ and $\hat{y} = vec(\mathbf{K}_y)$. A new matrix $\hat{\mathbf{X}} \in \mathbb{R}^{\hat{n} \times p}$ ($\hat{n} = n^2$) is defined as $\hat{\mathbf{X}} = (\hat{x}_1, \ldots, \hat{x}_p)$. Let $\mathbf{M}$ and a be defined as in Proposition (1). Then the solution v^* of the following non-negative least squares regression:*

$$\min_{v \geq 0} \|\hat{\mathbf{X}} v - \hat{y}\|^2 \tag{5}$$

is the same as the solution in optimization (4).

Proof.

$$||\hat{\mathbf{X}}\boldsymbol{v} - \hat{\boldsymbol{y}}||^2 = \boldsymbol{v}^T\hat{\mathbf{X}}^T\hat{\mathbf{X}}\boldsymbol{v} - 2\boldsymbol{v}^T\hat{\mathbf{X}}^T\hat{\boldsymbol{y}} + \hat{\boldsymbol{y}}^T\hat{\boldsymbol{y}}$$

$$= \sum_{i=1}^{p}\sum_{j=1}^{p}v_iv_j\hat{\boldsymbol{x}}_i^T\hat{\boldsymbol{x}}_j - 2\sum_{i=1}^{p}v_i\hat{\boldsymbol{x}}_i^T\hat{\boldsymbol{y}} + \hat{\boldsymbol{y}}^T\hat{\boldsymbol{y}}$$

$$= \sum_{i=1}^{p}\sum_{j=1}^{p}v_iv_j\langle\mathbf{K}_i,\mathbf{K}_j\rangle_F - 2\sum_{i=1}^{p}v_i\langle\mathbf{K}_i,\mathbf{K}_y\rangle_F + \hat{\boldsymbol{y}}^T\hat{\boldsymbol{y}}$$

$$= \boldsymbol{v}^T\mathbf{M}\boldsymbol{v} - 2\boldsymbol{v}^T\boldsymbol{a} + \hat{\boldsymbol{y}}^T\hat{\boldsymbol{y}}.$$

Since the term $\hat{\boldsymbol{y}}^T\hat{\boldsymbol{y}}$ is a constant, it is clear that this problem has the same objective as Optimization (4). The conversion of $\boldsymbol{v}$ back to the solutions of ALIGNF requires normalization as in (4).

Now we shortly review the TSMKL algorithm [13] and follow the notations in the original paper. It first transforms the original input and output space $\mathcal{X}$ and $\mathcal{Y}$ to a new space, named as K-space: $\{(z_{xx'}, t_{yy'})|((x,y),(x',y')) \sim P \times P\}$ where $z_{xx'} = (K_1(x,x'),\ldots,K_p(x,x'))$ and $t_{yy'} = yy'$ for binary classification and $t_{yy'} = 2\cdot1_{\{y=y'\}}-1$ for multi-class classification. $z_{xx'}$ is called K-example and $t_{yy'}$ is called K-label. Then a linear SVM restricted to the non-negative orthant is learned in the K-space and the solutions of this linear SVM is directly taken as the kernel weights.

The definition of K-labels coincides with the entries of target kernel matrix computed by $\mathbf{K}_y = \boldsymbol{y}\boldsymbol{y}^T$. To see this, let $q = (j-1)n + i$, K-label $t_{y_iy_j} = y_iy_j = (\mathbf{K}_y)_{ij}$, which is the qth entry of $vec(\mathbf{K}_y) = \hat{\boldsymbol{y}}$. K-example $z_{x_ix_j} = (K_1(x_i,x_j),\ldots,K_p(x_i,x_j))$ is a concatenation of entries on i^{th} row and j^{th} column of all the kernel matrices, i.e. $z_{x_ix_j} = ((\mathbf{K}_1)_{ij},\ldots,(\mathbf{K}_p)_{ij})$, which is the same as the qth row of $\hat{\mathbf{X}}$, denoted as $\hat{\boldsymbol{x}}_q = (vec(\mathbf{K}_1)_q,\ldots,vec(\mathbf{K}_p)_q)$.

As a result, when ℓ is fixed to be the hinge loss, the TSMKL [13] algorithm can be restated as:

$$\min_{\boldsymbol{v}\geq0}\frac{\lambda}{2}||\boldsymbol{v}|| + \frac{1}{\hat{n}}\sum_{q=1}^{\hat{n}}\ell(\boldsymbol{v};\hat{\boldsymbol{x}}_i,\hat{y}_i). \tag{6}$$

Now it is clear that, by vectorizing the input and target kernel matrices, ALIGNF solves a non-negative least squares and TSMKL solves a non-negative linear SVM with the same input and output. In this case, large scale linear solver and stochastic optimization will be needed since the number of data examples in the vectorized kernel space is quadratic to the number of original data points. In terms of worst case time complexity, solving ALIGNF in (4) including the computation of matrix $\mathbf{M}$ and vector $\boldsymbol{a}$ takes $O(n^2p^2+p^3)$ and solving ALIGNF in (5) takes $O(n^5)$. Therefore, when the number of kernels is much smaller than the number of data examples, solving ALIGNF in (4) is more efficient.

3   ALIGNF+

In this section, we first introduce a problem whose dual corresponds to (4). Then we extend ALIGNF with similar idea of soft margin SVM, adding slack variables in the constraints.

Proposition 3. *The dual of the the following problem coincides with* (4):

$$\min_{\mathbf{W} \in \mathbf{R}^{n \times n}} \quad \langle \mathbf{W}, \mathbf{W} \rangle_F \tag{7}$$

$$s.t. \quad 2\langle \mathbf{W}, \mathbf{K}_k \rangle_F \geq 2\langle \mathbf{K}_y, \mathbf{K}_k \rangle_F, \ k = 1 \ldots, p \tag{8}$$

Proof. Introducing Lagrangian multipliers $v \geq 0$ *for* (8) *and the Lagrangian becomes:*

$$L(\mathbf{W}, v) = \langle \mathbf{W}, \mathbf{W} \rangle_F - 2 \sum_{k=1}^{p} v_k \langle \mathbf{W}, \mathbf{K}_k \rangle_F + 2 \sum_{k=1}^{p} v_k \langle \mathbf{K}_y, \mathbf{K}_k \rangle_F. \tag{9}$$

The derivative of the Lagrangian with respect to $\mathbf{W}$ *is:*

$$\frac{\partial L}{\partial \mathbf{W}} = 2\mathbf{W} - 2 \sum_{k=1}^{p} v_k \mathbf{K}_k \tag{10}$$

Setting the derivative of Lagrangian with respect to $\mathbf{W}$ *to zero leads to:*

$$\mathbf{W}^* = \sum_{k=1}^{p} v_k^* \mathbf{K}_k. \tag{11}$$

With the same definition of $\mathbf{M}$ *and* a *as in Proposition 1, plugging* (11) *back to the Lagrangian we have the following dual problem:*

$$\min_{v \geq 0} v^T \mathbf{M} v - 2v^T a,$$

which is the same as (4).

We now call the problem defined in Proposition (3) as *primal ALIGNF* and the problem defined in Eq. (4) as *dual ALIGNF*. Strong duality holds between them. When we reach optimal in both primal and dual, the following complementary slackness can be derived:

$$(\langle \mathbf{W}^*, \mathbf{K}_k \rangle_F - \langle \mathbf{K}_y, \mathbf{K}_k \rangle_F) v_k^* = 0, \ k = 1, \ldots, p. \tag{12}$$

The equation above can be interpreted as follows: when the input kernel matrix correlates better with the solution matrix $\mathbf{W}$ than the correlation with the target kernel (indicates a bad alignment between the input kernel and target kernel), it will have no weight. When the input kernel matrix correlates to the target kernel as well as the solution matrix (indicates a good alignment between the input kernel and target kernel), it will receive a non-zero weight.

Now we pay attention to the constraints (8), which can be seen as similar constraints to the ones in the one-class SVM [18]. These constraints are not robust to outliers and noise. Some input kernels may happen to be well correlated with the target kernel but they may contain small amount of information. The estimate of the target kernel could also be noisy. With the same rationale of soft margin SVM, which embraces better stability than hard margin SVM as discussed in [10], we propose to add slack variables to the constraints (8), leading to the following problem.

Definition 1. *Primal problem of ALIGNF+:*

$$\min_{\mathbf{W},\boldsymbol{\xi}} \ \langle \mathbf{W}, \mathbf{W} \rangle_F + C \sum_{k=1}^{p} \xi_k \tag{13}$$

$$s.t. \ \ 2\langle \mathbf{W}, \mathbf{K}_k \rangle_F \geq 2\langle \mathbf{K}_y, \mathbf{K}_k \rangle_F - \xi_k, \ k = 1 \ldots, p, \tag{14}$$

$$\xi_k \geq 0, \ k = 1 \ldots, p. \tag{15}$$

By applying the standard Lagrangian technique, we obtain the following dual problem of ALIGNF+.

Definition 2. *Dual problem of ALIGNF+:*

$$\min_{\boldsymbol{v}} \ \boldsymbol{v}^T \mathbf{M} \boldsymbol{v} - 2 \boldsymbol{v}^T \boldsymbol{a} \tag{16}$$

$$s.t. \ \ 0 \leq v_k \leq C, \ k = 1 \ldots, p. \tag{17}$$

After the optimal $\boldsymbol{v}^*$ found, the kernel weights $\boldsymbol{\mu}$ can be computed as $\boldsymbol{\mu} = \boldsymbol{v}^* / \|\boldsymbol{v}^*\|$. Similar to soft-margin SVM, the ALIGNF+ algorithm just requires an upper bound on all the components of the solution.

In fact, by controlling the upper bound of ALIGNF+, we interpolate between ALIGNF and the simple uniform combination of kernels (UNIMKL), where every kernel receives the same weight. The ALIGNF+ reduces to ALIGNF when $C \to +\infty$ and it reduces to UNIMKL when $C \to 0$. We demonstrate this issue on the Spambase dataset (refer to the experiments section for more detail about this dataset) in the Fig. 1 where we plot the change of angles between the solutions of ALIGNF+ and UNIMKL against increasing upper bound in ALIGNF+ (blue line). We also plot the angle between the solutions of ALIGNF and UNIMKL (red dashed line). It is clear that the solutions of ALIGNF+ evolve from UNIMKL to ALIGNF when the upper bound is increased.

It is also possible to formulate ALIGNF+ as a bounded least squares problem in the vectorized kernel space. With the same notation as shown in Sect. 2.2, the soft ALIGNF (ALIGNF+) can be equivalently formulated as:

$$\min_{0 \leq \boldsymbol{v} \leq C} \|\hat{\mathbf{X}} \boldsymbol{v} - \hat{\boldsymbol{y}}\|^2. \tag{18}$$

When the number of kernels is much larger than the number of data examples, it is advantageous to solve ALIGNF+ in (18). When the number of kernels is much smaller than the number of data examples, it is better to use the formulation of ALIGNF+ given in Definition (2), which is the formulation used in all our experiments.

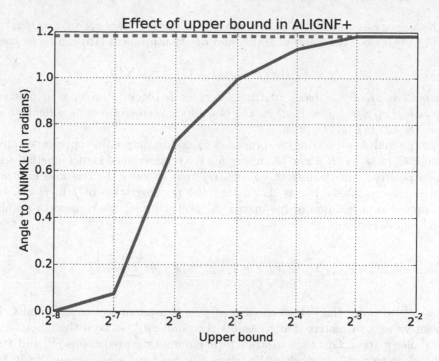

Fig. 1. For the Spambase dataset, the change of angles between the solutions of ALIGNF+ and UNIMKL (on the Y-axis) against increasing upper bound in ALIGNF+ (on the X-axis), shown in the blue line. The angle between the solution of ALIGNF and UNIMKL is shown in the red dashed line. (Color figure online)

3.1 Coupling ALIGNF+ with a Multilabel Method

Both ALIGNF and ALIGNF+ are designed for single label kernel target alignment. They can be adapted for multilabel dataset either by considering each label independently or by defining target kernel for multiple labels at the same time. We will test both in the experiments section. Let us denote the labels as a matrix $\mathbf{Y} \in \mathbb{R}^{n \times m}$ where m is the number of labels. We focus on the simplest target kernel defined as $\mathbf{K}_Y = \mathbf{Y}\mathbf{Y}^T$ for all the labels. Notice that by defining the target kernel this way, the learnt kernel weights are shared for all the labels.

To compare the weights learnt by ALIGNF and ALIGNF+, we use operator-valued kernel regression (OVKR) in the second stage. We briefly recap OVKR here. OVKR [16] extends the classical regularized kernel regression to the case of vector-valued functions. In this setting, the values of the input kernel are not scalar but operators on vectors in $\mathcal{Y}$. In the multilabel setting, the kernel values $\mathcal{K}(\boldsymbol{x}, \boldsymbol{x}')$ are matrices of size $m \times m$. Given a loss function $\mathcal{L} : \mathcal{Y} \times \mathcal{Y} \to \mathbb{R}$, the optimization problem of OVKR is the following:

$$\arg\min_{f \in \mathcal{H}} \sum_{i=1}^{n} \mathcal{L}(f(\boldsymbol{x_i}), \mathbf{y}_i) + \lambda \|f\|_{\mathcal{H}}^2, \ \lambda > 0. \tag{19}$$

In this paper we consider the least squares loss function: $\mathcal{L}(f(\boldsymbol{x}_i), \mathbf{y}_i) = \|f(\boldsymbol{x}_i) - \mathbf{y}_i\|_{\mathcal{Y}}^2$. In this case, the solution of the minimization problem (19) can be written as:

$$\forall \boldsymbol{x} \in \mathcal{X}, \; f(\boldsymbol{x}) = \mathbf{G}_x (\lambda \mathbf{I} + \mathbf{G})^{-1} \text{vec}(\mathbf{Y}^T),$$

where $\mathbf{G}$ is an $n \times n$ block matrix, where each block is an $m \times m$ matrix: $\mathbf{G}_{ij} = \mathcal{K}_x(\boldsymbol{x}_i, \boldsymbol{x}_j)$, $i, j = 1, \ldots, n$. $\mathbf{G}_x$ is a block vector of length n defined as $\mathbf{G}_x = (\mathcal{K}_x(\boldsymbol{x}, \boldsymbol{x}_1), \ldots, \mathcal{K}_x(\boldsymbol{x}, \boldsymbol{x}_n))$.

In the multilabel setting, [6] proposed to use decomposable operator-valued kernels: $\mathcal{K}_x(\boldsymbol{x}, \boldsymbol{x}') = K_x(\boldsymbol{x}, \boldsymbol{x}')\mathbf{A}$, where K_x is a scalar-valued kernel function and $\mathbf{A}$ is a positive semi-definite $m \times m$ matrix that encodes the relations existing between the m labels. It was shown that the regularization of f in $\mathcal{H}$ can be interpreted as a function of the matrix $\mathbf{A}$. Following [6], we consider a graph-based regularization:

$$\|f\|_{\mathcal{H}}^2 = \frac{\eta}{2} \sum_{i,j=1}^{m} \mathbf{C}_{ij} \|f^{(i)} - f^{(j)}\|^2 + (1 - \eta) \sum_{i=1}^{d} \mathbf{C}_{ii} \|f^{(i)}\|^2,$$

where $f^{(i)}$ denotes the i-th component of f, η is a parameter in $[0, 1]$ and $\mathbf{C}$ is a positive $m \times m$ matrix that measures the similarity between the labels, e.g. covariance matrix. This regularization term enforces the predictions $f^{(i)}$ and $f^{(j)}$ to be close to each other when the labels i and j have a high similarity in $\mathbf{C}$. When $\eta = 0$, the labels are predicted independently from each other. [6] showed that using this regularization term is equivalent to use a decomposable kernel with the matrix $\mathbf{A}$ defined as $\mathbf{A} = (\eta \mathbf{L_C} + (1 - \eta)\mathbf{I})^{-1}$, where $\mathbf{L_C}$ denotes the graph Laplacian of $\mathbf{C}$. This is the formulation used in our experiments with $\mathbf{C}$ as covariance matrix.

4   Experiments

We have two sets of experiments. The first set covers Sects. 4.2 and 4.3, where the individual labels in a multilabel dataset are handled independently in both stages, first for learning the kernel weights, and training SVM in the second stage. This is the main focus of the paper. The second set covers Sect. 4.4, where the labels for multilabel datasets are treated jointly and OVKR is used in the second stage to predict the multiple labels at once. We start with an introduction on the datasets used for evaluation.

4.1   Datasets

We used 17 datasets for evaluation: 4 public datasets, 5 real world bioinformatics datasets and 8 real world image annotation datasets. The public datasets contain one single label dataset Spambase, three multilabel datasets Emotions, Yeast, and Enron. The single-label dataset Spambase is downloaded from UCI Machine Learning Repository (http://archive.ics.uci.edu/ml/) and the others are from

Mulan library (http://mulan.sourceforge.net/datasets.html). We computed RBF kernels with different γ parameters. For multilabel datasets, γ are taken from the set $\{2^{-13}, 2^{-11}, \ldots, 2^1, 2^3\}$. For single-label dataset Spambase, the same set of γ parameters are used as stated in [2]. Both ALIGNF and ALIGNF+ are single label methods, so we handle each label independently and we do the same for all the multilabel datasets in this paper without mentioning otherwise.

The 5 bioinformatic datasets contain three multiclass datasets Psort$^+$, Psort$^-$ and Plant and two multilabel datasets FP^S and FP. The multiclass datasets are taken from [23] where we used all the 69 kernels as input including 64 sequence motif kernels, 3 kernels from BLAST E-values and 2 kernels based on phylogenetic trees. The target kernels for multiclass datasets are computed as in Sect. 2.2. We used the full datasets without any filtering. The FP^S dataset (superscript 'S' means small) consists of 1000 mass spectra as input and 101 molecular fingerprints as labels with balanced positive and negative examples. The 12 input kernels include one probability product kernels [9] and 11 fragmentation tree kernels as described in [21]. The FP dataset contains 24 kernels, 12 of them are the same as FP^S and the others are described in [5].

The 8 image annotation datasets are two versions of the datasets Corel5k, Espgame, Iaprtc12 and Mirflickr [8,15]. The linear kernel are computed on 15 features which are described in [8]. Many labels are extremely dominated by negative examples, so we only consider the labels with more than 2 % of positive labels. For these datasets, the upper bounds are selected on the training set and the learned weights are used to combine the kernels and the performances are evaluated on a fixed test set. They have 499, 2081, 1962 and 12500 fixed test examples respectively. We also sampled 1000 examples from the training set of all the image datasets and we denote these smaller ones by Corel5k^S, Espgame^S, Iaprtc12^S and Mirflickr^S. All the datasets are summarized in Table 1.

4.2   Comparison to ALIGNF

We first report the performance of ALIGNF+ comparing to original ALIGNF. The kernel weights are first learned by ALIGNF and ALIGNF+ independently for each label. Then, for each label, the kernels weights are normalized and the kernels are combined with the weights before being used by SVM in the second stage.

Table 1. Datasets summary. n and m are the numbers of training examples and labels. Psort$^+$, Psort$^-$ and Plant are multiclass datasets which have 4, 5 and 4 classes respectively.

	Spambase	Emotions	Yeast	Enron	Psort$^+$	Psort$^-$	Plant	FP	FP^S
n	1000	593	1500	1702	541	1444	940	4138	1000
m	1	6	14	53	1	1	1	505	101
	Corel5k	Espgame	Iaprtc12	Mirflickr	Corel5k^S	Espgame^S	Iaprtc12^S	Mirflickr^S	
n	4500	18689	17665	12500	1000	1000	1000	1000	
m	43	58	74	17	37	52	67	15	

For the full image datasets with fixed test sets, the upper bound in ALIGNF+ is selected within the training set and the performance reported are based on the fixed test set. The results for other datasets, except a 10-fold strategy as in the original publication for the FP dataset [5], are based on 5-fold cross-validation. The upper bound in ALIGNF+ is selected in an inner cross validation based on F1 scores on validation folds. The upper bound is chosen from the set $\{2^{-9}, 2^{-8}, \ldots, 2^5, 2^6\}$. For the multilabel datasets, we use a stratification setting of [19] to split to data in such a way that each fold has similar label distribution.

The macro-average F1, micro-average F1 and accuracy of ALIGNF and ALIGNF+ are recorded in Table 2. For single label dataset Spambase, the macro and micro average F1 scores are just F1 scores. The performance deviation comes from averaging the results of different test folds in cross validations and the full image datasets have zero deviation due to the fixed test sets.

From Table 2, in general the ALIGNF+ outperforms original ALIGNF especially in macro-average F1 and accuracy. This can be explained because the upper bound in ALIGNF+ is selected based on the F1 score for each label. Macro-average F1, as a summary statistic, reflects the label-wise difference better

Table 2. Performance comparison of ALIGNF and ALIGNF+. The performance deviation comes from averaging the results of different test folds. The full image datasets have deviation zero due to fixed test sets. The better method between the two is highlighted in bold font for each dataset.

Data	Macro-average F1		Micro-average F1		Accuracy	
	ALIGNF	ALIGNF+	ALIGNF	ALIGNF+	ALIGNF	ALIGNF+
Spambase	92.3±1.7	**93.1±1.6**	92.3±1.7	**93.1±1.6**	94.1±1.3	**94.7±1.2**
Emotions	**66.0±2.1**	65.9±1.7	**68.6±2.0**	68.4±1.9	81.8±0.7	81.8±0.5
Yeast	41.4±1.3	**41.8±1.0**	**65.8±0.6**	65.6±0.8	**79.8±0.2**	79.6±0.4
Enron	31.1±1.5	**33.2±2.1**	60.2±1.4	**60.6±1.5**	91.0±0.2	91.0±0.2
Psort$^+$	88.8±2.4	**90.5±2.4**	89.9±1.7	**91.3±1.7**	89.9±1.7	**91.3±1.7**
Psort$^-$	**91.2±2.2**	91.0±1.7	**91.8±2.1**	91.6±1.5	**91.8±2.1**	91.6±1.5
Plant	91.7±2.2	**92.4±1.1**	92.8±2.0	**93.5±1.1**	92.8±2.0	**93.5±1.1**
FP^S	84.5±0.6	**86.4±0.3**	84.8±0.5	**86.7±0.3**	84.8±0.4	**86.7±0.2**
Corel5k^S	49.1±4.7	**51.9±5.1**	54.0±3.7	**56.0±3.5**	96.1±0.3	**96.2±0.3**
Espgame^S	14.0±1.4	**15.7±1.6**	26.0±3.0	**27.2±2.2**	94.8±0.2	94.8±0.2
Iaprtc12^S	16.8±3.5	16.8±3.6	**27.1±2.4**	26.7±2.7	**94.6±0.2**	94.5±0.2
Mirflickr^S	15.2±7.8	**15.7±8.0**	**16.9±6.1**	16.6±6.8	97.2±0.3	**97.3±0.2**
FP	78.3±0.8	**79.3±0.7**	85.6±0.5	**86.5±0.4**	89.0±0.3	**89.7±0.3**
Corel5k	50.8±0.0	**51.5±0.0**	**53.1±0.0**	53.0±0.0	95.7±0.0	**95.8±0.0**
Espgame	28.8±0.0	**29.0±0.0**	33.7±0.0	**34.5±0.0**	93.4±0.0	93.4±0.0
Iaprtc12	34.9±0.0	**35.7±0.0**	37.7±0.0	**39.1±0.0**	93.6±0.0	**93.7±0.0**
Mirflickr	10.0±0.0	**10.5±0.0**	10.1±0.0	**10.3±0.0**	95.5±0.0	**95.6±0.0**

than the micro-average F1. We also observed that the performance of ALIGNF+ is very sensitive to the upper bound. Because ALIGNF is a special case of ALIGNF+, ideally, ALIGNF+ should be at least as good as ALIGNF when the right upper bound could be estimated from the training data. However, the estimation based on the training folds may not be optimal for testing folds due to lack of data or sub-optimal train-test split. We tested the methods difference significance level with Friedman test, the resulting p-value is 0.0027, 0.2253 and 0.0522 for macro F1, micro F1 and accuracy respectively.

With the slack variables introduced in the primal problem of ALIGNF+, it should be more robust to noise than ALIGNF. To demonstrate this advantage, we choose one dataset which has the most positive labels from each category and randomly flip the labels of Emotions, FP^S and Iaprtc12^S datasets with probabilities $\{0.5\%, 1.0\%\ 1.5\%, 2.0\%, 2.5\%\}$. We plot the macro and micro average F1 scores of ALIGNF+ (red line) and ALIGNF (blue line) in Fig. 2.

Clearly, for Emotions dataset, with increasing noise on the labels, the performance of ALIGNF+ drops slower than ALIGNF. For FP^S dataset, the performance of the two seems to drop at the same pace but ALIGNF+ is always better than ALIGNF on all noise levels. For Iaprtc12^S dataset, although ALIGNF+ is worse in the beginning (this may due to errors for upper bound estimation on the training data, more advanced method on choosing the hyper-parameters can be used such as [17]), its performance is more stable when more noise is presented. In Fig. 2, the F1 scores are not averaged over 5 folds but computed on the whole prediction, which explains why some F1 scores are different to those in Table 2.

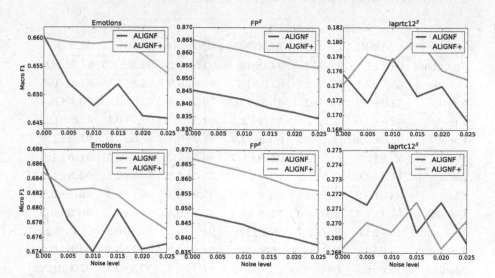

Fig. 2. The X-axis represents the probability to flip the labels (noise level). The Y-axis corresponds to the macro-average F1 (top) and micro-average F1 (bottom). Red line represents ALIGNF+ and blue line represents ALIGNF. (Color figure online)

4.3   Comparison to Other Two-Stage MKL Methods

We consider two competitors, the uniform combination of kernels UNIMKL and the TSMKL algorithm described in Sect. 2.2. In the experiments, the maximum number of iterations for TSMKL is set to 10^5 and a python implementation of pegasos [20] using stochastic gradient descent (https://github.com/ejlb/pegasos) is used to solve the linear SVM with projection onto non-negative orthant after each iteration. The regularization parameter λ is selected based on a random 80%–20% split of the training folds [13]. The 5 fold cross validated macro-average F1 and accuracy are reported in Table 3. From the table, ALIGNF+ outperforms UNIMKL almost consistently and the advantage stays when comparing to TSMKL in the accuracy measure. We tested the methods difference significance level with Friedman test, the resulting p-value is 0.0658 and 0.0074 for macro F1 and accuracy respectively.

As mentioned in Sect. 2.2, TSMKL is equivalent to solving a linear SVM in the vectorized kernel space. Here we also point out another issue of TSMKL: Even though large scale linear solver could be used to solve this problem, one should mind for the convergence issue while ALIGNF+ in the dual formulation (16) is much more simple. This QP can be solved very efficiently when the problem size is modest, say no more than couple of thousands.

Table 3. Performance comparison of UNIMKL, TSMKL and ALIGNF+. The performance deviation comes from averaging the results over different cross-validation folds. The best methods of the three are highlighted in bold font.

Data	Macro-average F1			Accuracy		
	UNIMKL	TSMKL	ALIGNF+	UNIMKL	TSMKL	ALIGNF+
Spambase	93.2±1.5	**93.3±1.9**	93.1±1.6	94.8±1.2	**94.9±1.5**	94.7±1.2
Emotions	65.4±2.1	65.9±2.1	65.9±1.7	81.5±0.9	80.8±1.5	**81.8±0.5**
Yeast	42.0±1.0	**44.5±1.4**	41.8±1.0	79.5±0.4	78.6±0.3	**79.6±0.4**
Enron	32.6±2.0	32.9±2.1	**33.2±2.1**	**91.1±0.2**	90.7±0.1	91.0±0.2
Psort$^+$	82.6±3.8	81.1±4.7	**90.5±2.4**	83.0±3.4	81.6±5.0	**91.3±1.7**
Psort$^-$	85.9±2.8	84.7±2.6	**91.0±1.7**	86.8±2.7	85.7±2.5	**91.6±1.5**
plant	71.4±2.0	83.6±1.9	**92.4±1.1**	75.0±1.7	86.0±2.1	**93.5±1.1**
FP^S	85.6±0.2	86.1±0.3	**86.4±0.3**	85.9±0.3	86.4±0.2	**86.7±0.2**
Corel5k^S	51.4±5.4	50.0±4.0	**51.9±5.1**	96.1±0.3	95.9±0.3	**96.2±0.3**
Espgame^S	15.7±1.6	15.7±1.7	15.7±1.6	94.8±0.2	94.4±0.3	94.8±0.2
Iaprtc12^S	16.6±3.2	**18.6±1.7**	16.8±3.6	94.4±0.2	94.1±0.3	**94.5±0.2**
Mirflickr^S	15.0±8.5	**16.9±7.5**	15.7±8.0	97.3±0.3	97.3±0.3	97.3±0.2

4.4   Multilabel Classification Experiments

In this section, we show the results using multilabel kernel target alignment in the first stage where the target kernel is computed as $\mathbf{K}_Y = \mathbf{Y}\mathbf{Y}^T$. In the second stage, we use operator-valued kernel regression (OVKR). The detail of the above two components can be found in Sect. 3.1. The parameters λ and η have been selected on the training folds from $\lambda \in \{0.0001, 0.001, 0.01, 0.1, 1, 10\}$ and $\eta \in \{0.1, 0.3, 0.5, 0.7, 0.9\}$.

We also tested the macro-average and micro-average F1 scores for multilabel datasets. There is no clear winner between ALIGNF+ and ALIGNF so we omit the results here. This could be partially explained by the fact that the target kernel matrix computed in this way is already robust. Let the entry on i^{th} row and j^{th} column of target kernel matrix be $(\mathbf{K}_Y)_{ij} = \sum_{t=1}^{m}(\mathbf{y}_i)_t(\mathbf{y}_j)_t$. If one label is corrupted, this value could be corrected by the other labels to certain extent. For this reason, we repeated the experiments on the noisy labels with OVKR for Emotions, FP^S and Iaprtc12^S datasets. The results are shown in Fig. 3. It is clear that the ALIGNF+ formulation almost outperforms, or is at least as good as ALIGNF on all noise levels.

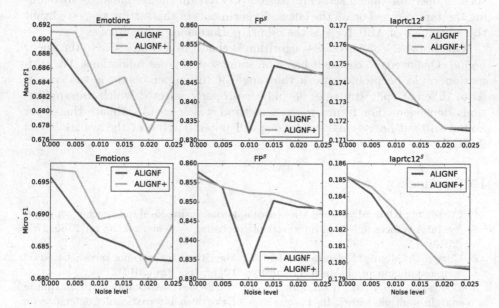

Fig. 3. Performance comparison between ALIGNF and ALIGNF+ with target kernel computed as $\mathbf{K}_Y = \mathbf{Y}\mathbf{Y}'$ and using OVKR in the second stage. The real valued prediction is binarized to 1 and -1 with threshold 0. The X-axis represents the probability to flip the labels (noise levels). The Y-axis corresponds to the macro-average F1 (top) and micro-average F1 (bottom). Red line represents ALIGNF+ and blue line represents ALIGNF. (Color figure online)

5   Conclusions

We have shown that through vectorizing the input and target kernel matrices, the two recently proposed two-stage MKL algorithms ALIGNF [2] and TSMKL [13] show correspondence to familiar machine learning problems: the first solves a non-negative least squares while the second solves a non-negative linear SVM with the same input and output. Based on centered kernel target alignment algorithm ALIGNF, we derived the corresponding primal problem, which is essentially a one-class SVM problem. Then we proposed a soft formulation, named ALIGNF+, where slack variables are included into the constraints, similar to soft-margin SVM, to make the algorithm more robust to noise. The dual of ALIGNF+ turned out to be as simple as ALIGNF, just with an upper bound on the solution. This upper bound was shown to interplay between UNIMKL and ALIGNF.

Experiments with SVM in the second stage showed that the ALIGNF+ algorithm outperforms the original ALIGNF and other two-stage MKL algorithms on several tasks. When noise was added to the labels, ALIGNF+ was less sensitive against the noise than ALIGNF. It is also possible to use multilabel method such as operator-valued kernel regression (OVKR) in the second stage by defining the target kernel on all the labels. Experiments in this setting suggested that the advantage of ALIGNF+ is the robust performance against noise.

One drawback of ALIGNF+ algorithm is that it is very sensitive to the upper bound. One needs to choose it based on some sort of cross validations. The estimation could be problematic if the range of the upper bound is too coarse. Also, different split strategies should be properly selected which are applications dependent. For future work, a method which could estimate the upper bound without heavy cross validations will potentially boost the availability of ALIGNF+.

References

1. Cortes, C., Kloft, M., Mohri, M.: Learning kernels using local rademacher complexity. In: Advances in Neural Information Processing Systems, vol. 26, pp. 2760–2768 (2013)
2. Cortes, C., Mohri, M., Rostamizadeh, A.: Algorithms for learning kernels based on centered alignment. J. Mach. Learn. Res. 13(1), 795–828 (2012)
3. Cortes, C., Mohri, M., Rostamizadeh, A.: Multi-class classification with maximum margin multiple kernel. In: Proceedings of the 30th International Conference on Machine Learning, pp. 46–54 (2013)
4. Cristianini, N., Kandola, J., Elisseeff, A., Shawe-Taylor, J.: On kernel-target alignment. In: Advances in Neural Information Processing Systems, vol. 14, pp. 367–373. MIT Press (2002)
5. Dührkop, K., Shen, H., Meusel, M., Rousu, J., Böcker, S.: Searching molecular structure databases with tandem mass spectra using CSI: fingerid. Proc. Nat. Acad. Sci. 112(41), 12580–12585 (2015)
6. Evgeniou, T., Micchelli, C.A., Pontil, M.: Learning multiple tasks with kernel methods. J. Mach. Learn. Res. 6, 615–637 (2005)

7. Gönen, M., Alpaydın, E.: Multiple kernel learning algorithms. J. Mach. Learn. Res. **12**, 2211–2268 (2011)
8. Guillaumin, M., Verbeek, J., Schmid, C.: Multimodal semi-supervised learning for image classification. In: IEEE Conference on Computer Vision & Pattern Recognition, pp. 902–909 (2010)
9. Heinonen, M., Shen, H., Zamboni, N., Rousu, J.: Metabolite identification and molecular fingerprint prediction through machine learning. Bioinformatics **28**(18), 2333–2341 (2012)
10. Herbrich, R.: Learning Kernel Classifiers: Theory and Algorithms. MIT Press, Cambridge (2001)
11. Jawanpuria, P., Varma, M., Nath, S.: On p-norm path following in multiple kernel learning for non-linear feature selection. In: Proceedings of the 31st International Conference on Machine Learning, pp. 118–126 (2014)
12. Kloft, M., Brefeld, U., Sonnenburg, S., Zien, A.: Lp-norm multiple kernel learning. J. Mach. Learn. Res. **12**, 953–997 (2011)
13. Kumar, A., Niculescu-Mizil, A., Kavukcuoglu, K., Daume III, H.: A binary classification framework for two-stage multiple kernel learning. In: Proceedings of the 29th International Conference on Machine Learning (2012)
14. Lanckriet, G.R., Cristianini, N., Bartlett, P., Ghaoui, L.E., Jordan, M.I.: Learning the kernel matrix with semidefinite programming. J. Mach. Learn. Res. **5**, 27–72 (2004)
15. Makadia, A., Pavlovic, V., Kumar, S.: Baselines for image annotation. Int. J. Comput. Vis. **90**(1), 88–105 (2010)
16. Micchelli, C.A., Pontil, M.A.: On learning vector-valued functions. Neural Comput. **17**, 177–204 (2005)
17. Moore, G., Bergeron, C., Bennett, K.P.: Model selection for primal SVM. Mach. Learn. **85**(1), 175–208 (2011)
18. Schölkopf, B., Platt, J.C., Shawe-Taylor, J., Smola, A.J., Williamson, R.C.: Estimating the support of a high-dimensional distribution. Neural Comput. **13**(7), 1443–1471 (2001)
19. Sechidis, K., Tsoumakas, G., Vlahavas, I.: On the stratification of multi-label data. In: Gunopulos, D., Hofmann, T., Malerba, D., Vazirgiannis, M. (eds.) ECML PKDD 2011, Part III. LNCS(LNAI), vol. 6913, pp. 145–158. Springer, Heidelberg (2011)
20. Shalev-Shwartz, S., Singer, Y., Srebro, N., Cotter, A.: Pegasos: primal estimated sub-gradient solver for SVM. Math. Program. **127**(1), 3–30 (2011)
21. Shen, H., Dührkop, K., Böcker, S., Rousu, J.: Metabolite identification through multiple kernel learning on fragmentation trees. Bioinformatics **30**(12), i157–i164 (2014)
22. Yamada, M., Jitkrittum, W., Sigal, L., Xing, E.P., Sugiyama, M.: High-dimensional feature selection by feature-wise kernelized lasso. Neural Comput. **26**(1), 185–207 (2014)
23. Zien, A., Ong, C.S.: Multiclass multiple kernel learning. In: Proceedings of the 24th International Conference on Machine learning, pp. 1191–1198. ACM (2007)

Unsupervised Anomaly Detection in Noisy Business Process Event Logs Using Denoising Autoencoders

Timo Nolle[(✉)], Alexander Seeliger, and Max Mühlhäuser

Technische Universität Darmstadt, Telecooperation Lab, Darmstadt, Germany
{nolle,seeliger,max}@tk.tu-darmstadt.de

Abstract. Business processes are prone to subtle changes over time, as unwanted behavior manifests in the execution over time. This problem is related to anomaly detection, as these subtle changes start of as anomalies at first, and thus it is important to detect them early. However, the necessary process documentation is often outdated, and thus not usable. Moreover, the only way of analyzing a process in execution is the use of event logs coming from process-aware information systems, but these event logs already contain anomalous behavior and other sorts of noise. Classic process anomaly detection algorithms require a dataset that is free of anomalies; thus, they are unable to process the noisy event logs. Within this paper we propose a system, relying on neural network technology, that is able to deal with the noise in the event log and learn a representation of the underlying model, and thus detect anomalous behavior based on this representation. We evaluate our approach on five different event logs, coming from process models with different complexities, and demonstrate that our approach yields remarkable results of 97.2 % F1-score in detecting anomalous traces in the event log, and 95.6 % accuracy in detecting the respective anomalous activities within the traces.

1 Introduction

Anomaly detection, or outlier detection, is an important topic for todays businesses. Companies all over the world are interested in anomalous executions within their process, as these can be indicators for fraud, or inefficiencies in their process.

More and more companies rely on process-aware information systems (PAISs) [7] to assist their employees in the execution. The increasing numbers of PAISs has generated a lot of interest in the data these systems are storing about the execution of a process. PAISs provide data analysts with a huge amount of information in form of log files. These log files can be used to extract the events that happened during the execution of the process, and hence create so called event log files. Event logs are comprised of activities that have occurred during the execution of a process. These event logs enable a process analyst to explore the process by which this event log has been generated. In other words, the event log is the leftover evidence of the process that produced it. Consequently, it is

T. Calders et al. (Eds.): DS 2016, LNAI 9956, pp. 442–456, 2016.
DOI: 10.1007/978-3-319-46307-0_28

possible to recreate the process model by evaluating its event log. This is known as process model discovery and is one of the main ideas in the domain of process mining [22].

Often, process models have been designed by experts some time in the past, but over the years the process has slowly mutated. These mutations can be caused by new employees, working slightly different than their predecessors, or the use of new technology, or simply changes in the business plan. Process changes usually happen in a very subtle way. In a procurement process, for example, an employee stops requesting the required approval of their advisor, as this speeds up the process. At first this will happen rarely, but over time it will start to overwhelm the designated behavior. In other words, these subtle changes start off as anomalies in the process, and therefore it is important to detect them early and take actions, before they can manifest.

Process mining provides methodologies to detect these changes, e.g., by discovering the as-is process model from the event log [1]; that is, generating a valid process model that is capable of producing the same event log. After the discovery of the as-is model, it can be compared to a reference model in order to detect the changes. This is known as conformance checking [18]. However, this approach requires the existence of such a reference model of some form (e.g., BMPN model, petri-net, rule set). Unfortunately, these reference models are often not well maintained by the company, or even non-existent.

The absence of a reference model is a big problem for conformance checking, and especially in the field of anomaly detection, as there is no model that defines what a normal execution of the process ought to look like. Thus, we can not define what an anomalous execution is either. Many of the current anomaly detection algorithms work by first learning what a normal example is, by training on a training set that solely consists of normal examples, and then using this knowledge to detect the anomalies, as they are much different from what they have learned about normal examples during training. However, this is not a valid assumption we can make when considering process event logs from PAISs, as they usually already contain anomalous behavior and other sorts of noise. What is a valid assumption, on the other hand, is that the anomalous executions are highly outnumbered by the normal ones, which we will take advantage of.

In this paper we propose a method to automatically split a noisy event log into normal and anomalous traces. The main contribution of this approach is that it does not require the existence of a reference model, nor prior knowledge about the underlying process. We train our system on the raw input from the event log, including the anomalous traces, and without the use of labels indicating which cases are, in fact, anomalous. The system has to deduce the difference between normal and anomalous traces purely based on the patterns in the raw data. Our approach is based on a special type of neural network, called an autoencoder, that is trained in an unsupervised fashion. We will demonstrate that our system is able to understand the underlying processes of five different event logs. Consequently, it can automatically analyze a given event log and filter out anomalous traces, based on the implicitly inferred model. Not only can we filter out anomalous traces, but we can also infer which specific activity in a trace is the cause for the anomaly.

2   Related Work

In the field of business processes, and especially process mining [22], anomaly detection is not very frequently researched. The most recent publication [4] describes an approach where a reference model is build through the use of discovery algorithms. Then, this reference model can be used to automatically detect anomalous traces. However, this approach relies on a clean dataset, that is, no anomalous traces must be present in the data set during the discovery. As we have described earlier, this is usually not the case, as the event logs from PAISs most likely already do contain these anomalies.

The approach within this paper is highly influenced by the works in [9,12], where they propose the use of replicator neural networks [10], i.e., networks that reproduce their input, which are based on the idea of autoencoders from [11]. The approaches from [9,12], however, do not work well with variable length input.

A review of novelty detection (i.e., anomaly detection) methodology can be found in [17], where they describe and compare many methods that have been proposed over the last decades. The authors differentiate between five different basic methods for novelty detection: probabilistic, distance-based, reconstruction-based, domain-based, and information-theoretic novelty detection.

Probabilistic approaches try to estimate the probability distribution of the normal class, and thus are able to detect anomalies as they were sampled from a different distribution. However, this approach also requires a clean dataset. Distance-based novelty detection (e.g., nearest neighbor, clustering) does not require a cleaned dataset, yet it is only partly applicable for process traces, as anomalous traces are usually very similar to normal ones.

Reconstruction-based novelty detection (e.g., neural networks) is similar to the aforementioned approaches in [9,12]. However, training a neural network usually also requires a cleaned dataset. Nevertheless, we will show that our approach works on the noisy dataset, by taking advantage of the skewed distribution of normal data and anomalies, as demonstrated in [8].

Domain-based novelty detection requires domain knowledge, which violates our assumption, that we do not require any prior knowledge, only the data. Information-theoretic novelty detection defines anomalies as the examples that most influence an information measure (e.g., entropy) on the whole dataset. Iteratively removing the data with the highest impact will yield a cleaned dataset, and thus a set of anomalies. With the exception of reconstruction-based approaches, this is the only approach that can, to a certain degree, handle noisy datasets.

Within this paper, we opted to use a neural network based approach, as the recent achievements in machine translation and natural language processing indicate that neural networks are an excellent choice when modeling sequential data. At last, we want to point out that one-class support vector machines (SVMs) [6] are usually very sensitive to outliers in the data [2], which is why we did not apply classic one-class SVMs in this setting.

3   Dataset

PAISs keep a record of almost everything that has happened during the execution of a business process. This information can be extracted in form of an event log. An event log consists of traces, each consisting of the activities that have been executed. Table 1 shows an excerpt of such an event log, in this case it has been generated by a procurement process model. Notice that an event log must consist of at least three columns: trace id, to uniquely assign an executed activity to a trace; a timestamp, to order the activities within a trace; and an activity label, to distinguish the different activities.

Table 1. Example event log of a procurement process

Trace ID	Timestamp	Activity
1	2015-03-21 12:38:39	PR Created
1	2015-03-28 07:09:26	PR Released
1	2015-04-07 22:36:15	PO Created
1	2015-04-08 22:12:08	PO Released
1	2015-04-21 16:59:49	Goods Receipt
2	2015-05-14 11:31:53	SC Created
2	2015-05-21 09:21:26	SC Purchased
2	2015-05-28 18:48:27	SC Approved
2	2015-06-01 04:43:08	PO Created

In order to create a test setting for our approach we randomly generated process models and then sampled event logs from those models. These event logs have been generated by PLG2 [5], a process simulation and randomization tool. PLG2 allows the user to randomly generate a process model and then simulate it to generate genuine event logs. It also comes with a feature to perturb the generated event log by, for example, skipping events, doubling events, or changing the sequence in which events have occurred. PLG2 was specifically designed for process mining researchers' needs, as the amount of publicly available datasets of reasonable sizes is minuscule.

We have used the PLG2 tool to generate random process models of different complexities. Table 2 shows the complexity of these models in terms of the number of distinct activities and the number of gateways in the model. The resulting models were then used to generate noisy event logs, i.e., event logs including anomalous traces. Each trace in the event log has the chance of being

Table 2. Overview over the four different randomly generated process models and the corresponding event logs

Model	Activities	Gateways	Traces	Unique	Anomalous
Small	12	2	10 000	240	978
Medium	32	12		398	973
Large	42	14		621	985
Huge	51	22	50 000	1 044	4 778
P2P	12	8	10 000	232	968

affected by any of the following mutations: skipping an event; swapping two activities, or duplicating an activity so that it appears twice in a row.

The probability that a mutation occurs has been set to 3.3 % for all the three mentioned mutations. Thus, the resulting event log will contain roughly ten percent anomalous traces, as shown in Table 2. When considering a real-life business process, ten percent anomalous traces in the event log is quite high, and thus we have chosen to set this as our upper bound. Notice that our approach ought to yield better results with less anomalous traces in the log, as it becomes easier to generalize to the normal traces. This is why we use a highly noisy event log as compared to real life event logs, where the number of anomalies is usually much smaller. Notice that we used a fixed size of 10 000 traces for the small, medium, and large dataset, and a bigger size of 50 000 traces for the huge dataset. The huge dataset required a bigger sample due to its higher complexity. The increasing complexity with every dataset is also illustrated by the number of unique traces in each log, as shown in Table 2.

Figure 1 shows a t-SNE [15] visualization of the four randomly generated datasets, depicting anomalous traces in red and normal ones in green. We can clearly recognize the clusters that are being formed by the normal traces. However, within these clusters there also lie anomalous traces, which is exactly what we want, as anomalies in real life are typically very similar to normal traces in terms of the sequence of activities.

In addition to the four randomly generated models, we also used a simplified version of a purchase to pay (P2P) process model as is depicted by the BPMN model shown in Fig. 2. This model was mainly created for the evaluation part within this paper, as it features interpretable activity names, unlike the randomly generated ones. The resulting event log for the P2P model was generated in the same fashion as those of the randomly generated models, using the same parameters as mentioned above. A trace in our P2P model can either start with the manual creation of a purchase request (PR) or the creation of a shopping cart (SC). In both cases, after the necessary approval of the SC and the release of the PR, a purchase order (PO) is created. This PO can be altered by increasing or decreasing the order quantity. After the PO has been released the orderer receives the goods, and usually in quick succession also the corresponding invoice.

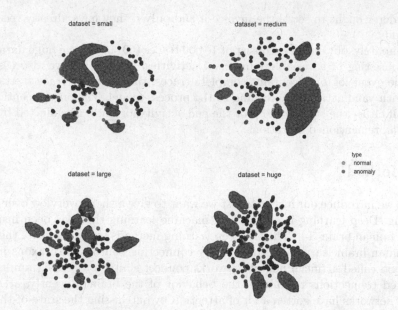

Fig. 1. t-SNE visualization of the randomly generated datasets from Table 2 (Color figure online)

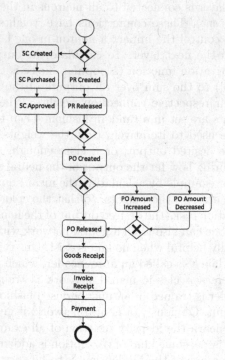

Fig. 2. BPMN model of a simplified purchase to pay process

The orderer ought to settle the invoice if and only if they have already received the goods.

Ultimately, our datasets consist of 10 000 traces (50 000 for the huge dataset), each consisting of a variable number of activities. Notice that we also assume that the event log only contains complete traces, that is, every trace starts and ends with valid activities according to the process model. The only exception to this is if either the start activity or the end activity, or both, are affected by one of the aforementioned mutations.

4   Method

Before we introduce our method, first we want to give a short overview over deep learning. Deep learning is a branch of machine learning that has been inspired by the human brain [14]. That is, deep learning methods try to replicate the way the human brain learns new concepts by connecting neurons with axons in the brain. So called artificial neural networks connect simple processing units with weighted connections to imitate the behavior of the brain. Recently, artificial neural networks have gotten a lot of attention by outclassing the state-of-the-art methods in many domains such as object recognition in images [13], or machine translation [3].

A neural network consists of multiple layers, each containing many neurons. Every neuron in one layer is connected to all neurons in the preceding and succeeding layers (if present). These connections have weights attached to them, which can be used to control the impact a neuron in one layer has on the activation of a neuron in the next layer. To calculate the output of a neuron we apply a non-linear activation function (a popular choice is the rectifier function $f(x) = \max(0, x)$ [16]) to the sum over all outputs of the neurons in the previous layer times their respective connection weights. When training a neural network all the weights are set in a random fashion. Then the backpropagation algorithm [19] can be used to iteratively tune the weights, so that the neural network produces the desired output, or a close enough approximation of it. This is done by measuring how far the output of the neural network differs from the desired output, for example by calculating the mean squared error, and then back-propagating the error to the weights, so that the error gets minimized.

In classic classification tasks the desired output of the neural network will be a class label. However, one can also train a neural network without the use of class labels. This is especially helpful when no labels exist. One type of neural network that does not rely on labels is called an autoencoder, which is what we deployed in our method. Whereas a classic neural network is trained in a supervised fashion, an autoencoder is trained in an unsupervised fashion, as it is trained to reproduce its own input. Obviously, a neural network, if given enough capacity and time, can simply learn the identity function of all examples in the training set. To overcome this issue, some kind of corruption is added to the autoencoder, for example, by forcing one of the hidden layers to be very small, therefore not allowing the autoencoder to simply learn the identity. Another very common way of adding corruption is to distribute additive gaussian noise over the input vector of the autoencoder. Thus, the autoencoder, even if repeatedly trained on the same trace, will always receive a different input. These types of autoencoders are known as denoising autoencoders, as they are basically producing a noise free version of their input. Our method is based on exactly this kind of autoencoder.

4.1 Setup

An autoencoder has a fixed size input; hence, we have to transform the variable sized traces from the event log. First, we force all traces to have the same length by repeatedly adding a special padding activity to the end until all traces have the same length (this can be set by hand or set to the maximum trace length encountered in the event log). Thereafter, we encoded the activity names using a one-hot encoding. Every activity is encoded by an n-dimensional vector, where n is the number of different activities in the event log, so that every activity is connected to exactly one dimension in the one-hot vector. To encode one activity we simply set the corresponding dimension of the one-hot vector to a fixed value of one, whilst setting all the other dimensions to zero. Notice that, instead of using zero, we opted to use -1 as this results in better distribution of the additive gaussian noise. Because we are using rectified linear units (i.e., units using the aforementioned rectifier function as their activation function), using -1 instead of 0 does not have a huge impact. This is done for every activity in every trace, including the special padding activity.

Consider the following example: let us assume an event log consists of 10 different activities and the maximum length of all traces in the event log is also 10. After the padding, every trace will have a fixed size of 10; and every activity is encoded by a 10-dimensional one-hot vector. Consequently, the resulting one-hot vector for every trace will have a size of 100.

Using the one-hot encoded event log we can train the autoencoder with stochastic gradient descent and backpropagation, using the event log both as the input, and the label. Figure 3 shows a simplified version of the architecture. Notice that Fig. 3 shows the event log with variable size traces for reasons of simplicity. In reality the event log is transformed before being fed into the autoencoder and then decoded afterwards. The special noise layer adds gaussian noise before feeding the input into the autoencoder, this layer is only active

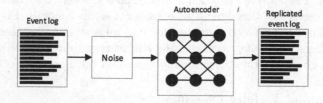

Fig. 3. Autoencoder is trained to replicate the traces in the event log after the addition of gaussian noise

Table 3. Overview of the hidden layer sizes

Model	Input/Output size	Hidden size
Small	264	286
Medium	340	374
Large	616	660
Huge	728	784
P2P	154	168

during training. Now the autoencoder is trained to reproduce its input, that is, to minimize the mean squared error between the input and its output.

We trained the autoencoder for a fixed number of 500 epochs using a mini batch size of 32. As the optimizer we used stochastic gradient descent with a learning rate of 0.01, a learning rate decay of 10^{-5}, and nesterov momentum [21] with a momentum factor of 0.9. Additionally, we used a maxnorm weight constraint of 0.5, as well as a dropout of 0.5, as suggested in [20]; the additive gaussian noise was sampled from a zero centric gaussian distribution with a standard deviation of 0.1. Each autoencoder consisted of an input and an output layer with linear units, and exactly one hidden layer with rectified linear units. These training parameters were used for each of the different event logs, but the size of the hidden layer was adapted depending on the event log. The number of neurons in the hidden layer was set to the size of the input plus one neuron for each possible activity in the event log. This was an arbitrary choice for the hidden layer size, but we found that it worked sufficiently good. However, choosing a hidden layer size smaller than the input layer size did not yield good results. The actual hidden layer sizes can be found in Table 3.

4.2 Anomalous Trace Classifier

After training the autoencoder, it can be used to reproduce the traces in the same event log it was trained on, but without applying the noise. Now, we can measure the mean squared error between the input vector and the output vector to detect anomalies in the event log. Because the distribution of normal traces and anomalous traces in the event log is one sided we can assume that the autoencoder will reproduce the normal traces with less reproduction error than the anomalies. Therefore we can define a threshold t, where if a traces reproduction error succeeds this threshold t we consider it as an anomaly. Figure 4 shows how to transform the trained autoencoder into an anomaly classifier by adding a threshold classifier. We found that using the average reproduction error on the event log is a good general choice for the threshold (cf. Fig. 5 in Sect. 5).

4.3 Anomalous Activity Classifier

We have described how to detect anomalous traces in the event log, now we want to refine this method. Not only can we detect that a trace is anomalous, but also

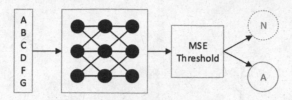

Fig. 4. Threshold classifier based on the mean squared error between the input vector and the output of the autoencoder

what activity in the trace influences the reproduction error the most. Therefore, we have to change our calculation of the reproduction error from trace based to activity based. Up until now, we calculated the reproduction error as the mean squared error between the entire one-hot encoded input and output sequence of the autoencoder. However, we can also consider the mean squared error for every activity in the sequence separately. After a trace has been reproduced by the autoencoder we split the input and the output vectors into subparts, so that every subpart contains the one-hot encoding for one single activity. Now we can compute the mean squared error for each activity separately and then apply the same threshold classifier method as before, only this time the classifier detect anomalous activities as apposed to anomalous traces.

5   Evaluation

We evaluated our approach on four different event logs coming from process models with different complexities, ranging from low to high complexity. In addition to those four event logs we also used an interpretable version with low complexity for demonstrative purposes.

After training one autoencoder for each event log, we evaluated the autoencoders on the same dataset, but without adding gaussian noise, as in the training phase. Therefore, we calculated the mean squared error for every trace in the training set and analyzed the resulting distribution. Figure 5 shows the distribution of the reproduction error for each dataset split into anomalous and normal traces. To indicate the variance of the distribution we used so called box-and-whisker plots. Figure 5 indicates that all five autoencoders are able to perfectly split the normal traces from the anomalous one solely based on the reproduction error. It also shows that the average reproduction error is a good threshold value for the anomaly classifier, albeit one would prefer a more pessimistically set value for real life scenarios, so that the anomaly class precision and the normal class recall are maximized. However, as we do not have the benefit of being provided labels in real life, we stick with the simple solution here. We plan on investigating more sophisticated methods of automatically setting the threshold.

Table 4 shows the respective precision, recall, and F1-score for the 5 different autoencoders in their classification report. Notice that using the average reproduction error as the threshold leads to the low anomaly precision class and

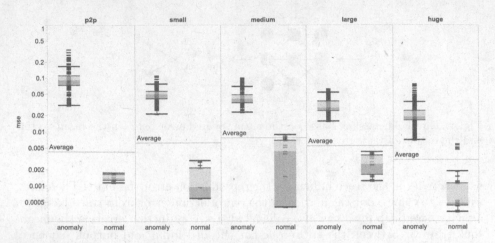

Fig. 5. The autoencoder succeeds in perfectly splitting the dataset into normal and anomalous traces solely based on the reproduction error

Table 4. Classification report for the anomalous traces detector

Dataset	Class	Precision	Recall	F1-score	Support
P2P	normal	1.00	1.00	1.00	9032
	anomaly	1.00	1.00	1.00	968
	average	1.00	1.00	1.00	10000
Small	normal	1.00	1.00	1.00	9022
	anomaly	1.00	1.00	1.00	978
	average	1.00	1.00	1.00	10000
Medium	normal	1.00	0.95	0.98	9027
	anomaly	0.70	1.00	0.83	973
	average	0.97	0.96	0.96	10000
Large	normal	1.00	1.00	1.00	9015
	anomaly	1.00	1.00	1.00	985
	average	1.00	1.00	1.00	10000
Huge	normal	1.00	0.87	0.93	45222
	anomaly	0.46	1.00	0.63	4778
	average	0.95	0.89	0.90	50000

normal class recall scores in the medium and huge event log. Even though the overall result is still remarkable, we still have room to improve the automatic adjusting of the threshold, as one can clearly see that setting the threshold optimally is indeed possible.

Next, we want to evaluate the activity based anomaly detection described earlier, but first we want to consider a few examples of classified traces from

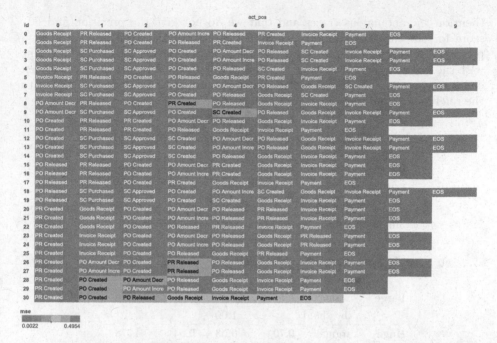

Fig. 6. Conformance check on a sample of the P2P dataset (Color figure online)

the P2P dataset autoencoder. Figure 6 shows a sample of 30 traces of the P2P event log, where every activity has been color coded according to the autoencoders reproduction error. The color coding simply linearly distributes 6 color patches across the range from zero to the maximum reproduction error in the event log. When comparing the traces to the reference procurement model from Fig. 2, we find that the autoencoder detects the anomalous activity in the traces remarkably well. However, we can also see that it has problems when certain activities are left out and the rest of the trace consequently gets shifted by one activity. Trace number 30 demonstrates this phenomenon. We can see that the necessary activity 'PR Released' has been skipped, but the system indicates that all activities after this point are anomalous, albeit the rest of the sequence being valid. The system has problems to handle shifted subsequences. Nevertheless, the autoencoder successfully detects the first activity that does not conform with the underlying model, that is the reproduction error of that activity is significantly high.

We have evaluated two versions of the anomalous activity classifier. The first produces only one position in the sequence by returning the index with the highest reproduction error. The second approach, similar to the anomalous trace classifier before, returns all indices where the reproduction error exceeds the average reproduction error on the whole event log. We shall call the former the argmax approach and the latter the threshold approach. Table 5 shows the classification reports for those two approaches. Notice that we only evaluated them on the anomalous traces and that we do not give the precision and F1-score for the threshold approach. The threshold approach will always yield a precision of one, hence we concentrate on the recall score, which is equivalent with the accuracy here. In case of the threshold approach, its prediction has been count as correct if the actual index of the first anomalous activity, according to the reference model, produced an above average reproduction error.

Table 5. Classification report for the anomalous activity detector

Dataset	Type	Precision	Recall	F1-score	Support
P2P	argmax	0.55	0.47	0.50	968
	threshold		0.85		
Small	argmax	0.76	0.66	0.68	978
	threshold		0.98		
Medium	argmax	0.68	0.54	0.56	973
	threshold		0.99		
Large	argmax	0.67	0.59	0.61	985
	threshold		1.00		
Huge	argmax	0.70	0.63	0.64	4 778
	threshold		0.96		

Table 5 shows that the argmax approach does not perform well. This is due to the fact that in most cases the reproduction error of the first anomalous activity in the trace is indeed significantly high, yet the overall highest reproduction error is found at a different index. Notice that this is also an effect of the reproduction error being carried through, as mentioned before. However, the threshold approach clearly shows that the anomalous activity almost always produces an above average reproduction error; hence, the autoencoder is capable of detecting them.

6   Conclusion and Future Work

Real-life event logs often contain anomalous traces. We have presented an approach that is capable of automatically filtering out anomalous traces in a noisy event log without any prior knowledge being fed into the system. The system learns to discriminate between normal and anomalous traces only from the present pattern in the data. We have evaluated the system on five different noisy event logs from randomly generated process models (one model was produced manually). Our evaluation has shown that our threshold based anomalous activity classifier is indeed capable of automatically detecting the anomalous activity in a trace with an accuracy of at least 85.0 % and 95.6 on average over all training sets. Especially for the more complex process models this is a remarkable result.

We want to point out that our approach is susceptible to anomalous behavior in the event log that is very frequent, that is the same anomalous trace is found multiple times. This is something we want to investigate in the future. However, as changes in a business process usually happen subtly, anomalous traces with the same sequence should be infrequent at first; thus, our approach will be able to detect them early, so that they do not have time to settle in. We also want to test our approach on a range of different noise levels in the event log (i.e., more anomalous traces), as well as include incomplete traces in the event log. As event logs consist of sequences of activities, it is also sensible to apply recurrent neural networks to the problem. Using recurrent networks could overcome the issue that our system is susceptible to skipped activities, which results in a shifted event sequence that is otherwise valid. Recurrent networks can learn these pattern regardless of where they exactly occur in the sequence, which is something the autoencoder in our approach is unable to do.

Our approach proves that neural networks are applicable within the domain of business processes. Moreover, we have shown that denoising autoencoders are capable of dealing with event logs that do already contain the anomalous traces, as opposed to training them on event logs that only contain normal traces. This approach is especially interesting, as it shows that an autoencoder can capture the underlying process of an event log, without being provided extra knowledge.

Acknowledgments. This project (HA project no. 479/15-21) is funded in the framework of Hessen ModellProjekte, financed with funds of LOEWE – Landes-Offensive zur Entwicklung Wissenschaftlich-ökonomischer Exzellenz, Förderlinie 3: KMU-Verbundvorhaben (State Offensive for the Development of Scientific and Economic Excellence) and by the LOEWE initiative (Hessen, Germany) within the NICER project [III L 5-518/81.004].

References

1. Van der Aalst, W., Weijters, T., Maruster, L.: Workflow mining: discovering process models from event logs. IEEE Trans. Knowl. Data Eng. **16**(9), 1128–1142 (2004)
2. Amer, M., Goldstein, M., Abdennadher, S.: Enhancing one-class support vector machines for unsupervised anomaly detection. In: Proceedings of the ACM SIGKDD Workshop on Outlier Detection and Description, pp. 8–15. ACM (2013)
3. Bahdanau, D., Cho, K., Bengio, Y.: Neural machine translation by jointly learning to align and translate (2014). arXiv preprint: arXiv:1409.0473
4. Bezerra, F., Wainer, J., Aalst, W.M.P.: Anomaly detection using process mining. In: Halpin, T., Krogstie, J., Nurcan, S., Proper, E., Schmidt, R., Soffer, P., Ukor, R. (eds.) BPMDS/EMMSAD 2009. LNBIP, pp. 149–161. Springer, Heidelberg (2009). doi:10.1007/978-3-642-01862-6_13
5. Burattin, A.: PLG2: multiperspective processes randomization and simulation for online and offline settings. CoRR abs/1506.0 (2015)
6. Cortes, C., Vapnik, V.: Support-vector networks. Mach. Learn. **20**(3), 273–297 (1995)
7. Dumas, M., Van der Aalst, W.M., Ter Hofstede, A.H.: Process-Aware Information Systems: Bridging People and Software Through Process Technology. John Wiley & Sons, Hoboken (2005)

8. Eskin, E.: Anomaly detection over noisy data using learned probability distributions. In: Proceedings of the International Conference on Machine Learning. Citeseer (2000)

9. Hawkins, S., He, H., Williams, G., Baxter, R.: Outlier detection using replicator neural networks. In: Kambayashi, Y., Winiwarter, W., Arikawa, M. (eds.) DaWaK 2002. LNCS, pp. 170–180. Springer, Heidelberg (2002). doi:10.1007/3-540-46145-0_17

10. Hecht-Nielsen, R.: Replicator neural networks for universal optimal source coding. Science **269**(5232), 1861 (1995)

11. Hinton, G.E.: Connectionist learning procedures. Artif. Intell. **40**(1), 185–234 (1989)

12. Japkowicz, N.: Supervised versus unsupervised binary-learning by feedforward neural networks. Mach. Learn. **42**(1), 97–122 (2001)

13. Krizhevsky, A., Sutskever, I., Hinton, G.E.: Imagenet classification with deep convolutional neural networks. In: Advances in Neural Information Processing Systems, pp. 1097–1105 (2012)

14. LeCun, Y., Bengio, Y., Hinton, G.: Deep learning. Nature **521**(7553), 436–444 (2015)

15. Maaten, L.V.D., Hinton, G.E.: Visualizing data using t-SNE. J. Mach. Learn. Res. **9**, 2579–2605 (2008)

16. Nair, V., Hinton, G.E.: Rectified linear units improve restricted boltzmann machines. In: Proceedings of the 27th International Conference on Machine Learning (ICML 2010), pp. 807–814 (2010)

17. Pimentel, M.A., Clifton, D.A., Clifton, L., Tarassenko, L.: A review of novelty detection. Sign. Process. **99**, 215–249 (2014)

18. Rozinat, A., van der Aalst, W.M.: Conformance checking of processes based on monitoring real behavior. Inf. Syst. **33**(1), 64–95 (2008)

19. Rumelhart, D.E., Hinton, G.E., Williams, R.J.: Learning representations by back-propagating errors. Cogn. Model. **5**(3), 1 (1988)

20. Srivastava, N., Hinton, G., Krizhevsky, A., Sutskever, I., Salakhutdinov, R.: Dropout: a simple way to prevent neural networks from overfitting. J. Mach. Learn. Res. **15**(1), 1929–1958 (2014)

21. Sutskever, I., Martens, J., Dahl, G., Hinton, G.: On the importance of initialization and momentum in deep learning. In: Proceedings of the 30th International Conference on Machine Learning (ICML 2013), pp. 1139–1147 (2013)

22. Van Der Aalst, W., Adriansyah, A., de Medeiros, A.K.A., Arcieri, F., Baier, T., Blickle, T., Bose, J.C., van den Brand, P., Brandtjen, R., Buijs, J., et al.: Process mining manifesto. In: Daniel, F., Barkaoui, K., Dustdar, S. (eds.) BPM 2011 Workshops, Part I. LNBIP, vol. 99, pp. 169–194. Springer, Heidelberg (2012). doi:10.1007/978-3-642-28108-2_19

DeepRED – Rule Extraction from Deep Neural Networks

Jan Ruben Zilke[✉], Eneldo Loza Mencía, and Frederik Janssen

Knowledge Engineering Group, Technische Universität Darmstadt,
Darmstadt, Germany
j.zilke@mail.de, {eneldo,janssen}@ke.tu-darmstadt.de

Abstract. Neural network classifiers are known to be able to learn very accurate models. In the recent past, researchers have even been able to train neural networks with multiple hidden layers (deep neural networks) more effectively and efficiently. However, the major downside of neural networks is that it is not trivial to understand the way how they derive their classification decisions. To solve this problem, there has been research on extracting better understandable rules from neural networks. However, most authors focus on nets with only one single hidden layer. The present paper introduces a new decompositional algorithm – *DeepRED* – that is able to extract rules from deep neural networks.

The evaluation of the proposed algorithm shows its ability to outperform a pedagogical baseline on several tasks, including the successful extraction of rules from a neural network realizing the XOR function.

1 Introduction

To tackle classification problems, i.e., deciding whether or not a data instance belongs to a specific class, machine learning offers a wide variety of methods. If the only goal is to accurately assign correct classes to new, unseen data, neural networks (NN) are able to produce very low error rates that yet could not be achieved by other machine learning techniques [8]. Sufficiently deep NNs, so-called deep neural networks (DNN), are even able to realize arbitrary functions. And lately, researchers improved the training of these structures such that they can generalize better and better from training data, for instance in the research fields of speech recognition and computer vision.

However, there is a major downside of NNs: For humans, it is not easy to understand how they derive their decisions [11]. However, understanding a NN's function can be essential. For instance, this is the case in safety-critical application domains such as medicine, power systems, or financial markets, where a hidden malfunction could lead to life-threatening actions or enormous economic loss. A better understanding of the way NNs derive their decisions could also push NN training research. Making NNs more transparent, for instance, could help to discover so-called hidden features that might be formed in DNNs while learning. Such features are not present in the plain input data, but emerge from combining them in a useful way.

© Springer International Publishing Switzerland 2016
T. Calders et al. (Eds.): DS 2016, LNAI 9956, pp. 457–473, 2016.
DOI: 10.1007/978-3-319-46307-0_29

In contrast to NNs, rule-based approaches like decision trees or simple IF-THEN rules are known to be better understandable. Since most humans tackle classification problems in a manner very similar to the one implemented by many rule-based learning techniques, i.e., listing simple conditions that need to be met, their models and decision processes are more comprehensible.

To overcome the weakness of NNs being black boxes, especially in the 1990s researchers have worked on the idea of extracting rules from them. Since then, a lot of rule extraction techniques have been developed and evaluated – and for many approaches quite good results have been reported. However, most algorithms to extract rules focus on small NNs with only one hidden layer. Surprisingly little work has been done on analysing the challenges of extracting rules from the more complex DNNs. Although some authors have presented rather general approaches, to the best of our knowledge, there does not exist any algorithm that has explicitly been tested on the task of extracting rules from DNNs. But indeed, having a large amount of layers makes the extraction of comprehensible rules more difficult.

In this paper, we introduce and evaluate *DeepRED*, a new algorithm that is able to extract rules from DNNs. The decompositional algorithm extracts intermediate rules for each layer of a DNN. A final merging step produces rules that describe the DNN's behaviour by means of its inputs. The evaluation shows *DeepRED*'s ability to successfully extract rules from DNNs and to outperform a pedagogical approach.

The work is structured as follows: First, we give an overview of the current state-of-the-art. Afterwards, we describe our proposed algorithm in more detail (Sect. 3). We evaluate *DeepRED* on several tasks (Sects. 4 and 5) before we summarize and conclude our work.

2   Related Work

It is commonly believed that "concepts learned by neural networks are difficult to understand because they are represented using large assemblages of real-valued parameters" [4]. To transform NNs into a form that is better comprehensible for humans, many concepts are conceivable. One famous choice is to extract rules from NNs. In [1], the authors have introduced a widely used taxonomy to distinguish between different rule extraction algorithms. The taxonomy, for instance, comprises the *translucency* dimension that defines the strategy used: decompositional (considering NN's inner structure), pedagogical (black box approach), or eclectic (mixture of both).

Examples of eclectic methods are the *MofN* algorithm [19] as well as the *FERNN* algorithm [15]. *MofN*'s main approach is to find NN connections with similar weights. *FERNN* focuses on identifying a NN's relevant inputs and hidden neurons.

The works dealing with pedagogical rule extraction, for instance, comprise the *VIA* method discussed in [17,18], different sampling-based approaches [5,13,14,16,21], and the *RxREN* algorithm [2]. *VIA* uses validity interval analysis to find provably correct rules. Sampling-based methods try to create extensive artificial training sets where rule learning algorithms later can extract rules from. *RxREN* provides interesting ideas to prune a NN before rules are extracted (cf. Sect. 3.4).

Examples for decompositional approaches are the *KT* method that heuristically searches for input combinations that let a neuron fire [7], a more efficient implementation of the latter [20], and an algorithm that transforms NNs to fuzzy rules [3]. However, the most important basis for this present work is the (decompositional) *CRED* algorithm presented in [12]. *CRED* uses decision trees to describe a NN's behaviour based on the units in its hidden layer. In a second step it builds up new decision trees to describe the split points of the first decision trees. Afterwards rules are extracted and merged.

Although we encounter a wide variety of interesting rule extraction approaches, there is a major shortcoming: Most decompositional algorithms introduced so far cannot deal with NNs with more than one hidden layer. Furthermore, to the best of our knowledge, there does not exist any algorithm that has explicitly been tested on the task of extracting rules from DNNs. This is the case even though most pedagogical approaches should be able to perform this task without any major modifications. However, due to the lack of pedagogical techniques to include the knowledge present in a DNN's inner structure, we focus on a decompositional method to extract rules from DNNs.

3 The DeepRED Algorithm

With *DeepRED* (Deep neural network Rule Extraction via Decision tree induction), in this paper we present an algorithm that is able to overcome the shortcoming of most state-of-the-art algorithms: *DeepRED* is applicable to DNNs. Since *DeepRED* is a decompositional approach, in contrast to pedagogical methods, the algorithm is able to easily extract hidden features from a DNN.

This section is structured as follows: First, we introduce our notation and assumptions. After describing the *CRED* algorithm in more detail, we present *DeepRED*'s way to extract rules from DNNs as well as the pruning technique used.

3.1 Preliminaries

We assume a multiclass classification problem with m training examples $x_1, \ldots, x_m$, each x_j associated with one class $y_j \in \{\lambda_1, \ldots, \lambda_u\}$. We represent the input values of a NN as $i = i_1, \ldots, i_n$ and the output values as $o = o_1, \ldots, o_u$, one for each possible class. The hidden layers are abbreviated as $h_i \in \{h_1, \ldots, h_k\}$ while the hidden layer h_i consists of the neurons $h_{i,1} \ldots, h_{i,H_i}$ (in the case of a single-hidden-layer NN, the hidden neurons are written as $h_1, \ldots, h_H$). For convenience, we set $h_0 = i$ and $h_{k+1} = o$ and let $h_i(x)$ denote the specific layer values for input instance x. *DeepRED* produces intermediate rule sets $R_{a \rightarrow b}$ that include rules that describe layer b by means of terms based on layer a.

A rule r : IF *body* THEN *head* is constituted by a set of terms in the body (conditions on attribute values, or, in our case, also conditions on activation values) and an assignment term in the head.

3.2   CRED as Basis

As mentioned earlier, *DeepRED* is based on *CRED* (Continuous/discrete Rule Extractor via Decision tree induction). In a first step, the original algorithm introduced by Sato and Tsukimoto [12] uses the well-known *C4.5* algorithm [10] to transform each output unit of a NN (with one hidden layer) into a decision tree. Such a tree's inner nodes are tests on the values of the hidden layer's units. The leaves represent the class such an example would belong to (cf. Fig. 1). Afterwards, intermediate rules are extracted directly from these trees. In the case illustrated in Fig. 1, if the task is to find rules for *class*$_1$, a single intermediate rule would be extracted, i.e. IF $h_2 > 0.5$ AND $h_1 \leq 0.6$ THEN $\hat{y} = \lambda_1$. This leads us to a rule set that describes the behaviour of a NN on the basis of the hidden layer's units.

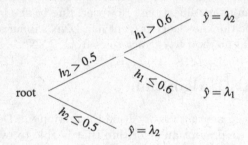

Fig. 1. Result of CRED's first step: a decision tree providing rules for the neural network's output based on its hidden layer (adapted from [12]).

Next, for each term used in these rules, another decision tree is created using split points on the NN's input layer. In these new decision trees, the leaves do not directly decide for an example's class, but rather on the tests used in the first tree's node, as exemplarily depicted in Fig. 2. Extracting rules from this second decision tree leads us to a description of the hidden neurons' state by terms consisting of input variables. In case of the state $h_2 > 0.5$, the following intermediate rules could be extracted: IF $i_0 > 0.5$ THEN $h_2 > 0.5$, IF $i_0 \leq 0.5$ AND $i_1 = a$ THEN $h_2 > 0.5$, and IF $i_0 \leq 0.5$ AND $i_1 = b$ THEN $h_2 > 0.5$. Further decision trees must be extracted for all the other split points found in the first tree (in our example for $h_1 \leq 0.6$).

As a last step, the intermediate rules that describe the output by means of the hidden layer and those that describe the hidden layer based on the NN's inputs are substituted and merged to build rules that describe the NN's output on the basis of its inputs. Note that the resulting rules from the decision tree are disjunctive, i.e., only one will fire at a time for any test instance.

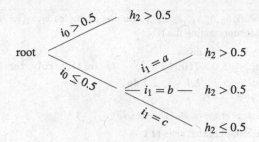

Fig. 2. Result of one iteration of CRED's second step: a decision tree providing rules for $h_2 > 0.5$ based on the neural network's inputs (adapted from [12]).

3.3   Extending CRED to DeepRED

Since *CRED* is a decompositional rule extraction approach, it is not possible to apply the original implementation directly to NNs with multiple layers of hidden neurons. However, the algorithmic approach can be extended relatively straight-forward by deriving additional decision trees and intermediate rules for every supplementary hidden layer.

The general process of *DeepRED* is the following: The algorithm extracts rules for each class/NN output one after another. For each class it processes every hidden layer in a descending order. *DeepRED* extracts rules for each hidden layer that describes its behaviour based on the preceding layer. In the end, all rules for one class are getting merged such that we arrive at the rule set $R_{i \to o}$.

Our modified version of the *CRED* algorithm – *DeepRED* – starts just like the original implementation by using *C4.5* to create decision trees consisting of split points on the activation values of the last hidden layer's neurons and the regarding classifications in the trees' leaves. Figure 3 provides the pseudo code for *DeepRED*. To exemplify *DeepRED*'s approach we consider, without loss of generality, a NN with k hidden layers. As a result of the first step we get a decision tree that describes the output layer by split points on values regarding h_k, i.e., the rule set $R_{h_k \to o}$. The data to run *C4.5* on is generated by feeding the training data to the NN and recording the outputs of the hidden neurons (line 8).

In the next step, in contrast to the algorithm presented by Sato and Tsukimoto, we do not directly refer to the input layer, but instead process the next shallower hidden layer, i.e. h_{k-1} (loop starting in line 1). For every term present in one of the rules in $R_{h_k \to o}$, we need to apply the *C4.5* algorithm to find decision trees that describe layer h_k by means of h_{k-1} and can be transformed to the rule set $R_{h_{k-1} \to h_k}$ (line 10). We also see that the proposed algorithm also implements a procedure to prevent itself from performing redundant runs of *C4.5* since we only learn a decision tree for terms which were not already extracted (line 6). Just like in the *CRED* example presented earlier, the terms in $R_{h_k \to o}$ are directly used to differentiate positive and negative examples for the regarding *C4.5* runs (line 9).

Input: Neural network $h_0, h_1, \ldots, h_k, h_{k+1}$, training examples $x_1, x_2, \ldots, x_m$
Output: Set of rules representing the NN

1 **foreach** *class* $\lambda_v \in \lambda_1, \ldots, \lambda_u$ **do**
2 $\quad R^v_{h_k \to o} \leftarrow$ IF $h_{k+1,v} > 0.5$ THEN $\hat{y} = \lambda_v$ `// initial rule for NN's prediction`
3 $\quad$ **foreach** *hidden layer* $j = k, k-1, \ldots 1$ **do**
4 $\quad\quad R^v_{h_{j-1} \to h_j} \leftarrow \emptyset$
5 $\quad\quad T \leftarrow$ `extractTermsFromRuleBodies` $(R^v_{h_j \to h_{j+1}})$
6 $\quad\quad T \leftarrow$ `removeDuplicateTerms` (T)
7 $\quad\quad$ **foreach** $t \in T$ **do**
8 $\quad\quad\quad x'_1, \ldots, x'_m \leftarrow h_j(x_1), \ldots, h_j(x_m)$
$\quad\quad\quad\quad$ `// obtain activation values for all neurons in the` $j-1$`-th`
$\quad\quad\quad\quad$ `layer for each training example`
9 $\quad\quad\quad y'_1, \ldots, y'_m \leftarrow t(h_{j+1}(x_1)), \ldots, t(h_{j+1}(x_m))$
$\quad\quad\quad\quad$ `// apply term t on respective activation in layer j and get`
$\quad\quad\quad\quad$ `binary outcome, the training signal, for each training example`
10 $\quad\quad\quad R^v_{h_{j-1} \to h_j} \leftarrow R^v_{h_{j-1} \to h_j} \cup$ `C4.5` $((x'_1, y'_1), \ldots, (x'_m, y'_m))$
$\quad\quad\quad\quad$ `// learn rules for t and add to rules representing the layer`
11 $\quad\quad$ **end**
12 $\quad$ **end**
$\quad$ `// merging to conjunctive rules and cleaning up`
13 $\quad$ **foreach** *hidden layer* $j = k, k-1, \ldots 1$ **do**
14 $\quad\quad R^v_{h_{j-1} \to o} =$ `mergeIntermediateTerms` $(R^v_{h_{j-1} \to h_j}, R^v_{h_j \to o})$
$\quad\quad\quad\quad$ `// dissolve rules of remaining two layers`
15 $\quad\quad R^v_{h_{j-1} \to o} =$ `deleteUnsatisfiableTerms` $(R^v_{h_{j-1} \to o})$
16 $\quad\quad R^v_{h_{j-1} \to o} =$ `deleteRedundantTerms` $(R^v_{h_{j-1} \to o})$
17 $\quad$ **end**
18 **end**
19 return $R^1_{i \to o}, R^2_{i \to o}, \ldots, R^u_{i \to o}$ `// describes output of NN by rules consisting of`
`terms on input values of first layer`

Fig. 3. Pseudo code of DeepRED.

We proceed in the same manner until we arrive at decision trees/rules that describe terms in the first hidden layer h_1 by terms consisting of the original inputs to the NN, i.e. $R_{i \to h_1}$.

Now we have rule sets that describe each layer of the NN by their respective preceding layer, i.e. we have the sets of intermediate rules $R_{i \to h_1}, R_{h_1 \to h_2}, \ldots,$ $R_{h_{k-1} \to h_k}$, and $R_{h_k \to o}$. To get a rule set $R_{i \to o}$ that describes the NN's outputs by its inputs, these rules need to be merged. This merging process proceeds layer-wise (block starting at line 13). First, *DeepRED* substitutes the terms in $R_{h_k \to o}$ by the regarding rules in $R_{h_{k-1} \to h_k}$ to get the rule set $R_{h_{k-1} \to o}$ (line 14). As mentioned in Fig. 3, unsatisfiable intermediate rules as well as redundant terms are deleted (lines 15 and 16).[1] This happens to reduce computation time and memory

---

[1] The merging may produce rules of the form $i_1 < 0.1$ AND $i_1 > 0.2$, or $i_1 > 0.4$ AND $i_1 > 0.5$.

usage drastically. Next, we merge the rule sets $R_{h_{k-1}\to o}$ and $R_{h_{k-2}\to h_{k-1}}$. Step by step we go through all layers until we arrive at rules that describe the classification/outputs according to the inputs to the NN, which is the result we were looking for.

3.4   Pruning

One way to facilitate the rule extraction process is to prune NN components that can be considered as less important for the classification. While the authors of *CRED* did not report on any of these approaches, other researchers have invented several techniques that help to extract comprehensible rules. As a pre-processing step before *DeepRED* is applied, we have borrowed the input pruning technique from *RxREN* [2]. It proceeds by testing the NN's performance while ignoring individual inputs. Those inputs that where not necessary to still produce an acceptable classification performance are getting pruned.

Table 1. Overview of deep neural networks used for evaluation, including the characteristics of the original data the NNs were trained on and the accuracies of the NN on the training and test set.

	#attributes	#training ex.	#test ex.	NN structure	acc(training)	acc(test)
MNIST	784	12056	2195	784-10-5-2	99.6 %	98.8 %
letter	16	1239	438	16-40-30-26	96.9 %	97.3 %
artif-I	5	20000	10000	5-10-5-2	99.5 %	99.4 %
artif-II	5	3348	1652	5-10-5-2	99.4 %	99.0 %
XOR	8	150	106	8-8-4-4-2-2-2	100 %	100 %

4   Experiments

Since we are not aware of any algorithm that has been tested on extracting rules from DNNs, we want to fill this gap with a *DeepRED* evaluation. Although *DeepRED* is able to extract rules from DNNs with an arbitrary number of output neurons, in this study we limit ourselves to two-class-problems. Before we present and discuss the results more extensively, we first take a look at the evaluation data base.

To evaluate *DeepRED*, we need suitable input data, i.e. trained NNs and training sets. Unfortunately, it is not trivial to find already trained NNs in literature or online. Although NN research is a lot about training, usually only network structures, training methods and performances are reported, while specific weights are not of particular interest for most researchers. Therefore, we trained DNNs for real-world and artificial problems. Table 1 summarizes our data basis.

Both, the *MNIST* and the *letter* data are taken from real-world problems. The basis for our *MNIST* dataset is the dataset presented in [9]. We picked a subset of the original data, reduced the problem to distinguish only between two different classes (*1* vs. rest), and trained a DNN on it. The *letter* recognition problem is derived from the dataset introduced in [6]. Here, we first trained the DNN on a multi-class training set. Afterwards we reduced the training and test set to a binary classification problem (accuracy rates reported are based on the training set reduced to two classes).

The datasets *artif-I* and *artif-II* are artificial ones that we have created to be able to compare original and extracted models. Both databases comprise examples with five attributes (discrete and continuous domains). A challenging characteristic of *artif-I* is that it contains greater-than relations on real-valued attributes. These functions cannot easily be modelled by decision trees. This is not the case with *artif-II*, however, the second artificial dataset is based on a model where the value of one attribute has no effect on the class, which as well might be challenging for rule extraction algorithms.[2]

As a fifth problem we manually constructed a DNN that solves the *XOR* problem with eight inputs. The parity function is well-known as a hard problem for rule learners. Top-down learner, for instance, need all possible input examples to correctly model *XOR*.

4.1 Rule Extraction Algorithms

To get an idea of how well the proposed algorithm performs the task of extracting rules from DNNs, in this evaluation we compare two variants of *DeepRED* – a version without and a version with *RxREN* pruning – with a pedagogical baseline.

To control the behaviour of *DeepRED*, we implemented two parameters that influence the included *C4.5* algorithm and one for the pruning mechanism. The first two parameters control when *C4.5* should stop further growing a decision tree. As an additional general setting we adjust *C4.5* to only perform binary splits and to produce a maximum decision tree depth of ten.

The first parameter – *class dominance* – is a threshold that considers the classes of the current data base. If the percentage of examples that belong to the most frequent class exceeds the value in the class dominance parameter, this class is getting predicted instead of further dividing the data base.

The second parameter is the minimum *data base size*. This value tries to stop *C4.5* further growing the decision tree when there is not enough data available to base dividing steps on. This parameter takes the number of training examples available in the first step as a reference value and defines a percentage of this size as the minimum data base size. If for the current step, there are less

---

[2] Input instances are drawn randomly from $x \in \{0, 0.5, 1\} \times \{0, 0.25, 0.5, 0.75, 1\} \times [0, 1]^3$. For *artif-I* $y = \lambda_1$ if $x_1 = x_2$, if $x_1 > x_2$ AND $x_3 > 0.4$, or if $x_3 > x_4$ AND $x_4 > x_5$ AND $x_2 > 0$, else $y = \lambda_2$, whereas for *artif-I* $y = \lambda_1$ if $x_1 = x_2$, if $x_1 > x_2$ AND $x_3 > 0.4$, or IF $x_5 > 0.8$.

examples available than the parameter requires, *C4.5* produces a leaf with the most frequent class at this point.

The *RxREN* pruning parameter, namely the maximum *accuracy decrease*, controls the pruning intensity. Step by step, the technique prunes the inputs that are considered as the least significant ones, as long as the NN's accuracy does not drop below the decrease allowed by this parameter.

As a pedagogical rule extraction baseline we choose to use the *C4.5* algorithm, that is also used as a part of *DeepRED* itself. Like the sampling-based approaches, our baseline is provided with the training examples as well as the NN's classification for these instances (instead of the real classes). Using these values, *C4.5* extracts a decision tree that, afterwards, is transformed to a rule set. The two introduced *C4.5* parameters do also apply to the baseline.

Note that no rule rewriting, rule pruning or any other rule optimization mechanisms are implemented by the *DeepRED* variants or the baseline (except *DeepRED*'s reported steps to delete redundant terms and unsatisfiable rules).

4.2 Evaluation Measures

To compare the results of the algorithms presented using the trained DNNs introduced, we derive two central measures from the extracted rule sets, that are used to assess their quality. First, we measure the comprehensibility by the number of terms in the extracted rule set. Second, we use the fidelity (ratio of prediction matches) to compare the mimicking performance of the extracted rules in contrast to the original NN's behaviour.

4.3 Evaluation Setup

Before we setup the evaluation, we collected our expectations of *DeepRED*'s performance. Our expectations were:

1. *DeepRED is able to extract rules from deep neural networks.*
   As stated earlier, to the best of our knowledge, no rule extraction algorithm has ever been tested on a DNN. We believed that in general the proposed algorithm can manage this task. However, we also expected *DeepRED*'s memory and computation time demands to sometimes make rule extraction processes intractable. We assumed that this was especially true for classification problems where an instance is described by a large number of attributes.
2. *DeepRED outperforms the baseline on more complex problems.*
   We believed that our algorithm can profit from a NN's structure when the training data model a concept that cannot be described/learned by a simple decision tree – including but not restricted to the *XOR* problem. On the other hand, for decision tree problems, i.e. classification tasks that can more easily be solved by a decision tree, our assumption was that the baseline could outperform *DeepRED*.

3. *RxREN pruning facilitates the extracted rules.*

   For problems where not all inputs are relevant, we expected the extracted rules to be more comprehensible when *RxREN* pruning has been applied. Depending on the problem, we even thought that the fidelity rates could be improved. However, since the optimal setting of the pruning parameters is not trivial, the extracted rules probably would not reach the possibly best tradeoff between comprehensibility and fidelity.

4. *DeepRED extracts more accurate rules if more data is available.*

   We believed that the *DeepRED* algorithm would need a certain amount of data to extract reasonable rules. However, we also thought that our algorithm is less dependant on a sufficiently large dataset than *C4.5* is, because *DeepRED* focuses on the setup of the NN's inner structure using instances, which is a more efficient representation compared to the pedagogical point of view.

To check these expectations, our evaluation setup includes three different pairs of *C4.5* parameter values, three different pruning settings, as well as four values that control the number of training examples visible to the rule extraction algorithm. Table 2 summarizes the different parameter settings chosen.

The number of experiments for the proposed algorithm sums up to 180 (or 36 per dataset) while the baseline is executed in 60 experiments (or 12 per dataset) – *RxREN* pruning never is applied here.

5   Evaluation

The results of *DeepRED*'s evaluation shows that most of our expectations are met. The main results are:

DeepRED can Successfully Extract Rules from Deep Neural Networks – Table 3 gives an overview of the general success of applying *DeepRED* to different DNNs. For every dataset, we list the number of experiments that were executed[3] as well as the number of successful and aborted attempts[4].

One observation is that for every DNN tested there is at least one parameter setting that enables *DeepRED* to successfully extract rules. Except for the *MNIST* dataset, this also holds true for every *RxREN* pruning variant evaluated. On the other hand is the high abortion rate of approximately 46 % an indication that the right parameters are important when using *DeepRED*. Especially for classification problems with a large number of inputs, i.e. *MNIST*, the success rate is very low.

---

[3] You might notice that, earlier, we mentioned that there are 36 experiments per dataset. However, to avoid sophisticating the outcomes, we discard those experiments where the *RxREN* pruning results in no pruned inputs at all.

[4] An abortion could either be the case if the experiment exceeds the allocated memory space (10000 MB) or if *DeepRED* needs more than the maximum execution time (24 h).

Table 2. The parameter settings used for the evaluation.

C4.5 parameters	RxREN pruning	Training set
92% / ≤ 2%	No pruning	10%
95% / ≤ 1%	5%	25%
99% / ≤ 0%	10%	50%
		100%

Table 3. Statistics of performed experiments for DeepRED.

	artif-I	artif-II	letter	MNIST	XOR
Executed	36	36	36	21	12
Successful	11	23	26	4	12
Aborted (memory)	24	13	10	7	0
Aborted (time)	1	0	0	10	0

A further investigation concerning the abortion reasons shows that most of the time memory limits are exceeded while merging if a lot of intermediate rules are present in the first two layers. The abortions due to violation of time constraints can all be ascribed to large training sets passed to the algorithm – the implemented *C4.5* is not efficient enough when dealing with large datasets.

DeepRED can Extract Comprehensible Rules for Complex Problems – Taking a closer look at the results on the datasets *artif-I* and *artif-II* lets us assess *DeepRED*'s performance on more complex problems. As described earlier, the data in *artif-I* model a function that cannot easily be realized by a decision tree. We consider this a difficult problem for the baseline. The *artif-II* data, on the other hand, can quite easily be modelled by a decision tree. However, please notice that the functions the DNNs have learned are not equivalent to the true models.

Figure 4 shows the results. We clearly see that the best *DeepRED* result outperforms the baseline on the DNN representing the *artif-I* dataset. Especially the comprehensibility of the rules extracted by our approach is much better. It seems like the proposed algorithm could use the inner structure of the DNN to create rules that efficiently model the NN's behaviour.

The *artif-II* dataset gives a different picture: The baseline produces much better results than *DeepRED* does. Regarding fidelity, the outcomes of our approach are more or less similar to those of the *artif-I* data. However, the baseline is able to find a more efficient way to model the NN's outputs – and does so much more accurate. For some reason, the inner structure of the DNN seems to confuse *DeepRED* such that it cannot extract high-quality rules.

Part of our expectation also was, that *DeepRED* can outperform the baseline by far when extracting rules from an *XOR* NN. Figure 5 gives an insight into the performance on this task. Please notice, that compared to Fig. 4 we needed to adjust the axes to draw a meaningful graph. As we expected it and it is

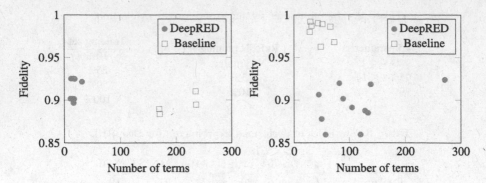

Fig. 4. Evaluation results for DeepRED and the baseline on artif-I (left) and artif-II (right).

well-known, *C4.5* cannot generalize from *XOR* training examples to correctly classify distinct test instances, i.e. it never gets better than random. However, as we see in the graph, *DeepRED* is able to use the knowledge embedded in the DNN, to extract rules that correctly classify unseen examples. As we will discuss in a moment, *DeepRED* needs a sufficiently large data basis to perform this task. But if it is given, an error rate of 0 % can be achieved (only having the training examples).

You might argue that *DeepRED* is not extracting comprehensible rules. In fact, rules in the form as we use them are not able to efficiently model the *XOR* problem. To correctly classify all possible 256 input combinations for this problem, 128 rules with 8 terms each are necessary to cover all 128 positive examples (the default rule covers the negative examples). In sum this leads to a number of 1024 terms in the rule set – which is the number of terms, the most accurate rules extracted by *DeepRED* have.

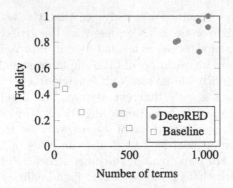

Fig. 5. Evaluation results for DeepRED and the baseline on XOR.

RxREN Pruning Helps DeepRED to Extract More Comprehensible Rules – *RxREN* pruning can facilitate the rule extraction process for *DeepRED*. Here, we want to analyse how the *RxREN* pruning affects a rule set's comprehensibility – and what it means for its fidelity. Figure 6 offers some insights for this question. We only show the datasets where not all inputs are important, i.e. *letter*, *artif-I*, and *MNIST*. The connected points represent experiments with the same *C4.5* parameters and training set sizes. Please note, that the figure only shows groups of experiments where at least two different pruning settings were successful.

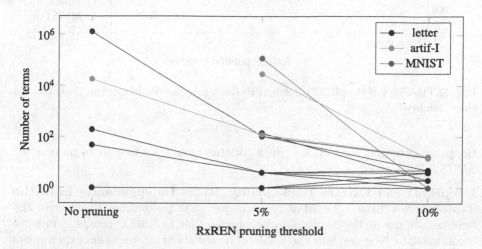

Fig. 6. The effect of DeepRED's different RxREN pruning parameters on the extracted rules' comprehensibility.

One can clearly see that for the vast majority of experiments, the number of terms gets smaller when the pruning intensity gets higher. This holds true for all experiments when we only consider no pruning versus pruning with a threshold of 5 %.

Figure 7 provides fidelity rates for exactly the same experiments as depicted in Fig. 6. For the *artif-I* problem, where the 10 % pruning setting produced the shortest rules, we also see the best fidelity rates for the same pruning setting. For the *MNIST* problem, instead, the threshold of 10 % is too greedy and leads to a much worse fidelity.

The results for the *letter* problem are less clear. We conclude that the pruning intensity needs to be adjusted in a more elaborate way to ensure the best achievable results. In total the results are better than expected since, although *RxREN* pruning overall increases the comprehensibility drastically, a clear trend towards lower fidelity rates cannot be recognized.

Another interesting observation is the fact that *RxREN* pruning in eight of 13 cases in the first place enables *DeepRED* to successfully extract rules. Without

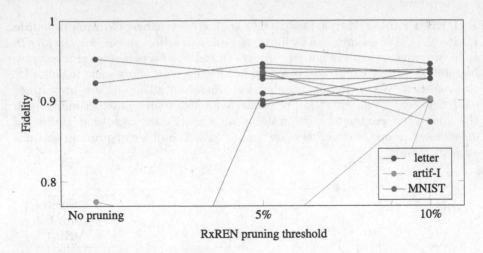

Fig. 7. The effect of DeepRED's different RxREN pruning parameters on the extracted rules' fidelity.

the pruning option, for instance, the algorithm would not be able to process the *MNIST* DNN at all.

DeepRED can Extract High-Quality Rules Independently from the Training Set Size – As a last measure we want to analyse the effect of the training set size on the results. Our belief was, that *DeepRED* profits from more available data. The graph in Fig. 8 shows the results of all successful experiments on datasets with less than 4000 training examples (with a minimum fidelity of 77 %).

Although we can clearly see that increasing the training set size from 10 % to 25 % has a positive influence on the fidelity, the overall results for the *artif-II* and *letter* datasets are surprising. Contrary to our expectations more data do not necessarily lead *DeepRED* to extract better rules. Often, the fidelity even gets worse when making more data available to the algorithm, while the baseline more often can profit from additional training data. A further investigation why this is the case is left for future research.

For the *XOR* dataset, Fig. 8 shows us exactly what we expected – a strong correlation between the number of training examples and the fidelity of the rule sets. While the results for the baseline are not shown in the extract of the graph (since the results are never better than 50 %), *DeepRED* manages to extract sensible rule sets with only 50 % of the training data (which is about 29 % of all possible input combinations). In our experiments with 100 % training data, the proposed algorithm extracted rules that perfectly mimic the NN's behaviour. Again, we want to clarify that the 100 % training data are approximately 59 % of all possible input combinations. The baseline would only be able to extract such a rule set, if all 256 possible instances are available for training.

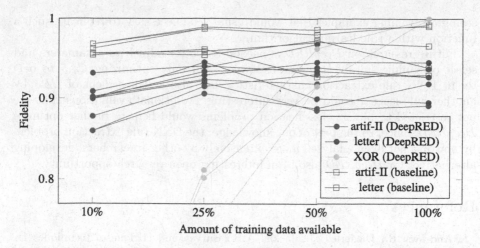

Fig. 8. The implications of different sizes of the available data on the extracted rules' fidelity.

This clearly shows the benefit of having access to the NN's structure. *DeepRED* does not need every input combination of all attributes in the training data – the examples must only contain the relevant input combinations anywhere in the training set. In the best case, for an *XOR* with four inputs, only four examples are needed[5], which is only 25 % of all 16 possible input combinations. The deeper a similar *XOR* network, the more *DeepRED* benefits from the NN's architecture.

6   Conclusion

In this paper, we presented and evaluated *DeepRED*, a new decompositional algorithm that solves the problem of extracting rules from DNNs to make their decision processes more comprehensible.

Since most state-of-the-art algorithms do not consider DNNs at all, we created a new rule extraction algorithm that is able to deal with DNNs. We proposed *DeepRED* which extends the *CRED* algorithm. Our approach uses *C4.5* to create rules that describe neurons on the basis of neurons in the preceding layer. *DeepRED* then merges these rules to produce a rule set that mimics the overall behaviour of the given DNN.

Our evaluation with different parameter settings on five datasets showed *DeepRED*'s advantages over a pedagogical baseline. The proposed algorithm even managed to successfully extract correct rules from an *XOR* NN despite not

---

[5] An example of a sufficient training set with the instance notation $x = x_1 x_2 x_3 x_4$ would be 0011, 1101, 1000, and 0110. It contains all combinations of x_1/x_2 and x_3/x_4.

seeing all training examples that would usually be necessary to replicate such a function with a rule learning algorithm.

Future research on *DeepRED* is needed to better adjust its parameters and select or sample suitable training examples. Not only optimizing *C4.5* to better fit DNN rule extraction characteristics, but also a replacement of *C4.5* by another rule learner could be very interesting. Additional evaluations that are not restricted to binary classification problems would help to further optimize *DeepRED*. Since, to the best of our knowledge, the DNN rule extraction problem by now has not been analysed more extensively by other researchers, developing alternatives for *DeepRED* also is an interesting open research opportunity.

References

1. Andrews, R., Diederich, J., Tickle, A.B.: Survey and critique of techniques for extracting rules from trained artificial neural networks. Knowl. Based Syst. **8**(6), 373–389 (1995)
2. Augasta, M.G., Kathirvalavakumar, T.: Reverse engineering the neural networks for rule extraction in classification problems. Neural Process. Lett. **35**(2), 131–150 (2012)
3. Benítez, J.M., Castro, J.L., Requena, I.: Are artificial neural networks black boxes? IEEE Trans. Neural Netw. **8**(5), 1156–1164 (1997)
4. Craven, M., Shavlik, J.W.: Using sampling and queries to extract rules from trained neural networks. In: ICML, pp. 37–45 (1994)
5. Craven, M.W., Shavlik, J.W.: Extracting tree-structured representations of trained networks. In: Advances in Neural Information Processing Systems, pp. 24–30 (1996)
6. Frey, P.W., Slate, D.J.: Letter recognition using Holland-style adaptive classifiers. Mach. Learn. **6**(2), 161–182 (1991)
7. Fu, L.: Rule generation from neural networks. IEEE Trans. Syst. Man Cybern. **24**(8), 1114–1124 (1994)
8. Johansson, U., Lofstrom, T., Konig, R., Sonstrod, C., Niklasson, L.: Rule extraction from opaque models-a slightly different perspective. In: 5th International Conference on Machine Learning and Applications, ICMLA 2006, pp. 22–27. IEEE (2006)
9. LeCun, Y., Bottou, L., Bengio, Y., Haffner, P.: Gradient-based learning applied to document recognition. Proc. IEEE **86**(11), 2278–2324 (1998)
10. Quinlan, J.R.: C4.5: Programs for Machine Learning, vol. 1. Morgan Kaufmann, San Francisco (1993)
11. Russell, S., Norvig, P.: Artificial Intelligence: A Modern Approach. Pearson Education, New York (1995)
12. Sato, M., Tsukimoto, H.: Rule extraction from neural networks via decision tree induction. In: Proceedings of the International Joint Conference on Neural Networks, IJCNN 2001, vol. 3, pp. 1870–1875. IEEE (2001)
13. Schmitz, G.P., Aldrich, C., Gouws, F.S.: ANN-DT: an algorithm for extraction of decision trees from artificial neural networks. IEEE Trans. Neural Netw. **10**(6), 1392–1401 (1999)
14. Sethi, K.K., Mishra, D.K., Mishra, B.: KDRuleEx: a novel approach for enhancing user comprehensibility using rule extraction. In: 2012 Third International Conference on Intelligent Systems, Modelling and Simulation (ISMS), pp. 55–60. IEEE (2012)

15. Setiono, R., Leow, W.K.: FERNN: an algorithm for fast extraction of rules from neural networks. Appl. Intell. **12**(1–2), 15–25 (2000)
16. Taha, I.A., Ghosh, J.: Symbolic interpretation of artificial neural networks. IEEE Trans. Knowl. Data Eng. **11**(3), 448–463 (1999)
17. Thrun, S.: Extracting provably correct rules from artificial neural networks. Technical report, University of Bonn, Institut für Informatik III (1993)
18. Thrun, S.: Extracting rules from artificial neural networks with distributed representations. In: Advances in neural information processing systems, pp. 505–512 (1995)
19. Towell, G.G., Shavlik, J.W.: Extracting refined rules from knowledge-based neural networks. Mach. Learn. **13**(1), 71–101 (1993)
20. Tsukimoto, H.: Extracting rules from trained neural networks. IEEE Trans. Neural Netw. **11**(2), 377–389 (2000)
21. Zhou, Z.H., Chen, S.F., Chen, Z.Q.: A statistics based approach for extracting priority rules from trained neural networks. In: Proceedings of the IEEE-INNS-ENNS International Joint Conference on Neural Networks, IJCNN 2000, vol. 3, pp. 401–406. IEEE (2000)

Ligand Affinity Prediction with Multi-pattern Kernels

Katrin Ullrich[1,2]([✉]), Jennifer Mack[1], and Pascal Welke[1]

[1] University of Bonn, Bonn, Germany
ullrich@iai.uni-bonn.de, mack@cs.uni-bonn.de, welke@uni-bonn.de
[2] Fraunhofer IAIS, Sankt Augustin, Germany

Abstract. We consider the problem of affinity prediction for protein ligands. For this purpose, small molecule candidates can easily become regression algorithm inputs if they are represented as vectors indexed by a set of physico-chemical properties or structural features of their molecular graphs. There are plenty of so-called molecular fingerprints, each with a characteristic composition or generation of features. This raises the question which fingerprint to choose for a given learning task? In addition, none of the standard fingerprints, however, systematically gathers all circular and tree patterns independent of size and the adjacency information of atoms. Since structural and neighborhood information are crucial for the binding capacity of small molecules, we combine the features of existing graph kernels in a novel way such that finally both aspects are covered and the fingerprint choice is included in the learning process. More precisely, we apply the Weisfeiler-Lehman labeling algorithm to encode neighborhood information in the vertex labels. Based on the relabeled graphs we calculate four types of structural features: Cyclic and tree patterns, shortest paths and the Weisfeiler-Lehman labels. We combine these different *views* using different multi-view regression algorithms. Our experiments demonstrate that affinity prediction profits from the application of multiple views, outperforming state-of-the-art single fingerprint approaches.

Keywords: Graph kernels · Molecular fingerprints · Multiple kernel learning · Support vector regression · Weisfeiler-lehman labeling

1 Introduction

In biological organisms small molecular compounds bind to large proteins with protein-ligand-specific affinities. If the real-valued affinity exceeds a given limit the compound is called a *ligand* of the protein. The ligand binding process typically triggers biochemical processes, for example, via a change of conformation or charge at the protein surface. *Ligand (affinity) prediction* is an important and challenging practical problem in the range of *quantitative structure-activity relationship* (QSAR) models as many drugs act as ligands. In practice,

© Springer International Publishing Switzerland 2016
T. Calders et al. (Eds.): DS 2016, LNAI 9956, pp. 474–489, 2016.
DOI: 10.1007/978-3-319-46307-0_30

for *drug discovery* and *design* millions of compounds can be tested in the laboratory already quite efficiently with *high-throughput-screening* (HTS) setups. Nevertheless, because of the expensive equipment and the quasi infinite number of synthesizable chemical compounds the process is still very time- and cost-consuming. Therefore, in chemoinformatics *similarity-based virtual screening* uses statistical ranking and machine learning methods in combination with molecular descriptors to train a binding model for the considered protein. Behind these approaches is the similarity assumption that similar compounds show similar binding behavior. As mentioned above, the protein-ligand-docking complex has a certain strength which can be measured as *binding affinity* K_i. For the prediction of affinities the well-known *support vector regression* (SVR) utilizing a vectorial feature representation of small molecules, the so-called *molecular fingerprints*, is the state-of-the-art method and was tested successfully (e.g., [1,9,17]). However, other machine learning algorithms like neural networks trained with molecular fingerprint data [10] were applied for the affinity prediction task as well. For a complete overview of approaches we point to the survey of Cherksov et al. [4].

Many publicly available or commercial fingerprint descriptors exist, just to mention a few: Maccs Keys, ECFP/C and FCFP/C fingerprints, GpiDAPH fingerprints, TGD or TGT. The fingerprints list (or count) diverse physico-chemical properties of the molecule, structural properties of their molecular graphs, or 3D information, and can be grouped according to their respective generation of features [2]. Additionally, even more molecular descriptors gathering graph structure information arise from the field of graph theory and have been proven beneficial (e.g., [7,11]). The variety of data descriptions is a blessing and a curse at the same time. On the one hand, a lot of different information sources for diverse learning tasks are available. On the other hand, this includes the necessity to choose a representation in order to obtain optimal results. Hence, we intend to overcome this problem by using multiple representations simultaneously. There are related QSAR approaches (e.g., [7,18]) which we would like to complement in this paper. Anyway, the method we will present is applicable for arbitrary graph data with multiple representations and real-valued labels.

Both structural and neighborhood information are crucial for the capacity of small molecules to be a ligand and for the strength of the bond (e.g., [12], [7]). For example, the presence of a benzene ring or that of an alcoholic group and their relative positions influence the chemical properties of the compound at hand. None of the existing fingerprints that collect structural information, however, captures both all circular and tree patterns of the molecular graph independent of size and the adjacency and connectivity information of atoms within the graph structure. Therefore, we propose to combine the feature set of the *cyclic pattern kernel* (CPK) [8] with that of shortest path (SP) kernels [3] and *Weisfeiler-Lehman* (WL) labels [14]. The WL algorithm assigns (new) labels to each vertex in the graph that depend on the surrounding vertices up to a certain distance h. CPK decomposes a graph into the set of contained cycles ($\mathcal{C}$) and remaining tree components ($\mathcal{T}$) of edges that do not belong to cycles.

Shortest path features ($\mathcal{P}$) collect the shortest paths from one vertex to another. Finally, we also consider the labels of the atoms ($\mathcal{L}$) themselves as features. Hence, for each depth h of the WL algorithm, we obtain four types of features ($\mathcal{C}_h, \mathcal{T}_h, \mathcal{P}_h, \mathcal{L}_h$) for each graph. Each of these $4 \cdot h$ feature sets can possibly be a (weighted) part of a resulting fingerprint. Additionally, a feature vector can either list or count features of a compound (binary or counting feature representation, see [12]). However, it is neither clear which of them to keep in the application scenario of affinity prediction, nor obvious which role the components play for the predictive process. Hence, we apply a systematic process to obtain an optimal combination of the proposed feature sets.

On an abstract level, these components can also be considered multiple views on molecular graphs. Including different views on data to learn one prediction function is known as multi-view learning. In general, a view or feature vector representation Φ is canonically related to a kernel function k via *Mercer's theorem*

$$k(x, x') = \langle \Phi(x), \Phi(x') \rangle \quad , \quad x, x' \in \mathcal{X}, \tag{1}$$

where $\mathcal{X}$ is the instance space of learning objects. In our case $\mathcal{X}$ are the potential ligands and $\Phi(x)$ would be the feature vector. We use the supervised kernel methods *multiple-kernel learning* (MKL) and its solution from Vishwanathan et al. [19] and *learning kernel ridge regression* (LKRR) by Cortes et al. [6]. Both algorithms learn a linear combination of kernel functions corresponding to the provided views. The linear combination of functions is included into a regularized empirical risk functional with ε-insensitive loss function (MKL) or squared loss function (LKRR). Hence, they virtually solve a multi-view SVR and multi-view *regularized least squares regression* (RLSR) problem. We will employ these multi-view kernel methods to find and use a combination of our features that suits the learning process best. Out of the rich set of patterns we aim at finding optimal compositions or combinations of views for the affinity prediction learning task. In the case of the linear kernel this is equivalent with utilizing a novel fingerprint representation with differently weighted pattern components, such that the identification of optimal weights is part of the learning process itself. Finally, this approach allows us to incorporate multiple feature representation for structural and neighborhood information with an automatic weighting that highlights the importance of the pattern group for the affinity prediction task.

2   Regression with Multiple Views

In the practical scenario of ligand affinity prediction we are looking for a real-valued predictor function f defined on an instance space $\mathcal{X}$ of molecules, i.e., $f : \mathcal{X} \rightarrow \mathbb{R}$. Here, f should be an element of a certain candidate space of functions $\mathcal{H}$ and the molecules in $\mathcal{X}$ can be regarded as graphs of atoms and bonds (see Fig. 1 below). A particular choice for $\mathcal{H}$ is a *reproducing kernel Hilbert space* (RKHS), a function space that is canonically related to a so-called kernel function $k : \mathcal{X} \times \mathcal{X} \rightarrow \mathbb{R}$. In the supervised scenario with training examples $(x_1, y_1), \ldots, (x_n, y_n) \in \mathcal{X} \times \mathbb{R}$, a common approach to find a good predictor is the principle of regularized *empirical risk minimization* (ERM) as a trade-off between empirical risk and function norm. Utilizing the ε-insensitive loss or the squared loss function for the empirical risk we obtain the SVR and RLSR formulation, respectively. For a thorough study of SVR and RLSR as well as their kernelized variants consult, for example, [16] or [5]. When we choose an RKHS $\mathcal{H}$ with reproducing kernel k as function space the *representer theorem* of Schölkopf et al. [15] implies a representation of the solution f^*. In fact, it turns out that f^* must be a linear combination of the kernel function k centered at training points that can be used to parameterize all considered optimization problems. Now let $\Phi_1, \ldots, \Phi_M$ be M views with corresponding kernels $k_1, \ldots, k_M$ according to (1), and respective RKHSs $\mathcal{H}_1, \ldots, \mathcal{H}_M$. Having multiple representations on data we want to apply SVR- and RLSR-related multi-view kernel algorithms. In contrast to using only one single view or kernel in the regularized ERM, the supervised multi-view approaches we consider incorporate a linear combination of kernel functions

$$k(x, x') = \sum_{v=1}^{M} b_v k_v(x, x') = \sum_{v=1}^{M} b_v \langle \Phi_v(x), \Phi_v(x') \rangle, \tag{2}$$

which itself is a kernel again for $b_v \geq 0$. In order to prevent overfitting, the linear factors $b = (b_1, \ldots, b_M)$ need to be regularized additionally. As the kernel expansion parameters from the representer theorem and linear factors b have to be learned simultaneously, the solution strategies of SVR- and RLSR-related multi-view approaches are different to the ones of single-view SVR and RLSR. On the one hand, the *multiple kernel learning* (MKL) algorithm presented in [19] utilizes the p-norm, $p > 1$, for the regularization of b. Hence, the MKL objective becomes $f_1^*, \ldots, f_M^* =$

$$\operatorname*{argmin}_{f_v \in \mathcal{H}_v, b_v \geq 0} \frac{1}{2} \sum_{v=1}^{M} \|f_v\|_{\mathcal{H}_v}^2 + C \sum_{i=1}^{n} \max\{0, |f(x_i) - y_i| - \varepsilon\} + \frac{\Lambda}{2} \left(\sum_{v=1}^{M} b_v^p \right)^{\frac{2}{p}},$$

where $f = \sum_{v=1}^{M} b_v f_v$ and $C, \Lambda, \varepsilon > 0$. Actually, there are other variants of MKL which we do not want to consider. On the other hand, the learning kernel ridge regression (LKRR) algorithm from [6] requires b being close to a constant vector b_0. The LKRR objective is

$$f_1^*, \ldots, f_M^* = \operatorname*{argmin}_{f_v \in \mathcal{H}_v} \sum_{v=1}^{M} \|f_v\|_{H_v}^2 + C \sum_{i=1}^{n} |f(x_i) - y_i|^2$$

$$\text{s.t. } \|b - b^0\| \leq \Lambda, b_v \geq 0,$$

where again $f = \sum_{v=1}^{M} b_v f_v$ as well as $b_v^0, \Lambda > 0$. For the solution of MKL and LKRR we refer to [6,19].

3   Patterns for Molecular Graphs

Our contribution in this paper is the intelligent combination of several graph patterns for the task of ligand prediction. To this end we now define four pattern sets or classes and corresponding kernels that incorporate structural and neighborhood information. Finally, we define a multi-pattern kernel based on these patterns that allows efficient combinations of patterns based on the methods presented in the previous section.

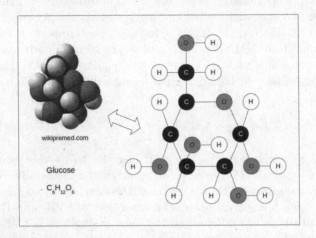

Fig. 1. Glucose molecule in its 3D (left) and graph representation (right). All bonds in glucose are single bonds, hence edge labels are omitted.

A *labeled, undirected graph* is a quadruple $G = (V, E, \Sigma, \lambda)$, with V being a finite set of vertices, $E \subseteq \binom{V}{2}$ a set of edges, Σ a finite linearly ordered set of labels, and $\lambda : V \cup E \to \Sigma$ a function assigning a label to each vertex and edge. For the computation of patterns we consider molecules as labeled, undirected graphs such that atoms correspond to vertices (labels: C, O, H, N, ...) and bonds to edges (labels: single, double, and aromatic). An example is shown in Fig. 1. A sequence $w = \{v_0, v_1\}, \{v_1, v_2\}, \dots, \{v_{k-2}, v_{k-1}\}, \{v_{k-1}, v_k\}$ of edges of a graph is called *simple path* if $v_i \neq v_j$ for all i, j with $1 \leq i < j \leq k$ (the vertex v should not be confused with the view index v). If additionally $v_0 = v_k$ holds true, the sequence is called *simple cycle*. Edges not belonging to any simple cycle are called *bridges*. A *forest* is an undirected graph that does not contain a cycle, a connected (i.e., where any two vertices are connected by a simple path) forest is called a *tree*. Two labeled, undirected graphs $G = (V, E, \Sigma, \lambda)$ and $G' = (V', E', \Sigma', \lambda')$ are *isomorphic*, if there is a bijection $\varphi : V \to V'$ that respects edges and labels, i.e., $\{v, w\} \in E$ if and only if $\{\varphi(v), \varphi(w)\} \in E'$, as well as $\lambda(v) = \lambda'(\varphi(v))$, and $\lambda(\{v, w\}) = \lambda'(\{\varphi(v), \varphi(w)\})$.

In the following we will review four types of graph patterns: Label patterns $\mathcal{L}$, cyclic patterns $\mathcal{C}$, tree patterns $\mathcal{T}$, and shortest path patterns $\mathcal{P}$. These patterns already appear in the definitions of the popular graph kernels Weisfeiler-Lehman

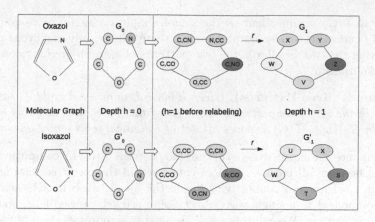

Fig. 2. WL labeling of two molecular graphs.

kernel (WLK), cyclic pattern kernel (CPK), and shortest path kernel (SPK). First, we introduce the *Weisfeiler-Lehman labeling* of a graph G's vertices as the basis for the WL test of graph isomorphism as well as the WLK of Shervashidze et al. [14], represented by the recursive labeling function $\lambda_G^h : V \rightarrow \Sigma^*$ for recursion depth h. Initially, it holds that $\lambda_G^0 = \lambda$. That means for depth $h = 0$ the WL labels are the original vertex labels. For each recursion step we append to each vertex label $\lambda_G^h(v)$, $v \in V$, a sorted list of all labels from adjacent vertices and obtain

$$\lambda_G^{h+1}(v) = r\left(\lambda_G^h(v), sort\left(\{\lambda_G^h(w) : (v, w) \in E\}\right)\right),$$

where a, b represents the concatenation of strings a and b (using a separating comma), $\{\cdot\}$ is a multiset, and r is a renaming function (see Shervashidze et al. [14]). Based on this definition we denote with $G_h = (V, E, \Sigma, \lambda_G^h)$ the graph G with WL labels of depth h. The edge labels remain unaffected in the sense that $\lambda_G^h(e) = \lambda(e)$ for all $h = 0, 1, \ldots$ and all $e \in E$. Two small examples can be found in Fig. 2.

Definition 1 (Label Patterns). *For a labeled, undirected graph G we define $\mathcal{L}(G)$ to be the set of all vertex labels of G.*

The above definition implies that $\mathcal{L}(G_h)$ are the WL labels of depth $h \geq 0$. Next, *cyclic* and *tree patterns* correspond to the features defined by Horváth et. al. [8] for the CPK.

Definition 2 (Cyclic Patterns). *Given a labeled, undirected graph G we define $S(G)$ to be the set of all simple cycles in G. Then $C(G)$ denotes the set of canonical representations[1] of $S(G)$.*

---

[1] See [8] for a definition of such a canonical representation.

That is, $\mathcal{C}(G)$ is equivalent to the set of simple cycles *up to isomorphism*. Hence, if G contains two isomorphic cycles, there will be only one cyclic pattern representing the two. Similarly, the tree patterns of a graph are only considered up to isomorphism.

Definition 3 (Tree Patterns). *Given a labeled, undirected graph G we define $T(G)$ to be the set of connected components of the forest that contains only the bridges in G. Then, $\mathcal{T}(G)$ denotes the set of canonical representations of $T(G)$.*

We enhance the expressiveness of the tree patterns $\mathcal{T}(G)$ by computing shortest paths between all pairs of vertices contained in the forest consisting of the bridges in G. To regain connectivity inside the forest, we contract each biconnected component to a single vertex and assign a fixed, unused label to each of those fusion vertices. We call this newly derived tree representation of the original graph *contracted graph* and define the shortest path patterns corresponding to the SPK of Borgwardt et al. [3] as follows.

Definition 4 (Shortest Path Patterns). *Given a labeled, undirected graph G we define $P(G)$ to be the set of shortest paths between all pairs of vertices in the contracted graph of G. With $\mathcal{P}(G)$ we denote the canonical representations of $P(G)$.*

Considering that a WL labeled graph G_h differs from its underlying original graph G only with respect to the labels, we can apply the definitions of cyclic, tree, and shortest path patterns to such graphs as well. This allows us to derive even more detailed patterns $\mathcal{C}(G_h)$, $\mathcal{T}(G_h)$, $\mathcal{P}(G_h)$, and of course $\mathcal{L}(G_h)$ for depths h greater than zero. With $\Phi_v : \mathcal{X} \to \mathbb{R}^{d_v}$, $v \in \{\mathcal{C}, \mathcal{T}, \mathcal{P}, \mathcal{L}\}$, we denote the binary or counting feature representation of the respective patterns. In practice, the feature space dimension d_v depends on the view v, the considered depth h, and the graph dataset at hand. In Sect. 4 we will use the term (set) intersection or (set) counting kernel when we refer to the linear kernel on binary or counting feature vectors, respectively. Analogous to the WLK in [14] we define cumulative pattern kernels and a non-cumulative version of them.

Definition 5 (Pattern Kernel). *Let $v \in \{\mathcal{C}, \mathcal{T}, \mathcal{P}, \mathcal{L}\}$ be a graph pattern class and Φ_v its binary or counting feature mapping. For two labeled, undirected graphs G and G' the cumulative pattern kernel k_v^h of depth h is defined as*

$$k_v^h(G, G') = \langle \Phi_v(G_0), \Phi_v(G_0') \rangle + \cdots + \langle \Phi_v(G_h), \Phi_v(G_h') \rangle, \tag{3}$$

whereas the non-cumulative pattern kernel of depth h is just

$$k_v^h(G, G') = \langle \Phi_v(G_h), \Phi_v(G_h') \rangle. \tag{4}$$

Obviously, (3) is a generalization of the WLK and (4) is an instance of its non-negatively weighted variant. Although we restrict to the linear view kernel $\langle \Phi_v(\cdot), \Phi_v(\cdot) \rangle$ in Definition 5 also other base kernels (compare WLK definition in [14]) could be applied. For the sake of convenience we do not use extra indices for cumulative/non-cumulative or intersection/counting kernels. This will be clear from the context in the practical part below. Finally, we define the multi-pattern kernel (MPK). Interestingly, with minor modifications the molecular fingerprint ECFPx corresponds to the cumulative WL labels of a molecular graph up to depth $h = x/2$.

Definition 6 (Multi-pattern Kernel). *We consider non-negative weights b_v, $v \in \{\mathcal{C}, \mathcal{T}, \mathcal{P}, \mathcal{L}\}$. The multi-pattern kernel k_{MPK} of two labeled, undirected graphs G and G' is defined as*

$$k_{MPK}(G, G') = \sum_{v \in \{\mathcal{C}, \mathcal{T}, \mathcal{P}, \mathcal{L}\}} b_v \cdot k_v^{h_v}(G, G') \quad , \quad b_v \geq 0,$$

where the WL depth h_v depends on the pattern v and $k_v^{h_v}$ can be a cumulative or non-cumulative pattern kernel.

Now we want to investigate MPKs with multi-view kernel approaches in the context of ligand affinity prediction.

4   Experiments

We provided a considerable number of representation variants for molecular graphs that can be used as views for single- and multi-view kernel approaches which themselves can be parameterized as well in several ways (regularization and kernel type). At first, in the preliminary experiments we intend to extract promising views or view combinations for the practical task of ligand affinity prediction using only the single view regression methods SVR and RLSR. For this purpose, the views are either individual graph pattern vectors or their concatenation. In a second step, we use the best patterns and check whether we can take profit from multi-view kernel methods for regression. We will compare our results with the performance of standard fingerprints.

4.1   Setup and Datasets

Our experiments will be performed on 20 datasets, each of which representing ligands of one of 20 human proteins. A set contains between 90 and 986 ligands of the respective protein gathered from BindingDB[2]. We ordered the protein datasets according to the number of contained ligands and renamed them (see *Number* in Table 1, DS = dataset). We divided the 20 datasets into two groups: The first group, consisting of odd ordinal numbers, was used for preliminary and parameter tuning experiments. The second group with even numbers was utilized for the main experiments with multi-view kernel methods. Every ligand is a single molecule in the sense of a connected graph and is labeled with its affinity value ($pK_i = -\log_{10} K_i$) towards the protein target. The ligands are given

---

[2] Binding database, https://www.bindingdb.org/bind/index.jsp.

Table 1. Datasets with name, ordinal number, number of ligands, and label range.

Target	P23946	Q99895	P09871	P25774	Q9Y5Y6
Number	DS 1	DS 2	DS 3	DS 4	DS 5
Ligands	90	91	92	104	125
Range	5.4-8.9	2.7-8.0	4.8-9.0	4.3-9.8	4.0-10.1
Target	P17655	P42574	P00740	P07384	P07339
Number	DS 6	DS 7	DS 8	DS 9	DS 10
Ligands	128	133	171	189	197
Range	4.8-10.8	4.9-11.9	3.9-8.7	3.1-10.7	4.1-11.0
Target	P08709	P43235	P00750	P07858	P29466
Number	DS 11	DS 12	DS 13	DS 14	DS 15
Ligands	249	252	264	278	310
Range	3.9-9.5	3.9-11.5	2.2-9.5	3.0-10.5	3.1-9.8
Target	P07711	P00747	P00749	P08246	P07477
Number	DS 16	DS 17	DS 18	DS 19	DS 20
Ligands	357	474	600	742	986
Range	3.9-10.6	1.9-11.0	0.3-11.1	2.7-11.2	2.0-10.6

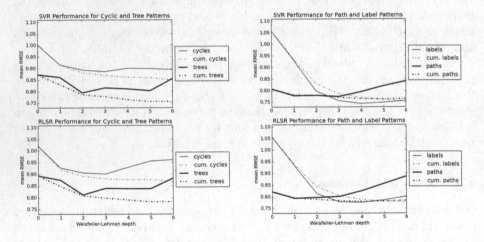

Fig. 3. SVR and RLSR results for the intersection kernel.

in SMILES-format (Simplified Molecular Input Line Entry Specification) from which the labeled graph structure can be deduced easily, e.g., with the chemistry toolbox Open Babel[3] and the structure data format (SDF). We introduced the edge label *aromatic* by hand using a Hückel's rule heuristic. Thus, for all ligands in all datasets the binary and counting feature vectors for all pattern types up to

---

[3] openbabel.org.

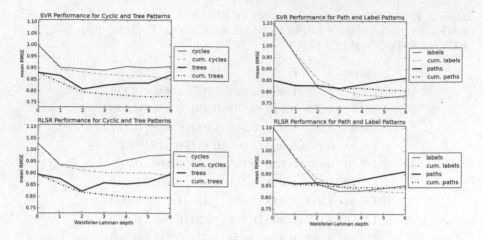

Fig. 4. SVR and RLSR results for the counting kernel.

WL depth $i = 6$ and the standard molecular fingerprints Maccs and ECFP6 were available for our experiments.

We use the SMO-MKL software[4] for efficient MKL that is based on libSVM[5] for both our MKL and SVR experiments. For LKRR we use our own implementation of Algorithm 1 in [6]. For a learned predictor function f we report root mean squared error $\text{RMSE}(f) = \frac{1}{\sqrt{m}} \|Y_{test} - f(X_{test})\|_2$, i.e., the root mean squared distance between the original label vector Y_{test} of m test instances X_{test} and the corresponding vector of prediction values $Y_{pred} = f(X_{test})$. In our experiments we perform 5-fold cross validation. In every training and parameter tuning fold we randomly sample 80 % training data and the remainder as test instances. We use the linear kernel on binary and counting feature vectors $k_v(G, G') = \langle \Phi_v(G), \Phi_v(G') \rangle$. For the reason of calculation stability of the used software, we normalized every kernel matrix initially with its Frobenius matrix norm. For the parameters we chose $C \in [50.0, 100.0]$ for all algorithms and $\Lambda = 1.0$. The trade-off parameter C was generally chosen quite large during the parameter tuning phase (which we account for the kernel normalization), whereas all algorithms seemed to be almost insensible to the choice of Λ. Leaned on the expert knowledge in chemoinformatics with affinity prediction (e.g., [1]) we used $\varepsilon = 0.1$.

4.2 Results

Preliminary Experiments: Initially, we considered the patterns *cycles, trees, shortest paths,* and *labels* individually applying SVR and RLSR for the prediction of ligand affinities. Therefore, we used a cumulative and non-cumulative feature

---

[4] Available at http://research.microsoft.com/en-us/um/people/manik/code/smo-mkl/download.html.

[5] https://www.csie.ntu.edu.tw/~cjlin/libsvm/.

Table 2. RMSE for standard fingerprints. A: SVR with Maccs, B: SVR with ECFP6, C: MKL with Maccs and ECFP6, D: RLSR with Maccs, E: RLSR with ECFP6, F: LKRR with Maccs and ECFP6.

Target	A	B	C	D	E	F
DS 2	1.039	**1.004**	1.010	1.007	1.000	**0.995**
DS 4	0.923	**0.719**	0.737	0.970	0.762	**0.746**
DS 6	1.052	**0.906**	0.915	1.063	0.903	**0.812**
DS 8	0.790	**0.655**	0.675	0.841	**0.618**	0.621
DS 10	0.897	**0.726**	0.755	1.083	**0.906**	0.914
DS 12	1.289	**1.077**	1.111	1.265	1.071	**1.053**
DS 14	1.254	**1.048**	1.078	1.271	1.073	**1.033**
DS 16	1.216	**0.948**	0.996	1.190	0.961	**0.925**
DS 18	1.068	**0.820**	0.834	1.073	0.842	**0.801**
DS 20	1.138	**0.869**	1.045	1.116	0.855	**0.816**

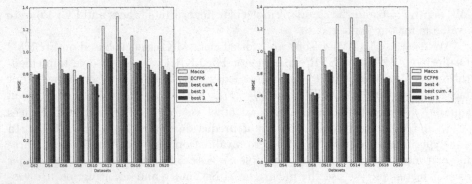

Fig. 5. Graphical visualization of main results for the counting kernel. Left: SVR/MKL results from Table 4, columns A, B, G, H, I. Right: RLSR/LKRR results from Table 6, columns A, B, F, G, H.

vector variant which we refer to with "cum. pattern" or "pattern", respectively. For the first variant, we use all features based on all WL depths up to some depth h in a concatenated feature vector. For the second, we only use features of a fixed depth h. The results can be found in Figs. 3 and 4. We report the mean RMSE with respect to all datasets with odd numbers. We observe that the qualitative performance trend is very similar for SVR and RLSR. Obviously, the non-cumulative patterns reach an optimal WL depth and decline for greater depths. The cumulative ones appear to converge to the optimal performance with increasing WL depth. Nevertheless, the best RMSE is very similar for cumulative and non-cumulative patterns. In general, for the predictive task at hand the performance of labels seems to be very high for an appropriate WL depth, whereas for all WL depths the one for cycles is quite low.

Table 3. RMSE for SVR/MKL experiments with intersection kernel. A: SVR with Maccs, B: SVR with ECFP6, C: SVR with cum. SPK $h = 6$, D: SVR with cum. WLK $h = 6$, E: SVR with cum. CPK $h = 6$, F: MKL with best depth of all patterns, G: MKL with best depth of all cum. patterns, H: MKL with best depths of 3 best patterns, I: MKL with best depths of 2 best patterns, "-" algorithm did not converge.

Target	A	B	C	D	E	F	G	H	I
DS 2	**1.028**	1.056	1.100	1.084	1.071	1.095	1.088	1.107	1.101
DS 4	0.932	**0.707**	0.726	0.752	0.948	0.777	0.769	0.762	0.743
DS 6	.068	0.894	**0.842**	0.882	1.041	0.891	0.864	0.881	0.860
DS 8	0.849	0.700	0.715	0.694	0.714	**0.693**	0.697	0.696	0.707
DS 10	0.935	0.744	0.747	0.757	0.819	0.742	0.752	**0.740**	0.745
DS 12	1.307	1.099	1.104	1.069	1.244	1.036	**1.030**	1.016	1.038
DS 14	1.261	1.102	0.931	0.956	1.118	0.932	**0.899**	0.913	0.917
DS 16	1.189	0.946	0.940	0.941	1.070	0.979	0.952	0.952	**0.931**
DS 18	1.103	0.846	0.869	0.838	0.804	**0.784**	0.785	0.787	0.832
DS 20	1.110	0.838	-	0.886	-	-	0.834	0.751	**0.718**

Table 4. RMSE for SVR/MKL experiments with counting kernel. A: SVR with Maccs, B: SVR with ECFP6, C: SVR with cum. SPK $h = 5$, D: SVR with cum. WLK $h = 6$, E: SVR with cum. CPK $h = 4$, F: MKL with best depth of all patterns, G: MKL with best depth of all cum. patterns, H: MKL with best depths of 3 best patterns, I: MKL with best depths of 2 best patterns, "-" algorithm did not converge.

Target	A	B	C	D	E	F	G	H	I
DS 2	0.818	**0.776**	0.815	0.787	0.805	0.802	0.796	0.794	0.807
DS 4	0.930	**0.673**	0.675	0.691	0.894	0.728	0.725	0.706	0.718
DS 6	1.037	0.841	**0.763**	0.820	0.995	0.826	0.808	0.805	0.808
DS 8	0.837	**0.751**	0.785	0.761	0.781	0.775	0.764	0.788	0.776
DS 10	0.896	0.728	0.698	0.715	0.807	0.698	0.705	**0.687**	0.710
DS 12	1.231	0.991	0.993	**0.973**	1.212	1.007	0.981	0.984	0.978
DS 14	1.317	1.130	0.990	0.989	1.205	0.978	0.994	0.959	**0.943**
DS 16	1.142	0.891	**0.860**	0.886	1.008	0.912	0.904	0.902	0.917
DS 18	1.106	0.880	0.893	0.877	0.846	0.822	0.834	0.816	**0.803**
DS 20	1.142	0.868	-	-	-	**0.767**	0.833	0.797	0.814

As the information for individual datasets is not apparent in the diagrams of Figs. 3 and 4 for each cumulative and non-cumulative pattern variant we chose the best WL depth and extracted the performance for every dataset with odd number. We omit the results for reasons of space, but in essence they show the same impact of the different pattern classes for individual datasets as for the average over the datasets.

Table 5. RMSE for RLSR/LKRR experiments with intersection kernel. A: RLSR with Maccs, B: RLSR with ECFP6, C: RLSR with cum. SPK $h = 4$, D: RLSR with cum. WLK $h = 6$, E: RLSR with cum. CPK $h = 2$, F: LKRR with best depth of all patterns, G: LKRR with best depth of all cum. patterns, H: LKRR with best depths of 3 best patterns, I: LKRR with best depths of 2 best patterns.

Target	A	B	C	D	E	F	G	H	I
DS 2	0.991	**0.990**	1.011	1.016	1.082	0.998	1.011	1.018	0.995
DS 4	0.966	0.839	0.827	0.879	1.039	0.807	0.815	0.797	**0.774**
DS 6	1.097	0.974	0.902	0.948	1.055	0.878	**0.870**	0.829	0.784
DS 8	0.820	0.678	0.715	0.684	0.741	0.669	0.688	**0.667**	0.689
DS 10	0.872	**0.697**	0.729	0.725	0.770	0.710	0.715	0.716	0.711
DS 12	1.175	1.028	1.000	0.985	1.210	0.976	0.973	0.954	**0.931**
DS 14	1.251	1.084	0.948	0.980	1.169	0.928	0.940	0.906	**0.887**
DS 16	1.186	0.958	0.933	0.952	1.088	0.918	0.910	0.901	**0.881**
DS 18	1.046	0.828	0.856	0.807	0.770	0.729	0.733	**0.723**	0.738
DS 20	1.160	0.888	0.810	0.890	0.961	0.764	0.767	0.753	**0.742**

Table 6. RMSE for RLSR/LKRR experiments with counting kernel. A: RLSR with Maccs, B: RLSR with ECFP6, C: RLSR with cum. SPK $h = 4$, D: RLSR with cum. WLK $h = 6$, E: RLSR with cum. CPK $h = 3$, F: LKRR with best depth of all patterns, G: LKRR with best depth of all cum. patterns, H: LKRR with best depths of 3 best patterns, I: LKRR with best depths of 2 best patterns.

Target	A	B	C	D	E	F	G	H	I
DS 2	0.985	**0.962**	1.040	1.021	1.034	1.005	0.997	1.024	0.997
DS 4	0.949	**0.766**	0.782	0.826	1.049	0.808	0.801	0.798	0.828
DS 6	1.078	0.918	0.844	0.886	1.034	0.847	0.857	**0.834**	0.860
DS 8	0.787	0.610	**0.599**	0.621	0.676	0.630	0.604	0.620	0.620
DS 10	1.013	0.876	0.838	0.870	0.971	0.826	0.832	**0.816**	0.851
DS 12	1.244	1.013	1.007	1.006	1.254	1.014	0.993	**0.986**	**0.986**
DS 14	1.300	1.096	0.961	0.994	1.130	0.940	0.943	0.926	**0.924**
DS 16	1.238	0.998	0.968	0.988	1.093	0.946	0.951	**0.933**	0.961
DS 18	1.090	0.872	0.899	0.841	0.805	0.751	0.763	0.754	**0.734**
DS 20	1.140	0.869	0.764	0.851	0.912	0.741	**0.719**	0.736	0.785

Main Experiments: At first, we tested whether we can already take profit from multi-view approaches using standard molecular fingerprints only. The results are presented in Table 2. We find that in the case of RLSR/LKRR there is a performance improvement in favor of the multi-view algorithm. This cannot be verified in the case of SVR/MKL as single-view SVR with fingerprint ECFP6 turns out to be the best method for all datasets. Subsequently, we investigated whether we can take advantage of different graph pattern features in the multi-view setting. Actually, there are too many pattern combinations to test all of them in the scope of multi-view algorithms. Therefore, we chose the most promising combinations and depths from the preliminary experiments (with respect to the RMSE) to compare them with baseline kernels and standard fingerprints. The results are shown in Tables 3, 4, 5 and 6, as well as Fig. 5. In columns A and B the single-view results with standard fingerprints are shown. The columns C to E present the performance of the popular graph kernels SPK, WLK, and CPK, each with its optimal depth taken from the preliminary experiments. Finally, the columns F - I obtain the results for optimal multi-view combinations from the preliminary experiments (see the respective table caption). In the case of SVR/MKL experiments we observe that MKL approaches show the best performances in 7 out of 10 cases (Table 3). If we utilize the counting kernel the MKL approaches outperform the single-view SVR variants with standard molecular fingerprints or standard graph kernels in the majority of cases (Table 4). In the RLSR/LKRR scenario the multi-view approaches exhibit the lowest RMSE for 8 out of 10 or 7 out of 10 datasets when we apply the intersection or counting kernel, respectively (Tables 5 and 6). We observe that the single-view approaches SVR and RLSR applying the standard graph kernel features of SPK, WLK, and CPK outperform the other algorithms in very few cases. Most interestingly, the more simple non-cumulative pattern combinations perform very well in comparison to cumulative combinations, even if we do not lift all pattern types cycles, trees, shortest paths, or labels, but rather only 2 or 3 of them. Apparently, it is sufficient to use non-cumulative pattern combinations of few pattern types.

5   Conclusion

We considered the problem of ligand affinity prediction with a variety of different feature vectors representing small molecular compounds and compared single- and multi-view regression approaches for this learning task. We showed that one can profit from the application of linear combinations of multiple views on molecular data in this practical scenario. It turned out that the multi-view approaches based on structural features and neighborhood information outperform the SVR and RLSR algorithm using standard molecular fingerprints or popular graph kernels. This effect was the more visible the greater the dataset sizes were (see Fig. 5). During our experiments we observed that the application of WL labels $\mathcal{L}$ and shortest path patterns $\mathcal{P}$ improved the prediction results particularly. In general, the squared loss-approaches RLSR and LKRR achieved better results than the analogue ε-insensitive loss algorithms SVR and MKL. Hence, using combinations of graph patterns based on WL labels of appropriate depths together with multi-view methods represents a noteworthy alternative to the application of SVR and standard molecular fingerprints which is the state-of-the-art approach for affinity prediction in the field of QSAR modeling.

Acknowledgements. We want to thank Dr. Martin Vogt from the Department of Life Science Informatics, B-IT, of the university of Bonn for preparing the protein dataset and making it available for us. Furthermore, we thank Dr. Martin Vogt and his colleagues for many valuable discussions on this topic. We would also like to thank Prof. Thomas Gärtner for guidance and advice.

References

1. Balfer, J., Bajorath, J.: Artifacts in support vector regression-based compound potency prediction revealed by statistical and activity landscape analysis. PLoS ONE **10** (2015)
2. Bender, A., Jenkins, J.L., Scheiber, J., Sukuru, S.C.K., Glick, M., Davies, J.W.: How similar are similarity searching methods? A principal component analysis of molecular descriptor space. J. Chem. Inf. Model. **49**, 108–119 (2009)
3. Borgwardt, K.M., Kriegel, H.-P.: Shortest-path kernels on graphs. In: Proceedings of ICDM, pp. 74–81 (2005)
4. Cherkasov, A., Muratov, E.N., Fourches, D., Varnek, A., Baskin, I., Cronin, M., et al.: QSAR modeling: where have you been? Where are you going to? J. Med. Chem. **57**, 4977–5010 (2014)
5. Christianini, N., Shawe-Taylor, J.: An Introduction to Support Vector Machines and Other Kernel-Based Learning Methods. Cambridge University Press, New York (2000)
6. Cortes, C., Mohri, M., Rostaminzadeh, A.: L_2 regularization for learning kernels. In: Proceedings of UAI, pp. 109–116 (2009)
7. Gaüzère, B., Brun, L., Villemin, D.: Treelet kernel incorporating cyclic, stereo and inter pattern information in Chemoinformatics. Pattern Recogn. **48**, 356–367 (2014)
8. Horváth, T., Gärtner, T., Wrobel, S.: Cyclic pattern kernels for predictive graph mining. In: Proceedings of KDD, pp. 158–167 (2004)
9. Liu, W., Meng, X., Xu, Q., Flower, D.R., Li, T.: Quantitative prediction of mouse class I MHC peptide binding affinity using support vector machine regression (SVR) models. BMC Bioinform. **7** (2006)
10. Myint, K.-Z., Wang, L., Tong, Q., Xie, X.-Q.: Molecular fingerprint-based artificial neural networks QSAR for ligand biological activity predictions. Mol. Pharm. **9**, 2912–2923 (2012)
11. Ning, X., Rangwala, H., Karypis, E.: Multi-assay-based structure-activity-relationship models: improving structure-activity-relationship models by incorporating activity information from related targets. J. Chem. Inf. Model. **49**, 2444–2456 (2009)
12. Ralaivola, L., Swamidass, S.J., Saigo, H., Baldi, P.: Graph kernels for chemical informatics. Neural Netw. **18**, 1093–1110 (2005)
13. Rogers, D., Hahn, M.: Extended connectivity fingerprints. J. Chem. Inf. Model. **50**, 742–754 (2010)
14. Shervashidze, N., Schweitzer, P., van Leeuwen, E.J., Mehlhorn, K., Borgwardt, K.M.: Weisfeiler-Lehman graph kernels. J. Mach. Learn. Res. **12**, 2539–2561 (2011)
15. Schölkopf, B., Herbrich, R., Smola, A.J.: A generalized representer theorem. In: Helmbold, D., Williamson, B. (eds.) COLT 2001. LNCS (LNAI), vol. 2111, pp. 416–426. Springer, Heidelberg (2001). doi:10.1007/3-540-44581-1_27
16. Smola, A.J., Schölkopf, B.: A tutorial on support vector regression. Stat. Comput. **14**, 199–222 (2004)

17. Sugaya, N.: Ligand efficiency-based support vector regression models for predicting bioactivities of ligands to drug target proteins. J. Chem. Inf. Model. **54**, 2751–2763 (2014)
18. Qiu, S., Lane, T.: Multiple kernel support vector regression for siRNA efficacy prediction. IEEE/ACM Trans. Comput. Biol. Bioinform. **4983**, 367–378 (2008)
19. Vishwanathan, S.V.N., Sun, Z., Theera-Ampornpunt, N., Varma, M.: Multiple kernel learning and the SMO algorithm. In: Proceedings of NIPS, pp. 2361–2369 (2010)

Author Index